physica status solidi (a)

applied research

Volume 110 Number 2 December 1988

AKADEMIE-VERLAG BERLIN
VCH PUBLISHERS INC. · NEW YORK, N.Y.

ISSN 0031-8965 phys. stat. sol. (a), Berlin 110 (1988) 2, 309–696, K61–K116, A13–A20

International Classification System for Physics*)

60. Condensed matter: structure, mechanical and thermal properties
    61. Structure of liquids and solids; crystallography
    (see also 68.20. Solid surface structure, 71. Electron states)
    62. Mechanical and acoustical properties of condensed matter
    (see also 61.70. Defects in crystals, 68.30. Surfaces and interfaces)
    63. Lattice dynamics and crystal statistics
    (see also 65. Thermal properties, 66.70. Thermal conduction, 68.30. Dynamics of surface and interface vibrations, 78.30. Infrared and Raman spectra)
    64. Equations of state, phase equilibria, and phase transitions
    65. Thermal properties of condensed matter
    (see also 63. Lattice dynamics; for thermodynamic properties of quantum fluids, see 67.40; for thermal properties of solid helium, see 67.80)
    66. Transport properties of condensed matter (nonelectronic)
    67. Quantum fluids and solids; liquid and solid helium
    68. Surfaces and interfaces; thin films and whiskers
    (for impact phenomena, see 79; for crystal growth, see 61.50)

70. Condensed matter: electronic structure; electrical, magnetic, and optical properties
    71. Electronic states
    (see also 63. Lattice dynamics, 73. Electronic structure and electrical properties of surfaces, interfaces, and thin films)
    72. Electronic transport in condensed matter
    (for surfaces, interfaces, and thin films, see 73)
    73. Electronic and electrical properties of surfaces, interfaces, and thin films
    74. Superconductivity
    75. Magnetic properties and materials
    76. Magnetic resonances and relaxation in condensed matter; Mössbauer effect
    77. Dielectric properties and materials
    (for conductivity phenomena, see 72.20 and 72.80)
    78. Optical properties and condensed-matter spectroscopy and other interactions of matter with particles and radiation
    (for phonon spectra, see 63)
    79. Electron and ion emission by liquids and solids; impact phenomena

85. Devices**)

*) Excerpt; reproduced with permission of International Council for Scientific and Technical Information (ICSTI).

**) Outside the ICSTI Classification for Physics.

(The Substance Classification is given on cover three)

physica status solidi (a)

applied research

Volume 110 · Number 2 · Pages 309 to 696, K61 to K116, and A13 to A20

December 16, 1988

AKADEMIE-VERLAG · BERLIN

and

VCH PUBLISHERS INC. · NEW YORK, N. Y.

For the USA and Canada:

physica status solidi (a), ISSN 0031-8965, is published monthly by VCH Publishers, Inc., 303 NW 12th Avenue, Deerfield Beach, FL 33442-1788; Telex 5101011104. VCH Pub. Telefax (305) 428-8201; Telephone (305) 428-5566 or 1-800-422-8824 (toll free).

Second class postage paid at Deerfield Beach, FL 33441.

Annual subscription price: US $ 800.00 including postage and handling charges.

U. S. Postmaster: Send address changes to VCH Publishers, Inc., 303 NW 12th Avenue, Deerfield Beach, FL 33442-1788.

Enquiries, subscriptions, and orders for single copies from customers in the USA and CANADA should be directed to:

VCH Publishers, Inc., 303 NW 12th Avenue, Deerfield Beach, FL 33442—1788, USA;

from all other countries to:

Akademie-Verlag Berlin, Leipziger Straße 3—4, DDR-1086 Berlin, GDR.

Editorial Note:

"physica status solidi (a)" undertakes that an original paper *accepted for publication* before the 23rd of any month will be published within 50 days of this date unless the author requests a postponement. In special cases there may be some delay between *receipt* and *acceptance* of a paper due to the review and, if necessary, revision of the paper.

Schriftleiter und verantwortlich für den Inhalt:
Professor Dr. E. Gutsche, Leipziger Straße 3—4, Postfach 1233, DDR-1086 Berlin.

Verlag:
Akademie-Verlag Berlin, Leipziger Straße 3—4, DDR-1086 Berlin; Fernruf: 2236221 und 2236229; Telex-Nr.: 114420; Bank: Staatsbank der DDR, Berlin, Kto.-Nr.: 6836-26-20712.

Chefredakteur:
Dr. H.-J. Hänsch.

Redaktionskollegium:
Prof. Dr. E. Gutsche, Dr. H.-J. Hänsch, Dr. H. Lange, Dr. S. Oberländer.

Anschrift der Redaktion:
Leipziger Straße 3—4, Postfach 1233, DDR-1086 Berlin; Fernruf: 2236279.

Veröffentlicht unter der Lizenznummer 1620 des Presseamtes beim Vorsitzenden des Ministerrates der Deutschen Demokratischen Republik.

Gesamtherstellung:
VEB Druckerei „Thomas Müntzer", DDR-5820 Bad Langensalza.

Erscheinungsweise:
Die Zeitschrift „physica status solidi (a)" erscheint jeweils am 16. eines jeden Monats. Jährlich erscheinen 6 Bände zu je 2 Heften.

Bestellnummer dieses Bandes: 1085/110.

phys. stat. sol. (a) **110**, No. 2 (1988)

Contents

Original Papers and Short Notes

Structure; crystallography

Defects; nonelectronic transport

Lattice properties

Surfaces, interfaces, thin films; lower-dimensional systems

Electronic transport; superconductivity

Magnetic properties; resonances

physica status solidi (a) is indexed in Current Contents/Physical, Chemical & Earth Sciences.

Systematic List

The Author Index of Volume 110 Begins on Page 697

(It will be delivered together with Volume 111, Number 1.)

Original Papers

phys. stat. sol. (a) **110**, 317 (1988)

Subject classification: 61.40; 64.70; S1.2

South West Centre for Reactor Engineering Research and Design, Chengdu (a), Institute of Solid State Physics, Academia Sinica, Hefei[1]) *(b), and Microstructure Research Centre, Academia Sinica, Chengdu*[2]) *(c)*

An Investigation of Fast Neutron Irradiation Effects on the Structure of $Cu_{50}Ti_{50}$ Metallic Glass

By

GUOAN WU (a), YUANDA DONG (b), KEQIN XIAO (b), YIZHEN HE (b), LONGSHU WANG (c), and JIAQING HE (c)

The effect of irradiation on the atomic structure of a $Cu_{50}Ti_{50}$ metallic glass is investigated by X-ray diffraction. The irradiation is seen to produce significant changes in the structure factor $S(Q)$ of this glass, and the changes observed are quite opposite to those caused by low-temperature annealing. Changes are also observed in the reduced radial distribution function $G(r)$. All these changes are finally discussed in conjunction with the topological short range ordering (TSRO) and chemical short range ordering (CSRO) of the metallic glass.

Bestrahlungseffekte infolge Röntgenbeugung auf die atomare Struktur von metallischen $Cu_{50}Ti_{50}$-Glas werden untersucht. Die Bestrahlung ruft signifikante Änderungen des Strukturfaktors $S(Q)$ dieses Glases hervor und die beobachteten Änderungen stehen im völligen Gegensatz zu denen der Niedertemperaturausheilung. Änderungen werden auch für die reduzierte radiale Verteilungsfunktion $G(r)$ beobachtet. Alle diese Änderungen werden im Zusammenhang mit der topologischen Nahordnung (TSRO) und der chemischen Nahordnung (CSRO) der Metallgläser diskutiert.

1. Introduction

There has been a significant growth of interest in the study of irradiation effects on metallic glasses in recent years as, in terms of technology, there is some hope for the use of metallic glasses as device and structural materials in nuclear reactors. A few investigations have been devoted to neutron radiation effects on their physical properties, such as crystallization behaviour [1, 2], Young's modulus [2], electrical resistivity [3], and ductility [4] and atomic structure, such as structure factors [5, 6] and microscopic morphology [2, 7], and very interesting results have been obtained. However, the interpretation of these measurements is not always without difficulty, because the changes in short-range and medium-range orders induced by irradiation, which might be fundamental to an understanding of some of the properties, are still practically unknown. On the other hand, Fukunaga et al. [8] and Sakata et al. [9] studied systematically the structure of Cu–Ti metallic glasses using X-ray and neutron diffraction techniques and gave a picture of the chemical short-range order (CSRO) for $Cu_{50}Ti_{50}$. On this basis, we undertake the study of neutron irradiation effects on the structure of $Cu_{50}Ti_{50}$ metallic glass by means of X-ray diffraction for comparison as an attempt to cast fresh light on clarifying the irradiation effects.

The present paper gives a description of the experimental work of resulting structure factors and reduced radial distribution functions, and discusses the local atomic structure of the $Cu_{50}Ti_{50}$ glass on the basis of these results.

---

[1]) Hefei, People's Republic of China.

[2]) Chengdu, People's Republic of China.

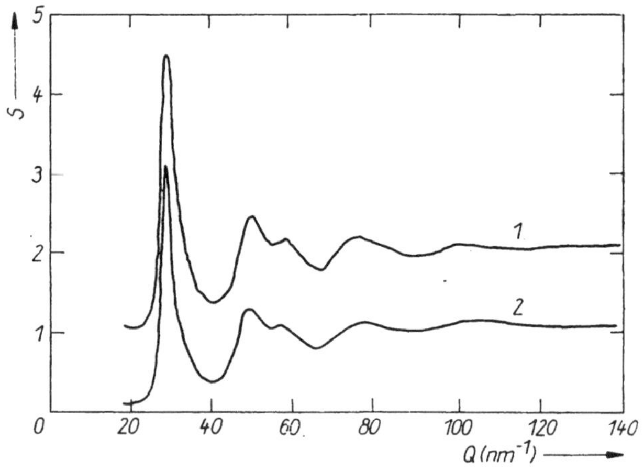

Fig. 1. Structure factor curves for $Cu_{50}Ti_{50}$ metallic glass (1) before and (2) after irradiation

2. Experimental

A master ingot of composition $Cu_{50}Ti_{50}$ was prepared by vacuum induction melting of pure Cu (99.99%) and Ti (99.5%). Weight loss on melting was negligible. The alloy was conventionally melt-spun on to a copper wheel to produce metallic glass ribbons approximately 3 mm wide and 30 μm thick. The absence of crystallization was confirmed by X-ray diffraction.

The samples were sealed in aluminium cans, filled with water to conduct away the heat generated by radiation. The cans were placed in a hole inside the high flux engineering testing reactor (HFETR), located in the South West Centre for Reactor Engineering Research and Design of China. After one week irradiation, the estimated fast neutron fluence and displacement per atom were about 2.5×10^{19} neutrons/cm² ($E > 1$ MeV) and 0.04, respectively. The sample temperature throughout irradiation was kept below 320 K. After waiting for radioactivity to die down to acceptable levels, samples were then taken out from the cans for further measurements.

X-ray diffraction experiments were performed using a Rigaku D/Mas-rA diffractometer in conjunction with a molybdenum target and a graphite curved crystal monochromator. A range of scattering angles $5° < 2\theta < 135°$ was covered in 0.5° steps with a pre-set time of 40 s per step. The data were processed using an extensively modified version of a computer program by Wagner [10]. The structure factors $S(Q)$ obtained for the unirradiated and irradiated specimens are shown in Fig. 1.

Table 1

The position of peaks in the structure factors for $Cu_{50}Ti_{50}$ metallic glass before and after irradiation, Q_1, Q_2, etc. are given with corresponding peak position in $G(r)$s r_1, r_2, etc. The width of the first peaks in $S(Q)$ is described by ΔQ_1

sample state	peak positions in $S(Q)$ (nm⁻¹)				coordination distances in $G(r)$ (nm)			
	Q_1	Q_2	Q_3	ΔQ_1	r_{11}	r_{12}	r_{21}	r_{22}
unirradiated	29.2	50.0	59.0	4.00	0.263	—	0.450	0.520
irradiated	29.0	49.5	58.0	4.80	0.257	0.270 to 0.290	0.450	0.525

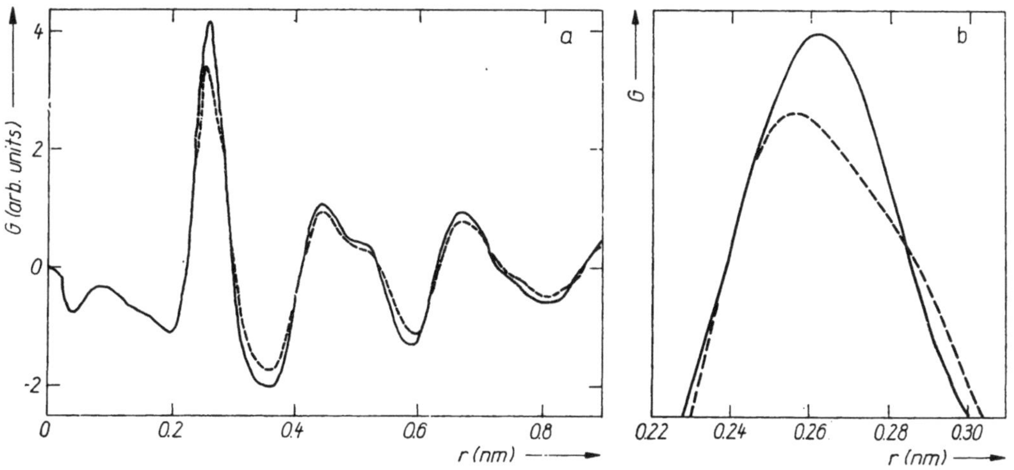

Fig. 2. a) The reduced radial distribution function $G(r)$ curves for $Cu_{50}Ti_{50}$ metallic glass. The solid line is for the specimen before irradiation and the dashed line is for the same glass after irradiation. b) The first peak of the $G(r)$ curve on a finer r scale

3. Results

It can be seen from Fig. 1 that the $S(Q)$ curves obtained before and after irradiation are rather similar. From a comparison of the curves in Fig. 1 it follows that the structure factor of the unirradiated specimen shows characteristic features common to most metallic glasses, namely a sharp first peak, and a second peak with a shoulder on the high Q side and well developed oscillations out to Q_{max}. The structure factor of the irradiated specimen shows the same basic features, but the height of the first peak is noticeably reduced by about 8%, and its position is shifted to a slightly lower Q value. The subsequent peaks in $S(Q)$ are also slightly reduced, but their positions in Q remain roughly unchanged, while the second peak shoulder becomes less distinct. A quantitative description of these changes is given in Table 1. In general, the changes in peak height are consistent with those observed by Itoh et al. [5].

The changes of structure in real space, as expressed by $G(r)$ curves which are the Fourier transforms of the structure factors $S(Q)$, are of a nature rather similar to the changes in $S(Q)$. This is illustrated by Fig. 2a. The irradiation is seen to cause a slight reduction in the magnitude of the oscillations in $G(r)$ and to cause a noticeable reduction in height and a broadening of the first peak. The $G(r)$s in Fig. 2b are the Fourier transforms of $S(Q)$ on a finer r scale than those shown in Fig. 2a. Table 1 gives some numerical values of the $G(r)$ parameters.

One interesting point is that the irradiation appears to have an exactly opposite effect on the structure to that produced by low-temperature annealing. Both X-ray [11] and neutron [12] diffraction experiments show that annealing produces an increase in the peak heights and a general sharpening of the features in $S(Q)$.

4. Discussion

For an unirradiated $Cu_{50}Ti_{50}$ specimen the average nearest neighbour distance was found to be $r_1 = 0.263$ nm. According to the Faber-Ziman relation, the total reduced radial distribution function $G(r)$ can be expressed in terms of the three partial pair correlation functions in X-ray diffraction as follows:

$$G(r) = 0.323\, G_{CuCu}(r) + 0.491\, G_{CuTi}(r) + 0.186\, G_{TiTi}(r)\,, \tag{1}$$

strictly at $Q = 0$. From the weighting factors it follows that the contribution of the Ti–Ti correlation function is small. From this and from the fact that the atomic diameter of titanium is larger than that of copper we conclude that the first peak of the unirradiated $Cu_{50}Ti_{50}$ specimen consists of an average of Cu–Cu and Cu–Ti neighbours. After irradiation the first peak splits into a main peak, positioned at $r = 0.257$ nm, and a shoulder ranging from 0.270 to 0.290 nm. From the known Goldschmidt radii one can assume that Cu–Cu correlations contribute to the main peak, while Cu–Ti and Ti–Ti correlations contribute to the shoulder.

Fukunaga et al. [8] studied the structure of $Cu_{50}Ti_{50}$ metallic glass by pulsed neutron scattering and obtained the correlation distances for this alloy. They concluded that the atomic arrangement of $Cu_{50}Ti_{50}$ alloy glass is similar to the CSRO in the crystalline $Cu_{50}Ti_{50}$ compound (γ-CuTi type). It is of interest to know whether the CSRO in $Cu_{50}Ti_{50}$ metallic glass changes after irradiation. Our observations suggest that after irradiation the CSRO in $Cu_{50}Ti_{50}$ is closer to crystalline γ-CuTi than the CSRO before irradiation.

Itoh et al. [5] and Kayano et al. [2] studied the neutron effects on the structure of MoSi and PdSi metallic glasses and held that, due to the destruction of the short-range order by neutron irradiation, the atomic structure of metallic glasses is changed to a more random state. This is also the case in our experiments.

In Fig. 1 it can be seen that the widths of the first peaks in $S(Q)$ before and after irradiation are different. The correlation length ξ of the topological short-range ordering (TSRO) can be calculated from the width ΔQ according to the formula [13]

$$\xi = \frac{2\pi}{\Delta Q}\,. \tag{2}$$

With the experimental data from Table 1 we obtain the correlation length $\xi_{unirr}^{TSRO} \approx 1.6$ nm for the unirradiated specimen, and $\xi_{irr}^{TSRO} \approx 1.3$ nm for the irradiated specimen. The means that the range of correlation in an irradiated $Cu_{50}Ti_{50}$ glass is smaller than that in an unirradiated $Cu_{50}Ti_{50}$ glass. It suggests that irradiation induces topologically atomic rearrangements that diminish the order in the unirradiated glass. This point is confirmed by the fact that the heights of the first and second peaks in $S(Q)$ are reduced and the positions of the two peaks in $S(Q)$ are slightly shifted towards lower Q values after irradiation. We believe that it is this type of change in TSRO that is responsible for the retardation of crystallization [14] and an increase in internal friction [15] at the crystallization temperature in $Cu_{50}Ti_{50}$ metallic glass.

After calculating the probabilities for cascades containing a certain number of displaced atoms, Gerling and Wagner [7] pointed out that most of the fast neutron induced damage cascades contain more than ten displaced atoms; almost 39% of the damage cascades even consist of more than hundred displaced atoms. According to this view, we may suggest that the rearrangement of atoms in irradiated $Cu_{50}Ti_{50}$ metallic glass in our experiments involves an extended region. As a result the CSRO and TSRO would be altered.

5. Conclusion

X-ray diffraction experiments have been made on specimens of $Cu_{50}Ti_{50}$ metallic glass before and after irradiation in order to study the nature of the structural changes produced by irradiation. Irradiation appears to cause significant changes in the structure factors. The changes observed are opposite to those caused by low-temperature annealing. These facts indicate that the irradiated specimens suffer an increase in structural disorder. The reduced radial distribution functions $G(r)$ are obtained by Fourier transformation of the structure factors. The changes in the radial distribution function $G(r)$ show that the irradiation enhances the chemical short-range order.

Acknowledgements

The authors would like to thank Zihou Zhao and Liangyou He for their assistance in sample irradiation arrangement and irradiated sample handling.

References

[1] J. Weis and I. Červeň, phys. stat. sol. (a) **76**, K49 (1983).
[2] H. Kayano, T. Masumoto, S. Tomizawa, and S. Yajima, Sci. Rep. Res. Inst. Tôhoku Univ. **26**, 240 (1977).
[3] S. Takamura and M. Kobigama, Radiat. Eff. **86**, 43 (1984).
[4] R. Gerling, P. Schimanski, and R. Wagner, Acta metall. **35**, 1001 (1987).
[5] F. Itoh, S. Ikeda, M. Ikebe, M. Hasegawa, T. Honda, T. Fukunaga, H. Fujimori, and K. Suzuki, Proc. 4th Internat. Conf. Rapidly Quenched Metals, 1982 (p. 763).
[6] K. Doi, Proc. 4th. Internat. Conf. Rapidly Quenched Metals, 1982 (p. 339).
[7] R. Gerling and R. Wagner, Proc. 4th Internat. Conf. Rapidly Quenched Metals, Jap. Inst. of Metals, Sendai 1982 (p. 767).
[8] T. Fukunaga, K. Kai, M. Naka, N. Watanabe, and K. Suzuki, Proc. 4th Internat. Conf. Rapidly Quenched Metals, 1982 (p. 347).
[9] M. Sakata, N. Cowlam, and H. A. Davies, Proc. 4th Internat. Conf. Rapidly Quenched Metals, Jap. Inst. of Metals, Sendai 1982 (p. 347).
[10] C. N. J. Wagner, Tech. Rep. Nat. Sci. Foundation Grant GL 3213 (1968).
[11] T. Egami, J. appl. Phys. **50**, 1564 (1979).
[12] N. Cowlam, K. Dini, and H. A. Davies, Metallic Glasses: Science and Technology, Kultura, Budapest 1981 (p. 267).
[13] E. Nassif, P. Lamparter, W. Sperl, and S. Steeb, Z. Naturf. **38a**, 142 (1983).
[14] Dong Yuanda, Wu Guoan, Cheng Lifang, Xiao Keqin, Li Xiaoguang, Zhu Zhaoqin, and He Yizhen, Chinese Phys. Letters **12**, 541 (1987).
[15] Dong Yuanda, Wu Guoan, Xiao Keqin, Li Xiaoguang, and He Yizhen, J. appl. Phys., to be published.

(Received February 12, 1988; in revised form August 19, 1988)

phys. stat. sol. (a) **110**, 323 (1988)

Subject classification: 61.10; 61.70; S5

Laboratory of Solid State Microstructure, Nanjing University[1])

The Simulation of Modulated I_3 Fringes in an X-Ray Section Topograph of a Stacking Fault

By

S. S. JIANG, Y. QIU, and G. S. GREEN[2])

Computer simulations are performed of the fading phenomena visible in the fringe systems of a section topograph image of a stacking fault obtained with an unpolarized incident X-ray beam. Periodic fading of the I_3 fringes is seen in a section topograph of a stacking fault in an otherwise nearly perfect diamond taken under high absorption conditions. Simulations based on dynamical diffraction theory are performed separately for the σ and π polarization states. The image formed by the superposition of these simulations also shows I_3 fringe modulation, and is in excellent agreement with the experiment.

Für das Schwundphänomen in den Streifensystemen des Sektionstopographdiagramms eines Stapelfehlers mit einem unpolarisiert einfallenden Röntgenstrahl werden Computersimulationen durchgeführt. Periodisches Schwinden der I_3-Streifen wird in einer Sektionstopographie eines Stapelfehlers in einem sonst nahezu perfekten Diamant unter Hochabsorptionsbedingungen beobachtet. Simulationen auf der Grundlage der dynamischen Beugungstheorie werden getrennt für σ- und π-Polarisationszustände durchgeführt. Das durch Überlagerung dieser Simulationen gebildete Diagramm zeigt ebenfalls I_3-Streifenmodulation und befindet sich in ausgezeichneter Übereinstimmung mit dem Experiment.

1. Introduction

The fringe pattern in an X-ray section topograph of a stacking fault is complicated in comparison to that produced by transmission electron microscopy. These complications are due to several factors: the spherical wave nature of the incident X-ray beam under usual experimental conditions combined with the greater specimen thickness and larger Bragg angles, and the possibility of polarization effects in the X-ray case.

The image contrast was first analyzed by Kato et al. [1] for the zero absorption case. Authier [2] later developed the theory for an absorbing crystal. The experimental dynamical image of a stacking fault in dolomite was reported by Authier [3], and the theoretical analysis was confirmed by Wonsiewicz and Patel [4] using computer simulations. Both K_0 and K_h high resolution images of a stacking fault in natural diamond were obtained by Jiang and Lang [5], who demonstrated that the nature of the stacking fault may be determined from the contrast of the first I_3 fringes in both types of image. The closely related dynamical images of Brazil twin boundaries in natural quartz have been investigated by Yang et al. [6], who also studied both K_0 and K_h images.

The periodic modulation of pendellösung fringes can be interpreted in terms of the dynamical theory of X-ray diffraction by taking the unpolarized nature of the incident X-ray beam into account [7 to 9]. For a crystal containing a stacking fault,

[1]) Nanjing, People's Republic of China.

[2]) Permanent address: Department of Physics, University of Durham, South Road, Durham DH1 3LE, Great Britain.

fringe fading phenomena have been observed by section topography [5], but were not simulated at that time. Kowalski and Lang [10] later performed simulations which included the polarization effect for low absorption conditions. In this paper, simulations of the image of a stacking fault in a diamond crystal which take polarization into account are presented for the high absorption case, where the I_3 fringes are predominant. The modulated fringe pattern thereby produced is compared to that in the experimental section topographs obtained previously.

2. Brief Summary of the Theory

Based upon the spherical wave theory of pendellösung interference in perfect crystals developed by Kato [11], the contrast of a stacking fault fringe pattern was first calculated by Kato et al. [1]. Authier [2] developed the analysis to consider explicitly the effect of absorption.

In the general case shown schematically in Fig. 1, a spherical wave is incident upon the specimen at the point A. The whole dispersion surface is therefore excited, and directions of energy flow from A fill the triangle BAC. The wave field which propagates in a given direction is formed from the interference of two waves, corresponding to a conjugate pair of tie points on the two branches of the dispersion surface.

The crystal is split into two regions by a stacking fault which intersects the plane of incidence along the line B_1C_1. The ray AQ is a typical wave field propagating in region I of the crystal. At the fault plane, interbranch scattering occurs, resulting in the creation of a new wave field QP, while QO contains the fraction of energy in AQ not scattered by the stacking fault. The image at the point P is therefore due to the interference between the new wave field QP and the unscattered wave field AP. In an otherwise perfect crystal, all the new wave fields formed by interbranch scattering at the stacking fault are brought to a focus at the point A′, which is a corner of the parallelogram $AB_1A'C_1$.

If the stacking fault is inclined within the crystal in such a way that it intersects both entrance and exit surfaces, it is clear from the geometry that there will exist a plane of incidence in which the focus of the new wave fields, A′, lies on the exit surface, thus giving rise to the characteristic hour-glass shape of the section topograph image.

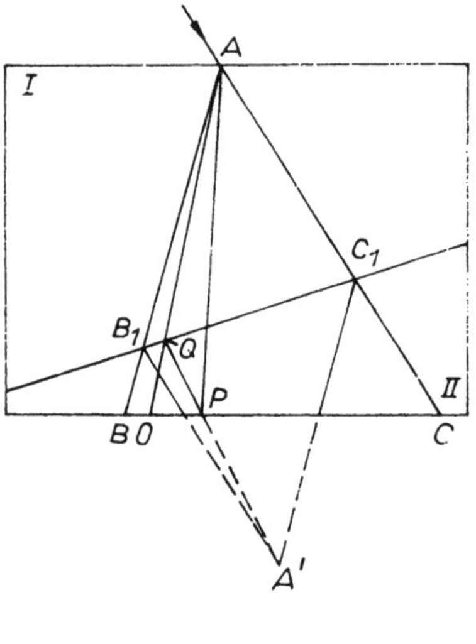

Fig. 1. Paths of the X-ray wave fields in a single plane of incidence

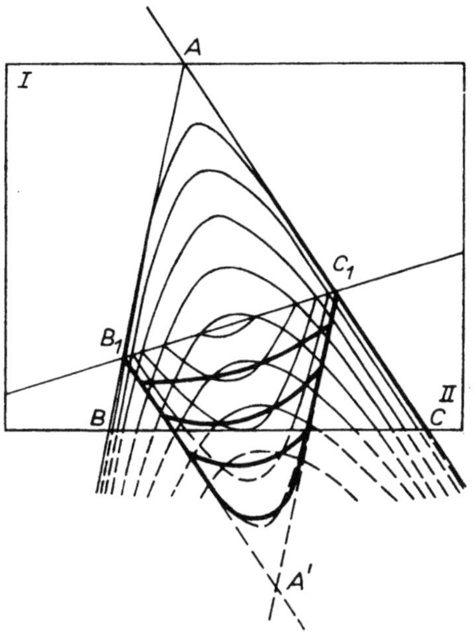

Fig. 2. The formation of the fringe patterns in the crystal

The formation of interference fringes in the stacking fault image is shown in Fig. 2. The hyperbolae shown, extending throughout the whole crystal, are lines of constant phase for the old wave fields which are not scattered by the fault. The hyperbolae which lie within the triangle $B_1A'C_1$ in region II of the crystal represent the fringe system of the new wave fields created by interbranch scattering at the fault. The thick lines in region II of the crystal represent the fringes due to interference between the old and new wave fields.

The intensity at a point along BC on the exit surface may be written as $I = I_1 + I_2 + I_3$,

where

$$I_1 = I_p(1 - A \sin^2 \tfrac{1}{2} \delta) ,$$

$$I_2 = B \sin^2 \tfrac{1}{2} \delta ,$$

$$I_3 = C \sin^2 \tfrac{1}{2} \delta + D \sin \delta .$$

δ is the phase shift introduced when a wave crosses the fault and I_p is the intensity of a perfect crystal. I_1 is the term due to the old wave fields and is the same as that of a perfect crystal, but reduced by the factor $(1 - A \sin^2 \frac{1}{2} \delta)$. I_2 is due to interference between new interbranch scattered waves, and I_3 is due to interference between the old and new wave fields. Complete expressions for A, B, C, and D are given by Authier [2].

The relative magnitude of these constants is dependent on the absorption. In particular, D is zero in the case of zero absorption and the intensity I_1 is usually quite low. Under conditions of low absorption, the stacking fault image is dominated by the I_2 fringes, whereas when absorption is high, D is relatively large and the I_3 fringes predominate.

Under usual experimental conditions in the laboratory, the incident X-ray beam is unpolarized. It therefore contains two components, with the electric field vector lying perpendicular to or within the plane of incidence. The extinction distance of pendellösung fringes is given by

$$\xi_h = \frac{(\gamma_0 \gamma_h)^{1/2}}{KP(\chi_h \chi_{\bar{h}})^{1/2}},$$

where $\gamma_0 = \boldsymbol{s}_0 \cdot \boldsymbol{n}_e$ and $\gamma_h = \boldsymbol{s}_h \cdot \boldsymbol{n}_e$, $\boldsymbol{n}_e$ is the unit vector normal to the entrance surface, $\boldsymbol{s}_0$ and $\boldsymbol{s}_h$ are unit vectors in the directions $\boldsymbol{K}_0$ and $\boldsymbol{K}_h$, respectively, $K = 1/\lambda$, and the polarization factor P is given by

$$P = 1 \qquad (\sigma \text{ polarization}),$$

$$P = |\cos 2\theta_B| \qquad (\pi \text{ polarization}).$$

Clearly the extinction distances for the σ and π polarization states are different, leading to the possibility of observing fading phenomena due to beating between the two interference fringe systems.

3. Results

The stacking fault fringe patterns observed in high resolution X-ray section topographs have been described previously [5]. Simulations were presented of the images observed in the low absorption cases using AgK_α and MoK_α radiation. In both cases the Bragg angle for the $1\bar{1}1$ reflection studied is small, so the simulations were performed using a mean value for the polarization factor. For this reason they did not show the periodic fading of the I_2 fringes observed experimentally. Kowalski and Lang [10] have since produced a simulation of an image taken with MoK_α radiation which takes the polarization effect into account. The fading of the I_2 fringes in the simulation is in good agreement with the experiment.

The image under consideration here is that of the asymmetric $1\bar{1}1$ reflection taken using CuK_α radiation. The sample is a nearly perfect natural diamond cut and polished on $\langle 001 \rangle$ surfaces to form a plate 1013 μm thick, so that $\mu t = 1.5$, which represents a moderately high absorption condition. The Bragg angle in this case is 21.9° and so, using the following value for the Fourier component of the susceptibility,

$$\chi_h = 8.27 \times 10^{-6} + i\, 2.774 \times 10^{-8},$$

the extinction distances for the σ and π polarization states are found to be

$$\xi_h^\sigma = 16.3\ \mu\text{m} \quad \text{and} \quad \xi_h^\pi = 22.6\ \mu\text{m}.$$

The difference between these extinction distances is considerable, leading to fading of the interference fringes over a short period. In order to produce an accurate simulation of the image, it is clearly necessary to calculate both σ and π components separately.

The simulations for the two polarization states are shown in Fig. 3. They are photographs of dot matrix prints produced on an IBM PC-AT with an M2024 printer using the TOPEMP3 program developed by Dr. Y. Epelboin. The simulations are 480 points wide, representing a resolution of 1.5 μm between pixels on the topograph image. Fig. 4a shows the image obtained by combining the intensities of the simulations for the two polarization states. It should be compared with the corresponding experimental topograph which is shown in Fig. 4b.

The top half of the experimental image does not form a triangle whose width fills the image, as in the simulation. This is because the stacking fault does not completely

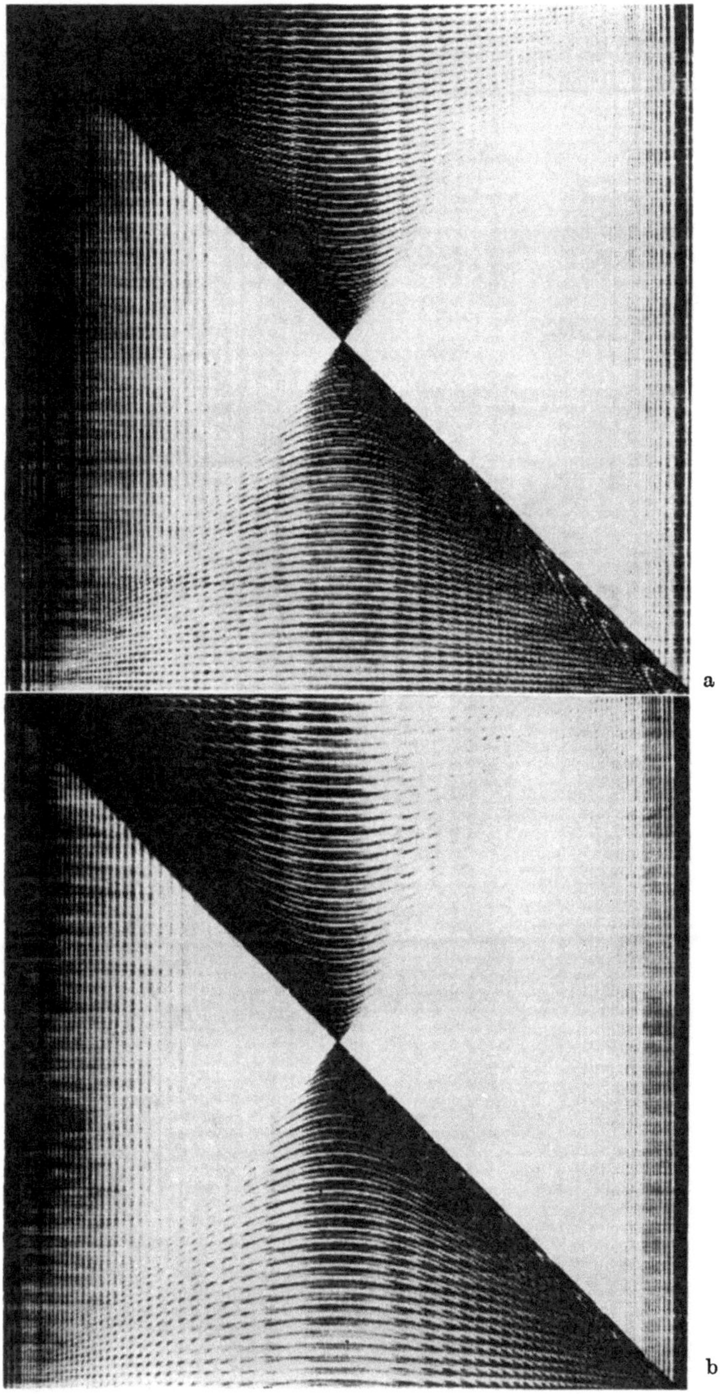

Fig. 3. Computer simulations of the stacking fault image; a) σ polarization state, b) π polarization state

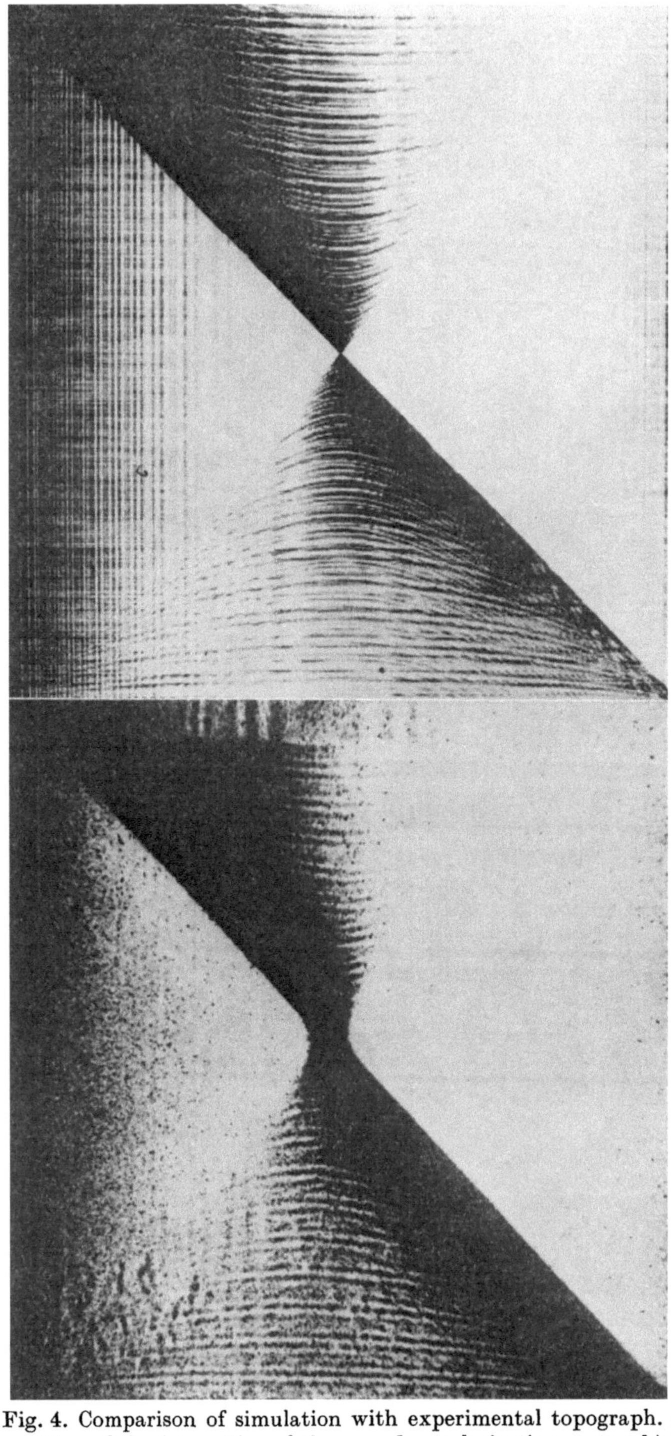

a

b

Fig. 4. Comparison of simulation with experimental topograph. a) Simulation formed by superposition of the intensities of the σ and π polarization states. b) Section topograph of a stacking fault in diamond. CuK_α radiation, $1\bar{1}1$ reflection. The image width is 0.72 mm

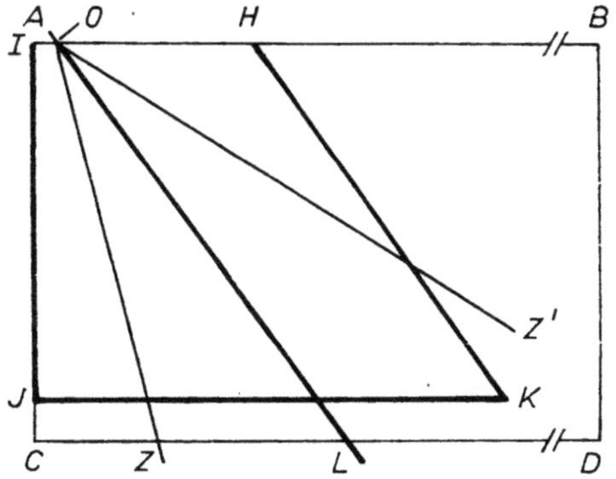

Fig. 5. Scale drawing of the projection of the stacking fault onto a plane of incidence in the specimen

intersect the Borrmann triangle. Fig. 5 shows the projection of the stacking fault onto a plane of incidence. The fault was nucleated at the point K, and within the crystal it has edges KH and KJ which lie in the [011] and [$\bar{1}$10] directions, respectively. ZOZ' defines the energy flow triangle, with the incident wave taking the direction OZ for the 1$\bar{1}$1 reflection.

The extinction distance is the period of pendellösung oscillations parallel to the Bragg planes, so that the total number of oscillations visible in one half of the image is given by

$$N = \frac{t}{\xi_h \cos \psi},$$

where ψ is the angle between the Bragg planes and the normal to the exit surface, which is 35.3° in the present case. The period of oscillation of the I_3 fringes is twice that of the I_2 and I_1 systems, so that the number of I_3 fringes $N_3 = \frac{1}{2} N$, and so the numbers of fringes in the two polarization cases are

$$N_3^\sigma = 38.1 , \qquad N_3^\pi = 27.5 .$$

These values are in good agreement with the measured values obtained from the simulations which are 39 and 28, respectively.

The mean number, $\bar{n}$, of fringes in a repetition period of the fading pattern in the combined image is given by

$$\bar{n} = \frac{1 + P}{1 - P} = 3.1$$

and so the total number, N', of fading periods is given by

$$N' = \frac{\bar{N}_3}{\bar{n}} = \frac{\frac{1}{2}(38.1 + 27.5)}{3.1} = 10.6$$

which is in excellent agreement with the experiment and the simulation which both contain eleven fading periods.

4. Discussion

In contrast to the low absorption case using MoK_α and AgK_α radiation, in which fading was only visible in the I_2 fringe system, the high absorption case studied in, this paper exhibits extensive fading phenomena. In both experiment and simulation the most obvious fading occurs in the I_3 fringes, but it is also clear in the I_1 fringes and it may just be seen in the I_2 fringes by viewing the photographs at a grazing angle along the I_2 fringe asymptotes. The experimental image also shows periodic fading in the perfect crystal pendellösung fringes at the top of the figure.

The most obvious difference between the experimental fringe patterns and the simulations is the imperfect focussing of the interbranch scattered waves in the experiment. This is due to slight curvature of the lattice, and has been discussed previously [5].

The simulations show a fringe pattern on the right-hand side of the lower half of the hour-glass image which is not observed experimentally. This is a Moiré fringe pattern which is an artefact of the simulation conditions.

The fading phenomena discussed in this paper serve to highlight the care that must be taken when using pendellösung fringes in the determination of structure factors [8].

Acknowledgements

The authors thank Dr. Y. Epelboin (Laboratoire de Minéralogie-Cristallographie, University of Paris VI) for making his TOPEMP3 simulation program available. One of the authors (G.S.G.) thanks the British Council for financial assistance. The financial support for this work is from the National Natural Science Foundation of China.

References

[1] N. Kato, K. Usami, and T. Katagawa, Adv. X-Ray Analysis **10**, 46 (1967).
[2] A. Authier, phys. stat. sol. **27**, 77 (1968).
[3] A. Authier, Adv. X-Ray Analysis **10**, 24 (1967).
[4] B. C. Wonsiewicz and J. R. Patel, J. appl. Phys. **47**, 1837 (1976).
[5] S. S. Jiang and A. R. Lang, Proc. Roy. Soc. **A388**, 249 (1983).
[6] P. Yang, S. S. Jiang, H. N. Zhou, and D. Feng, phys. stat. sol. (a) **97**, 411 (1986).
[7] H. Hattori, H. Kuriyama, and N. Kato, J. Phys. Soc. Japan **20**, 1047 (1965).
[8] M. Hart and A. R. Lang, Acta cryst. **19**, 73 (1965).
[9] S. S. Jiang, Acta phys. Sin. **32**, 1497 (1983).
[10] G. Kowalski and A. R. Lang, J. appl. Cryst. **19**, 224 (1986).
[11] N. Kato, Acta cryst. **13**, 349 (1960).

(Received August 2, 1988)

phys. stat. sol. (a) **110**, 331 (1988)

Subject classification: 61.55; S1.2

Physical Metallurgy Division (a) and Radiometallurgy Division (b), Bhabha Atomic Research Centre, Trombay[1])

Computer Simulation of the Early Stages of Ordering in Ni–Mo Alloys

By

U. D. KULKARNI (a), S. MURALIDHAR (b), and S. BANERJEE (a)

The nature of the short range ordered (SRO) state and the mechanism of evolution of the long range ordered (LRO) structure in the Ni–Mo system are subjects of considerable controversy. Attempts are made to simulate the state of SRO and the transitional states between SRO and LRO at the atomistic level. The two descriptions of the SRO state, namely, the concentration wave model and the cluster model are discussed in the light of the results of the present simulation.

Die Natur des nahgeordneten (SRO) Zustandes und der Mechanismus der Entwicklung von ferngeordneten (LRO) Strukturen im System Ni–Mo sind derzeit Gegenstand einer Kontroverse. Es wird ein Versuch unternommen, auf atomistischer Ebene den Nahordnungszustand und die Übergangszustände zu simulieren, die beim Übergang von Nahordnung zur Fernordnung auftreten können. Elektronenbeugungsdiagramme werden berechnet und mit experimentellen Aufnahmen verglichen. Zwei Beschreibungen des nahgeordneten Zustandes, nämlich das Konzentrationswellen-Modell und das Cluster-Modell, werden im Rahmen der Ergebnisse dieser Untersuchungen diskutiert.

1. Introduction

It is generally believed that the SRO state of an alloy is a prefiguration of its structure in the LRO state. Information about the LRO state is expected to be present in the SRO state in an embryonic form. In fact, there is a school of thought which interprets a short-range ordered alloy as composed of microdomains of LRO structures [1 to 3]. The ordering behaviour of the Ni–Mo system, however, does not seem to follow this pattern. Samples of Ni-rich alloys belonging to this system, when quenched from the single phase f.c.c. region exhibit SRO intensity maxima at $\{1\frac{1}{2}0\}$ and equivalent positions in the reciprocal space. A complete absence of intensity at {210} positions is observed. The fact that such a diffraction geometry is not consistent with the positions of superlattice reflections of any of the coherent LRO structures appearing in this system [4 to 6] suggests that the structure represented by the SRO state is not the same as that of any of its LRO counterparts. This anomaly has been a subject of a great deal of discussion and of numerous experimental investigations [1 to 11].

Of the several models proposed in order to explain the nature of the SRO state and the evolutionary states between SRO and LRO, the most attractive ones are the cluster model proposed by Amelinckx and coworkers [3, 10, 11] and the static concentration wave approach of de Fontaine [7]. While the former is based on the assumption that SRO is characterised by small locally ordered atomic configurations representing

[1]) Trombay, Bombay 400085, India.

motifs of the coherent LRO structure(s), the latter ascribes the occurrence of the $\{1\frac{1}{2}0\}$ SRO maxima to the formation, in the solid solution, of low amplitude concentration waves with $\langle 1\frac{1}{2}0\rangle$ $\boldsymbol{k}$-vectors. The microdomain model due to Thomas and coworkers [1, 2] envisages the sizes of the locally ordered regions to be of the dimension of several unit cells of the LRO structures. In a recent publication [3], van Tendeloo et al. have jointly concluded that the seemingly contradictory concepts, namely, the concentration wave and the microdomains are, in fact, two aspects of the same physical reality or two languages to describe the same phenomenon. The present work seeks to resolve the controversies related to the structures of the SRO state and the SRO to LRO transition states via a computer simulation of these states on the atomistic scale.

Ordering processes in the Ni–Mo system show several remarkable features. Firstly, alloys containing 15 to 28 at% Mo, when quenched from the single phase f.c.c. field exhibit SRO characterised by the presence of intensity maxima at $\{1\frac{1}{2}0\}$ positions in the reciprocal space [6]. Secondly, the transformation from the SRO state to the equilibrium/metastable coherent LRO structures appears to take place by a continuous mechanism at relatively low temperatures of ageing [6, 9, 12]. Three different coherent LRO structures, namely, the equilibrium Ni_4Mo phase ($D1_a$ structure) and the metastable phases Ni_3Mo (DO_{22} structure) and Ni_2Mo (Pt_2Mo-type) have been reported to evolve from the short-range ordered alloy, depending upon the ageing treatment and the composition of the alloy [1 to 3, 5, 6, 9, 12]. These structures belong to the "$1\frac{1}{2}0$ family" of ordered structures and can be described by different stacking sequences of all Ni or all Mo layers on {420} sets of planes of the disordered f.c.c. lattice [1, 7, 8]. The initial stages of evolution of these structures are marked by the appearance of streaks of diffuse intensity in the electron diffraction patterns, Some of these patterns have been presented in the later part of this paper (in Fig. 11). These streaks are continuous lines/circular arcs which run from the positions of the SRO maxima to the superlattice positions of the emerging LRO structures or in certain cases, from the superlattice positions of one LRO structure to those of another. The pattern of diffuse intensity has been found to depend upon the alloy composition and the ageing treatment and to vary continuously during the SRO to LRO transition [1, 3, 5, 6, 9, 12]. Amelinckx and coworkers [3, 10, 11] have attributed the appearance of such streaks to the formation, in the solid solution, of atomic clusters of certain kind during the SRO to LRO transition.

In the present work, two-dimensional (2D) diffraction gratings representing a unit cell thick slab of two adjacent (200) layers of the f.c.c. structure have been used to simulate the SRO and the SRO to LRO transitional states. Computed [100] diffraction patterns from such gratings have been generated for comparison with the experimentally obtained selected area diffraction (SAD) patterns. These simulations are similar, in principle, to the optical simulations of the SRO state by Amelinckx and coworkers [3, 11]. The generation of the computed diffraction patterns (CDPs) requires a great deal of computational effort. This is the main difficulty in attempting a simulation using a 3D lattice as a diffraction grating, although such a simulation is expected to be more realistic. It has been possible, however, to simulate many of the features of the SRO and the transitional states in the 2D simulation carried out in the present work.

2. Ordering in the Ni–Mo System

2.1 *Ordering below the $1\frac{1}{2}0$ spinodal*

The $1\frac{1}{2}0$ ordering spinodal in Ni–Mo alloys owes its existence to the minima of the $\boldsymbol{k}$-space potential energy function, $\boldsymbol{V}(\boldsymbol{k})$, at $\langle 1\frac{1}{2}0\rangle$ special points [7, 8]. This spinodal

Table 1

Clusters used in the simulation of various transitional states

structure and stoichiometry	arrangement on {420} planes	description of the structure	cluster size used in simulation in No. of molecules
N_2M_2	NNMM	hypothetical structure generated by a single $\langle 1\frac{1}{2}0\rangle$ concentration wave [7, 8]	1, 4, 16, 44
Pt_2Mo	NNM	b.c. orthorhombic	1, 4, 16
$DO_{22}(N_3M)$	NNNM	b.c. tetragonal	1, 4, 16
$D1_a(N_4M)$	NNNNM	b.c. tetragonal	1, 4, 16
N_3M	—	hypothetical; result of superposition of two perpendicular $\langle 1\,\frac{1}{2}\,0\rangle$ waves	1

Two orientational variants as illustrated in Fig. 2 were used in each of the first four cases. 32 different unimolecular configurations from a single variant of N_3M were chosen in the last case.

defines a region in the phase diagram wherein the f.c.c. solid solution becomes unstable with respect to concentration waves with $\boldsymbol{k}$-vectors equal to $\langle 1\frac{1}{2}0\rangle$. On quenching from the single phase f.c.c. region into the region of the $1\frac{1}{2}0$ spinoidal, the solid solution comes under the influence of the spinodal, leading to an amplification of $\langle 1\frac{1}{2}0\rangle$ concentration waves. This phenomenon known as $\langle 1\frac{1}{2}0\rangle$ spinodal ordering results in the formation of the "$1\frac{1}{2}0$ state of short-range order" characterised by diffuse intensity maxima at $\{1\frac{1}{2}0\}$ and equivalent positions in the reciprocal space. The structural interpretation of a state of order of this kind has been a subject of controversy [1, 3, 7, 10, 11]. However, it has been pointed out in a previous publication [8] that such a state of order would be marked by the presence of clusters of a partially ordered, off-stoichiometric N_2M_2 structure — a hypothetical structure proposed by Khachaturyan [13]. This structure consists of an Ni–Ni–Mo–Mo sequence of atoms on a {420} set of planes of the f.c.c. lattice (Table 1) and gives rise to superlattice reflections at $\frac{1}{4}\{420\}$, $\frac{3}{4}\{420\}$, and equivalent positions as shown schematically in Fig. 1. A stacking sequence of this kind is equivalent to a fully amplified $\langle 1\frac{1}{2}0\rangle$ concentration wave in an equiatomic alloy. On subsequent ageing at temperatures

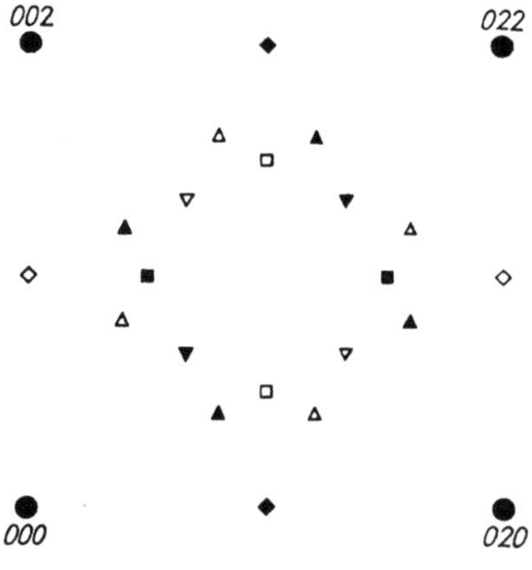

Fig. 1. A quadrant of a [100] diffraction pattern (schematic). ● f.c.c.; △ $D1_a$; □, ◇ DO_{22}; □ N_2M_2 (and SRO maxima); ▽ Pt_2Mo reflections. Open and solid symbols refer to two different variants

below the $1\frac{1}{2}0$ spinodal, the short-range ordered alloy transforms into different equilibrium/metastable coherent LRO phases depending upon the alloy composition and the ageing treatment. It has been suggested in a previous publication [8] that the SRO to LRO transition involves interaction of two perpendicular $\langle 1\frac{1}{2}0\rangle$ concentration waves. Such an interaction would result in an "N_3M structure" which comprises a periodic arrangement of unimolecular subunit cell clusters belonging to two variants each of the Ni_4Mo and the Ni_2Mo structures.

2.2 SRO to LRO transition in Ni_4Mo alloy

On ageing, a spinodally ordered alloy of Ni_4Mo composition transforms into the equilibrium Ni_4Mo ($D1_a$) phase. The $D1_a$ structure consists of an Ni–Ni–Ni–Ni–Mo stacking sequence of atoms on a {420} set of planes (Table 1) and gives rise to superlattice reflections at $\frac{1}{5}$, $\frac{2}{5}$, $\frac{3}{5}$, and $\frac{4}{5}$ {420} and equivalent positions (Fig. 1). At relatively lower ageing temperatures, this structural transformation appears to proceed in a continuous manner during the initial stages [1, 12], as evidenced by a continuous transfer of intensity from the SRO intensity maxima to the superlattice positions of the $D1_a$ structure. This transfer of intensity takes place along circular arcs emanating from the SRO maxima and reaching the $D1_a$ superlattice positions.

2.3 SRO to LRO transition in Ni_3Mo alloy

The SRO to LRO transition in the Ni_3Mo alloy is more complex in that, on ageing, the SRO alloy decomposes into a phase mixture of two coherent LRO structures, namely: the Ni_4Mo phase with a $D1_a$ structure and the Ni_2Mo phase with a Pt_2Mo-type structure [2, 5, 6, 9]. The latter structure consists of an Ni–Ni–Mo stacking sequence of atoms on a {420} set of planes of the f.c.c. lattice and it gives rise to superlattice reflections at $\frac{1}{3}${420}, $\frac{2}{3}${420}, and equivalent positions as shown in Fig. 1. The onset of this phase separation reaction is marked by the appearance of pseudocircles of diffuse intensity passing through the four $\{1\frac{1}{2}0\}$ maxima in the {020} square in the [100] diffraction pattern (Fig. 1). The pattern of diffuse intensity changes continuously during ageing before resolving itself into distinct superlattice reflections of the Ni_2Mo and the Ni_4Mo structures. The relevant diffraction patterns have been presented in the later part of this paper (in Fig. 11).

During electron irradiation at low temperatures [9], a secondary tendency for the formation of an Ni_3Mo phase (DO_{22} structure) which possesses the same stoichiometry as the alloy, manifests itself in the form of faint ⟨100⟩ streaks joining the SRO maxima and the {100} positions. This structure which consists of an Ni–Ni–Ni–Mo stacking sequence of atoms on a {420} set of planes of the f.c.c. lattice would yield superlattice reflections at $\frac{1}{4}${420}, $\frac{2}{4}${420}, $\frac{3}{4}${420}, and equivalent positions as shown in Fig. 1. Interestingly, such a tendency for the evolution of the DO_{22} structure is also noticed during the evolution of the LRO Ni_2Mo (Pt_2Mo) structure in a rapidly solidified Ni_2Mo alloy [14].

3. Computed Diffraction Patterns

3.1 Simulated diffraction gratings

Two-dimensional diffraction gratings consisting of 168 × 168 points were used to simulate the various states of order in Ni–Mo alloys. These gratings represent the [100] projection of two successive (200) layers of the f.c.c. lattice as shown in Fig. 2. The occupancy of lattice points (by an N or an M atom) was determined with the

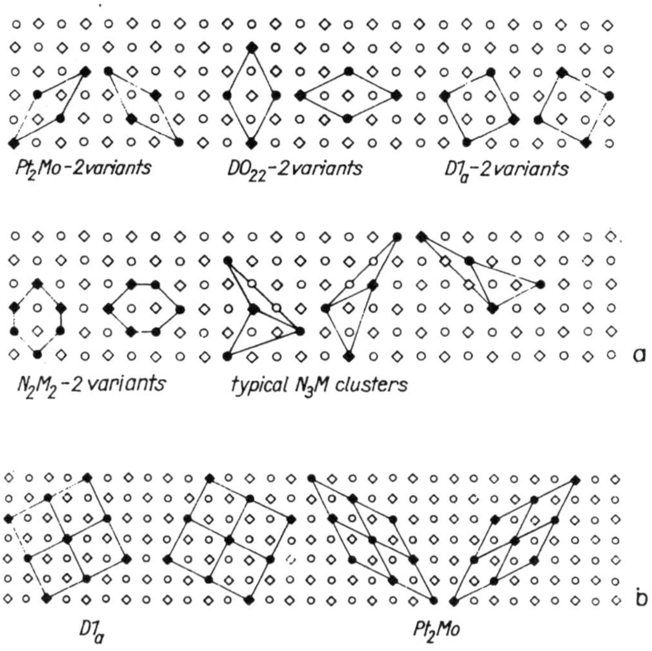

Fig. 2. A [100] projection of f.c.c. lattice showing typical clusters used in the simulation. a) Unimolecular clusters, b) clusters of 4-molecule size. 1/2th layer (◇,◆), 0th layer (○, ●); open symbols for Ni atoms, full symbols for Mo atoms

help of two different computational algorithms for the two cases considered in the present simulation, viz. the concentration wave approach and the cluster approach.

3.1.1 Simulation using the concentration wave approach

In this approach, the probability of occupation of a lattice site by an atom of a particular species is a function of the average alloy composition and of the amplitude of the concentration wave in question. Two different situations were considered in the present simulation. In the first case, a single $\langle 1\frac{1}{2}0\rangle$ wave was generated over the entire grating, while in the second case, two perpendicular $\langle 1\frac{1}{2}0\rangle$ waves were introduced over the grating. The overall composition was assumed to be N_3M and the amplitude of the wave was varied between zero (the completely disordered state) to 0.25 (the maximally ordered state). Atoms were placed at lattice sites by following the Monte Carlo approach, wherein random events are simulated by the generation of random numbers. Thus, an M-atom was placed at a given lattice site if the random number generated, R, satisfied the condition, $0 < R < P_M$, where P_M is the probability of occupation of the site by an M-atom. An N-atom was placed at the lattice site if the condition was not satisfied. A representative portion of a grating depicting the superposition of two perpendicular $\langle 1\frac{1}{2}0\rangle$ waves is shown in Fig. 3. Such an arrangement of atoms would represent a partially ordered (concentration amplitudes = 0.175) N_3M structure [8]. Square-shaped and lozenge-shaped atomic configurations of Mo atoms representing subunit cell clusters of the $D1_a$ and the Pt_2Mo structures, respectively, have been delineated.

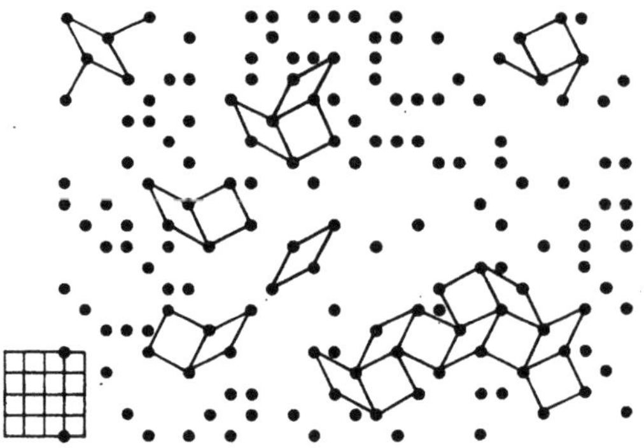

Fig. 3. A [100] projection of a partially ordered N_3M structure generated by a superposition of two perpendicular $\langle 1\frac{1}{2}0\rangle$ concentration waves. Points show the positions of Mo atoms. Square shaped Ni_4Mo and lozenge-shaped Ni_2Mo clusters have been delineated

3.1.2 Simulation using the cluster approach

The cluster approach is based on the assumption that the SRO state is composed of locally ordered atomic configurations (clusters) which have structural motifs representative of the coherent LRO structures in the Ni–Mo system. In the present simulation work, a variety of clusters belonging to the different LRO structures in the "$1\frac{1}{2}0$ familiy" were used as building blocks to simulate the various states of order. Clusters of various sizes each containing an integral number of stoichiometric units or "molecules" of the afore-mentioned LRO structures were used in most cases. The 2D gratings described in Section 3.1 were prepared by a random placement of such clusters at the lattice points. Table 1 gives details of the clusters employed in the simulation. Some of these are shown in Fig. 2.

The clusters chosen in these simulation experiments are expected to be atomic configurations of low internal energy since they maximise the number of third nearest neighbour Mo–Mo pairs [8]. As a matter of fact, the unimolecular as well as the larger clusters of the Pt_2Mo, the DO_{22}, and the $D1_a$ structures (Fig. 2) are bounded by $1\frac{1}{2}\frac{1}{2}$ vectors which are the position vectors of the f.c.c. third nearest neighbours. Like third nearest neighbour bonds such as these play a vital role in the occurrence of the $1\frac{1}{2}0$ instability and in stabilising the ordered structures belonging to this family [8]. Some of the clusters used in this simulation work have actually been observed in the high resolution electron microscopy (HREM) images of SRO Ni–Mo alloys [3, 11].

In the computer program written for constructing the diffraction gratings, the location of a cluster is decided by generating two random numbers between 1 and 168 for the X and the Y coordinates, respectively. In the event of encroachment of a new cluster into an existing one, the new cluster is allowed to erase parts of the old

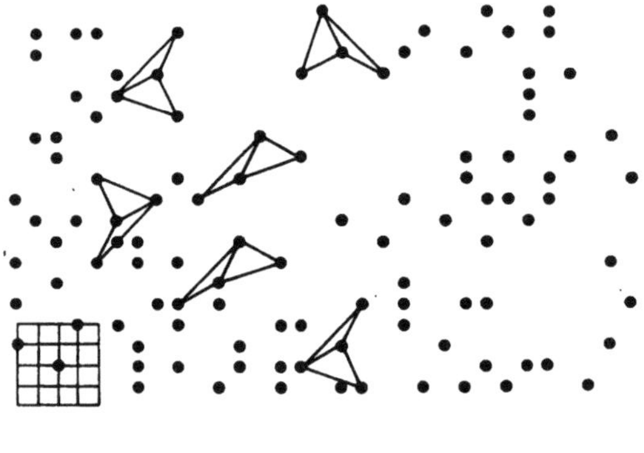

Fig. 4. A portion of a grating based on 16 different unimolecular N_3M clusters. Clusters which have remained intact during the generation of the grating have been delineated

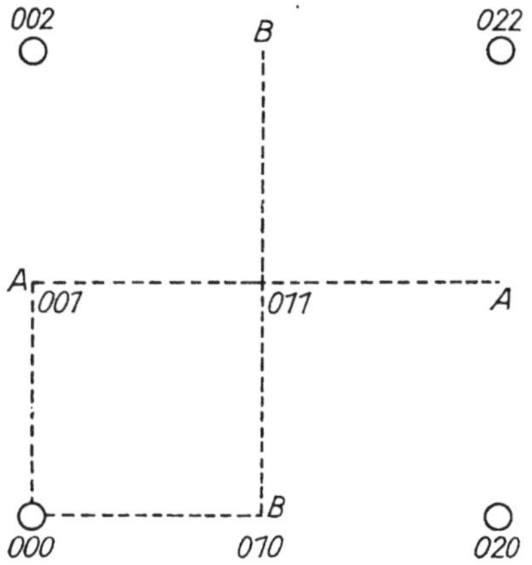

Fig. 5. Generation of the 000–002–022–020 square of a computed diffraction pattern from the smaller square by reflection of intensity values across the mirrors AA and BB

cluster and the number of lattice points so modified is counted. The process of placing clusters is terminated when the total number of points thus modified exceeds a preset value. As an illustration, a representative portion of a grating based on a cluster combination of sixteen different unimolecular N_3M clusters is shown in Fig. 4. Cluster collisions result in a melange in which individual clusters lose their identity. Some N_3M clusters that have remained intact are delineated.

3.2 Generation of diffraction patterns

The 2D diffraction gratings prepared with the help of the computational algorithms described in Section 3.1 were used to generate computed [100] reciprocal lattice sections. This procedure involves computation of structure factors (F) at 84 × 84 points over the 000–010–011–001 square in the reciprocal space (Fig. 5). In order to save computer time, it was assumed that only the minority Mo atoms contribute to scattering. In effect, f_{Ni} was assumed to be equal to zero. This simplification results in an increase in the intensity of the diffuse scattering and the superlattice reflections, vis-a-vis that of the matrix reflections. Intensity values (FF^*) over the larger 000–020–022–002 square were obtained by reflection across the two mirror planes AA and BB as shown in Fig. 5. In order to get a good averaging of the intensity values, they were averaged over five different diffraction gratings. Images of diffraction patterns were generated from the intensity data using an image processor. The screen of the image processor used consists of 640 × 512 square-shaped pixels. The intensity of each pixel can be controlled over 256 levels of illumination. The image processor permits a dynamic change of image contrast, while the constructed image is on the screen. As the contrast is decreased, more and more pixels illuminate the screen. When the intensity of a given pixel increases beyond the maximum limit during contrast reduction, the pixel turns dark. In the computed diffraction patterns in Fig. 11, the points close to the exact (000), (020), (022), and (002) positions of the fundamental reflections appear dark as their intensity exceeded the cutoff limit.

4. Results and Discussion

4.1 Diffraction from gratings based on the wave approach

The computed diffraction patterns (CDPs) obtained from gratings which were generated using the method outlined in Section 3.1.1. were found to contain only sharp diffraction spots. These patterns were found to be devoid of streaks and of diffuse

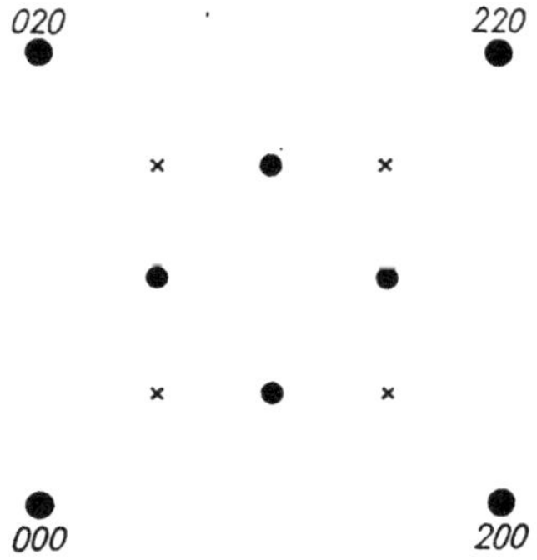

Fig. 6. A schematic CDP from the structure shown in Fig. 3. ● $1\frac{1}{2}0$ and equivalent reflections, × $\frac{1}{2}\frac{1}{2}0$ and equivalent reflections, ● f.c.c. reflections

intensity distributions of any kind. The diffraction effects arising from a single $\langle 1\frac{1}{2}0\rangle$ wave or a superposition of two $\langle 1\frac{1}{2}0\rangle$ waves can be explained with reference to the schematic quadrant of a [100] diffraction pattern as shown in Fig. 1 and 6. The single $\langle 1\frac{1}{2}0\rangle$ wave gives rise to superlattice reflections corresponding to only one set of SRO maxima — $0\frac{1}{2}1$ and $0\frac{3}{2}1$ or $01\frac{1}{2}$ and $01\frac{3}{2}$ depending upon its $\boldsymbol{k}$-vector ($[0\frac{1}{2}1]$ or $[01\frac{1}{2}]$). A superposition of two such waves, however, results in superlattice reflections at all the four SRO positions in addition to weaker reflections at four $\{\frac{1}{2}\frac{1}{2}0\}$ and equivalent positions as shown in Fig. 6. In either case, the intensity of the superlattice reflections increased as the square of the concentration amplitude (Fig. 7). This is the familiar S^2 dependence of the intensity of a superlattice reflection. The fact that all the four SRO maxima are invariably present in a (020) square of the SAD patterns implies that both the variants of $\langle 1\frac{1}{2}0\rangle$ waves are simultaneously present in the half a micron diameter diffraction area of the sample. However, the absence of intensity at $\{\frac{1}{2}\frac{1}{2}0\}$ and equivalent positions suggests that the "N_3M structure" [8] resulting from the interaction of these waves does not form over a long range during the ordering process.

In the wave approach, the probability of occupation of a lattice site by an atom of a particular species is fractional. However, in reality a lattice site is occupied by an M or an N atom. As a consequence, a partially amplified wave would be expected to contain some locally ordered clusters characteristic of the structure it represents. For instance, unimolecular clusters of both Pt_2Mo and $D1_a$ structures can be seen in the grating comprising two superimposed $\langle 1\frac{1}{2}0\rangle$ waves (Fig. 3). The probability of occurrence of such clusters is expected to increase with an increasing wave amplitude.

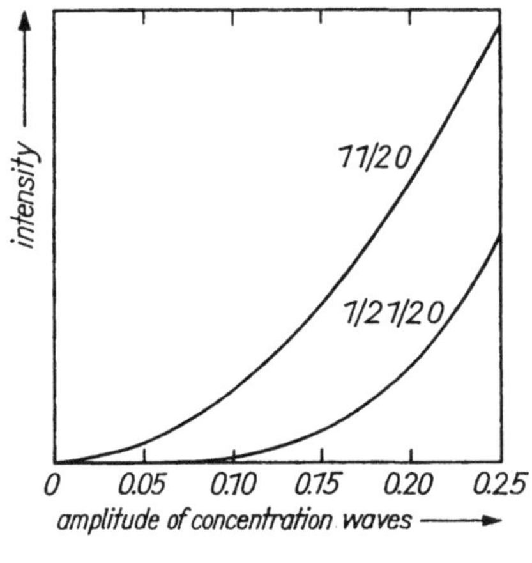

Fig. 7. A plot showing the nature of dependence of the intensity of superlattice reflections on the amplitude of the relevant concentration wave(s)

In the limiting case, when the amplitude of the waves reaches its saturation value, a perfectly long-range ordered "N_3M structure" results.

However, there is an important difference between such gratings characterised by concentration waves (Fig. 3) and those containing degenerate clusters (such as the one based on unimolecular N_3M clusters (Fig. 4). In the former case, the clusters are all in phase with the given concentration wave and contribute to the diffracted intensity of the same superlattice reflections. While in the latter case, the clusters are randomly distributed and are separated from other clusters of the same type by f.c.c. lattice translation vectors rather than the respective superlattice translation vectors. In such a situation, there is no overall concentration wave as there is, on an average, an equal number of clusters which are antiphase to one another. It will be demonstrated in the next section that the latter situation leads to the occurrence of various patterns of diffuse intensity in the SAD patterns.

4.2 Diffraction from gratings based on the cluster approach

4.2.1 Gratings based on identical clusters

Fig. 8 and 9 show CDPs from gratings containing one or four molecule clusters of different ordered structures in the $1\frac{1}{2}0$ family, respectively. In each case, clusters belonging to the two different variants (as described in Table 1) were used. A striking feature of these CDPs was that each cluster-type gave rise to a characteristic diffuse

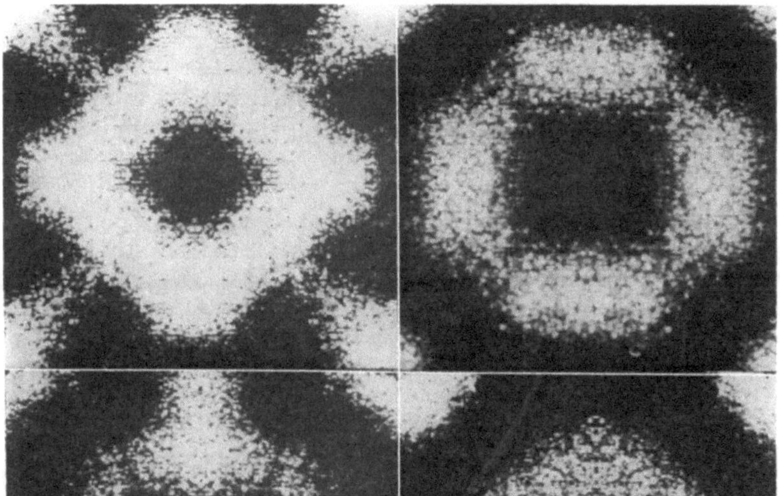

Fig. 8. Computed diffraction patterns from one-molecule clusters of a) Pt_2Mo, b) Dl_a, c) DO_{22}, d) N_2M_2 structures

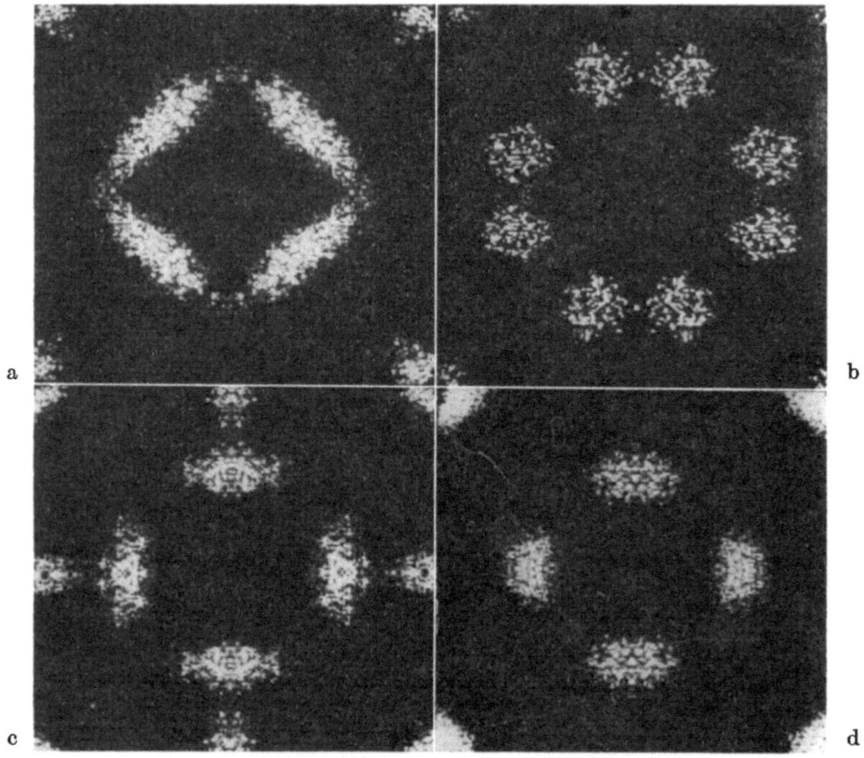

Fig. 9. Computed diffraction patterns from four-molecule clusters. Description of structures as in Fig. 8

intensity distribution which was intimately related to the diffraction pattern characteristic of the corresponding combination of variants of the LRO structures (Fig. 1).

The CDPs corresponding to the unimolecular cluster types (Fig. 8) exhibited diffuse intensity at $1\frac{1}{2}0$ and equivalent positions and also a pattern of diffuse intensity which formed a closed figure around the four SRO maxima. Additional features characteristic of the individual cluster types could be seen in the different CDPs. Unimolecular DO_{22} gave rise to a CDP (Fig. 8c) which was characterised by $\langle 100 \rangle$ streaks extending from the $\{1\frac{1}{2}0\}$ to the $\{100\}$ positions. CDP of the Pt_2Mo cluster type, on the other hand, showed $\langle 110 \rangle$ streaks extending from the fundamental reflections to the $\frac{1}{3}\{220\}$ or equivalent superlattice positions of the Pt_2Mo structure.

On increasing the cluster size from one molecule to four molecules, the streaks became discontinuous showing, thereby, a tendency to turn into superlattice reflections (Fig. 9). The streak pattern in the CDP of the DO_{22} cluster type (Fig. 9c) resolved itself into broad $1\frac{1}{2}0$-type and 100-type intensity maxima, which are the superlattice positions of the DO_{22} structure. Fig. 9a, which corresponds to the Pt_2Mo cluster type of four molecule size shows elongated streaks at the $\frac{1}{3}\{220\}$ and equivalent positions of the Pt_2Mo structure. The $\langle 110 \rangle$ streaks have now shrunk into patches of intensity near the fundamental reflections. Such an intensity distribution which is akin to a sideband in a spinodally decomposed alloy is indeed observed during the early stages of phase separation in Ni_3Mo-based alloys [9].

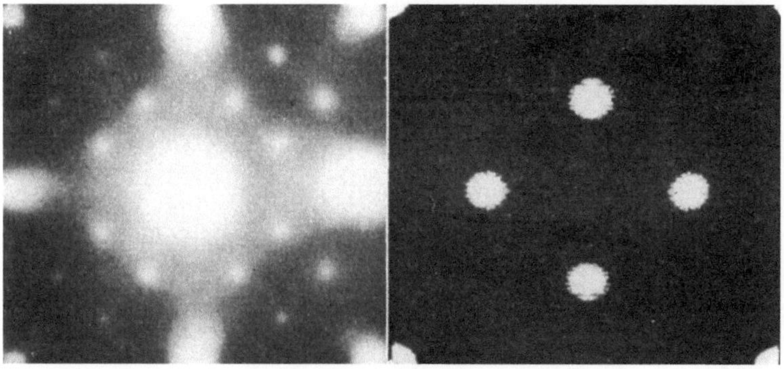

a b

Fig. 10. a) SAD pattern from an as-quenched Ni_4Mo alloy. b) (020) square of a CDP based on 44-molecule N_2M_2 clusters in two different orientational variants. Note the four nearly circular SRO maxima

The CDP based on four molecule N_2M_2 clusters (Fig. 9d) showed intensity maxima at $\{1\frac{1}{2}0\}$ and equivalent positions which are the positions of the SRO maxima. The CDP, however, differs from the SAD patterns obtained from an as-quenched alloy (Fig. 10a) in that the intensity maxima in the CDP have an elongated appearance which is a consequence of the geometry of the N_2M_2 clusters used in the simulation. Circular intensity maxima resembling the actual SRO maxima could be obtained from a grating containing nearly square-shaped 44-molecule N_2M_2 clusters (Fig. 10b). Interestingly, the position and the circular shape of these maxima were found to remain unchanged when the degree of order within such clusters was varied from perfect order to near disorder. N_2M_2 clusters of such dimensions and of a variable degree of order may be thought of as "wave packets" which represent a localisation, in space, of a $\langle 1\frac{1}{2}0\rangle$ concentration wave. Such wave packets, however, differ from the "static concentration wave packets" proposed by Chevalier and Stobbs [15] in that the latter are characterised by diffuse boundaries.

4.2.2 Diffraction from gratings based on clusters of different types

In an attempt to find a reasonably good match between the CDPs and the actual SAD patterns (which is lacking in the case of CDPs based on clusters of the same type) nearly fifty CDPs based on various combinations of clusters belonging to different LRO structures were generated. Fig. 11a to f show [100] SAD patterns depicting different stages of the SRO to LRO transition. (020) squares of CDPs showing features similar to those in the SAD patterns are shown on the right-hand side. In certain cases it was not possible to simulate all the features of an SAD pattern in a single CDP. In such cases, two different CDPs, each showing some of the features of the SAD pattern are presented. The details of the cluster combination used in the computation of each of these CDPs are listed in Table 2. The main features of the observations are summarised in the following:

The onset of the SRO to LRO transition in an Ni_3Mo alloy is marked by a ring of diffuse intensity passing through the four SRO maxima. The corresponding CDP (Fig. 11a) is generated using a cluster combination of sixteen different N_3M clusters of one-molecule size. It should be mentioned here that this structure in its LRO state yields four $\frac{1}{2}\frac{1}{2}0$-type reflections in addition to the $1\frac{1}{2}0$-type reflections as mentioned in Section 4.2.1. It should be emphasised that the CDP was produced with N_3M

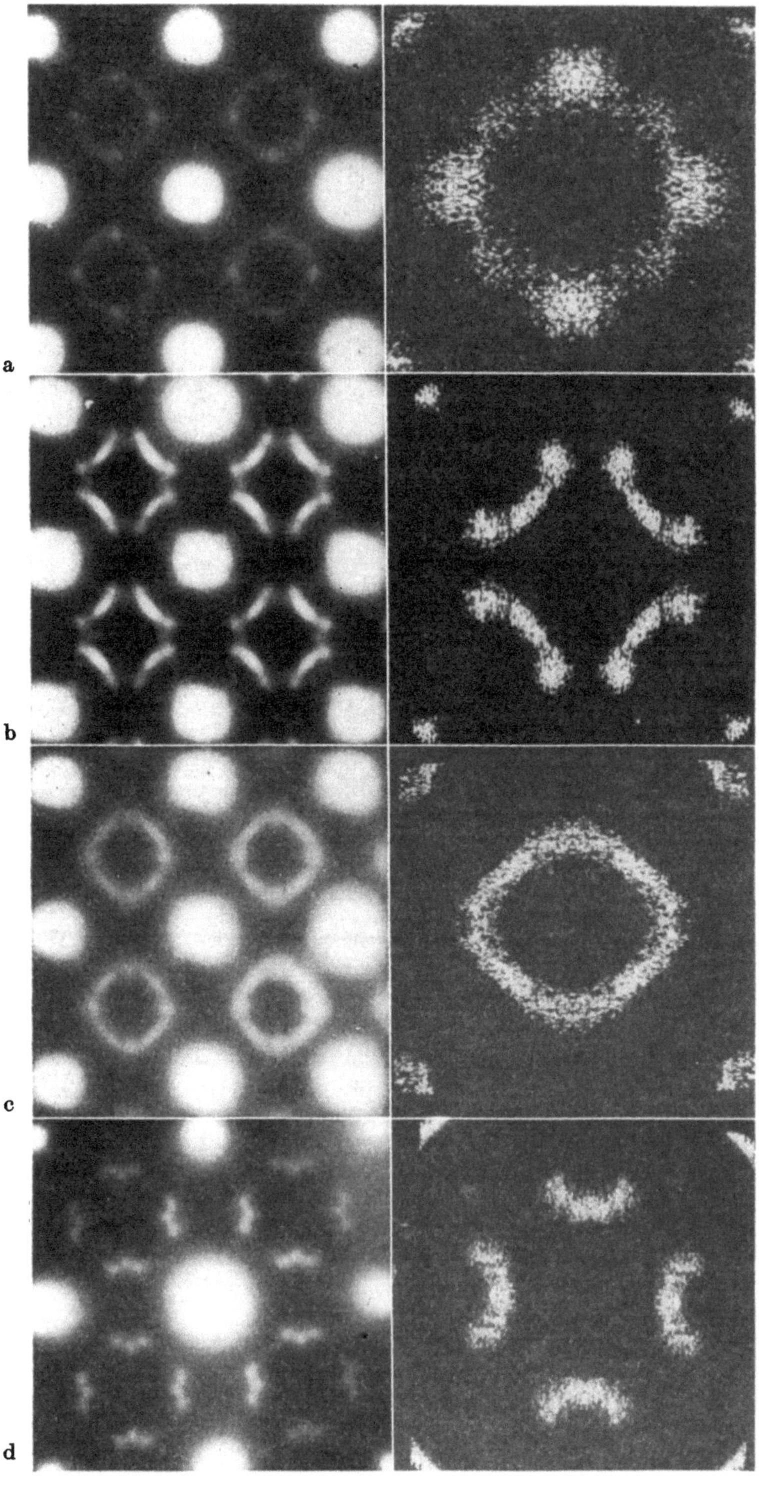

Fig. 11

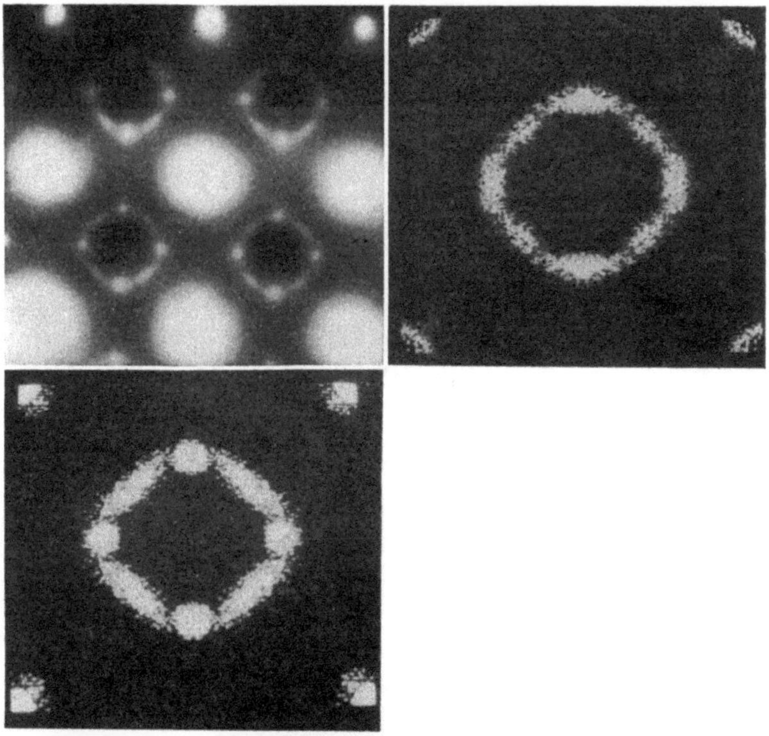

e

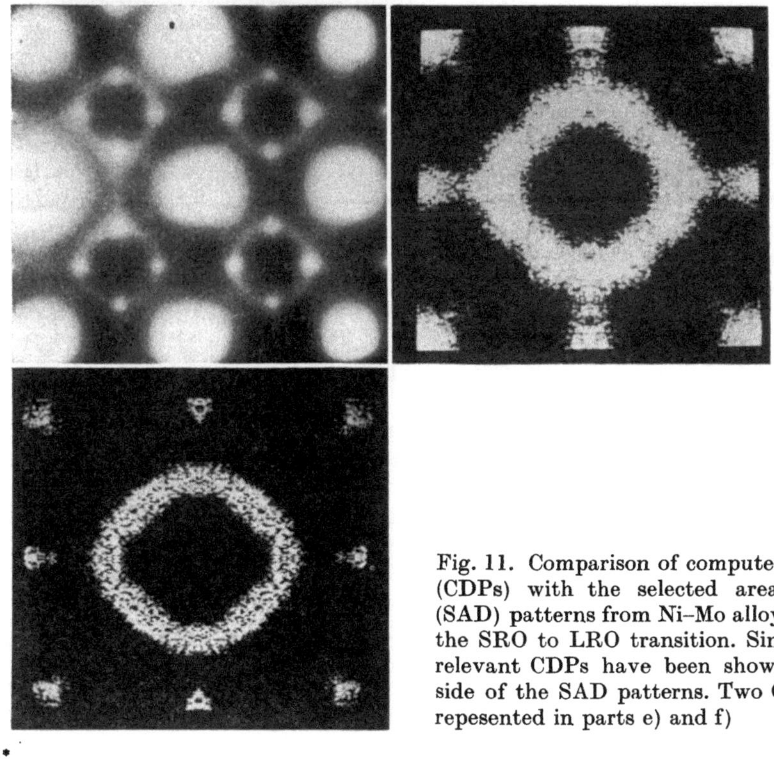

f

Fig. 11. Comparison of computed diffraction patterns (CDPs) with the selected area electron diffraction (SAD) patterns from Ni–Mo alloys in various stages of the SRO to LRO transition. Single quadrants of the relevant CDPs have been shown on the right-hand side of the SAD patterns. Two CDPs each have been repesented in parts e) and f)

Table 2

Details of cluster combinations used in the generation of computed diffraction patterns in Fig. 11

computed diffraction pattern	clusters used in simulation, figures in brackets indicate size in number of molecules	ratio
a	16 different N_3M clusters	equal proportion
b	$Dl_a(16)$, $Pt_2Mo(16)$, $Dl_a(4)$, $Pt_2Mo(4)$	1:2:2:4
c	$N_2M_2(4)$, $Pt_2Mo(4)$, $Pt_2Mo(16)$	2:4:2
d	$N_2M_2(16)$, $Dl_a(16)$, $Dl_a(4)$	1:1:4
e	$N_2M_2(16)$, $Pt_2Mo(16)$, $Pt_2Mo(4)$	1:1:4
e	$N_2M_2(44)$, $Pt_2Mo(16)$, $Pt_2Mo(4)$	1:3:12
f	$DO_{22}(4)$, $Pt_2Mo(4)$	1:1
f	$DO_{22}(16)$, $Pt_2Mo(4)$, $Pt_2Mo(16)$	1:4:1

clusters of identical sizes, rather than clusters of assorted sizes. More uniform rings with brighter arcs at $\frac{1}{2}\{220\}$ and equivalent positions (Fig. 11e) could be obtained with an assortment of N_2M_2 and Pt_2Mo clusters of various sizes (Table 2). This corresponds to the next evolutionary stage in the ordering sequence of the Ni_3Mo alloy. A different combination of N_2M_2 and Pt_2Mo clusters yields a subsequent transitional state, wherein the ring of diffuse intensity changes into a diamond-like figure (Fig. 11c). Faint $\langle 100 \rangle$ streaks in addition to the diffuse ring are evidenced during the transition under electron irradiation at lower temperatures (Fig. 11f). These streaks which are a precursor to the formation of DO_{22} microdomains [6, 9] could be simulated along with the ring by replacing the N_2M_2 clusters by DO_{22} clusters. The second CDP in this set depicts more uniform rings. The $\langle 100 \rangle$ streaks have, however, degenerated into $\{100\}$ spots.

Arcs of intensity connecting $\frac{1}{5}\{420\}$ spots belonging to two different Dla variants, through $\frac{1}{3}\{420\}$ positions are seen during the afore-mentioned phase separation in the Ni_3Mo alloy (Fig. 11b). Such a pattern compares favourably with the CDP based on Pt_2Mo and Dl_a clusters. A patch of intensity along the $\langle 110 \rangle$ directions near the fundamental reflections (resembling a sideband or a satellite reflection in the case of spinodally decomposed alloys) which arises from four-molecule Pt_2Mo clusters can also be seen in the computed as well as the SAD patterns. The SRO to LRO transition in the Ni_4Mo alloy is characterised by "wings" of diffuse intensity extending from the SRO maxima to the Dl_a positions (Fig. 11d). A CDP which answers this description has been generated with a suitable combination of N_2M_2 and Dl_a clusters.

It should be emphasised here that the comparison between the experimental and the computed diffraction patterns is purely qualitative and the sizes of the clusters chosen and the proportions in which they are mixed can only serve as approximate descriptions of the transitional states which presumably comprise a much wider spectrum of cluster sizes and cluster configurations.

5. Summary

On the basis of the results of the simulation work and the foregoing discussion, the following points have emerged regarding the nature of the SRO state and the transitional states between SRO and LRO.

The two models, the concentration wave model and the cluster/microdomain model appear to be relevant to two different regimes of the ordering process. While the former is more appropriate for the SRO state which is characterised by $\{1\frac{1}{2}0\}$ intensity maxima, the latter describes adequately the nature of the transitional states which give rise to the various patterns of diffuse intensity. The concentration waves are, however, expected to be present as "wave packets" rather than as concentration waves extending over a long range. The SRO to LRO transition involves the formation of degenerate clusters of N_3M, $D1_a$, Pt_2Mo, and DO_{22} structures. Such clusters can arise from secondary ordering processes [8] in the SRO state consisting of $\langle 1\frac{1}{2}0\rangle$ concentration waves. While the first three of these result from the interaction of two perpendicular $\langle 1\frac{1}{2}0\rangle$ waves, the DO_{22} clusters result from the coupling of a $\langle 1\frac{1}{2}0\rangle$ wave and an appropriate $\langle 100\rangle$ wave. The SRO to LRO transition involves a continuous increase in the numbers and the sizes of such clusters.

The cluster concept used here differs from the multiple microdomain approach of Das et al. [1] in one important respect. The present simulation permits interactions among various clusters. Indeed, it is the atomic arrangements resulting from collisions of clusters rather than their mere presence that gives rise to the various patterns of diffuse intensity.

References

[1] S. K. Das, P. R. Okomoto, P. M. J. Fischer, and G. Thomas, Acta metall. **21**, 913 (1973).
[2] S. K. Das and G. Thomas, phys. stat. sol. (a) **21**, 177 (1974).
[3] G. van Tendeloo, S. Amelinckx, and D. de Fontaine, Acta cryst. **B41**, 281 (1974).
[4] E. Ruedl, P. Delavignette, and S. Amelinckx, phys. stat. sol. **28**, 305 (1968).
[5] G. van Tendeloo, Mater. Sci. Engng. **26**, 209 (1976).
[6] P. L. Martin and J. C. Williams, Acta metall. **32**, 1681 (1984).
[7] D. de Fontaine, Acta metall. **23**, 553 (1975).
[8] U. D. Kulkarni and S. Banerjee, Acta metall. **36**, 413 (1988).
[9] S. Banerjee, U. D. Kulkarni, and K. Urban, Acta metall., in the press.
[10] R. de Ridder, G. van Tendeloo, and S. Amelinckx, Acta cryst. **A32**, 216 (1972).
[11] S. Amelinckx, G. van Tendeloo, and J. van Landyut, Mater. Sci. Forum **3**, 123 (1985).
[12] S. Banerjee, K. Urban, and M. Wilkens, Acta metall. **32**, 299 (1984).
[13] A. G. Khachaturyan, phys. stat. sol. (b) **60**, 9 (1973).
[14] U. D. Kulkarni, G. K. Dey, and S. Banerjee, Scripta metall. **22**, 437 (1988).
[15] J. P. Chevalier and W. M. Stobbs, Acta metall. **27**, 1197 (1979).

(Received May 17, 1988)

phys. stat. sol. (a) **110**, 347 (1988)

Subject classification: 61.25; S9.11

Department of Chemistry, Jadavpur University, Calcutta[1])

Structure Factor Computation of RbCl and RbBr

By

R. V. GOPALA RAO and RATNA DAS

The generalized mean spherical model approximation is used to obtain the total direct correlation function using a solution similar to that of Waisman-Lebowitz (WL) inside the core. Outside the core a sum of Coulomb and Yukawa potential is taken as a perturbation in the MSM approximation. The partial structure functions are computed by means of Pearson-Rushbrooke relations. Using these PSF's the radial distribution functions are calculated. All these partial structure functions have equal positive values at the long wave limit and the oscillations of the total direct correlation functions are found to cease with increasing k as expected. This new method is a generalization of previous MSM results in the sense that more general forms for the direct correlation function are considered. In both the cases, the coefficients of all terms can be independently adjusted. The partial structure functions and the radial distribution functions so computed gives better agreement than the mean spherical model approximation with experimental and also with molecular dynamics calculations.

Es wird die verallgemeinerte „Mean-Spherical-Model-Approximation" benutzt, um die direkte Gesamtkorrelationsfunktion zu erhalten, mittels einer Lösung die der von Waisman-Lebowitz (WL) im Rumpfinneren ähnlich ist. Außerhalb des Rumpfes wird eine Summe des Coulomb- und Yukawa-Potentials als Störung der MSM-Näherung angenommen. Die Partial-Strukturfunktionen werden mittels Pearson-Rushbrooke-Beziehungen berechnet. Mit diesen PSF werden dann die radialen Verteilungsfunktionen berechnet. Alle Partial-Strukturfunktionen haben gleiche positive Werte im Grenzfall langer Wellen und es wird gefunden, daß die Oszillationen der direkten Gesamtkorrelationsfunktionen, wie erwartet, mit wachsendem k geringer werden. Diese neue Methode ist eine Verallgemeinerung früherer MSM-Ergebnisse in dem Sinne, daß allgemeinere Formen für die direkte Korrelationsfunktion betrachtet werden. In beiden Fällen lassen sich die Koeffizienten aller Terme unabhängig anpassen. Die so berechneten Partial-Strukturfunktionen und radialen Verteilungsfunktionen ergeben bessere Übereinstimmung mit experimentellen und molekulardynamisch berechneten Werten als die „Mean-Spherical-Model-Approximation".

1. Introduction

It is pointed out [1, 2] that thermodynamic self-consistency is achieved within the framework of the self-consistent Ornstein-Zernike approximation. In the generalized mean spherical model approximation (GMSMA) use is made of some thermodynamic properties like internal energy, compressibility, and contact values of the radial distribution function $[g(\sigma)]$ to obtain the so-called free parameters that arise in the direct correlation function (DCF) outside the core.

In a preliminary investigation it has been found that $C_{ij}(r)$ in general can be written in the form

$$C_{ij}(r) = \alpha_1 - \beta_1\left(\frac{r}{\sigma}\right) + \gamma_1\left(\frac{r}{\sigma}\right)^3 \quad \text{for} \quad r \leqq \sigma\,; \tag{1}$$

$\alpha_1, \beta_1, \gamma_1$ have been defined later (cf. Section 2).

---

[1]) Calcutta 700032, West Bengal, India.

This is the type of form that Waisman and Lebowitz obtained for $C_{ij}(r)$ for ions of equal diameters of charged hard spheres [1, 2]. This form has a close resemblance to that obtained for neutral hard spheres by Wertheim [3] and Theile [4]. Even for ions of unequal diameters one can write for $C_{ij}(r)$, as shown in [5, 6],

$$C_{ij}(r) = \frac{\delta}{r} + \alpha_2 + \beta_2 r + \gamma_2 r^3 , \tag{2}$$

where δ, α_2, β_2, and γ_2 are functions of the charge on the ions. The term δ/r can be included in the Coulombic term outside the core. The recent solution of Blum [7] obtained for diameters of unequal size can also be put in the form given by (2). The first term in (2) can be taken care of when integrating $C_{ij}(r)$ outside the core where the $(1/r)$ term comes up in the Coulomb term as well as in the second term that comes up in $C_{ij}(r)$ outside the core (cf. (2)). Further Harada et al. [8] found that the configurational integral does not change if the ion size ratio varies between $0.75 < \sigma^+/\sigma^- < 1.5$. Just as the configurational integral is not sensitive to the ion-size ratio, so is also the form of $C_{ij}(r)$ for fused salts.

Thus we put forward the working hypothesis that the form of $C_{ij}(r)$ can be taken as that given in (1). Thus by choosing a suitable σ_{ij} one can generate $S_{ij}(k)$ [9 to 12].

2. Formulation of the Problem

We write the DCF inside the core as

$$C_{ij}(r) = C_{ij}^{\mathrm{WL}}(r) = \alpha_1 - \beta_1\left(\frac{r}{\sigma}\right) + \gamma_1\left(\frac{r}{3}\right)^3 \quad \text{for} \quad r \leqq \sigma . \tag{3}$$

Here $C_{ij}^{\mathrm{WL}}(r)$ is the direct correlation function inside the core due to Waisman and Lebowitz, where

$$\alpha_1 = \alpha - 2K ,$$

$$\alpha = -\frac{(1+2\eta)^2}{(1-\eta)^4},$$

$$K = \frac{q_i q_j B_{ij}}{K_{\mathrm{B}} T \varepsilon \sigma_{ij}},$$

$$\beta_1 = \frac{-6\eta(1+\eta/2)^2}{(1-\eta)^4} + K B_{ij} ,$$

$$B_{ij} = \frac{1}{\xi}[1 + \xi - (1+2\xi)^{1/2}] ,$$

$$\xi = K_{\mathrm{D}}\sigma ,$$

$$K_{\mathrm{D}} = \frac{4\pi}{k_{\mathrm{B}} T \varepsilon} \sum_i \varrho_i q_i^2 ,$$

$$\gamma_1 = -\frac{\eta}{2}\,\frac{(1+2\eta)^2}{(1-\eta)^4}.$$

Further along with others [13, 14] we use the GMSMA for $C_{ij}(r)$ outside the core and write

$$C_{ij}^{\mathrm{lr}}(r) = \frac{\beta q_i q_j}{\varepsilon r} + D_{ij} \frac{\mathrm{e}^{-q_{ij} r}}{r} \quad \text{for} \quad r > \sigma . \tag{4}$$

"lr" means the long-range potential outside the core with $\beta = (k_{\mathrm{B}}T)^{-1}$ and $D_{+-} = D_{-+}$ for obvious symmetry properties of $C_{ij}(r)$. Equation (4) properly constitutes the present generalised MSM approximation. In the limit $D_{ij} \to 0$ one recovers the standard MSM approximation. Now the total DCF in k-space can be obtained by Fourier transformation of $C_{ij}(r)$ as

$$\tilde{C}_{ij}(k) = \frac{4\pi}{k} \int_0^{\sigma_{ij}} r \sin kr [C_{ij}^{\mathrm{WL}}(r)] + \frac{4\pi}{k} \int_{\sigma_{ij}}^{\infty} r \sin kr\, C_{ij}^{\mathrm{lr}}(r)\, \mathrm{d}r , \tag{5}$$

where $C_{ij}^{\mathrm{lr}}(r)$ is given by (4). Then one immediately writes $\varrho \tilde{C}_{ij}(k)$ as

$$\begin{aligned}
\varrho \tilde{C}_{ij}(k) &= 24\eta_{ij}(\alpha_1)_{ij} \left[\frac{k\sigma_{ij} \cos k\sigma_{ij} - \sin k\sigma_{ij}}{k^3\sigma_{ij}^3}\right] + \\
&+ 24\eta_{ij}\beta_{ij} \left[\frac{k^2\sigma_{ij}^2 \cos k\sigma_{ij} - 2k\sigma_{ij} \sin k\sigma_{ij} - 2 \cos k\sigma_{ij} + 2}{k^4\sigma_{ij}^4}\right] + \\
&+ 24\eta_{ij}\gamma_{ij} \left[\frac{(k^4\sigma_{ij}^4 \cos k\sigma_{ij} - 4k^3\sigma_{ij}^3 \sin k\sigma_{ij} - 12k^2\sigma_{ij}^2 \cos k\sigma_{ij} + {} \atop {} + 24k\sigma_{ij} \sin k\sigma_{ij} + 24 \cos k\sigma_{ij} - 24}{k^6\sigma_{ij}^3}\right] + \\
&+ 4\eta_{ij} \frac{\beta(q_i q_j)}{\varepsilon} \frac{B_{ij}^2}{\sigma_{ij}} \left[\frac{12 \cos k\sigma_{ij}}{k^4\sigma_{ij}^4} - \frac{12}{k^4\sigma_{ij}^4} + \frac{12 \sin k\sigma_{ij}}{k^3\sigma_{ij}^3} - \frac{6 \cos k\sigma_{ij}}{k^2\sigma_{ij}^2}\right] + \\
&+ 48\eta_{ij} \frac{q_i q_j \beta}{\varepsilon} \left(\frac{B_{ij}}{\sigma_{ij}}\right) \left[\frac{\cos k\sigma_{ij}}{k^2\sigma_{ij}^2} - \frac{\sin k\sigma_{ij}}{k^3\sigma_{ij}^3}\right] - 24\eta_{ij} \left(\frac{q_i q_j \beta}{\varepsilon}\right) \times \\
&\times \frac{1}{\sigma_{ij}} \left(\frac{\cos k\sigma_{ij}}{k^2\sigma_{ij}^2}\right) - 4\pi\varrho D_{ij} \exp(-q_{ij}\sigma_{ij}) \left[\frac{k \cos k\sigma_{ij} + q_{ij} \sin k\sigma_{ij}}{k(q_{ij}^2 + k^2)}\right],
\end{aligned} \tag{6}$$

where

$$B_{ij} = -6\eta_{ij} \frac{(1 + \eta_{ij}/2)^2}{(1 - \eta_{ij})^4},$$

$$\alpha_{ij} = -\frac{(1 - 2\eta_{ij})^2}{(1 - \eta_{ij})^4},$$

$$\gamma_{ij} = \frac{1}{2} \eta_{ij} \frac{(1 + 2\eta_{ij})^2}{(1 - \eta_{ij})^2}.$$

Here k is the wave vector.

The partial structure factors can be simply obtained from the Pearson-Rushbrooke equation which can be written as

$$S_{ii}(k) = \frac{1 - \varrho \tilde{C}_{ii}(k)}{D} \tag{7}$$

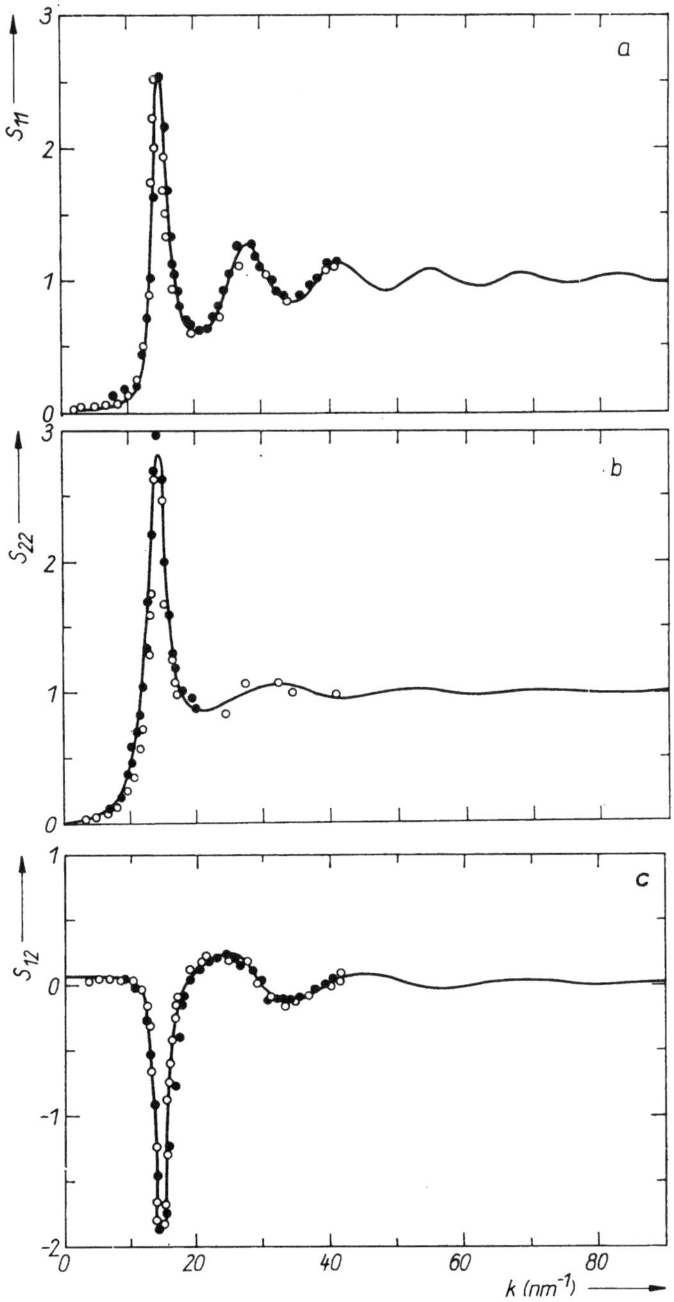

Fig. 1. Partial structure functions $S_{ij}(k)$ for RbBr; a) Rb^+, b) Br^-, and c) Rb–Br. Present calculated results (———), experimental results (○), and MSM results (●)

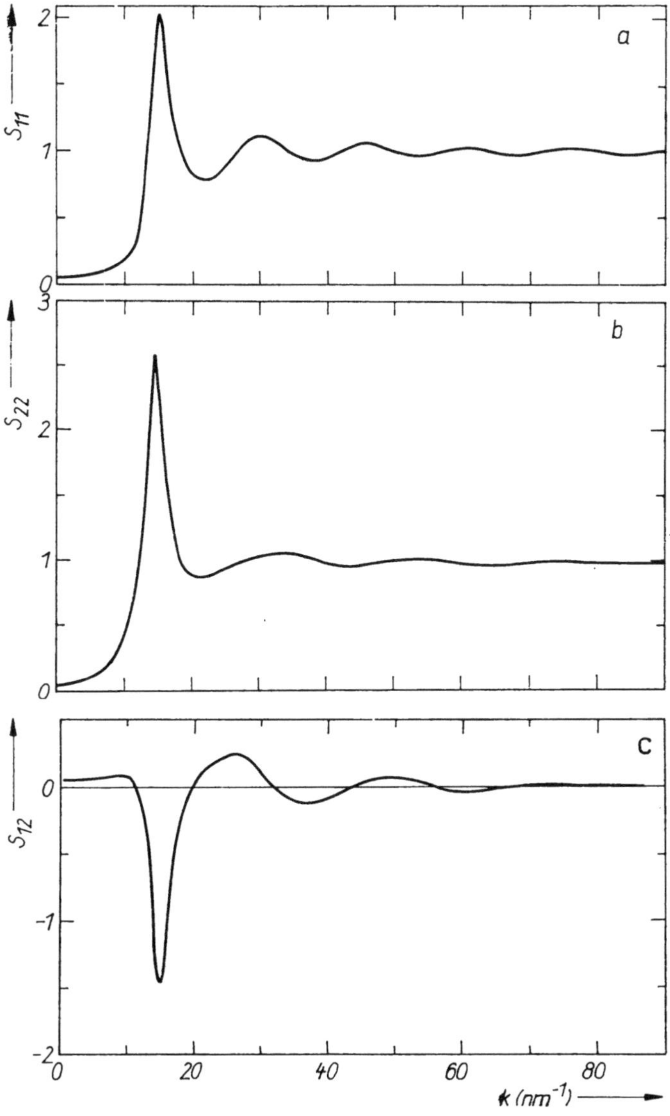

Fig. 2. Partial structure functions $S_{ij}(k)$ for RbCl; a) Rb^+, b) Cl^-, and c) Rb–Cl. Present calculated results (———)

Table 1
Molecular parameters and other input data

salts	temperature (K)	ionic diameters (nm)			dielectric constant	number density (nm^{-3})	Yukawa constant	
		σ_{11}	σ_{12}	σ_{22}			D	q
RbCl	1023.0	0.300	0.255	0.415	2.75	11.05	5.00	3.15
RbBr	958.0	0.310	0.275	0.460	2.50	9.94	5.00	3.00

and

$$S_{12}(k) = \frac{\varrho \tilde{C}_{12}(k)}{D}\,, \tag{8}$$

where

$$D = [1 - \varrho \tilde{C}_{11}(k)]\,[1 - \varrho \tilde{C}_{22}(k)] - \varrho^2 \tilde{C}_{12}^2(k)\,. \tag{9}$$

Here $i = 1$ and $j = 2$ represent the cation and anion, respectively. These equations have been applied to determine the partial structure factors of RbBr and RbCl and are given in Fig. 1 and 2 and are compared with other results. The dielectric constant values have been taken from the literature [15]. The σ_{ij}'s have been treated as parameters around the literature values [16, 17] so as to obtain correct values of the maximum. The molecular parameters are given in Table 1 and the structural features are given in Table 2.

From the $S_{ij}(k)$ values we obtain through Fourier inversion the partial radial distribution functions of RbBr and RbCl, using the equation

$$g_{ij}(r) - 1 = \frac{1}{2\pi^2 (\varrho_i \varrho_j)^{1/2}\, r} \int_0^\infty k \sin kr [S_{ij}(k) - \delta_{ij}]\, dk\,. \tag{10}$$

Here δ_{ij} is the usual Kronecker delta. The $g_{ij}(r)$ so obtained are given in Fig. 3 and 4. The $C_{ij}(k)$ values computed from (6) are shown in Fig. 5 and 6. For uni-univalent electrolytes we have the well-known relation

$$\lim_{k \to 0} C_{11}(k) = C_{22}(k) = -C_{12}(k)\,. \tag{11}$$

Table 2

Values of partial structure functions $S_{ij}(k)$

salts	temperature (K)	functions	present calculations position	present calculations height	experimental or MD results position	experimental or MD results height	M.S.M. results position	M.S.M. results height
RbBr	958.0	$S_{11}(k)$	14.5 (first max.)	2.50	14.5	2.49	15.0	2.47
			20.5 (first min.)	0.59	20.5	0.58	21.0	0.60
		$S_{22}(k)$	14.5 (first max.)	2.80	14.5	2.70	14.5	2.95
			20.5 (first min.)	0.85	20.5	0.70	20.0	0.87
		$S_{12}(k)$	14.5 (first max.)	−1.84	14.5	−1.83	15.0	−1.90
RbCl	1023.0	$S_{11}(k)$	15.0 (first max.)	2.02	—	—	14.0	2.18
		$S_{22}(k)$	14.5 (first max.)	2.56	—	—	14.0	2.89
		$S_{12}(k)$	15.0 (first max.)	−1.45	—	—	14.0	−1.87

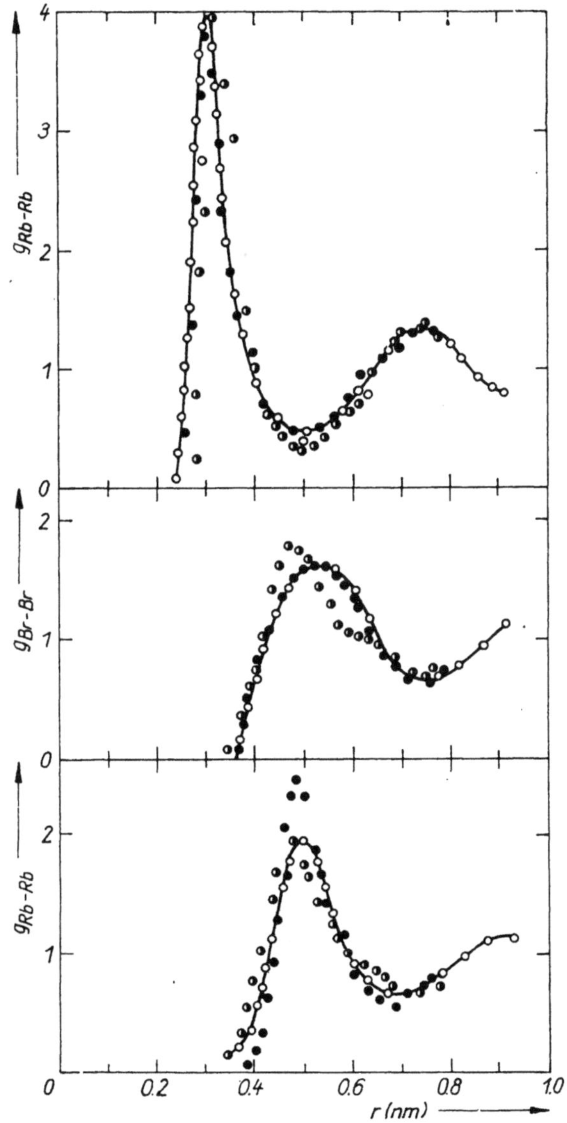

Fig. 3. Present calculated results (○), MSM results (●) and MD results (◑) of radial distribution functions of RbBr

Further

$$S_{11}(0) = S_{22}(0) = S_{12}(0) \tag{12}$$

should be positive. The σ_{ij} have been chosen from literature values and varied slightly so as to satisfy the criteria given in (11) and (12) and the dielectric constants are taken from the literature [15].

3. Results and Discussion

The input parameters and structural features are given in Tables 1 and 2 while those of the radial distribution functions are shown in Table 3.

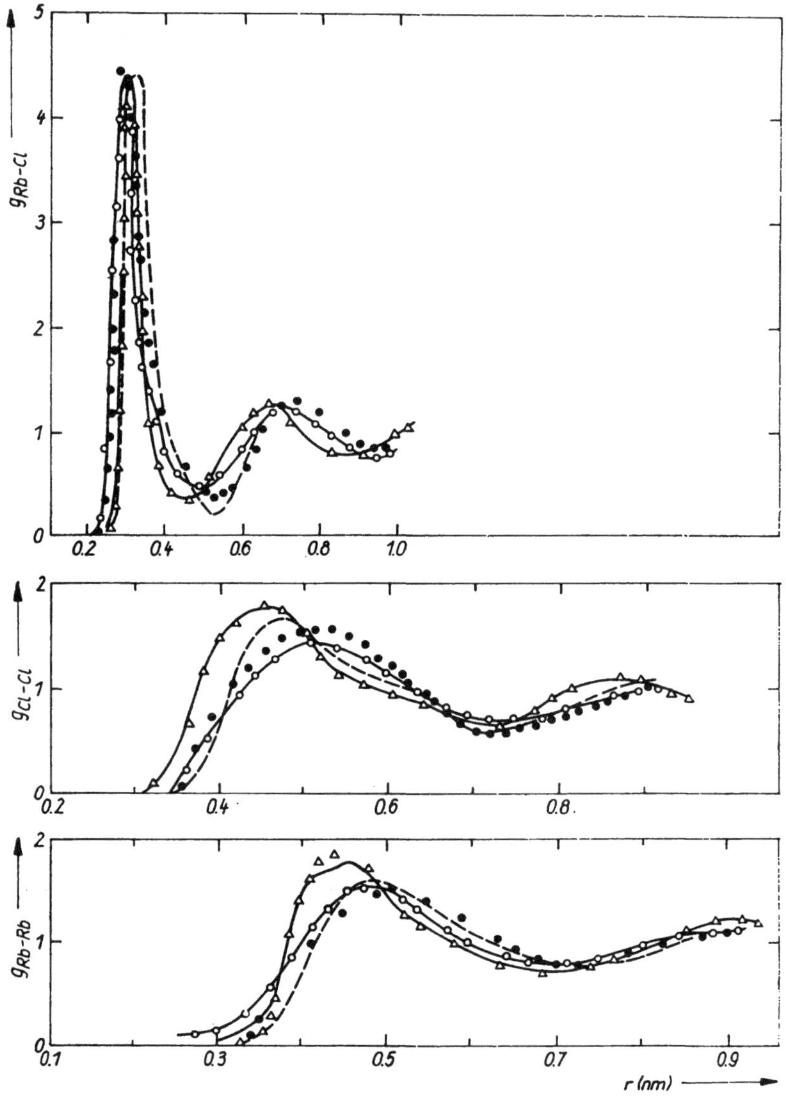

Fig. 4. Present calculated results (○), MSM results (●), experimental results (– – – –), rigid-ion model results (△), and polarizable-ion model results (———) of radial distribution function of RbCl

The $C_{i=j}(k)$ curves start with large negative values and show a positive peak corresponding to the first maximum in the $S_{ij}(k)$ curves, while in the case of $C_{i\neq j}(k)$ curves the values start with a large positive value and show a minimum at about the minimum of the $S_{i\neq j}(k)$ curves. All the $C_{ij}(k)$ vanish at large k.

In both cases of RbBr and RbCl, $S_{11}(k)$ and $S_{22}(k)$ become unity very fast after showing a principal peak. $S_{Br-Br}(k)$ and $S_{Cl-Cl}(k)$ do not show a sharp second peak. This may be due to the small size of the ion. The $S_{12}(k)$ curves show a very pronounced

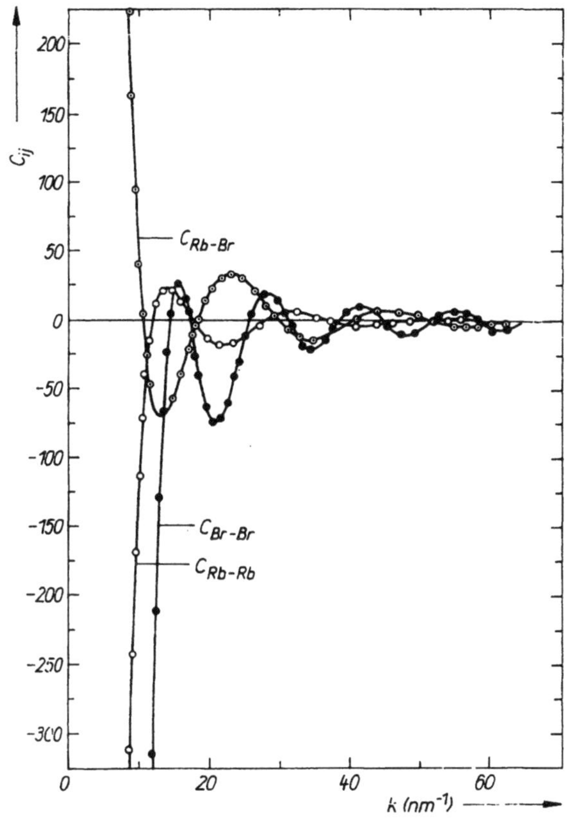

Fig. 5. Present calculated results of the direct correlation function $C_{ij}(k)$ for RbBr

minimum and vanish at large k. Further in both cases we find a positive maximum before they become zero.

The computed partial RDF results for RbCl are shown in Fig. 4. Not much improvement can be seen from the GMSMA calculations if the experimental results are accurate. Thus there is a phase shift in the partial RDF of $g_{Rb-Cl}(r)$. The same trend can be observed in the simulation studies made by Dixon and Sangster [18] using rigid-ion as well as polarization models. Their values of $g_{Rb-Cl}(r)$ on the right-hand side of the principal peak show exactly the same trend as the present calculations. The second peak maximum in the present calculations and that of Dixon et al. coincide with the observed values of Mitchell et al. [19]. Further $g_{Cl-Cl}(r)$ and $g_{Rb-Rb}(r)$ also show in all cases a phase shift. Experimental artefacts may be the cause for the phase shift in all cases. It may be pointed out that the present method gives the correct peak height as the experiment even though there is a phase shift. The results of Dixon et al. [18], while showing an agreement on the left-hand-side of the principal peak, also show a phase shift on the right-hand side. Further the height they obtained is much less than the experimental one.

Recent GMSMA calculations made by Abramo et al. [20] for RbCl to obtain the radial distribution function $g_{Rb-Rb}(r)$ give almost a flat portion around the principal peak. Their calculated values also show exactly the same trend as the present calculations. Thus their calculated values show a phase shift to the left with respect to the experimental values at the principal peak as in the present case.

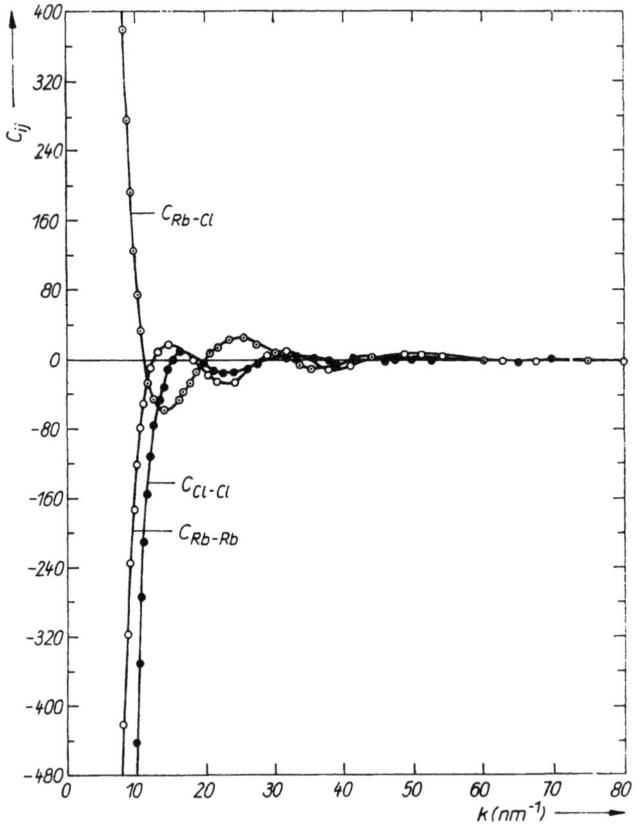

Fig. 6. Present calculated results of the direct correlation function $C_{ij}(k)$ for RbCl

Table 3
Structural features of the RDF's of RbCl and RbBr

salts	temperature (K)	functions	present calculations		experimental or MD results		M.S.M. results	
			position	height	position	height	position	height
RbBr	958.0	$g_{Rb-Rb}(r)$	0.480 (first max.)	1.96	0.480	1.93	0.490	2.46
		$g_{Br-Br}(r)$	0.520 (first max.)	1.60	0.480	1.75	0.520	1.60
		$g_{Rb-Br}(r)$	0.305 (first max.)	3.95	0.315	3.95	0.300	3.80
RbCl	1023.0	$g_{Rb-Rb}(r)$	0.485 (first max.)	1.53	0.486	1.58	0.530	1.49
		$g_{Cl-Cl}(r)$	0.500 (first max.)	1.50	0.480	1.71	0.530	1.67
		$g_{Rb-Cl}(r)$	0.300 (first max.)	4.44	0.318	4.44	0.280	4.40

In the case of RbBr the agreement between the MD values of Copley and Rahman [21] and the present GMSMA results are in better agreement than the MSM values. However, the simulation results are not expected to agree completely with the present results since in the simulation calculations a Fumi-Toshi potential was used, while in the present calculations we used a simple Coulomb potential only outside the core. Thus it may be seen from Fig. 4 that $g_{\mathrm{Rb-Rb}}(r)$ calculated by GMSMA gives better peak values than the MSM result [11].

Thus the present calculations show that the potential parameters chosen for GMSMA are good. At this juncture we may point out that even when sufficient data is not available it is possible to calculate the partial structure factors [12] using the various checks indicated above (cf. (11) and (12)). Thus it may be concluded that the present working hypothesis is reasonable in arriving at the partial structure factors in fused salts and hence the associated distribution functions.

Acknowledgements

The authors wish to acknowledge the financial support received from the DST, Govt. of India for this work. One of the authors (Ratna Das) is thankful to the same for the award of a fellowship and R.V.G.R. is thankful to the U.G.C. for financial assistance.

References

[1] E. Waisman and J. L. Lebowitz, J. chem. Phys. **56**, 3086 (1972).
[2] E. Waisman and J. L. Lebowitz, **56**, 3093 (1972).
[3] M. Wertheim, Phys. Rev. Letters **8**, 321 (1963).
[4] E. Thiele, J. chem. Phys. **39**, 474 (1963).
[5] R. V. Gopala Rao and Sobhen Kumar Pal, phys. stat. sol. (a) **83**, 429 (1984).
[6] R. V. Gopala Rao and Sobhen Kumar Pal, Indian J. pure appl. Phys. **22**, 696 (1984).
[7] L. Blum, Mol. Phys. **30**, 1529 (1975).
[8] M. Harada, M. Tanigaki, M. Yao, and M. Kinoshita, J. Chem. Soc. Faraday Trans. **78**, 1985 (1981).
[9] R. V. Gopala Rao and S. K. Pal, Pramana **22**, 523 (1984).
[10] R. V. Gopala Rao and U. Bandyopadhyay, phys. stat. sol. (a) **104**, 575 (1987).
[11] R. V. Gopala Rao and B. Dasgupta, Indian J. pure appl. Phys. **25**, 143 (1987).
[12] R. V. Gopala Rao and S. Kumar Pal, phys. stat. sol. (a) **86**, 43 (1984).
[13] M. C. Abramo, C. Caccamo, and G. Pizzimenti, J. chem. Phys. **78**, 357 (1983).
[14] M. C. Abramo, C. Caccamo, and G. Pizzimenti, Nuovo Cimento **30**, 297 (1981).
[15] G. J. Janz, Molten Salts Handbook, Academic Press, New York 1967 (p. 356).
[16] L. Pauling, The Nature of the Chemical Bond, 3rd ed., Oxford and IBH Publ. Co., 1975.
[17] F. G. Fumi and M. P. Tosi, J. Phys. Chem. Solids **25**, 31 (1964).
[18] M. Dixon and M. J. L. Sangster, Phil. Mag. **35**, 1049 (1977).
[19] E. W. J. Mitchell, P. F. J. Poncet, and R. J. Stewart, Phil. Mag. **34**, 721 (1976).
[20] M. C. Abramo, C. Caccamo, and G. Pizzimenti, Nuovo Cimento **3D**, 464 (1984).
[21] J. R. D. Copley and A. Rahman, Phys. Rev. C **9**, 3381 (1976).

(Received April 6, 1988; in revised form August 22, 1988)

phys. stat. sol. (a) **110**, 359 (1988)

Subject classification: 61.10

Department of Solid State Physics, Erevan State University[1])

Interpretation of Interference Patterns Obtained from X-Ray Laue Interferometers in Primary Plane Waves[2])

By

P. H. BEZIRGANYAN, V. G. ASLANYAN, and S. E. BEZIRGANYAN

The situations leading to the emergence of Moiré patterns in X-ray interferometers and the dependence of the period of these Moiré patterns on the wavelength and the order of reflection are investigated. The definite conditions of the emergence of Moiré patterns are elaborated. It is demonstrated, that not all the rotations of the reflecting planes and not all the differences in interplanar spacings in interferometers lead to the formation of Moiré patterns. It is shown also, that the period of Moiré patterns inside the crystal does not depend on the wavelength, whereas outside the crystal, this dependence is of a periodic character. The period is reduced along with an increase in the order.

Исследованы ситуации, приводящие к возникновению муаровых картин в рентгеновских интерферометрах и зависимость периода этих муаровых картин от длины волны и порядка отражения. Разработаны однозначные признаки возникновения муаровых картин. Показано, что не все повороты отражающих плоскостей и не все разности межплоскостных расстояний в интерферометрах приводят к возникновению муаровых картин. Показано также, что период муаровых картин внутри кристалла не зависит от длины волны, а вне кристалла — эта зависимость имеет периодический характер. С увеличением порядка отражения период уменьшается.

1. Introduction

Ulrich Bonse is truly considered to be the founder of X-ray interferometry and a number of our investigations [8 to 17] as well as the present paper have been inspired by the works of U. Bonse [1, 2, 18 to 30].

The intensity distribution on interference patterns obtained from ideal X-ray interferometers in primary plane waves is uniform, whereas real interferometers produce interference patterns either of periodical distribution of intensity or of the image of defects, in addition to a combination of both patterns.

Let us agree to consider the interferometer to be ideal, if its crystals are exactly plane-parallel plates of equal thickness, the reflecting planes of which are strictly parallel and possess equal interplanar spacings, the distances between adjacent crystals being either equal or satisfying certain conditions.

The number of various factors leading to a non-uniform distribution of intensity of the interference fields of interferometers is sufficiently high. We shall here confine ourselves to only those, which lead to a periodic intensity distribution of the interference fields.

---

[1]) Mravyan 1, SU-375049 Erevan, USSR.

[2]) Dedicated to Prof. U. Bonse on the occasion of his 60th birthday.

Among the above-mentioned factors we may enumerate the following known at present: difference in interplanar spacings, non-parallelism of the reflecting planes, unequal thickness, non-parallelism of the entrance and exit surfaces of the interferometer crystals and, finally, a disturbance in the conditions of the superimposed waves in the space between adjacent crystals. Thus, for example, beginning from the period when the first electronic and X-ray Moiré patterns were obtained from a system of crystals, up to the present time it was assumed that the reason for the formation of such patterns is a difference between the crystals of the given system or in the interplanar spacings or in the directions of the reflecting planes, or else in these and the others simultaneously [1 to 3]. Apparently, this assumption has been adopted from optics [4], where Moiré patterns are usually created by means of two gratings-model gratings and standard gratings. If the periods of these gratings are equal, then it is necessary to rotate one with respect to the other about their planes in order to obtain Moiré patterns. If their periods differ from each other, then Moiré patterns are also obtained in the case, when they are equally oriented (the lines of the gratings are parallel). In the latter case, when one grating is rotated with respect to the other, mixed Moiré patterns are obtained.

It appears, however, that X-ray Moiré patterns are formed also in the case when the crystals differ neither in interplanar spacings, nor in the directions of the reflecting planes. On the other hand, there are often also cases where differences in interplanar spacings and rotations of the reflecting planes do not lead to the formation of Moiré patterns. Even in such a fundamental work as that of Authier et al. [5] and in a later publication [6], it is presumed that the reasons for the emergence of Moiré patterns are only the rotational and dilatational distortions of the reflecting planes.

Hence, for an unambiguous interpretation of X-ray diffraction patterns of the imperfection of crystals, it is necessary to conduct additional investigations in this field. The detailed analysis of the reasons for the emergence of various periodic intensity distributions of X-ray interferometric patterns, to which the present paper is devoted, is an urgent problem in the field of X-ray diffraction diagnosis of the imperfection of crystals used in science, technology, and industry.

For further investigations, it is useful to examine first the interference of plane waves in vacuum.

2. Interference of Two Plane Monochromatic Waves in Vacuum

Let us consider first the dependence of the distribution of intensity in the interference field of two plane X-ray waves on their differences in the amplitudes, the initial phases, and the directions of distribution (the angle between their wave vectors).

In the field of superposition of the two vector waves,

$$\boldsymbol{E}_1 = \boldsymbol{E}_{01} \exp \{i[\omega t - (\boldsymbol{K}_1 \boldsymbol{r})] + \varphi_1\}$$

and

$$\boldsymbol{E}_2 = \boldsymbol{E}_{02} \exp \{i[\omega t - (\boldsymbol{K}_2 \boldsymbol{r})] + \varphi_2\} ,$$

the intensity distribution, in the case of general polarization, may be expressed by [7]

$$I = E_{01}^2 + E_{02}^2 + 2E_{01}E_{02} \cos \alpha \cos [(\boldsymbol{K}_1 - \boldsymbol{K}_2) \boldsymbol{r} + (\varphi_1 - \varphi_2)] , \tag{1}$$

where α is the angle between the amplitudes $\boldsymbol{E}_{01}$ and $\boldsymbol{E}_{02}$.

In case of σ-polarization, when the amplitudes are perpendicular to the plane of the wave vectors $\boldsymbol{K}_1$ and $\boldsymbol{K}_2$, i.e., when $\alpha = 0$, (1) may be brought to the form

$$I = (E_{01} - E_{02})^2 + 4E_{01}E_{02} \cos^2 \left[\left(\frac{\boldsymbol{K}_1 - \boldsymbol{K}_2}{2}\right) \boldsymbol{r} + \frac{\varphi_1 - \varphi_2}{2}\right]. \tag{2}$$

From this expression, for the period of planes of maximum intensities Δ and the visibility of the bands V, we obtain

$$\Delta = \frac{2\pi}{|\boldsymbol{K}_1 - \boldsymbol{K}_2|}, \tag{3}$$

$$V = \frac{2E_{01}E_{02}}{(E_{01} - E_{02})^2 + 2E_{01}E_{02}}. \tag{4}$$

As is obvious from (3) and (4), the period of the bands does not depend on the differences in the initial phases φ_1 and φ_2, it is unambiguously determined by the value of the vector difference of the wave vectors $\boldsymbol{K}_1$ and $\boldsymbol{K}_2$, and the lower the visibility V, the greater is the difference in the amplitudes; it is equal to one when the amplitudes are equal.

In case of equal given absolute values of wave vectors, the period of the intensity distribution of the total wave depends only on the size of the angle between these vectors. The observability of this intensity distribution (the observability of the bands) also depends on the angle between the wave vectors: in case of large angles, the period of the distribution is small and they are practically non-observable, while in case of very small angles this period is larger than the section of the X-ray beam and the bands appear outside the visual field.

Thus, in order to obtain the observed distribution of the total intensities (to obtain Moiré patterns) it is important that the angle between the wave vectors of the superimposed waves be neither too small nor too large. Therefore, in order to study the situation of the emergence and the observation of Moiré patterns, we must mainly investigate the angle between the wave vectors of the interference waves, which is rather simple using a graphical method in reciprocal space.

Let us mention one of the obtained results and its proof, upon which all our subsequent investigations will be based. This important result may be formulated as follows: for obtaining X-ray Moiré patterns in a plane-wave approximation, it is necessary to illuminate the reflecting planes of the crystal from two sides at such angles (within the limits of the angular region of reflection), the sum of which differs from the double Bragg angle. When this sum is equal to the double Bragg angle, then Moiré patterns do not appear.

The proof of this assertion may be divided into two parts. First of all, let us be convinced of the fact that when the sum of the angles of incidence θ_1 and θ_2 is equal to the double Bragg angle ($\theta_1 + \theta_2 = 2\theta_B$, Fig. 1), then Moiré patterns are not formed. In fact, the sum of the angles θ_1 and θ_2 may be equal to the double Bragg angle only in those cases, when $\theta_1 = \theta_2 = \theta_B$, or the larger one of them, the smaller the other is compared with the Bragg angle ($\theta_1 = \theta_B + \Delta\theta$; $\theta_2 = \theta_B - \Delta\theta$).

In this figure, $\boldsymbol{K}_1^i$ and $\boldsymbol{K}_2^i$ are the wave vectors of the incident waves; L_1 is a double point, i.e. the point of incidence for the waves having a wave vector $\boldsymbol{K}_1^i$ and the point of reflection for the waves having a wave vector $\boldsymbol{K}_2^i$; L_2 is also a double point, i.e. the point of incidence for waves having a wave vector $\boldsymbol{K}_2^i$ and the point of reflection for waves having a wave vector $\boldsymbol{K}_1^i$. Because of the condition $\theta_1 + \theta_2 = 2\theta_B$, the points L_1 and L_2 are situated on one and the same normal and, therefore, the wave vectors $\boldsymbol{K}_{10}$ and $\boldsymbol{K}_{2h}$ of the waves, propagating in the direction of the point 0 of the reciprocal lattice behind the crystal are strictly parallel. For the same reason, the vectors $\boldsymbol{K}_{20}^t$ and $\boldsymbol{K}_{1h}^t$ of the waves, propagating in the direction of the point H of the reciprocal lattice behind the crystal are also strictly parallel. Consequently, Moiré patterns do not arise in the case under consideration. It should be added that if the

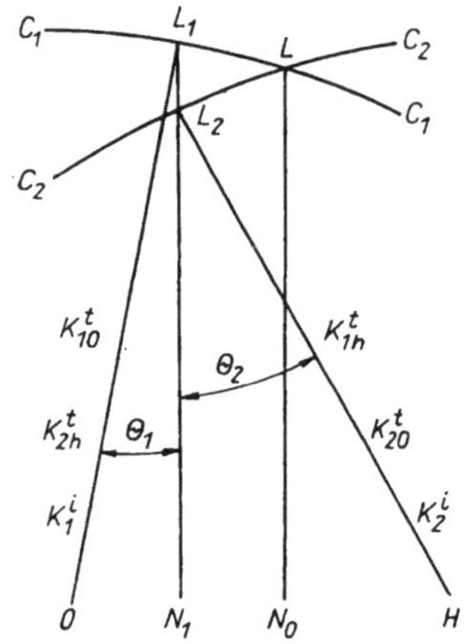

Fig. 1

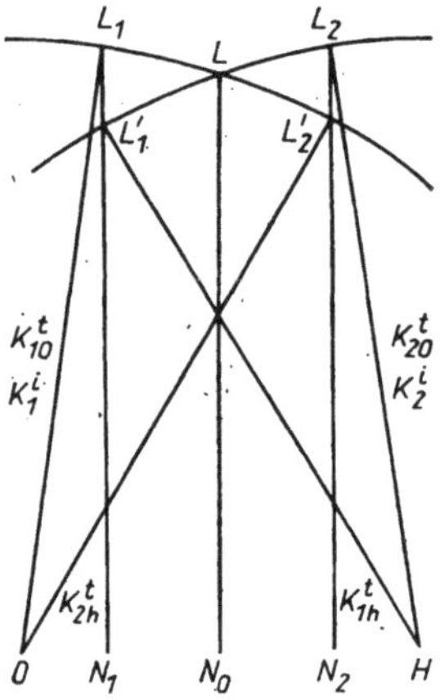

Fig. 2

Fig. 1. The sum of the angles of incidence θ_1 and θ_2 is equal to the double Bragg angle ($\theta_1 + \theta_2 = 2\theta_B$)

Fig. 2. The case when the condition $\theta_1 + \theta_2 \neq 2\theta_B$ is fulfilled

waves are incident exactly at the Bragg angle ($\theta_1 = \theta_2 = \theta_B$), then the points L_1 and L_2 coincide with the Laue point L and once again Moiré patterns are not formed.

Let us consider now the proof of the second part of the assertion, in other words, let us demonstrate that if the sum of the angles of incidence differs from the double Bragg angle, Moiré patterns are produced. The disposition of the wave vectors in this case is shown in Fig. 2. Now the points of incidence L_1 and L_2 are not situated on one normal and, therefore, they are not double points, but only points of incidence. Let us draw from these points the normals N_1 and N_2 and we obtain the points of reflection L_1' and L_2', which, connected with the points H and O of the reciprocal lattice, give the wave vectors $\boldsymbol{K}_{1h}^t$ and $\boldsymbol{K}_{2h}^t$ of the reflected waves. Thus, in the case under consideration, the waves having wave vectors $\boldsymbol{K}_{2h}^t$ and $\boldsymbol{K}_{1h}^t$, diffracted in the directions of incidence do not coincide with the waves having wave vectors $\boldsymbol{K}_{2h}^t$ and $\boldsymbol{K}_{1h}^t$, diffracted in the directions of reflection. Consequently, when the condition $\theta_1 + \theta_2 \neq 2\theta_B$ is fulfilled, Moiré patterns arise behind the crystal, the period of which is determined by the expression

$$\Delta_0 = \frac{1}{|\boldsymbol{K}_{10}^t - \boldsymbol{K}_{2h}^t|}; \qquad \Delta_h = \frac{1}{|\boldsymbol{K}_{20}^t - \boldsymbol{K}_{1h}^t|}.$$

3. Determination of the Value of the Reflection Angle in Case of a Given Angle of Incidence in Laue Geometry

For further investigations it is necessary to examine first the problem of the dependence of the angle of reflection upon the angle of incidence in Laue geometry.

As is well-known [7] in Laue geometry, in a double-wave approximation of the dynamic theory of the scattering of X-rays, the angle of incidence of the maximum reflection coincides with the exact Bragg angle of the kinematic theory and the angles of incidence and reflection are equal.

It is interesting to investigate the dependence of the angle of reflection on the angle of incidence in Laue geometry, when the primary wave is incident on the reflecting surface at an angle differing from the exact Bragg angle within the limits of the angular region of reflection.

Let us consider, for simplicity, the case of an absorbing crystal in which the first field is completely absorbed and the absorption of the second field may be disregarded.

Fig. 1 shows the case when the angle of incidence differs from the exact Bragg angle, the angle of incidence θ_1 in this figure being smaller than the Bragg angle,

$$\sin \theta_1 = \overline{\mathrm{ON}_1}/K_0^{\mathrm{i}} < \overline{\mathrm{ON}_0}/K_0^{\mathrm{i}} = \sin \theta_{\mathrm{B}} .$$

We can easily convince ourselves that when the angle of incidence is smaller than the Bragg angle, then the angle of reflection is larger than it and, on the contrary, when the angle of incidence is larger than the Bragg angle, then the angle of reflection is smaller.

It is interesting to note that the smaller the angle of incidence, the larger the angle of reflection is compared with the Bragg angle and, on the contrary, the larger the angle of incidence, the smaller the angle of reflection is compared with the Bragg angle. We may easily be convinced of this if we bear in mind that the spheres of incidence ($C_H C_H$) and of reflection ($C_0 C_0$) have equal radii K_0^{i}, if angle of incidence is equal to the Bragg angle, the point L_1 coincides with the Laue point L and the angle of reflection also coincides with the Bragg angle and, finally, an increase or a decrease of the angles of incidence and reflection is determined by one and the same arc L_1L of the above-mentioned spheres.

Thus, we come to an important conclusion: in Laue geometry a change in the direction of incidence of the primary beam in the angular region of reflection leads to such a change in the direction of the reflected beam that the sum of the angles of incidence and reflection remains equal to the double Bragg angle,

$$\theta_{01} + \theta_{\mathrm{h}}^{(1)} = \theta_{02} + 2\theta_{\mathrm{B}} . \qquad (5)$$

It is not difficult to show that the obtained results remain valid also for non-absorbing crystals. In fact, the patterns of the wave vectors in a thin (non-absorbing) crystal will differ from those shown in Fig. 1 only by the number of wave vectors inside the crystal — in a non-absorbing crystal, the waves of the first field having wave vectors $\boldsymbol{K}_{01}$ and $\boldsymbol{K}_{\mathrm{h}1}$ also propagate and which, behind a parallelepiped crystal, assume the values $\boldsymbol{K}_{01}^{\mathrm{t}}$ and $\boldsymbol{K}_{\mathrm{h}1}^{\mathrm{t}}$ and coincide with the wave vectors $\boldsymbol{K}_{02}^{\mathrm{t}}$ and $\boldsymbol{K}_{\mathrm{h}2}^{\mathrm{t}}$, respectively.

Thus, the absence of absorption in a parallelepiped crystal does not alter the direction of the wave vectors outside the crystal, therefore, our conclusions hold true for the transparent crystal as well.

It is appropriate here to note the following: in the general case waves having wave vectors $\boldsymbol{K}_{01}$, $\boldsymbol{K}_{02}$, $\boldsymbol{K}_{\mathrm{h}1}$, and $\boldsymbol{K}_{\mathrm{h}2}$ are propagating in the crystal with varying, but little differing, velocities, so that a difference in phases arises between them till their emergence from the crystal, i.e., outside the crystal $\boldsymbol{K}_{01}^{\mathrm{t}} \parallel \boldsymbol{K}_{02}^{\mathrm{t}}$ and $\boldsymbol{K}_{\mathrm{h}1}^{\mathrm{t}} \parallel \boldsymbol{K}_{\mathrm{h}2}^{\mathrm{t}}$, but their phases are not equal.

4. The Case that Rotations of the Reflecting Surfaces Do not Lead to the Formation of Moiré Patterns

Let us examine rotations of the reflecting surfaces of the second crystal in a two-crystal interferometer and of the third crystal in a three-crystal interferometer with respect to the reflecting surfaces of the first crystals.

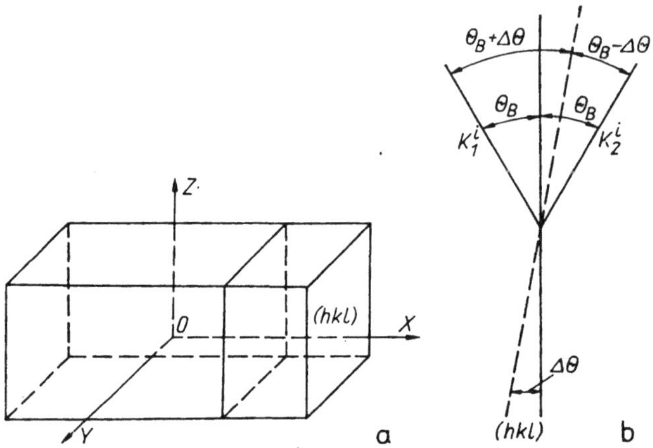

Fig. 3. a) Introduction of a coordinate system. The axis 0X is parallel to the reciprocal lattice vector of the reflecting planes; the axis 0Z is perpendicular to the entrance surface of the crystal; the reflecting planes (*hkl*) are parallel to the plane Y0Z. b) The angles ($\theta_B + \Delta\theta$) and ($\theta_B - \Delta\theta$) between the wave vectors $\boldsymbol{K}_1^i$ and $\boldsymbol{K}_2^i$ and the reflecting planes, respectively

Let us introduce a coordinate system, the axis 0X of which is parallel to the reciprocal lattice vector of the reflecting surfaces, 0Z is perpendicular to the entrance surface of the crystal, and the reflecting surfaces (*hkl*) are parallel to the plane Y0Z (Fig. 3a). Let us assume the case when the reflecting surfaces of the last crystal of the system are perpendicular to the exit surface, i.e. when the reflecting surfaces of the first and the last crystals of the system are strictly parallel to each other, the waves having wave vectors $\boldsymbol{K}_1^i$ and $\boldsymbol{K}_2^i$ fall on the reflecting surfaces of the last crystal at the Bragg angle (Fig. 3b, the trace of the plane is a continuous line) and when these planes are rotated about the axis 0Y at some angle $\Delta\theta$, then with the new positions of the planes (the trace of the plane is a dotted line) the wave vectors $\boldsymbol{K}_1^i$ and $\boldsymbol{K}_2^i$ make the angles $\theta_B + \Delta\theta$ and $\theta_B - \Delta\theta$, respectively.

Thus, rotations of the reflecting surfaces of one of the crystals in a two-crystal or three-crystal interferometer about the axis 0Y (about the axis perpendicular to the scattering surface) lead to the case that the waves fall on the reflecting planes from two sides at such angles of incidence that their sum is equal to the double Bragg angle. As we already know, Moiré patterns do not arise in such cases. Rotations of the reflecting planes about the axis 0Z lead to the formation of Moiré patterns, the periods of which are determined by the expression $\Lambda = d/\Delta\theta$ (Fig. 3), where d is the interplanar spacing of the reflecting planes and $\Delta\theta$ is the angle of rotation.

5. The Case that a Difference in Interplanar Spacings Does not Lead to the Formation of Moiré Patterns

In the Section 2 we have formulated a general feature of the emergence of Moiré patterns and, based on the foregoing, we have examined in Section 4 the case of such a rotation of the reflecting planes which does not lead to the formation of Moiré patterns. We shall describe here such a case of differences in interplanar spacings, which does not lead to the emergence of Moiré patterns either.

Let us examine a four-crystal, or rather a three-crystal interferometer, the second crystal of which consists of two parts (Fig. 4). Let us designate the interplanar spacings

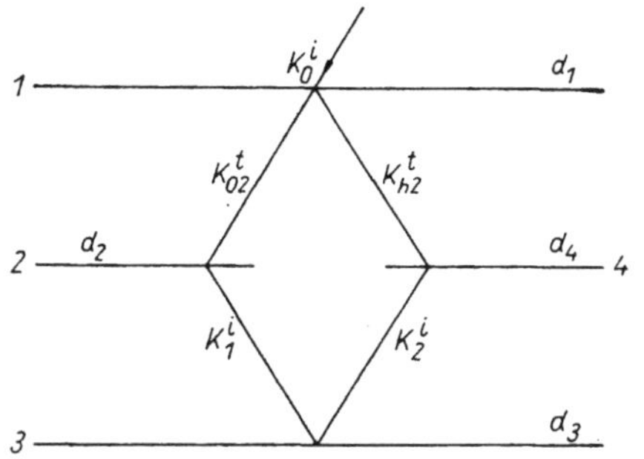

Fig. 4. A three-crystal interferometer the second crystal of which consists of two parts (see text)

of these crystals by d_1, d_2, d_3, and d_4, respectively, and their Bragg angles by θ_{B1}, θ_{B2}, θ_{B3}, and θ_{B4}, respectively. We assume the interplanar spacings d_1, d_2, d_3, and d_4 such that their Bragg angles comply with the following relationships:

$$\theta_{B1} = \theta_{B3} = \theta_B , \qquad \theta_{B2} = \theta_B + \Delta\theta_1 , \qquad \theta_{B4} = \theta_B - \Delta\theta . \tag{6}$$

If the primary wave having a wave vector $\boldsymbol{K}_0^i$ (Fig. 4) falls on the first crystal exactly at the Bragg angle θ_B, then the waves having wave vectors $\boldsymbol{K}_{02}^t$ and $\boldsymbol{K}_{h2}^t$, emerging from the first crystal fall on the reflecting plane of the second and the fourth crystals at an angle $\theta_{B1} = \theta_B$, i.e., at angles, differing from the Bragg angles of these crystals ($\theta_{B2} \neq \theta_{B1}$; $\theta_{B4} \neq \theta_{B1}$). Considering the latter and condition (5) obtained in Section 3 for the angles of incidence of the vectors $\boldsymbol{K}_1^i$ and $\boldsymbol{K}_2^i$, we find

$$\theta_1 = \theta_B + \Delta\theta ; \qquad \theta_2 = \theta_B - \Delta\theta ; \qquad \theta_1 + \theta_2 = 2\theta_B .$$

As is obvious, the reflecting planes of the third crystal are illuminated from two sides at such angles of incidence, that their sum is equal to the double Bragg angle of the reflecting planes of this crystal. Consequently, although the interplanar spacings of the interferometer crystals, shown in Fig. 4, differ from one another, nevertheless Moiré patterns do not arise.

It is clear that instead of the four-crystal interferometer we could examine such a three-crystal interferometer, the interplanar spacings of the second crystal of which satisfy the condition (6) in the sections of incidence of the waves having wave vectors $\boldsymbol{K}_{02}^t$ and $\boldsymbol{K}_{h2}^t$ (Fig. 4).

6. New Mechanism of the Formation of Moiré Patterns

We have already pointed out in the Introduction that until our publications the differences in the interplanar spacings and the directions of the reflecting planes were considered to be the reasons for the formation of X-ray Moiré patterns. It is possible to demonstrate, however, that Moiré patterns may arise even in case of strictly equal interplanar spacings and directions of reflecting planes as a result of the non-parallelism of the crystal surfaces of the interferometric system.

Let us examine two-crystal and three-crystal Laue interferometers, the reflecting planes of which are strictly parallel and their interplanar spacings equal.

Let us only study in detail the case, when the exit surface of the second crystal of a three-crystal interferometer deviates from the remaining surfaces, as is shown in Fig. 5. Fig. 6 shows the path of rays in a three-crystal interferometer, and Fig. 7 shows the direction of the wave vectors between the first and the second ($\boldsymbol{K}_{02}^t$, $\boldsymbol{K}_{h2}^t$), between the second and the third ($\boldsymbol{K}_{h2h}^t$, $\boldsymbol{K}_{02h}^t$) crystals, and behind the interferometer.

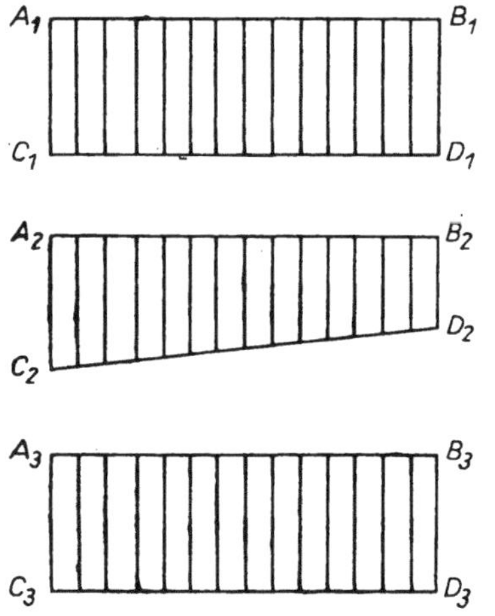

Fig. 5. The exit surface of the second crystal of a three-crystal interferometer deviating from the other surfaces (see text)

The waves $\boldsymbol{K}^{\mathrm{t}}_{02}$ and $\boldsymbol{K}^{\mathrm{t}}_{\mathrm{h}2}$ emerge at the Bragg angle from the first crystal and fall on the reflecting planes of the second crystal at the same angle. In order to find the direction of the wave vectors between the second and the third crystals, it is necessary to draw the normal N_4 through the propagation point A_2 to the exit surface (C_2D_2) of the second crystal. Joining the points of intersection of this normal L_1 and L_2 on the spheres C_0C_0 and C_HC_H with the points 0 and H of the reciprocal lattice, we find

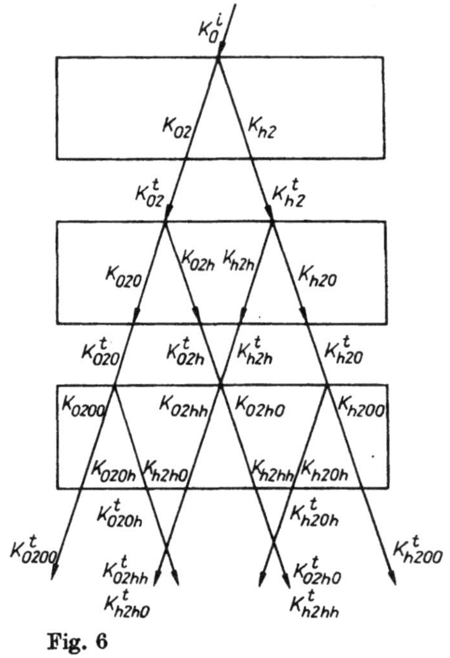

Fig. 6

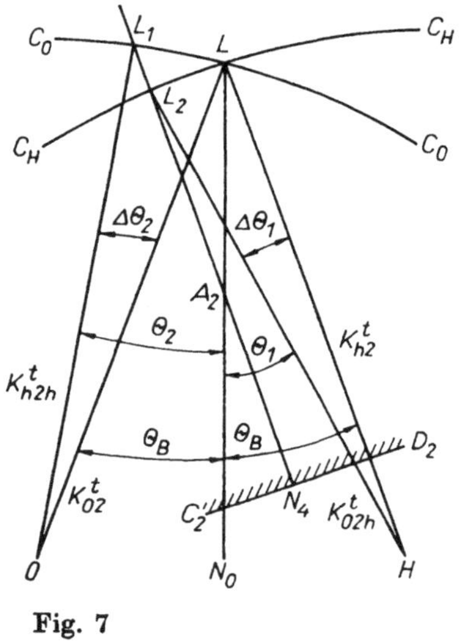

Fig. 7

Fig. 6. The path of rays in a three-crystal interferometer (see text)

Fig. 7. The wave vectors between the first and the second, between the second and the third crystals, and behind the interferometer (see text) the wave vectors $\boldsymbol{K}^{t}_{02h}$ and $\boldsymbol{K}^{t}_{h2h}$ of the waves, illuminating the reflecting planes of the third crystal from two sides (Fig. 6). As is obvious from Fig. 6, the wave vector $\boldsymbol{K}^{t}_{02h}$ makes the angle $\theta_1 = \theta_B + \Delta\theta_1$ with the reflecting planes, while the vector $\boldsymbol{K}_{h2h}$ makes the angle $\theta_2 = \theta_B - \Delta\theta_2$ with these same planes, consequently

$$\theta_1 + \theta_2 = 2\theta_B + (\Delta\theta_1 - \Delta\theta_2) \neq 2\theta_B\,, \tag{7}$$

i.e. the sum of these angles differs from the double Bragg angle and the condition for the formation of Moiré patterns is fulfilled. We may convince ourselves of the correctness of the condition (7) by taking into account that $\Delta\theta_2 > \Delta\theta_1$, since the arc L_1L is larger than the arc L_2L.

We have demonstrated on the basis of the general principle for the formation of X-ray Moiré patterns elaborated by us, that even in case of strictly equal interplanar spacings and directions of the reflecting planes, when the exit surface of the second crystal in a three-crystal interferometer deviates from the remaining surfaces, Moiré patterns are formed in it.

Finally, the problem of the formation of Moiré patterns is settled by finding the directions of the waves emerging from the last crystal of the interferometer. As we were persuaded, however, in order to settle this problem it is sufficient to find the sum of the incidence angles of waves, falling from two sides on the reflecting planes of the last crystal.

We are convinced that these two points of view lead to one and the same result. Let us find, therefore, the directions of the wave vectors outside the system.

We find the points of incidence on the entrance surface of the third crystal on the spheres of incidence, if we draw through the points L_1 and L_2 (Fig. 7) the normals N_6' and N_6'' to this surface (Fig. 8). Since the third crystal is a plane-parallel plate, the normals N_0' and N_0'' (Fig. 8) will be simultaneously normals to the exit surface as well. Therefore, the points of intersection of these normals L_3 and L_4 with the spheres C_HC_H and C_0C_0 will be, simultaneously with the points L_1 and L_2, points of incidence for the exit surface of the third crystal. Joining the points L_1, L_2, L_3, and L_4 with the points 0 and H of the reciprocal lattice, we find the wave vectors $\boldsymbol{K}^{t}_{02hh}$, $\boldsymbol{K}^{t}_{h2h0}$, $\boldsymbol{K}^{t}_{02h0}$, and $\boldsymbol{K}^{t}_{h2hh}$. As is obvious from Fig. 8, Moiré bands form both in the direction of incidence and the direction of reflection and their periods are determined by the following expressions:

$$\Delta_0 = \frac{1}{|\boldsymbol{K}^{t}_{02hh} - \boldsymbol{K}^{t}_{h2h0}|}\,; \qquad \Delta_h = \frac{1}{|\boldsymbol{K}^{t}_{02h0} - \boldsymbol{K}^{t}_{h2hh}|}$$

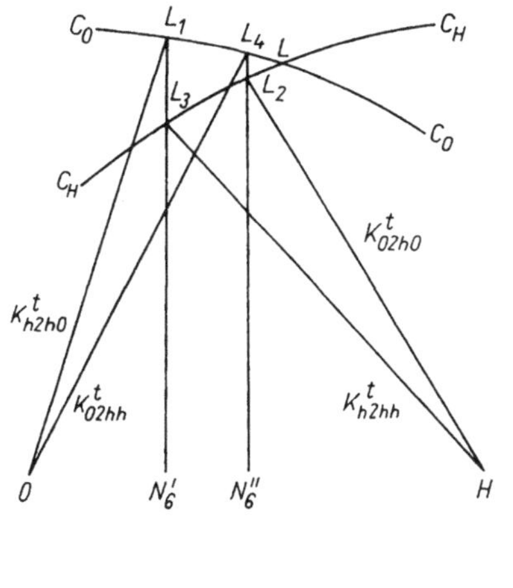

Fig. 8. Definition of the position of the points of incidence on the entrance surface of the third crystal (see text)

or

$$\Delta_0 = \frac{1}{|L_1L_4|}; \qquad \Delta_h = \frac{1}{|L_3L_2|}.$$

Using the elaborated method, an extensive investigation has been conducted also for the other cases of the examined field in plane-wave approximation. In order not to increase the volume of the paper, we mention only the main results of the remaining investigations:

Moiré patterns may be formed in two-crystal interferometers in case of parallel reflecting surfaces and equal interplanar spacings, when the exit surface of the first crystal (from the side of incidence of the primary beam) deviates from the remaining three surfaces. A deviation of any one of the remaining surfaces does not lead to the formation of Moiré patterns.

Besides the above-mentioned case, Moiré patterns are formed in three-crystal interferometers, when the exit surface of the first crystal deviates from the remaining five surfaces. A deviation of any other surface does not produce Moiré patterns.

The features of the formation of Moiré patterns due to the deviation of crystal surfaces may be briefly expressed as follows:

1. The deviating surface must be the exit surface of the crystal from the side of incidence (the second surface of the two-crystal interferometer, the second and the fourth surfaces of the three-crystal interferometer).

2. The deviating surfaces must differ in direction from their adjacent surfaces.

The study of the formation of Moiré patterns, depending on the directions of the crystal surfaces of the interferometer shows the enormous difficulty in the unambiguous interpretation of the observed diffraction patterns. As we have already seen, there is ambiguity even in the interpretation of such Moiré patterns formed as a result of differences in the interplanar spacings and in the directions of the reflecting planes.

First, our investigations have demonstrated the limitations of the possibilities of the simple Moiré patterns in the interpretation of imperfections of the crystal systems; they have shown that it is not possible to discover all the differences in the interplanar spacings, the rotations of the reflecting planes, and the deviations of the working surfaces of the crystals by means of Moiré patterns obtained by the usual methods. Second, it became clear from these investigations that various disturbances in the crystal lattice may create seemingly equal Moiré patterns, a fact which sharply reduces the unambiguous interpretation of Moiré patterns.

It follows from these statements that it is necessary to work out in the future more precise Moiré methods which would offer the possibility to discover and to unambiguously interpret disturbances of any kind in the crystal lattice.

7. The Dependence of Periods of X-Ray Moiré Patterns on the Wavelength

It is usual to consider the period of Moiré patterns not to be dependent on the wavelength [3]. Apparently, this has also been adopted from optics, where mainly geometrical (mechanical) Moiré patterns are examined, the periods of which do not actually depend on the wavelength [4]. In the X-ray range of wavelengths the artificial gratings are not efficient for the formation of Moiré patterns, therefore, the obtained X-ray Moiré patterns are, in practice, exclusively of interference character.

The interference character of X-ray Moiré patterns requires a more detailed examination of the problem of the dependence of the periods of these patterns on the wavelength.

Let us consider the dependence of the periods of the observed Moiré patterns upon the wavelength, taking as an example the two-crystal interferometer, where the values of the interplanar spacings of the reflecting planes of the first and the second crystals differ from each other. Let us investigate two different waves ($\lambda_1 > \lambda_2$; $K_1 = 1/\lambda_1 < K_2 = 1/\lambda_2$). Let $d_1 > d_2$ represent the interplanar spacings of the reflecting planes of the first and the second crystals, respectively, θ_{11} and θ_{12} the Bragg angles of the first and second crystals for the waves $\lambda_1 (\theta_{11} < \theta_{12})$, θ_{21} and θ_{22} the Bragg angles of the first and second crystals for the waves $\lambda_2 (\theta_{21} < \theta_{22})$. The first indices in the designations of the angles show the wave number and the second ones indicate the crystal number. Let two plane, monochromatic waves having wave vectors $\boldsymbol{K}_1$ and $\boldsymbol{K}_2$ fall on the first crystal at the exact Bragg angles θ_{11} and θ_{21}, respectively. Let us assume that the interferometer crystals are thick and the first fields are completely absorbed in them. Then, from the first crystal wave having wave vectors $\boldsymbol{K}_{02}^{t1}$ and $\boldsymbol{K}_{h2}^{t1}$ emerge at an angle θ_{11}, the modules of which are equal to $1/\lambda_1$, and waves having wave vectors $\boldsymbol{K}_{02}^{t2}$ and $\boldsymbol{K}_{h2}^{t2}$ emerge at an angle θ_{21}, the modules of which are equal to $1/\lambda_2$. All these four waves fall on the second crystal at angles, differing from its Bragg angles θ_{12} and θ_{22}. The wave vectors above and below the second crystal of the interferometer are shown in Fig. 9. In this figure, L_{11} and L_{12} are the Laue points of the first and the second crystals for the λ_1 wave, L_{21} and L_{22} are the Laue points of the first and the second crystals for the λ_2 wave, P_{11} and P_{12} are the points of incidence, and P_{11h} and P_{12h} are the points of reflection of the second crystal for the λ_1 wave, P_{21} and P_{22} are the points of incidence and P_{21h} and P_{22h} are the points of reflection of the second crystal for the λ_2 wave, $\overrightarrow{O_1H_1}$ and $\overrightarrow{O_2H_2}$ are the reciprocal lattice vectors of the first and the second crystals.

The waves $\boldsymbol{K}_{020}^{t1}$, $\boldsymbol{K}_{02h}^{t1}$, $\boldsymbol{K}_{h20}^{t1}$, and $\boldsymbol{K}_{h2h}^{t1}$ excited by the first wave and the waves $\boldsymbol{K}_{020}^{t2}$, $\boldsymbol{K}_{02h}^{t2}$, $\boldsymbol{K}_{h20}^{t2}$, and $\boldsymbol{K}_{h2h}^{t2}$ excited by the second wave emerge from the second crystal. To make things clear, we have shown in Fig. 10 only the wave vectors of the waves excited by the first wave λ_1 and in Fig. 11 only the wave vectors directed to the point O_2 of the reciprocal lattice. In these figures, N is the common normal to the entrance and exit surfaces of the interferometer crystals.

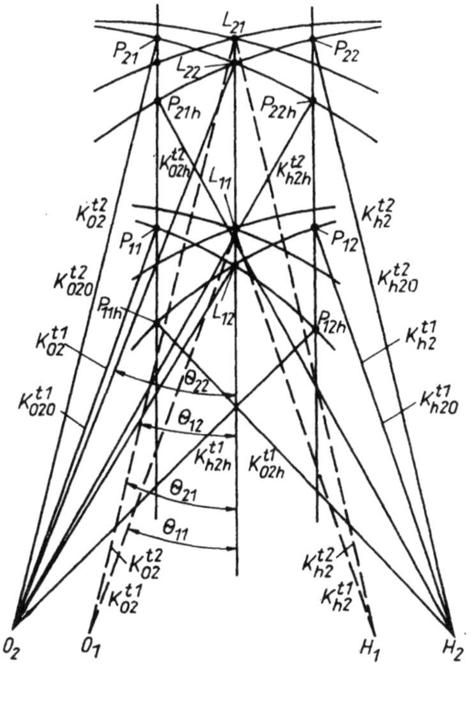

Fig. 9. The wave vectors above and below the second crystal of the interferometer (see text)

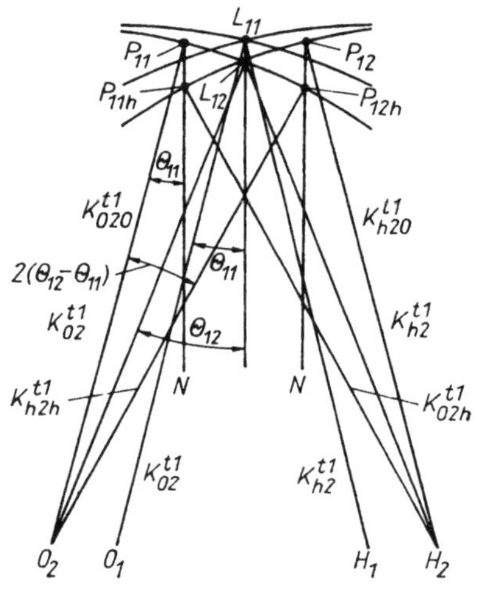

Fig. 10

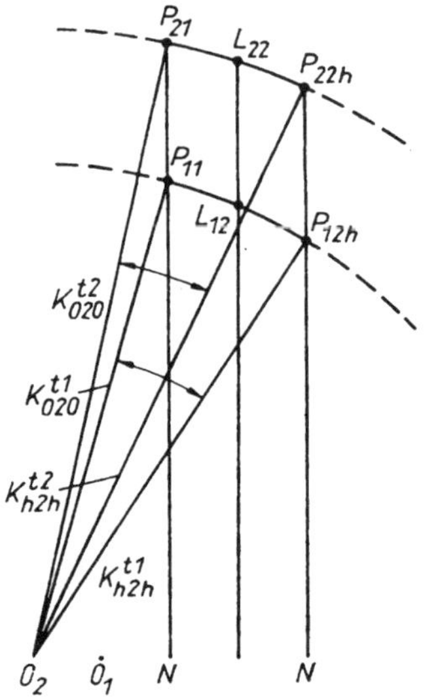

Fig. 11

Fig. 10. The wave vectors of the waves excited only by the first wave λ_1 (see text)

Fig. 11. The wave vectors directed to the point O_2 of the reciprocal lattice (see text)

With the help of Fig. 9 to 11 it is not difficult to prove the correctness of the following relationships:

$$\begin{aligned} \sphericalangle P_{11}O_2P_{12h} &= 2(\theta_{12} - \theta_{11}) = \sphericalangle P_{11h}H_2P_{12}\,, \\ \sphericalangle P_{21}O_2P_{22h} &= 2(\theta_{22} - \theta_{21}) = \sphericalangle P_{21h}H_2P_{22}\,. \end{aligned} \tag{8}$$

As is obvious from these figures, we can write the following expressions for the periods of Moiré beams:
for the λ_1 wave

$$\Delta_{10} = \Delta_{1h} = \frac{1}{|\overrightarrow{P_{11}P_{12h}}|} = \frac{1}{|\overrightarrow{P_{11h}P_{12}}|} = \frac{1}{|\boldsymbol{K}^{t1}_{020} - \boldsymbol{K}^{t1}_{h2h}|} = \frac{1}{|\boldsymbol{K}^{t1}_{h20} - \boldsymbol{K}^{t1}_{02h}|}\,, \tag{9}$$

for the λ_2 wave

$$\Delta_{20} = \Delta_{2h} = \frac{1}{|\overrightarrow{P_{21}P_{22h}}|} = \frac{1}{|\overrightarrow{P_{21h}P_{22}}|} = \frac{1}{|\boldsymbol{K}^{t2}_{020} - \boldsymbol{K}^{t2}_{h2h}|} = \frac{1}{|\boldsymbol{K}^{t2}_{h20} - \boldsymbol{K}^{t2}_{02h}|}\,. \tag{10}$$

Taking into consideration (8) and the insignificant difference of $\theta_{12} - \theta_{11}$ and $\theta_{22} - \theta_{21}$, these expressions may be rewritten for the periods in the following form:

$$\begin{aligned} \Delta_{10} &= \Delta_{1h} = \frac{1}{2K_1(\theta_{12} - \theta_{11})} = \frac{\lambda_1}{2(\theta_{12} - \theta_{11})}\,, \\ \Delta_{20} &= \Delta_{2h} = \frac{1}{2K_2(\theta_{22} - \theta_{21})} = \frac{\lambda_2}{2(\theta_{22} - \theta_{21})} \end{aligned} \tag{11}$$

Taking advantage of the Wulff-Bragg conditions for the reflecting surfaces of the first and the second crystals for λ_1 and λ_2, (11) may be transformed to

$$\begin{aligned} \Delta_{10} &= \Delta_{1\mathrm{h}} = \frac{d_1 d_2}{d_1 - d_2} \cos \theta_{11} = \Delta_0 \cos \theta_{11} \,, \\ \Delta_{20} &= \Delta_{2\mathrm{h}} = \frac{d_1 d_2}{d_1 - d_2} \cos \theta_{21} = \Delta_0 \cos \theta_{21} \,, \end{aligned} \tag{12}$$

where $\Delta_0 = d_{12}(d_1 - d_2)$ is the period of the Moiré beams in the second crystal. It is evident from (12), that

a) the period of the Moiré beams depends on the wavelength by the use of $\cos \theta$ only in vacuum,

b) the period of Moiré patterns Δ_0 itself inside the crystal is not dependent on the wavelength,

c) since $\lambda_1 > \lambda_2$ and $\theta_{11} > \theta_{21}$, $\cos \theta_{11} < \cos \theta_{21}$, consequently the periods of Moiré patterns in vacuum decrease along with an increase in the wavelength in case of a given difference in interplanar spacings.

All the above-stated facts may be clearly shown graphically. In fact, with the help of Fig. 11 we may show that

$$|\overrightarrow{\mathrm{P}_{11}\mathrm{P}_{12\mathrm{h}}}| = |\overrightarrow{\mathrm{P}_{11\mathrm{h}}\mathrm{P}_{12}}| > |\overrightarrow{\mathrm{P}_{21}\mathrm{P}_{22\mathrm{h}}}| = |\overrightarrow{\mathrm{P}_{21\mathrm{h}}\mathrm{P}_{22}}| \,. \tag{13}$$

Then, we obtain from (9), (10), and (13): $\Delta_{10} = \Delta_{1\mathrm{h}} < \Delta_{20} = \Delta_{2\mathrm{h}}$ since the arcs $(\mathrm{P}_{21}\mathrm{L}_{22}\mathrm{P}_{22\mathrm{h}})$ and $(\mathrm{P}_{11}\mathrm{L}_{12}\mathrm{P}_{12\mathrm{h}})$ and their corresponding chords $\mathrm{P}_{21}\mathrm{P}_{22\mathrm{h}}$ and $\mathrm{P}_{11}\mathrm{P}_{12\mathrm{h}}$ are limited by two normals, situated on one side of the diameter of concentric circles. It is evident that in such cases that chord is the greater, the smaller the corresponding radius is.

It is interesting to investigate also the dependence of the periods of Moiré patterns on the order of reflection.

In the general case, the Wulff-Bragg equation may be written in the form $2d \sin \theta_n = n\lambda$, where n is the order of the reflection and θ_n is the Bragg angle of the n-th order of reflection. As may be seen from Fig. 12, the absolute magnitude of the vector $\boldsymbol{H}_n = \boldsymbol{S}_{1n}/\lambda - \boldsymbol{S}_{0n}/\lambda$ increases with increasing order of reflection, where $\boldsymbol{S}_{1n}$ and $\boldsymbol{S}_{0n}$ are the unit vectors in the directions of incidence and reflection of the first and the second orders ($n = 1; 2$), hence, the relation between the interplanar spacing and the vector $\boldsymbol{H}_n$, which coincides with the reciprocal lattice vector only in the first order of reflection, may be expressed, in the general case, as follows:

$$d = \frac{n}{|\boldsymbol{H}_n|} \,.$$

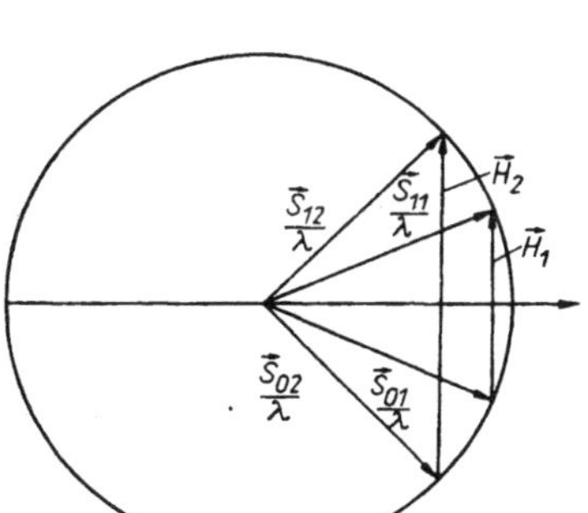

Fig. 12. Investigation of the dependence of the periods of Moiré patterns on the order of reflection (see text)

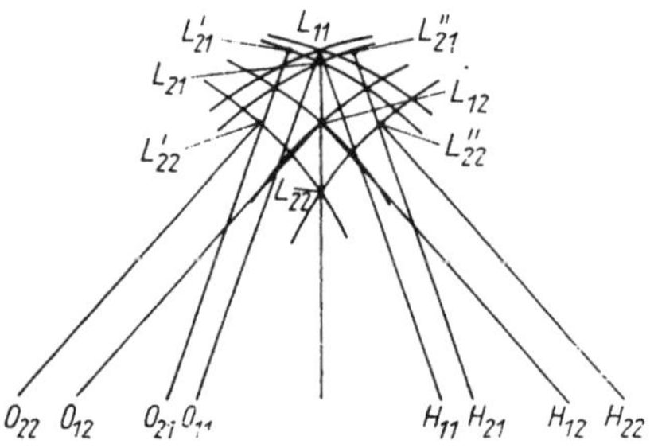

Fig. 13. The wave vectors and reciprocal lattice vectors. L_{11}, L_{12}, L_{21}, and L_{22} are the Laue points of the first and second crystals for the first and second orders of reflection (see text)

Fig. 13 shows the disposition of the wave vectors and of the reciprocal lattice vectors. L_{11}, L_{12}, L_{21}, and L_{22} are the Laue points of the first and the second crystals for the first and second orders of reflection.

Let us examine the formation of Moiré patterns in a two-crystal interferometer, the interplanar spacings of the reflecting planes of which differ from each other by Δd. Let us investigate the dependence of the periods of Moiré bands on the order of reflection. The reciprocal lattice vectors of the first crystal for reflectitons of first and second orders will be

$$|\boldsymbol{H}_{11}| = 1/d_1 ; \qquad |\boldsymbol{H}_{12}| = 2/d_1 .$$

The same vectors for the second crystal will be

$$|\boldsymbol{H}_{21}| = 1/d_2 = 1/(d_1 - \Delta d) = |\boldsymbol{H}_{11}| (1 + \Delta d/d) ,$$
$$|\boldsymbol{H}_{22}| = 2/d_2 = 2/(d_1 - \Delta d) = 2 |\boldsymbol{H}_{11}| (1 + \Delta d/d) ,$$

where the first indices show the crystal number and the second the order of reflection.

As is obvious from Fig. 13,

$$O_{22}O_{12} = H_{12}H_{22} = L'_{22}L_{12} = L_{12}L''_{22} = 2O_{21}O_{11} =$$
$$= 2H_{11}H_{21} = 2L'_{21}L_{11} = 2L_{11}L''_{21} ,$$

therefore, it is possible to affirm that the period of Moiré bands in the second order of reflection half that in the first order of reflection.

References

[1] U. BONSE and M. HART, Appl. Phys. Letters **6**, 155 (1965); **7**, 99 (1965); **7**, 239 (1965).
[2] U. BONSE and M. HART, Z. Phys. **188**, 154 (1965); **194**, 1 (1966).
U. BONSE and M. HART, Acta cryst. **A24**, 240 (1968).
[3] Z. G. PINSKER, Rentgenovskaya kristallooptika, Izd. Nauka, Moscow 1982 (p. 390).
[4] A. DURELLI and V. PARKS, Analiz deformatsii s ispolzovaniem moira, Izd. Mir, Moscow 1974 (p. 359).
[5] A. AUTHIER, A. D. MILNE, and M. SAUVAGE, phys. stat. sol. **26**, 469 (1968).
[6] Z. G. PINSKER, Problemi sovremennoi kristallografii, Izd. Nauka, Moscow 1975 (p. 172).
[7] R. W. JAMES, Solid State Phys. **15**, 53 (1963).
[8] P. H. BEZIRGANYAN, F. O. EIRAMDSHYAN, and K. G. TROUNI, phys. stat. sol. (a) **20**, 611 (1973).
[9] F. O. EIRAMDSHYAN and P. H. BEZIRGANYAN, Izv. Akad. Nauk Arm. SSR, Fiz. Mat. **7**, 215 (1972).
[10] A. O. ABOYAN, F. O. EIRAMDSHYAN, and P. H. BEZIRGANYAN, Izv. Akad. Nauk Arm. SSR, Fiz. Mat. **9**, 193 (1974).

[11] P. H. Bezirganyan, phys. stat. sol. (a) **40**, K77 (1977).
[12] P. H. Bezirganyan, R. Z. Gabrielyan, and F. O. Eiramdshyan, phys. stat. sol. (a) **43**, K183 (1977).
[13] P. H. Bezirganyan, Kh. R. Drmeyan, and G. M. Aladzadzhyan, phys. stat. sol. (a) **54**, K729 (1979).
[14] G. M. Aladzadzhyan, P. H. Bezirganyan, A. K. Kocharyan, and K. G. Truni, phys. stat. sol. (a) **58**, 65 (1980).
[15] P. H. Bezirganyan and V. G. Aslanyan, phys. stat. sol. (a) **78**, 415 (1983).
[16] P. H. Bezirganyan and V. G. Haroutyunyan, phys. stat. sol. (a) **99**, 37 (1987).
[17] P. H. Bezirganyan, L. G. Gasparyan, K. G. Truni, and V. P. Mkrtchyan, phys. stat. sol. (a) **104**, 549 (1987).
[18] U. Bonse and M. Hart, Z. Phys. **190**, 455 (1966).
[19] U. Bonse and E. Kaat, Z. Phys. **214**, 16 (1968).
[20] U. Bonse and M. Hart, Acta cryst. **A24**, 240 (1968).
[21] U. Bonse and H. Hellkötter, Z. Phys. **223**, 345 (1969).
[22] U. Bonse and E. Kaat, Z. Phys. **243**, 14 (1971).
[23] P. Becker and U. Bonse, J. appl. Cryst. **7**, 593 (1974).
[24] W. Graeff and U. Bonse, Z. Phys. B **27**, 19 (1977).
[25] W. Bauspiess, U. Bonse, and W. Graeff, J. appl. Cryst. **9**, 68 (1976).
[26] U. Bonse and G. Materlik, Z. Phys. **253**, 232 (1972).
[27] U. Bonse and W. Graeff, Acta cryst. **A31**, 254 (1975).
[28] U. Bonse and G. Materlik, Z. Phys. B **24**, 189 (1976).
[29] U. Bonse, G. Materlik, and W. Schroeder, J. appl. Cryst. **9**, 223 (1976).
[30] U. Bonse and W. Graeff, Appl. Phys. **22**, 94 (1977).

(Received April 29, 1988)

phys. stat. sol. (a) **110**, 375 (1988)

Subject classification: 61.60; 61.16; S1.63; S9

Ashigara Research Laboratories, Fuji Photo-Film Co., Ltd., Kanagawa[1]) *(a)*
and Institute for Chemical Research, Kyoto University, Uji[2]) *(b)*

High Resolution Electron Microscopic Study of Silver Sulfide Microcrystals Formed on Silver Bromide Emulsion Grains

By

T. SHIOZAWA (a) and T. KOBAYASHI (b)

Silver sulfide specks formed on the (111) plane of AgBr are examined by high resolution electron microscopy and the relationships in atom arrangement at their interface are revealed on the basis of the lattice images observed for the silver sulfide specks as well as their orientation relative to the direction of AgBr crystal axis. Silver sulfide are in contact with AgBr (111) plane with one of the four different planes, (100), (102), (104), and (10$\bar{3}$). In every interfaces, the *b*-axis of the microcrystal of silver sulfide is parallel with the [112] axis of AgBr. The atom arrangements of both the crystals are similar to each other at the interface and the lattice misfits between them are so small that the formation of the silver sulfide specks is considered to occur epitaxially on the AgBr (111) plane.

Silbersulfidflecken auf (111)-Ebenen von AgBr werden mittels hochauflösender Elektronenmikroskopie untersucht und die Beziehungen der Atomanordnung an ihren Grenzflächen werden anhand der für Silbersulfidflecken beobachteten Gitterdiagramme sowie deren Orientierung bezüglich der AgBr-Kristallachsenrichtungen gefunden. Silbersulfid befindet sich mit der AgBr (111)-Ebene und einer der vier verschiedenen Ebenen (100), (102), (104) und (10$\bar{3}$) in Kontakt. In jeder der Grenzflächen ist die *b*-Achse des Silbersulfid-Mikrokristalls parallel zur [112]-Achse von AgBr. Die Atomanordnungen beider Kristalle sind einander ähnlich an der Grenzfläche und die Gitterfehlanpassung zwischen ihnen ist so gering, daß angenommen wird, daß die Bildung der Silbersulfidflecken epitaktisch auf der AgBr (111)-Ebene erfolgt.

1. Introduction

Formation of sensitivity specks on photographic grains such as silver bromide and other silver halides prior to exposing them to light is a necessary step for obtaining high-sensitive conventional photographic materials. Such sensitivity specks are usually made by the following method. A small amount of the sensitizer is added to the emulsion (a suspension of silver halide grains in gelatin solution) and digested for a certain period at a constant temperature. As the sensitizer, sodium thiosulfate ($Na_2S_2O_3 \cdot 5\,H_2O$) and/or chloroauric acid ($HAuCl_4 \cdot 4\,H_2O$) are usually used as sulfur sensitization or gold plus sulfur sensitization, respectively. The resulting sensitivity specks are said to be Ag_2S [1] in the former case and $AuAg_3S_2$ [2] in the latter case. The knowledge of the positions and the numbers of such specks formed at the surface of the silver halide grains has been demanded in production of commercial photographic materials.

However, they are very small, 1.5 nm or less in size [3], and they form on the silver halide grain whose thickness is beyond the penetrating power of electron used in

[1]) 210 Nakanuma, Minamiashigara, Kanagawa 250-01, Japan.
[2]) Kyoto-fu 611, Uji, Japan.

a conventional electron microscope. The grains are sensitive to a damage due to the electron irradiation needed for a high resolution observation by electron microscope. Therefore, the information about the specks is very poor till now. In order to observe the sensitivity specks, some attempts have been tried and reported [4].

In the present work the orientational relationships between Ag_2S sensitivity specks and AgBr tabular grain will be revealed on the basis of the lattice images of Ag_2S observed by high resolution electron microscope and the atom arrangements at their interface are to be discussed.

2. Experimental

Preparation of AgBr tabular grains and Ag_2S specks was done as follows. Simultaneously adding of the same amount of water solution of silver nitrate (0.1 g/ml) and potassium bromide (0.12 g/ml) to an inert gelatin solution which was kept at 60 °C produced the AgBr tabular crystals in the emulsion form. The AgBr grains have a mean size of 1 μm square and thickness of 0.15 μm. The emulsion was digested for 60 min at the same temperature under the presence of sodium thiosulfate of the mole ratio of 0.003 to AgBr thus produced. The amount of sodium thiosulfate added was about 30 times more compared with that needed for obtaining an optimum sensitivity.

The AgBr emulsions were diluted with a large amount of water and separated from the gelatin solution by centrifuge and decantation.

A drop of the suspension of the grains in pure water put on a thin carbon substrate supported on a copper grid and most of grains of AgBr were removed by soaking the specimen into 10% sodium thiosulfate solution and washing again with water. All the operations were carried out under a safe light illumination.

The specimens were examined with a JEOL-200C electron microscope of a top-entry type equipped with a high resolution pole piece for the objective lens. The theoretical point-to-point resolution of the microscope is 0.20 nm and the lattice resolution is better than 0.10 nm.

3. Results

The sensitivity specks formed on the (111) face of AgBr grains are shown in Fig. 1. The crystal habit of the original AgBr grains can also be clearly observed as a replica made of thin gelatin film adsorbed on the grain. By digesting the AgBr with a solution of thiosulfate in a higher concentration than those in ordinary use, web-like specks

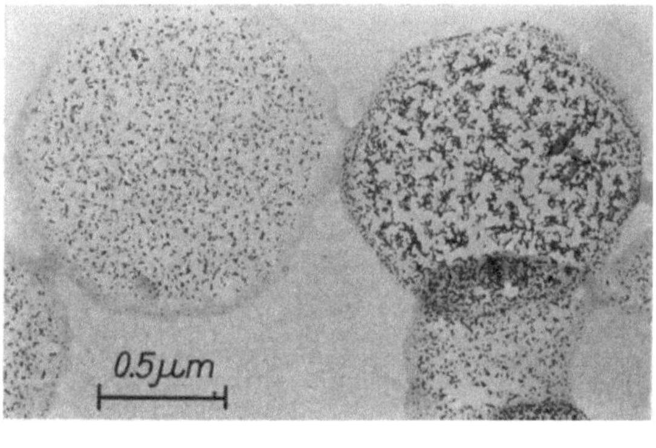

Fig. 1. Total aspect of Ag_2S formed at the surface of AgBr tabular grains

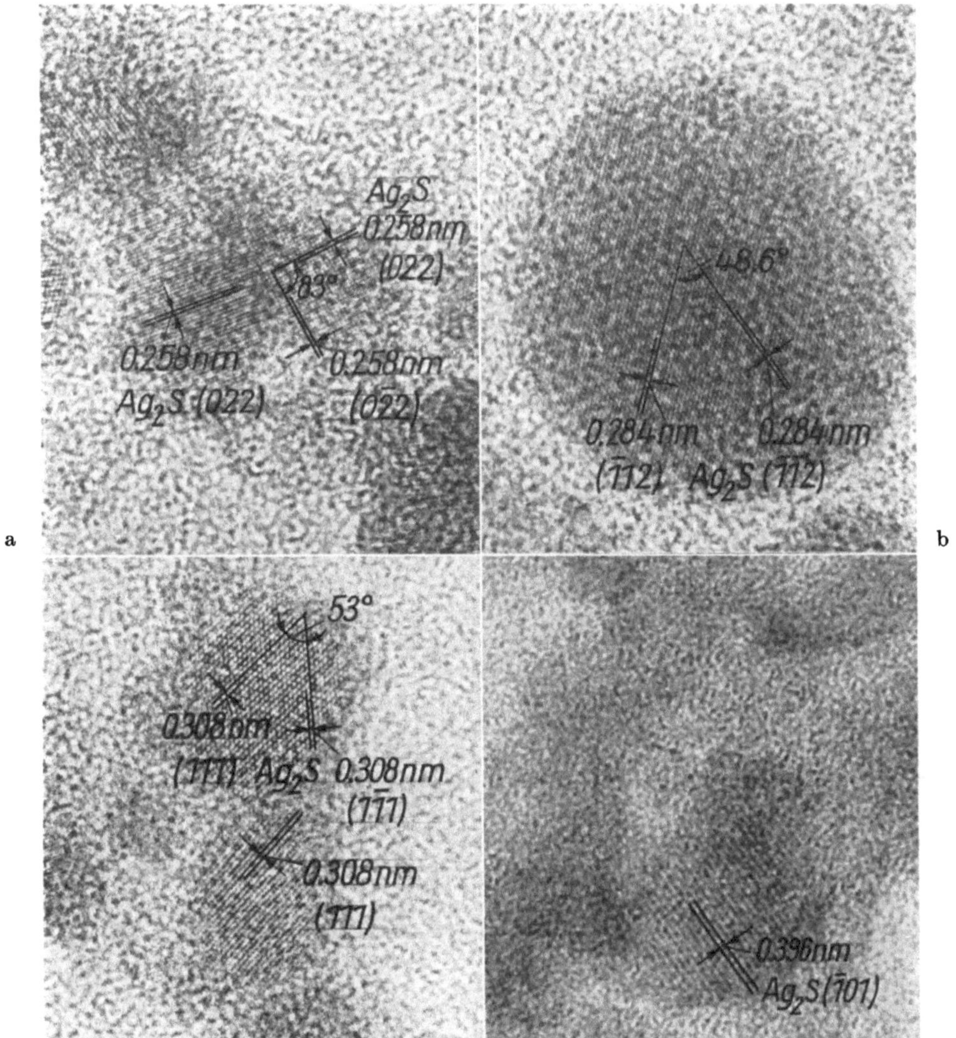

Fig. 2. Typical high resolution electron micrographs of the sensitivity specks showing the lattice mages with different intervals

are sometimes produced together on the grain as shown in Fig. 1 (right grain). No sufficient electron diffraction pattern for the analysis of the specimen was obtained even from these web-like specks because of their smallness. However, high resolution electron micrographs can reveal the crystallographic features of such small grains as shown in Fig. 2 a to d. These micrographs represent the lattice images whose lattice intervals are 0.258, 0.284, 0.308, and 0.396 nm. The orientation of these crystal lattices relative to the AgBr matrix is summarized in Fig. 3 schematically after the correction for image rotation due to the change in electric currents of the lenses needed for the magnification change.

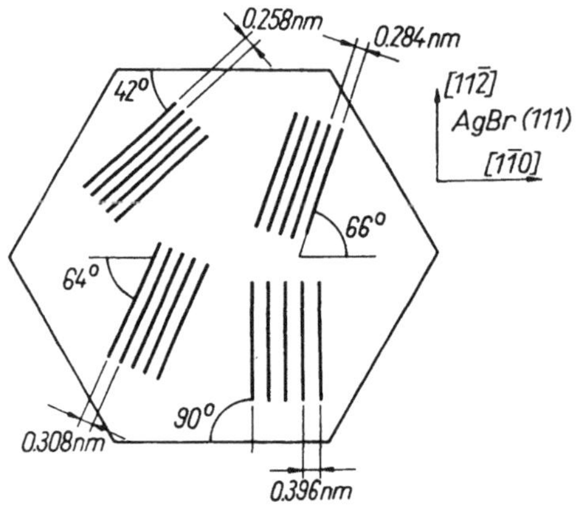

Fig. 3. Schematical representation of the orientation of the lattice images shown in Fig. 2 relative to the principal axes of AgBr crystals

4. Discussion

Crystals of Ag_2S have three polymorphs, two of which are low temperature phases and the other appears at a temperature higher than 170 °C. One of the low temperature phases is a monoclinic crystal having the following lattice constants [5]: $a_0 = 0.423$, $b_0 = 0.691$, $c_0 = 0.787$ nm, and $\beta = 99.6°$.

The crystals formed on the AgBr tabular crystals are identified to be this form of Ag_2S by comparing the observed lattice intervals and with those calculated by using these lattice constants. The lattice images having spacings of 0.258, 0.284, 0.308, and 0.396 nm can accordingly be assigned to (022), ($\bar{1}$12), (111), and ($\bar{1}$01) planes, respectively. The packing manner of atoms (ions) in a unit cell of the Ag_2S in monoclinic form is shown in Fig. 4.

The right part of the grain shown in Fig. 2a exhibits two sets of lattice fringes with the same intervals of 0.258 nm crossing almost perpendicularly. They correspond to (022) and (0$\bar{2}$2) planes of Ag_2S in monoclinic form. These planes should be almost parallel with the incident electron beam and, therfore, perpendicular to the substrate in

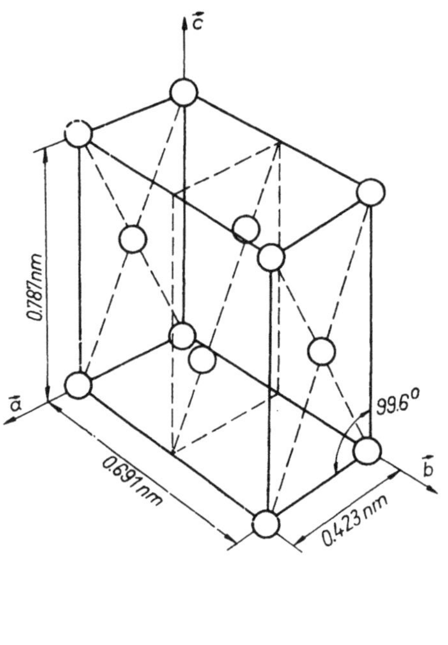

Fig. 4. Structure of monoclinic Ag_2S crystal. Only the positions of sulfur atoms are drawn

order to be observed as lattice images. The substrate in the present case is the (111) plane of an AgBr tabular grain. Accordingly a plane of Ag_2S perpendicular to these two planes is parallel to the substrate and forms an interface between the Ag_2S and AgBr.

The (111) plane of the AgBr is in principle composed of ions of the same sign, Ag^+ or Br^-, arranging in a triangular network because the crystal has a rocksalt-type cubic structure. However, it is said that the surface of the AgBr exhibiting (111) planes is covered partly with Ag^+ ions and the rest is covered with Br^- ions and that the ratio of the respective area is 3:4 according to a paper by Hamilton and Brady [6]. The existence of both kinds of areas is possible due to surface steps at an atomic level. The (100) plane of the Ag_2S shown in Fig. 5a is a plane perpendicular to both (022) and $(0\bar{2}2)$ and contacts with such (111) surface of AgBr forming the interface. On the basis of this fact one can deduce a mutual arrangement of ions belonging to crystals at both sides of the interface and the result is shown in Fig. 5b. The figure represents (022) and $(0\bar{2}2)$ planes of Ag_2S with their normal plane, (100), in Fig. 5 a and b the projection of the ions onto the (100) plane of Ag_2S and AgBr (111) plane are superimposed where the underlying AgBr is drawn by thinner lines. The circles represent Ag^+ and S^{2-} ions in the figure. The axial correlation between the two crystals is also written in the figure which was confirmed by observing the direction of the lattice relative to the crystal axis of the AgBr matrix detected as a replica, as shown in Fig. 3. In order to draw this figure the principle is adopted that the stable positions of negative ions, sulfur ions, are the centres of triangles composed of positive silver ions at the (111) face of AgBr. In this positional relationship $[1\bar{1}0]$ and $[\bar{1}\bar{1}2]$ directions of the AgBr coincide with the *c*- and *b*-axes of Ag_2S, respectively.

In Fig. 2b two sets of lattice fringes are crossing with an angle of $\approx 49°$. The lattice intervals are 0.284 nm in both directions and, therefore, the planes are assigned to be $(\bar{1}12)$ and $(\bar{1}\bar{1}2)$ planes of Ag_2S. A (102) plane of Ag_2S is the plane which makes an almost right angle (86.4°) with both $(\bar{1}12)$ and $(\bar{1}\bar{1}2)$ planes as illustrated in Fig. 6a.

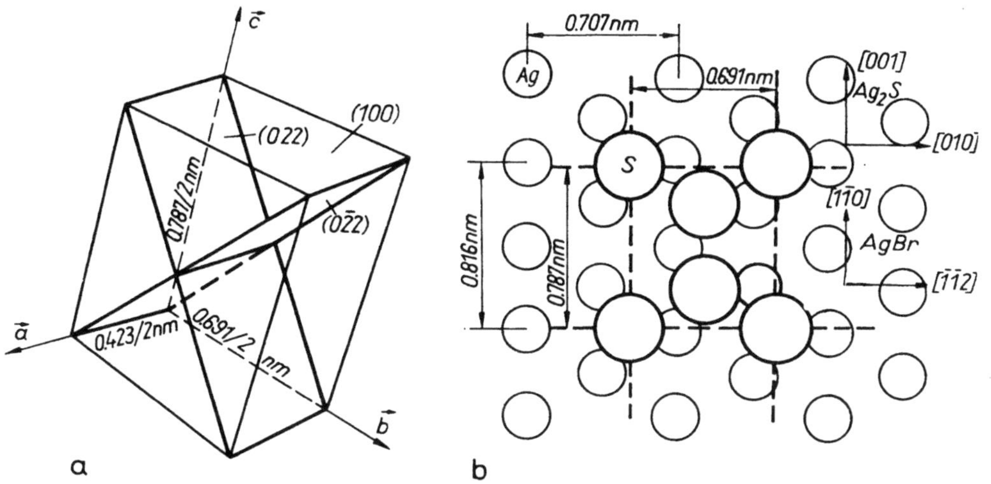

Fig. 5. a) Schematical illustration of the geometrical relationship among the (022), $(0\bar{2}2)$, and (100) planes of Ag_2S crystal. b) Plausible atom arrangement at the interface between Ag_2S and AgBr (111) planes. Positions of S atoms are projected on the (100) plane and superimposed on the (111) plane of AgBr. The large and small circles represent S atoms in Ag_2S and Ag in AgBr, respectively

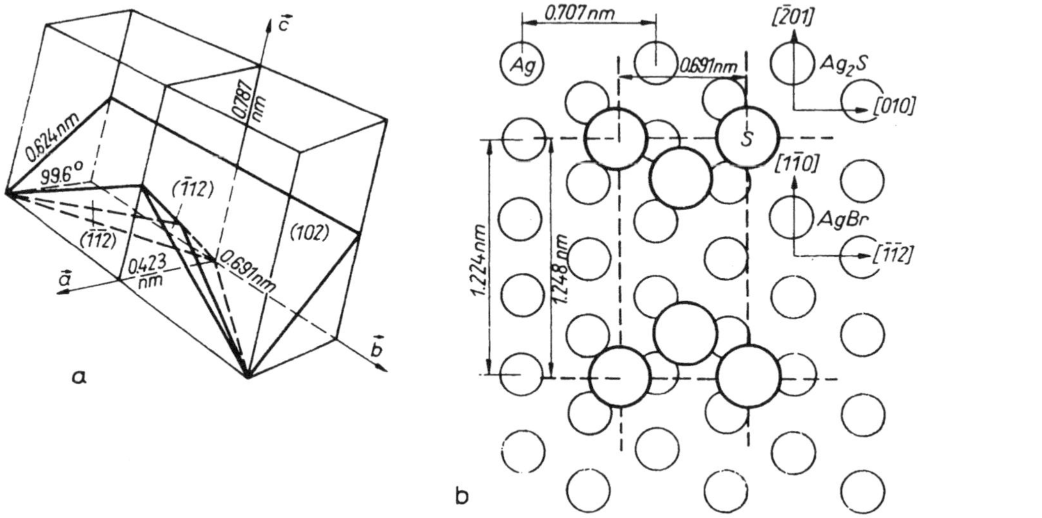

Fig. 6. a) Geometry of the planes $(\bar{1}12)$, $(\bar{1}\bar{1}2)$, and (102) of Ag_2S; b) atom arrangement at the interface composed of Ag_2S (102) and AgBr (111) planes

The (102) plane is considered to form the interface with the (111) plane of the AgBr and the ion arrangement derived from these facts can be represented as shown in Fig. 6b where the axial correlations obtained experimentally are also taken into account.

The $[\bar{2}01]$ and [010] axes of Ag_2S coincide with the $[1\bar{1}0]$ and $[\bar{1}\bar{1}2]$ axes of AgBr, respectively. At this interface the sulfur ions fit well with the centres of triangles of Ag^+ ions at the (111) face of AgBr.

The third example is the case that (111) and $(1\bar{1}1)$ planes are observed together in a single grain of Ag_2S as shown in Fig. 2c. The planes of Ag_2S which cut these two sets of planes at right angles are $(10\bar{3})$ planes as illustrated in Fig. 7a. The sulfur ion

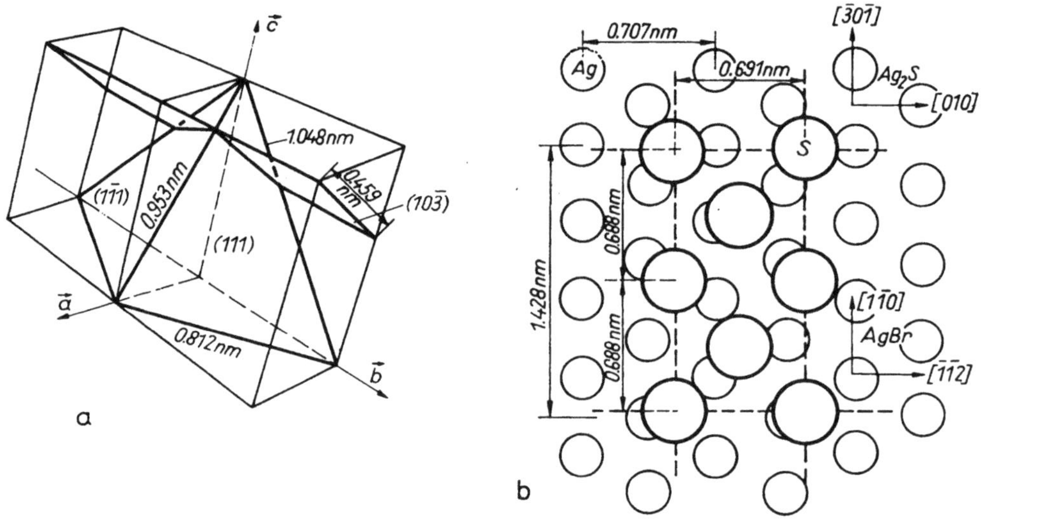

Fig. 7. a) (111) and $(1\bar{1}1)$ planes of Ag_2S which intersect the $(10\bar{3})$ plane at right angle; b) superposition of the S atoms of Ag_2S on AgBr (111) plane showing the structure of an interface

arrangement in this plane is superimposed on the (111) plane of AgBr in Fig. 7b which assumes the configuration of the interface between Ag_2S and AgBr. The axial correlations obtained from the observed lattice images agree well with that deduced from this figure and indicated in the illustration.

In these three cases the *b*-axes of the Ag_2S crystals lie in the interface, that is, the axis is parallel with the (111) plane of AgBr, and their directions always agree with the $[\bar{1}\bar{1}2]$ direction of the AgBr. On the other hand, in the $[1\bar{1}0]$ direction of AgBr, [001], $[\bar{2}01]$, and $[\bar{3}0\bar{1}]$ axes of Ag_2S are in parallel orientations.

In order to interpret these facts which are concerned with the growth mechanism of the sensitivity specks on a silver halide, it seems appropriate to introduce the misfit theory which is partly valid for the interpretation of the epitaxial growth of some metal films on a single crystalline surface. According to this theory the oriented overgrowth easily takes place when the lattice misfit is small.

The interplanar spacings of Ag_2S (010) planes are 0.691 nm along the *b*-axis while those of the AgBr (112) planes are 0.236 nm in the direction parallel with the *b*-axis of Ag_2S. Therefore, three times the interval of AgBr corresponds to one period of Ag_2S with a lattice misfit of 2%. Here the value of the lattice misfit is defined by

$$\left(\frac{1 - nd_{Ag_2S}}{md_{AgBr}}\right) \times 100\,,$$

where d_{Ag_2S} and d_{AgBr} denote the interplanar distance of Ag_2S and AgBr in the same direction and n and m are integers which are selected so as to minimize the misfits. The (110) interplanar distance of AgBr is 0.408 nm while those of the Ag_2S along the

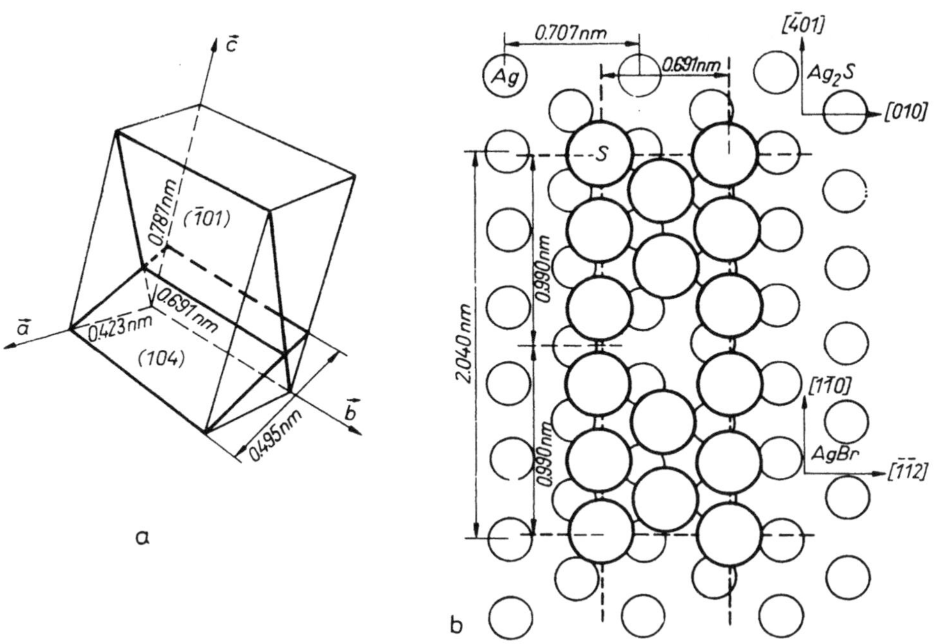

Fig. 8. a) An illustration showing that the (104) plane is one of the planes perpendicular to the $(\bar{1}01)$ plane; b) the (104) plane is superimposed on the AgBr (111) plane

c-axis and $[\bar{2}01]$ and $[\bar{3}0\bar{1}]$ directions are 0.787, 0.624, and 0.459 nm, respectively. Accordingly the lattice misfits along these axes are 4, 2, and 7% at the interface with the AgBr (111) plane.

The Ag_2S grain shown in Fig. 2d shows only a lattice image running in one direction. The spacing is 0.396 nm and the planes are considered to be the $(\bar{1}01)$ planes. The plane perpendicular to $(\bar{1}01)$ cannot be determined equivocally. However, from the facts described above we may assume that in this case also the b-axis of Ag_2S specks lies in the interface (Fig. 8a). Then the (104) plane can be deduced as the interface between this crystal and the matrix crystal. The arrangement of ions in this interface is represented in Fig. 8b. The lattice misfit is 1% along the $[\bar{4}01]$ axis.

5. Conclusions

1. The sensitivity specks formed on a AgBr grain by sulfur sensitization are Ag_2S microcrystals belonging to a monoclinic crystal form.
2. Most of the specks grown on the grains of AgBr make with their (100) or (102) planes the interface between AgBr and Ag_2S.
3. The formation of the specks takes place epitaxially and the b-axis of the Ag_2S microcrystals is always parallel with the [112] axis of the AgBr.
4. The lattice misfits are small at these interfaces.
5. Ag_2S crystals of cubic or pseudo-cubic type have not been found to form in the present work.

References

[1] S. E. SHEPPARD, Photogr. J. **65**, 380 (1925).
[2] P. FAELENS, Photogr. Korresp. **104**, 137 (1968).
[3] J. A. AZNAREZ, Photogr. Sci. Engng. **25**, 219 (1981).
[4] J. A. AZNAREZ, J. photogr. Sci. **25**, 53 (1977).
G. C. FARNELL, R. B. FLINT, and D. C. BIRCH, J. photogr. Sci. **25**, 203 (1977).
F. A. HAMM and J. J. COMER, J. appl. Phys. **24**, 1495 (1953).
[5] A. J. FRUEH, Z. Krist. **110**, 136 (1958).
[6] J. F. HAMILTON and L. E. BRADY, Surface Sci. **23**, 389 (1970).

(Received July 18, 1988)

phys. stat. sol. (a) **110**, 383 (1988)

Subject classification: 61.16 and 61.70

Department of Physics, Aristotle University of Thessaloniki[1])

Study of Special Triple Junctions and Faceted Boundaries by Means of the CSL Model[2])

By

E. G. DONI and G. L. BLERIS

Sufficient work has been made until now on the grain boundaries (GBs) multi-junctions in different polycrystalline cubic materials by using electron microscopy. In the present work the main characteristics of the majority of the junctions reported are theoretically investigated. In fact triple junctions (TJs) show a very systematic geometrical relationship, as far as the thermodynamically favorable grain boundaries are concerned. Thus the possible intersection of three grain boundaries is examined by using the analytical expression of the CSL rotation matrix and the CSL symmetry for cubic system. The general properties obtained are applied on TJs containing low Σ value grain boundaries and the resulting junctions are examined according to their stability.

Mittels Elektronenmikroskopie wurden bisher eine Reihe von Untersuchungen an Korngrenzen (GB)-Multiübergängen in verschiedenen polykristallinen kubischen Materialien durchgeführt. Hier werden die Hauptcharakteristiken der Mehrzahl der angegebenen Übergänge theoretisch untersucht. Tatsächlich zeigen Tripel-Übergänge (TJ) eine äußerst systematische geometrische Beziehung, insoweit die thermodynamisch bevorzugten Korngrenzen berücksichtigt werden. So wird die mögliche Durchschneidung der drei Korngrenzen mit dem analytischen Ausdruck der CSL-Rotationsmatrix und der CSL-Symmetrie für ein kubisches System untersucht. Die erhaltenen allgemeinen Eigenschaften werden auf TJ angewendet, die Korngrenzen mit niedrigem Σ-Wert enthalten, und die resultierenden Übergänge werden entsprechend ihrer Stabilität untersucht.

1. Introduction

Triple junctions (TJs) and faceted grain boundaries in polycrystalline materials have been subjects of great interest during the last few years. A lot of work has been carried out [1 to 17], which leads to the elucidation of the properties of TJs and faceted grain boundaries. Obviously the main reason for the generation and the stability of a TJ in a polycrystalline material is that it creates a thermodynamically stable configuration. Consequently, the main purpose of every experimental work on TJs is to give an answer to three main topics under investigation. Firstly to find the geometrical characteristics of a TJ, secondly to discover the macroscopic and microscopic reasons for the generation of a particular TJ, and thirdly to examine a TJ from the energetic point of view. Thus, grain boundary TJs for instance in Al [6, 7, 13, 16], Si [5, 12], and Cu [14] have been widely investigated from the experimental point of view during the last few years. Also in a very recent review Sutton and Bulluffi [17] gave a complete overview concerning TJs and faceted GBs which can be considered as the coexistence of two or more TJs in a variety of materials which have been already experimentally studied, such as for instance Au, Ag, Ni, and Cu.

[1]) 54006 Thessaloniki, Greece.

[2]) Work supported by the EEC no. ST2J-0289-C stimulation contract.

The experimental investigation of TJs has been strongly supported by a parallel theoretical investigation. The main results of this investigation can be summed up into the following statements:

(i) In general there exists a relationship of the type

$$R_3 = R_2 R_1 \tag{1}$$

between the three boundaries described by the coincidence site lattice (CSL) rotation matrices R_3, R_2, R_1, respectively (Warrington's idea [18]).

(ii) There is an isomorphism between the interfaces appearing in a TJ and the description of each interface by means of the CSL model, since a relation of the form (1) can always be established.

The first statement provides the main tool for the investigation of the interfaces in a TJ and the characterization of the grain boundaries according to the CSL model. The second statement leads to the prediction of all possible cases for the formation of a TJ from the CSL point of view, by making use of the CSL model for the description of a grain boundary.

In this work we shall make use of the previous two statements in order to complete a full theoretical study of TJs in polycrystalline materials belonging to the cubic system. Firstly, we shall analytically investigate (1). Secondly, by using the properties obtained we shall predict all possible cases of TJs in the cubic system in which the intersection axis is a symmetry axis. Finally, by using the symmetry properties of the CSL which have already been given [19], we shall examine some TJs concerning the most common cases in the cubic system.

2. Use of the CSL Model in the Study of TJs

Warrington [18] examined the case of a TJ between three crystals K_1, K_2, K_3 which is shown in Fig. 1, where the three rotation matrices R_1, R_2, R_3 are related by (1). In this case if the two matrices R_1 and R_2 are CSL rotation matrices, then the third matrix R_3 has also to be a CSL rotation matrix. These matrices have the general form

$$R = \frac{1}{\Sigma}[r_{ij}], \qquad j = 1, 2, 3, \tag{2}$$

where Σ and r_{ij} are integers without a common divisor as it has been proved elsewhere [20]. The integer Σ expresses the multiplicity of the CSL. The corresponding relation for the multiplicities Σ is

$$\Sigma_3 = \Sigma_2 \Sigma_1, \qquad \Sigma_3 > \Sigma_2, \Sigma_1,$$

or (3)

$$\Sigma_3 < \Sigma_2 \Sigma_1.$$

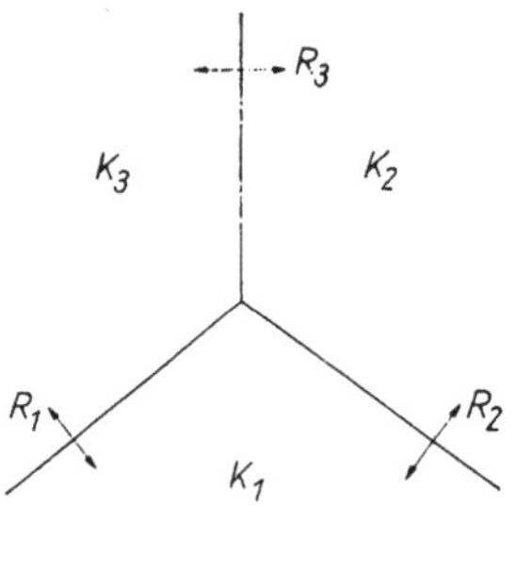

Fig. 1. The junction between three crystals K_1, K_2, K_3

Let Λ_1, Λ_2, Λ_3 be the lattices of the three crystals K_1, K_2, K_3, respectively. If we express the rotation matrices, R_1 and R_2 in the coordinate system of K_1, we have

$$R_1\Lambda_1 = \Lambda_3 ,$$

$$R_2\Lambda_1 = \Lambda_2 ,$$

that is

$$R_1R_2^{-1}\Lambda_2 = \Lambda_3 ,$$

and then

$$R_3 = R_1R_2^{-1} , \tag{4}$$

which allows us to express R_3 as a function of R_1 and R_2. Relation (4) is similar to (1), but it has the advantage that it gives the three matrices expressed in the same coordinate system. The relations between the corresponding multiplicities Σ_3, Σ_2, Σ_1 are given by (3). Moreover (4) has the same meaning as the relation proposed by Pond for the chatacterization of degenerated defects, i.e. defects in which their structures are related by symmetry [21]. In fact the rotation matrix R_1 expresses the operational relationship describing GB_2 in the frame of crystal K_1. The same is also true for the operator R_2 for GB_2. Then the operator $R_1R_2^{-1}$ will give the description of the existing degenerated defect. Since we are dealing with rotation operations only, the third operator R_3 will describe a disclination [21].

In order to determine the rotation angle and the rotation axis of the rotation operator R_3 given by (4) we follow the procedure given below. It is known that the analytical expression of the rotation matrix R which expresses a rotation around an axis $[uvw]$, through an angle θ is given for the cubic system [20] as the function

$$R(uvw, m, n, \alpha) , \tag{5}$$

where

$$R = \frac{1}{\Sigma}\left[\frac{r_{ij}^*}{\alpha}\right] = \frac{1}{\Sigma}\,[r_{ij}] , \qquad i, j = 1, 2, 3,$$

$$\Sigma = \frac{m^2 + dn^2}{\alpha}, \qquad (m, n) \div 1^{3)} , \tag{6}$$

$$d = u^2 + v^2 + w^2 ,$$

$$\tan\frac{\theta}{2} = \left[\frac{n^2 d}{m^2}\right]^{1/2} ,$$

and $\alpha = 1, 2$, or 4.

By substituting the expressions of the operators R_1 and R_2,

$$R_1(u_1v_1w_1, m_1, n_1, \alpha_1) ,$$

$$R_2(u_2v_2w_2, m_2, n_2, \alpha_2) ,$$

into (4) we can analytically find the matrix elements of the operator R_3. The analytical expressions of these matrix elements are given in the Appendix.

For the rotation angle θ_3 we find

$$\cos^2\frac{\theta_3}{2} = \frac{S_3 + \Sigma_3}{4\Sigma_3} , \tag{7}$$

[3]) $(m, n) \div 1$ means that the common divisor between the integers m, n is 1.

where S_3 is the trace of the rotation matrix R_3. By substituting in (7) the expression of S_3 as we can find it from the analyitcal expressions of the matrix elements (see Appendix) we deduce that

$$\cos\frac{\theta_3}{2} = \frac{m_1m_2 + (u_1u_2 + v_1v_2 + w_1w_2)\, n_1n_2}{(\alpha_1\alpha_2\Sigma_1\Sigma_2)^{1/2}} \tag{8a}$$

or

$$\cos\left(\frac{\theta_3}{2}\right) = \cos\left(\frac{\theta_1}{2}\right)\cos\left(\frac{\theta_2}{2}\right) + \sin\left(\frac{\theta_1}{2}\right)\sin\left(\frac{\theta_2}{2}\right)\cos\,(\bar{r}_1, \bar{r}_2)\,. \tag{8b}$$

Relation (8a) gives the general expression for the rotation angle in the case that R_3 is given by (4).

For the rotation axis computation we use the relations [22]

$$u_3 = \frac{(r_{32})_3 - (r_{23})_3}{2}\,,$$
$$v_3 = \frac{(r_{13})_3 - (r_{31})_3}{2}\,, \tag{9}$$
$$w_3 = \frac{(r_{21})_3 - (r_{12})_3}{2}\,,$$

and by substituting in (9) the general expressions of the elements of the rotation matrix R_3 we find

$$u_3 = (w_1v_2 - w_2v_1)\, n_1n_2 - u_2m_1n_2 + u_1m_2n_1\,,$$
$$v_3 = (u_1w_2 - u_2w_1)\, n_1n_2 - v_2m_1n_2 + v_1m_2n_1\,, \tag{10}$$
$$w_3 = (v_1u_2 - v_2u_1)\, n_1n_2 - w_2m_1n_2 + w_1m_2n_1\,.$$

So the general expression which gives the rotation axis of the operator R_3 for which (4) is valid, is

$$\boldsymbol{r}_3 = (\boldsymbol{r}_2 \times \boldsymbol{r}_1)\, n_1n_2 - \boldsymbol{r}_2m_1n_2 + \boldsymbol{r}_1m_2n_1\,. \tag{11}$$

Relations (8) and (11) give all the necessary information, that is the rotation angle and the rotation axis, for the determination of the characteristics of the operator R_3 given by (4). This relation rexpresses a purely geometrical relationship between the lattices of the three crystals. For any couple of operators R_1 and R_2 given we have to examine if there exists a relation of the form (4). If such a relation exists, then a TJ is described and the characteristics of the third description, that is the rotation angle and the rotation axis, are given by (8) and (11), respectively. By this sense, (4) in addition to (8) and (11) can be used in order to be able someone to predict all possible TJ in the cubic system.

3. Properties of TJs Defined by the Characteristics of the Two Grain Boundaries

In (4) the operators R_1 and R_2 express any CSL rotation around any CSL rotation axis. By making use of the general expressions of the rotation angle and the rotation axis we can find conditions for some special cases. So, in case that R_1 and R_2 express a rotation around the common intersection axis [uvw], (8a) becomes

$$\cos\frac{\theta_3}{2} = \frac{m_1m_2 + dn_1n_2}{(\alpha_1\alpha_2\Sigma_1\Sigma_2)^{1/2}}\,. \tag{12}$$

By making use of (6) we obtain

$$\cos\frac{\theta_3}{2} = \cos\frac{\pm(\theta_2 - \theta_1)}{2}$$

which means that

$$\theta_3 = \pm(\theta_2 - \theta_1) + 2\pi k\,. \tag{13}$$

Additionally for the rotation axis we find from (10) that

$$\left.\begin{aligned} u_3 &= u(-m_1n_2 + m_2n_1) \\ v_3 &= v(-m_1n_2 + m_2n_1) \\ w_3 &= w(-m_1m_2 + m_2n_1) \end{aligned}\right\} \Rightarrow [uvw]_3 = \langle uvw\rangle\,. \tag{14}$$

Furthermore if the operators R_1, R_2 express 180° rotations around axes perpendicular to the grain boundaries, then, since in this case we have [20]

$$R_1 = R_1(u_1, v_1, w_1, 0, 1, \alpha_1)\,,$$
$$R_2 = R_2(u_2, v_2, w_2, 0, 1, \alpha_2)\,,$$

(8b) and (9) give

$$\cos\frac{\theta_3}{2} = \cos\varphi \tag{15}$$

and

$$\boldsymbol{r}_3 = \boldsymbol{r}_2 \times \boldsymbol{r}_1\,. \tag{16}$$

Thus in this special case the rotation angle of the third CSL is two times the angle of the axes $[uvw]_1$ and $[uvw]_2$ and the intesection axis is defined by (16).

Relations (13) to (16) form a direct and easy way for the characterization of a TJ, if some one has already recognized some 180° rotation operators describing the CSLs of the GBs of the junction. For example, if the intersection of the three GBs is a CSL rotation axis of the two GBs, then the third GB should have a CSL rotation around the same axis. It is also obvious that if the two GBs are described by 180° CSL rotations then the third GB should necessarily have a CSL description with rotation angle θ_3 given by (15).

Although the above arguments are the necessary conditions for TJs having the special property of the common CSL axis as the intersection axis, or two GBs with 180° descriptions, they are not sufficient conditions for the study of any type of TJs. The characterization of a TJ by using the standard techniques [23] gives information for every GB independently. Then the use of (4) for the identification of the GB of the TJ according to the CSL model needs necessarily one of the rotation operators R_1 or R_2 to be transformed to the coordinate system of the other. Then the third CSL is described in the same system and the full identification of the TJ is possible.

We shall consider as a frame of reference Λ_1 the lattice of crystal K_1 and let R_2 be the operator describing the GB_2 in crystal K_2. Then (4) has to be written as

$$R_3 = R_1 S R_2^{-1} S^{-1}\,, \tag{17}$$

where S is the transformation operator relating the two coordinate systems. Now, for the operator R_3 to describe a CSL rotation relationship, the matrix

$$R_2' = S R_2^{-1} S^{-1} \tag{18}$$

should be of the same form as the CSL matrix given by (2). This is possible if and only if S is a symmetry element of the parent lattice. This remark implies that the two coordinate systems are related by a symmetry operator which makes the examination of TJs from the symmetry point of view to be an obvious necessity.

4. Symmetry Consideration of TJs

It is well known that the existence of a CSL means that a sublattice Λ_1^1 of the parent lattice exists with unit cell Σ times greater than the unit cell of the parent lattice. It is also known that the symmetry of the CSL is the symmetry of this sublattice and it is given by the group $G(\Sigma)$,

$$G(\Sigma) = H + RH\,, \tag{19}$$

where H is a subgroup of the point group G of the parent lattice and R is a second-order symmetry element of Λ_1^1 but not of the parent lattice. As has been already mentioned elsewhere the element R is a 180° CSL rotation of the examined CSL and for the majority of the CSLs such an operator always exists [19].

Let us consider now a given CSL grain boundary and its corresponding symmetry. It is obvious that if Λ_1^1 is the corresponding sublattice, then

$$R\Lambda_1^1 = \Lambda_1^1 \tag{20}$$

and its rotation axis belongs to Λ_1^1. For every $g \in G$ it is obvious that

$$gR\Lambda_1^1 = \begin{cases} \Lambda_1^1 & \text{if} \quad g \in H\,, \\ g\Lambda_1^1 = \Lambda_{1g} & \text{if} \quad g \notin H\,. \end{cases} \tag{21}$$

The operator $R_g = gR$ is a CSL operator also, describing one and the same CSL if and only if $g \in H$. Thus the second part of (21) shows that there are different variants [24] of the same sublattice in the parent lattice given by the left coset representatives g_i, where

$$G = H + g_1H + \dots + g_pH \tag{22}$$

and $g_i \in G$ but $g_i \notin H$ [19]. Let us consider now the coordinate system and the CSL description given on this system by their standard classification, for example in the cubic system by $u \geqq v \geqq w$. Another description of the same CSL may be on the same variant if its transformation by a symmetry element which belongs to the symmetry of the variant leads to a case where $u \geqq v \geqq w$. If this is not true the description belongs to another variant and by using the corresponding coset representative, (17) has to be established in the form

$$R_3 = R_1g^{-1}R_gg\,; \qquad g \notin H(R_g)\,. \tag{23}$$

Let us now consider the case of a special TJ, i.e. a TJ in which the intersection axis is a symmetry axis. In this particular case the symmetry element of the intersection axis is a common symmetry element of the symmetry groups of the two CSLs. By making use of the analytical expressions of the different CSL descriptions given elsewhere [25] and the fact that the symmetry axes for the cubic system are of the type ⟨100⟩, ⟨110⟩, and ⟨111⟩ we may easily conclude that a CSL described around such an axis has always at least one 180° description. The inverse is also true, i.e. if a CSL has a 180° description, then it has one CSL axis of the symmetry type. Thus the third CSL whose the characteristics are obtained by (15) and (16), has a symmetry group of the type (19), which means that it contains a 180° description. Moreover since the third CSL has a description around a symmetry axis, the symmetry element of this axis is a symmetry element of this CSL.

The above symmetry properties allow an obvious generalization: TJs whose intersection is a symmetry axis are composed by CSL grain boundaries which are characterized by the existence of 180° descriptions and they have a common symmetry element. Taking into account that GBs with 180° descriptions are thermodynamically most favorable, since the coherence of the structure is preserved as much as possible, we focus our attention to this case. In fact the great majority of TJs or faceted GBs which have been reported until now belong to this category. Thus by making use of the fact that if one of the 180° CSL GBs has a description around the intersection axis, then the three BGs have a description around this axis, we have found all the CSLs described around the ⟨100⟩, ⟨110⟩, and ⟨111⟩ axes in the cubic system. The results are tabulated in Tables 1, 2, and 3, respectively. The usage of these tables is obvious. If one has identified the common intersection axis as a symmetry axis, then the possible CSLs may also immediately be predicted.

5. Some Applications

Based mainly on the electron microscope observations there are many TJs or faceted GBs which have been reported. Since the faceted GBs can be analyzed on the basis of the idea of the coexistence of more than one TJs in close neighbourhood, we shall treat only TJs. The great majority of the TJs reported in the literature are composed

Table 1

The CSLs of the cubic system, produced by rotations around the ⟨100⟩-axis, up to $\Sigma = 97$

Σ	(m, n, α)	$\theta(°)$	Σ	(m, n, α)	$\theta(°)$
5	(1, 2, 1)	126.86	61	(5, 6, 1)	100.39
5	(2, 1, 1)	53.13	61	(6, 5, 1)	79.61
5	(1, 3, 2)	143.13	61	(1, 11, 2)	169.61
5	(3, 1, 2)	36.87	61	(11, 1, 2)	10.39
13	(2, 3, 1)	112.62	65	(1, 8, 1)	165.75
13	(3, 2, 1)	67.38	65	(4, 7, 1)	120.51
13	(1, 5, 2)	157.38	65	(7, 4, 1)	59.49
13	(5, 1, 2)	22.62	65	(8, 1, 1)	14.25
17	(1, 4, 1)	151.93	65	(3, 11, 2)	149.49
17	(4, 1, 1)	28.07	65	(7, 9, 2)	104.25
17	(3, 5, 2)	118.07	65	(9, 7, 2)	75.75
17	(5, 3, 2)	61.93	65	(11, 3, 2)	30.51
25	(3, 4, 1)	106.26	73	(3, 8, 1)	138.89
25	(4, 3, 1)	73.74	73	(8, 3, 1)	41.11
25	(1, 7, 2)	163.74	73	(5, 11, 2)	131.11
25	(7, 1, 2)	16.26	73	(11, 5, 2)	48.89
29	(2, 5, 1)	136.39	85	(2, 9, 1)	154.94
29	(5, 2, 1)	43.60	85	(6, 7, 1)	98.79
29	(3, 7, 2)	133.60	85	(7, 6, 1)	81.20
29	(7, 3, 2)	46.34	85	(9, 2, 1)	25.06
37	(1, 6, 1)	161.07	85	(1, 13, 2)	171.20
37	(6, 1, 1)	18.92	85	(7, 11, 2)	115.06
37	(5, 7, 2)	108.92	85	(11, 7, 2)	64.94
37	(7, 5, 2)	71.07	85	(13, 1, 2)	8.79
41	(4, 5, 1)	102.68	89	(5, 8, 1)	115.98
41	(5, 4, 1)	77.32	89	(8, 5, 1)	64.01
41	(1, 9, 2)	167.32	89	(3, 13, 2)	154.01
41	(9, 1, 2)	12.68	89	(13, 3, 2)	25.99
53	(2, 7, 1)	148.11	97	(4, 9, 1)	132.07
53	(7, 2, 1)	31.89	97	(9, 4, 1)	47.92
53	(5, 9, 2)	121.89	97	(5, 13, 2)	137.92
53	(9, 5, 2)	58.11	97	(13, 5, 2)	42.07

Table 2
The CSLs of the cubic system, produced by rotations around the ⟨110⟩-axis, up to $\Sigma=99$

Σ	(m, n, α)	$\theta(°)$	Σ	(m, n, α)	$\theta(°)$
3	(1, 1, 1)	109.47	51	(10, 1, 2)	16.10
3	(2, 1, 2)	70.53	57	(5, 4, 1)	97.05
9	(1, 2, 1)	141.05	57	(7, 2, 1)	44.00
9	(4, 1, 2)	38.94	57	(4, 7, 2)	157.16
11	(2, 3, 2)	129.52	57	(8, 5, 2)	148.41
11	(3, 1, 1)	50.48	59	(3, 5, 1)	134.02
17	(3, 2, 1)	86.63	59	(10, 3, 2)	45.98
17	(4, 3, 2)	93.37	67	(7, 3, 1)	62.44
19	(1, 3, 1)	153.47	67	(6, 7, 2)	117.56
19	(6, 1, 2)	26.52	73	(1, 6, 1)	166.56
27	(5, 1, 1)	31.59	73	(12, 1, 2)	13.44
27	(2, 5, 2)	148.41	81	(7, 4, 1)	77.88
33	(1, 4, 1)	159.95	81	(8, 7, 2)	102.11
33	(5, 2, 1)	58.99	83	(9, 1, 1)	17.86
33	(4, 5, 2)	121.00	83	(2, 9, 2)	162.14
33	(8, 1, 2)	20.05	89	(9, 2, 1)	34.89
41	(3, 4, 1)	124.12	89	(4, 9, 1)	145.11
41	(8, 3, 2)	55.88	97	(5, 6, 1)	118.98
43	(5, 3, 1)	80.63	97	(12, 5, 2)	61.02
43	(6, 5, 2)	99.36	99	(1, 7, 1)	168.46
51	(1, 5, 1)	163.90	99	(7, 5, 1)	90.58
51	(7, 1, 1)	22.84	99	(10, 7, 2)	89.42
51	(2, 7, 2)	157.16	99	(14, 1, 2)	11.54

by $\Sigma = 3$ CSL GBs. The $\Sigma = 3$ cubic CSL has a symmetry group [19] given by (19), where

$$\mathrm{H} = \{E, C_{31}^+, C_{31}^-, C_{2b}, C_{2e}, C_{2f}\} \tag{24}$$

and

$$R = \frac{1}{3}\begin{pmatrix} -1 & 2 & 2 \\ 2 & -1 & 2 \\ 2 & 2 & -1 \end{pmatrix} \tag{25}$$

is a second-order symmetry element describing a CSL rotation around the [111] axis. The symmetry axes are

$$[111], [\bar{1}10], [\bar{1}01], [0\bar{1}1] . \tag{26}$$

Since the order °H of H is 6 there are four variants which can be easily found by making use of the variant generating groups (V.G.G.) [26], which are

$$\mathrm{V.G.G.}_1 = \{E, C_{2x}, C_{2y}, C_{2z}\} , \tag{27}$$

$$\mathrm{V.G.G.}_2 = \{E, C_{2x}, C_{4x}^+, C_{4x}^-\} .$$

By using the elements of the V.G.G.s and (23) we may find the possible TJs in which the two GBs are twins. The result is always an intersection symmetry axis of the type ⟨110⟩ and the TJ is composed of $\Sigma = 3$, 3, 9 GBs. Their geometrical configuration is given in Fig. 2 and the two cases are complementary. The common intersection axis $[0\bar{1}1]$ is the symmetry axis of the C_{2e} symmetry element. The angle φ

Table 3

The CSLs of the cubic system, produced by rotations around the ⟨111⟩-axis, up to $\Sigma = 97$

Σ	(m, n, α)	$\theta(°)$	Σ	(m, n, α)	$\theta(°)$
3	(3, 1, 4)	60.0	49	(13, 3, 4)	43.57
3	(0, 1, 1)	180.00	57	(3, 4, 1)	133.17
7	(2, 1, 1)	81.78	57	(9, 7, 4)	106.83
7	(1, 3, 4)	158.21	57	(15, 1, 4)	13.17
7	(5, 1, 4)	38.21	61	(7, 2, 1)	52.66
13	(1, 2, 1)	147.79	61	(1, 9, 4)	172.66
13	(5, 3, 4)	92.20	61	(13, 5, 4)	67.34
13	(7, 1, 4)	27.79	67	(8, 1, 1)	24.43
19	(4, 1, 1)	46.83	67	(5, 9, 4)	144.43
19	(1, 5, 4)	158.21	67	(11, 7, 4)	95.57
19	(7, 3, 4)	73.17	73	(5, 4, 1)	108.36
21	(3, 2, 1)	98.21	73	(7, 9, 4)	131.63
21	(3, 5, 4)	141.78	73	(17, 1, 4)	11.63
21	(9, 1, 4)	21.78	79	(2, 5, 4)	153.99
31	(2, 3, 1)	137.89	79	(13, 7, 4)	86.01
31	(7, 5, 4)	102.10	79	(17, 3, 4)	33.99
31	(11, 1, 4)	17.89	91	(4, 5, 1)	130.42
37	(5, 2, 1)	69.43	91	(8, 3, 1)	66.01
37	(1, 7, 4)	170.57	91	(1, 11, 4)	173.99
37	(11, 3, 4)	50.57	91	(11, 9, 4)	109.58
39	(6, 1, 1)	32.20	91	(17, 5, 4)	53.99
39	(3, 7, 4)	152.20	91	(19, 1, 4)	10.42
39	(9, 5, 4)	87.79	93	(9, 2, 1)	42.10
43	(4, 3, 1)	104.82	93	(3, 11, 4)	162.10
43	(5, 7, 4)	135.18	93	(15, 7, 4)	77.89
43	(13, 1, 4)	15.18	97	(7, 4, 1)	89.41
49	(1, 4, 1)	163.57	97	(5, 11, 4)	150.59
49	(11, 5, 4)	76.43	97	(19, 3, 4)	30.59

between the two {111}-type GBs is 70.53° and the third GB has a description around the [$\bar{1}$01] axis of 141.06° as (15) predicts and it corresponds to $\Sigma = 9$ CSL (Fig. 2a). We use the symmetry element C_{2y} of the V.G.G.$_{1}$ in (23) and have

$$R_3 = RC_{2y}\, RC_{2y} = R(\bar{1}01, 4, 1, 2)\,. \tag{28}$$

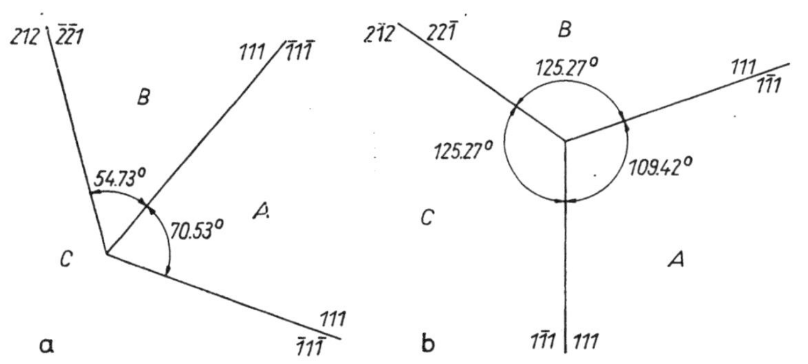

Fig. 2. TJs of $\Sigma = 3$ {111}-type GBs

It is obvious that every second-order element of the two V.G.G.s will give the same configuration. This type of TJ has already been reported [12]. By using the C_{4x}^{+} symmetry element we get the configuration of Fig. 2b, where the third GB is the $\Sigma = 9$ CSL around the $[10\bar{1}]$ axis and the corresponding rotation angle is 141.06°, i.e.

$$R_3 = RC_{4x}^{-}RC_{4x}^{+} = R(10\bar{1}, 4, 1, 2)\,. \tag{29}$$

These are the only cases if we use the {111} first-order twins of $\Sigma = 3$. But there is also the $\Sigma = 3$ {211} GB We shall examine two cases, the first is the coexistence of two {211} GBs and the second is the coexistence of {111} and {211} GBs. By using the same analysis for the symmetry group of $\Sigma = 3$ as in (24), we can see that there are three symmetry elements of the same variant around the ⟨211⟩-type axis, i.e.

$$\begin{aligned} R(11\bar{2}, 0, 1, 2) &= RC_{2b}\,, \\ R(1\bar{2}1, 0, 1, 2) &= RC_{2e}\,, \\ R(2\bar{1}\bar{1}, 0, 1, 2) &= RC_{2f}\,. \end{aligned} \tag{30}$$

Since they describe second-order twins inside the same variant we may directly apply (4). There are three cases, two of them being distinct, which are given in Fig. 3. It is rather obvious that the analysis of (4) by means of the theory given in Section 2 cannot predict these cases. Moreover one of them has been observed in Si [5], but it was not possible to be identified in that time.

In Fig. 4 we give the combination of the first- and second-order twins of $\Sigma = 3$. Its description is

$$R_3 = R(111, 0, 1, 1)\, C_{2x} R(211, 0, 1, 2)\, C_{2x} = C_{2f}\,. \tag{31}$$

In Fig. 5 we have TJs of $\Sigma = 3, 9, 27$ CSLs and in Fig. 6 the well-known faceted case [12] which can be considered as the coexistence of the cases of Fig. 5. Their descriptions are

$$\begin{aligned} R_3 &= R(\bar{4}11, 0, 1, 2)\, C_{2e} R(211, 0, 1, 2)\, C_{2e}\,, \\ R_3 &= R(111, 0, 1, 1)\, C_{2x} R(411, 0, 1, 2)\, C_{2x}\,, \\ R_3 &= R(111, 0, 1, 1)\, C_{2e} R(221, 0, 1, 1)\, C_{2e} = R(01\bar{1}, 5, 1, 1)\,. \end{aligned} \tag{32}$$

This case of faceted boundaries has been already remarked and studied from the energetic point of view by Sutton and Balluffi [17].

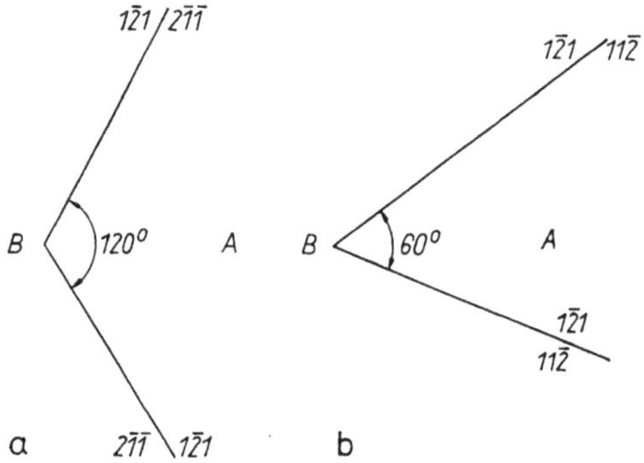

Fig. 3. The faceted $\Sigma = 3$ {211}-type GBs

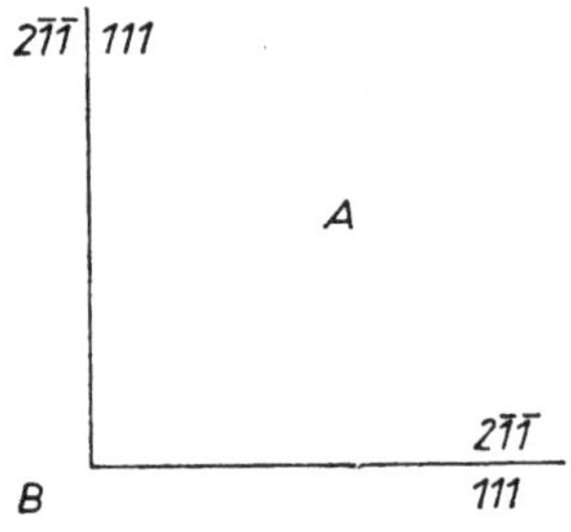

Fig. 4. The faceted $\Sigma = 3$ {111}- and {211}-type GBs

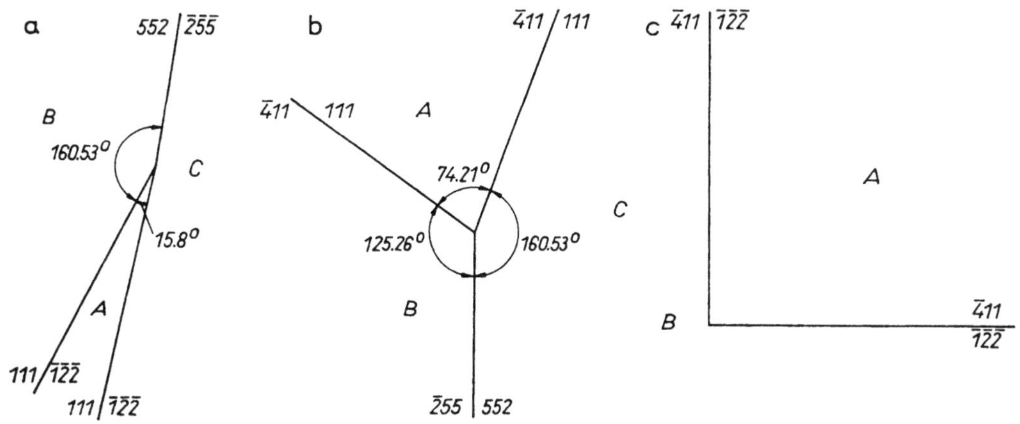

Fig. 5. TJ of a) $\Sigma = 3$, b) 9, c) 27 CSLs

6. Discussion

The previous analysis shows that many TJs which have been already reported might be described by the symmetry of the CSL and their identification is unique. This relation with the symmetry has to be taken into account on the microscopic level. It is rather obvious that there is a possibility that at least one common symmetry element exists between the symmetry groups of the parent lattice and the three CSLs and this element is related with the intersection axis, the axis around of which the three grains are bilt.

Moreover this symmetrical relationship between thermodynamically favourable GBs, i.e. twins of first or second order should be taken into consideration. A favourable GB may exist everywhere in the matrix. Thus the coexistence of two favourable GBs is equally probable. The experimental information and the examples presented here have shown that this coexistence is strongly supported by the symmetry elements of the parent lattice, which are absent in the composite lattice of the bicrystal. Since any kind of junction may be thought as a cooperation of at least two GBs, their coexistence should follow the classical symmetry laws. Considering one of them as most favourable from thermodynamical point of view, the second one existing should be compatible with the symmetry implied by the first. The third also must be compatible with the symmetry of the previous two. Thus the creation of a special triple junction may be established according to a group-subgroup property. From this point of view we cannot consider that the less thermodynamically favourable GB is decomposed to some other more favourable one, because such a procedure implies an increase of symmetry. This procedure may not be considered as a thermodynamical process compatible with the symmetry principles.

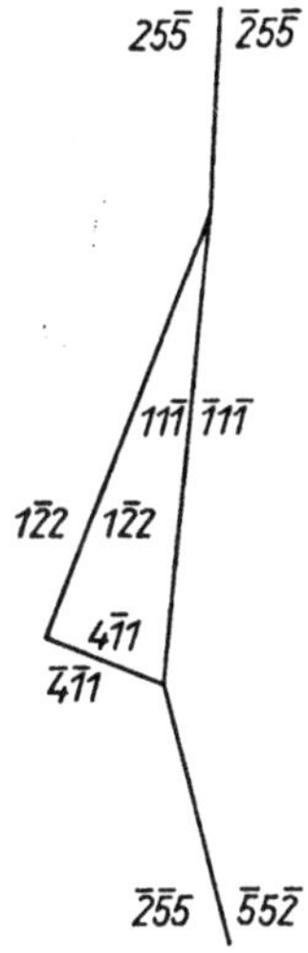

Fig. 6. The faceted GB of $\Sigma = 3, 9, 27$ CSLs

Appendix

Analytical expressions of the matrix elements of the description R_3 for which (4) is valid.

$$r_{11} = \frac{1}{\alpha_1\Sigma_1\alpha_2\Sigma_2} \{(4u_1^2u_2^2 - 2u_1^2d_2 - 2d_1u_2^2 + d_1d_2 + 4u_1v_1u_2v_2 + $$
$$+ 4u_1w_1u_2w_2)\, n_1^2n_2^2 + 4u_1\,(-v_1w_2 + w_1v_2)\, n_1^2m_2n_2 +$$
$$+ 4u_2(-w_1v_2 + v_1w_2)\, m_1n_1n_2^2 + 4(w_1w_2 + v_1v_2)\, m_1n_1m_2n_2 +$$
$$+ (2u_1^2 - d_1)\, n_1^2m_2^2 + (2u_2^2 - d_2)\, m_1^2n_2^2 + m_1^2m_2^2\}\ ,$$

$$r_{12} = \frac{1}{\alpha_1\Sigma_1\alpha_2\Sigma_2} \{2(u_1^2u_2v_2 + 2u_1v_1v_2^2 + 2v_1w_1v_2w_2 - u_2v_2d_1 -$$
$$- 2u_1v_1d_2)\, n_1^2n_2^2 + 2(2u_1^2w_2 - w_2d_1 - 2v_1w_1u_2)\, n_1^2m_2n_2 +$$
$$+ 2(-2w_1v_2^2 - w_1d_2 + 2u_1v_2w_2)\, m_1n_1n_2^2 - 4u_1u_2m_1n_1m_2n_2 +$$
$$+ 2u_2v_2m_1^2n_2^2 + 2u_1v_1n_1^2m_2^2 - 2w_1m_1n_1m_2^2 + 2w_2m_1^2n_2m_2\}\ ,$$

$$r_{13} = \frac{1}{\alpha_1\Sigma_1\alpha_2\Sigma_2} \{2(u_1^2u_2w_2 + 2u_1v_1v_2w_2 + 2u_1w_1w_2^2 - u_2w_2d_1 -$$
$$- u_1w_1d_2)\, n_1^2n_2^2 + 2(-2u_1^2v_2 + v_2d_1 + 2u_1v_1u_2)\, n_1^2m_2n_2 +$$
$$+ 2(-2w_1v_2w_2 + 2v_1w_2^2 - v_1d_2)\, m_1n_1m_2^2) - 4w_1u_2m_1n_1m_2n_2 +$$
$$+ 2u_2w_2m_1^2n_2^2 + 2u_1w_1n_1^2m_2^2 - 2v_2m_1^2m_2n_2 + 2v_1m_1n_1m_2^2\}\ ,$$

$$r_{21} = \frac{1}{\alpha_1\Sigma_1\alpha_2\Sigma_2} \{2(v_1^2u_2v_2 + 2u_1v_1u_2 + 2v_1w_1u_2w_2 - u_2v_2d_1 -$$
$$- 2u_1v_1d_2)\, n_1^2n_2^2 + 2(-2v_1^2w_2 + w_2d_1 + 2v_1w_1v_2)\, n_1^2m_2n_2 +$$
$$+ 2(2w_1u_2^2 - w_1d_2 - 2u_1u_2w_2)\, m_1n_1n_2^2 - 4u_1v_2m_1n_1m_2n_2 +$$
$$+ 2u_2v_2m_1^2n_2^2 + 2u_1v_1n_1^2m_2^2 + 2w_1m_1n_1m_2^2 - 2w_2m_1^2m_1n_2\}\ ,$$

$$r_{22} = \frac{1}{\alpha_1\Sigma_1\alpha_2\Sigma_2} \{(4v_1^2v_2^2 - 2v_1^2d_2 - 2d_1v_2^2 + d_1d_2 + 4u_1v_1u_2v_2 +$$
$$+ 4v_1w_1v_2w_2)\, n_1^2n_2^2 + 4v_1(-w_1u_2 + u_1w_2)\, n_1^2m_2n_2 +$$
$$+ 4v_2(-u_1w_2 + w_1u_2)\, m_1n_1n_2^2 + 4(w_1w_2 + u_1u_2)\, m_1n_1m_2n_2 +$$
$$+ (2v_1^2 - d_1)\, n_1^2m_2^2 + (2v_2^2 - d_2)\, m_1^2n_2^2 + m_1^2m_2^2\}\ ,$$

$$r_{23} = \frac{1}{\alpha_1\Sigma_1\alpha_2\Sigma_2} \{2(2u_1v_1u_2w_2 + 2v_1w_1w_2^2 - v_1w_1d_2 + 2v_1^2v_2w_2 -$$
$$- v_2w_2d_1)\, n_1^2n_2^2 + 2(-2u_1v_1v_2 + 2v_1^2u_2 - u_2d_1)\, n_1^2m_2n_2 +$$
$$+ 2(2w_1u_2w_2 + 2u_1w_2^2 + u_1d_2)\, m_1n_1n_2^2 - 4w_1v_2m_1n_1m_2n_2 +$$
$$+ 2v_1w_1n_1^2m_2^2 + 2v_2w_2m_1^2n_2^2 - 2u_1m_1n_1m_2^2 + 2u_2m_1^2m_2n_2\}\ ,$$

$$r_{31} = \frac{1}{\alpha_1\Sigma_1\alpha_2\Sigma_2} \{2(w_1^2u_2w_2 + 2w_1v_1v_2u_2 + 2u_1w_1u_2^2 - u_2w_2d_1 -$$
$$- u_1w_1d_2)\, n_1^2n_2^2 + 2(2w_1^2v_2 - v_2d_1 - 2v_1w_1w_2)\, n_2^2m_2n_2 +$$
$$+ 2(2u_1u_2v_2 + 2v_2w_1^2 + v_1d_2)\, m_1n_1n_2^2) - 4u_1w_2m_1n_1m_2n_2 +$$
$$+ 2u_2w_2m_1^2n_2^2 + 2u_1w_1n_1^2m_2^2 + 2v_2m_1^2m_2n_2 - 2v_1m_1n_1m_2^2\}\ ,$$

$$r_{32} = \frac{1}{\alpha_1 \Sigma_1 \alpha_2 \Sigma_2} \{2(2u_1w_1u_2v_2 + 2v_1w_1v_2^2 - v_1w_1d_2 + 2w_1^2v_2w_2 - \\ - v_2w_2d_1)n_1^2n_2^2 + 2(2u_1w_1w_2 - 2w_1^2u_2 + u_2d_1)\, n_1^2m_2n_2 + \\ + 2(-2v_1u_2v_2 + 2u_1v_2^2 - u_1d_2)\, m_1n_1n_2^2 - 4v_1w_2m_1n_1m_2n_2 + \\ + 2v_1w_1n_1^2m_2^2 + 2v_2w_2m_1^2n_2^2 + 2u_1m_1n_1m_2^2 - 2u_2m_1^2m_2n_2\}\ ,$$

$$r_{33} = \frac{1}{\alpha_1 \Sigma_1 \alpha_2 \Sigma_2} \{(4w_1^2w_2^2 - 2w_1^2d_2 - 2d_1w_2^2 + d_1d_2 + 4u_1w_1u_2w_2 + \\ + 4v_1w_1v_2w_2)\, n_1^2n_2^2 + 4w_1(-u_1v_2 + v_1u_2)\, n_1^2m_2n_2 + \\ + 4w_2(-v_1u_2 + u_1v_2)\, m_1n_1n_2^2 + 4(u_1u_2 + v_1v_2)\, m_1n_1m_2n_2 + \\ + (2w_1^2 - d_1)\, n_1^2m_2^2 + (2w_2^2 - d_2)\, m_1^2n_2^2 + m_1^2m_2^2\}\ .$$

References

[1] D. Vaughan, Phil. Mag. 22, 1003 (1970).
[2] P. I. Goodhew, Met. Sci. 13, 108 (1979).
[3] V. Andreeva, G. I. Salnikov, and L. K. Fionova, Acta metall. 26, 1331 (1980).
[4] L. K. Fionova, A. V. Andreeva, and T. I. Zhukova, phys. stat. sol. (a) 67, K15 (1981).
[5] F. Komninou, Th. Karakostas, G. L. Bleris, and N. A. Economou, J. Physique 43, C1–9 (1982).
[6] V. N. Perevezentzev, M. Y. Shcherban, E. A. Soldatov, and N. V. Belov, Soviet Phys. — Doklady 27, 93 (1982).
[7] T. I. Zhukova and L. K. Fionova, Soviet Phys. — Solid State 25, 472 (1983).
[8] G. D. Sukhomlin and A. V. Andreeva, phys. stat. sol. (a) 78, 333 (1983).
[9] M. D. Vaudin, B. C. Cunningham, and D. G. Ast, Scripta metall. 17, 191 (1983).
[10] C. T. Forwood and L. M. Clarebrough, Acta metall. 32, 757 (1984).
[11] L. C. Lim and R. Raj, Acta metall. 32, 1177 (1984).
[12] F. Komninou and Th. Karakostas, Proc. MRS Symp. Strasbourg (France), June 1984 (p. 95).
[13] L. K. Fionova, Phys. Chem. Mech. Surfaces 2, 3370 (1985).
[14] Ch. V. Kopetskii and L. K. Fionova, Phys. Chem. Mech. Surfaces 3, 313 (1985).
[15] Ch. V. Kopetskii and L. K. Fionova, Phys. Chem. Mech. Surfaces 3, 2004 (1985).
[16] S. Hasimoto, T. K. Fujii, and S. Miura, Scripta metall. 21, 169 (1987).
[17] A. P. Sutton and R. W. Balluffi, Acta metall. 35, 2177 (1987).
[18] D. H. Warrington, Grain Boundary Structure and Kinetics. Milwaukee, Wisconsin 1979.
[19] E. G. Doni, G. L. Bleris, Th. Karakostas, J. G. Antonopoulos, and P. Delavignette, Acta cryst. A41, 440 (1985).
[20] G. L. Bleris and P. Delavignette, Acta cryst. A37, 779 (1981).
[21] R. C. Pond, Line Defects in Interfaces, to be published in: Dislocations in Solids, Ed. F. R. Nabarro, North-Holland Publ. Co., Amsterdam 1988.
[22] G. L. Bleris, J. G. Antonopoulos, Th. Karakostas, and P. Delavignette, phys. stat. sol. (a) 67, 249 (1981).
[23] Th. Karakostas, G. Nouet, G. L. Bleris, S. Hagege, and P. Delavignette, phys. stat. sol. (a) 50, 703 (1978).
[24] G. van Tendeloo and S. Amelinckx, Acta cryst. A30, 431 (1974).
[25] G. L. Bleris, E. G. Doni, Th. Karakostas, J. G. Antonopoulos, and P. Delavignette, Acta cryst. A41, 445 (1985).
[26] G. van Tendeloo, Ph. D. Thesis, University of Antwerpen, 1974.

(Received September 23, 1988)

phys. stat. sol. (a) **110**, 397 (1988)

Subject classification: 61.70 and 78.20; 61.10; S11.1

Chair of Chemistry, Moscow Institute of Electronic Technology[1])

Properties of Proton Exchanged Optical Waveguiding Layers in $LiNbO_3$ and $LiTaO_3$

By

V. A. GANSHIN, YU. N. KORKISHKO, and T. V. MOROZOVA

Complex studies of proton exchange doping of $LiNbO_3$ and $LiTaO_3$ single crystals are carried out to investigate the regularities and specific features of structural changes and optical parameters of waveguiding layers. In fact, depending on doping melt type and crystalline cut H:$LiNbO_3$ waveguides can involve two to six and H:$LiTaO_3$ up to three layers with different properties. A qualitative description of the proton exchange mechanism is suggested.

Mit dem Ziel der Untersuchung von Gesetzmäßigkeiten und Besonderheiten von Strukturumwandlungen und Eigenschaften planarer Lichtleiter werden Komplexuntersuchungen zum Protonen–Lithium-Austausch an $LiNbO_3$ und $LiTaO_3$ Kristallen durchgeführt. Es wird gezeigt, daß in Abhängigkeit vom Schmelzentyp und der kristallographischen Orientierung der Plattenoberfläche H:$LiNbO_3$-Lichtleiter aus zwei bis sechs und bei H:$LiTaO_3$ aus bis zu drei Schichten mit verschiedenen Eigenschaften bestehen können. Es wird eine qualitative Beschreibung des Protonen–Lithium-Austausch gegeben.

1. Introduction

A number of methods of planar waveguide fabrication with lithium niobate (LN) and lithium tantalate (LT) substrates have been developed and tested to date, including out-diffusion of Li_2O, diffusion of metal ions from films, ion implantation, epitaxy, and ion exchange (IE) diffusion. The analysis of these technologies and properties of the resulting structures discloses that metal diffusion and IE techniques provide the best results [1]. This is especially true of IE, which is easy to implement in engineering and technology and provides the fabrication of LN waveguides with a wide range of refractive index (RI) increments for the design of certain unique optic elements, in particular planar lenses [2 ot 5].

Our paper deals with some new results on the IE process $H^+ \rightleftarrows Li^+$, or proton exchange (PE), in LN and LT crystals.

2. PE Waveguides in $LiNbO_3$

Ag^+, Tl^+, Cu^+, and K^+ ions can take part in the IE process forming waveguides in LN [6 to 8], and also protons [9]. A considerable number of works on PE has been published. The analysis of all known doping melts indicates that they can be classified into two types. Melts of type 1 (1-melts) are those that facilitate H:$LiNbO_3$ waveguide formation with Y-cut crystals featuring more than one mode, and without surface destruction. On the other hand, stongly acidic melts with a high intensity of PE are of type 2 (2-melts). Direct Y-cut waveguide formation in 2-melts is impossible because of high H^+ saturation of the surface and its destruction.

[1]) SU-103498 Moscow, USSR.

Type 1 includes already known melts with some organic acids as palmetic, stearic [10], lauric, myristic, etc., along with melts in which H^+ ions are generated by hydrolysis ($ZnCl_2$–KCl, $Ca(NO_3)_2$–KNO_3 systems, and others [11] in the presence of water), and also the mixture of benzoic acid with lithium benzoate [12]. Benzoic acid is a well known 2-melt.

All these melts are inconvenient in practice. The processes with organic compounds have to be carried out in evacuated and sealed off ampoules to exclude sublimation and dissimilation; hydrolysing melts are gradually depleted if the process is carried out in an open crucible, thus leading to a gradual reduction of the PE rate.

We have developed 2-melts of new compositions based on acidic inorganic salts, for example $KHSO_4$ [5], $NH_4H_2PO_4$, and others, featuring high air stability in the liquid phase and practically constant doping activity. Thanks to intensive PE the average time of waveguide fabrication in these melts is in the order of 10 min at temperatures from 200 to 300 °C. However, 2-melts have also some disadvantages. They are: impossibility to manufacture guides in *Y*-cut LN plates only by treating samples in melts, relatively low optical quality [13], lower thermal stability [10], and acousto-optical parameters [14] of waveguides. Nevertheless there exists the possibility to remove these drawbacks by subsequent annealing of samples, or by treating them in Li^+-containing melts after PE [15]. For example, in order to fabricate H:LN waveguides on *Y*-cut plates one can use the multistage process including treatment in melt and short-term annealing of samples at approximately 300 °C. Such technique allows to form H:LN(*Y*) waveguides with properties identical to those of specimens produced with 1-melts.

To study H:LN layers and waveguides produced in 2-melts we used, as previously [8] with 1-melts, several methods including X-ray structure analysis, rocking curves method, IR spectroscopy, and phase-contrast microscopy. The main features of these studies are:

a) Direct X-ray investigations of probes proved that a thin layer containing lithium pyroniobate, $LiNb_3O_8$, in its monoclinic phase is always formed when LN is doped with 2-melts as well as with 1-melts.

b) Analysis of mode spectra discloses that the most probable location of this layer is at the diffusion front between LN substrate and doped H:LN region. The existence of special ordinary modes (o-modes) with unusually low effective indices (EI) from 2.05 to 2.2 proves that the $LiNb_3O_8$ layer does not display any noticeable anisotropy of the biaxial crystal. Then there exist also unusual extraordinary modes (e-leaky-modes) with minimal EI's of the same values. The longer the doping PE process the lower are EI's of these leaky o- and e-modes and the thicker is the $LiNb_3O_8$ layer. The typical value of its thickness is in the order of 0.1 μm.

Similar qualitative results have been obtained with *X*-, *Z*-, and (014)-cut LN plates. Probably that is true also for *Y*-LN. However, it is impossible to form H:LN(*Y*) waveguides without annealing (see above) and thus to use waveguiding spectroscopy for specimens just after treatment of LN plates in 2-melts.

c) In contrast to samples formed in type 1 melts there is a strong absorption near 3240 cm^{-1} in IR spectra for all specimens treated in 2-melts.

d) In contrast to published data [16] no $HNbO_3$ in solid phase has been detected in the samples.

e) Depending on the crystal cut the doping region may contain from two to six layers differing from one another both in structure and optical properties. The most interesting process is with (014)-cut: at increased doping time six layers are formed in sequence and coexist. Some of their characteristics are listed in Table 1 along with those for *Z*- and *Y*-cut samples.

Table 1

cut	max. number of layers	layers and compositions	RI n_0 and n_e	crystal warping (%)
Y*)	2	$H_xLi_{1-x}NbO_3$	$n_e = 2.33$ $n_0 = 2.25$	$S_y = +3$, $S_z = 0$
		$H:LiNbO_3 + LiNb_3O_8$	$n_e = 2.2$ $n_0 = 2.27$ to 2.21	not studied
Z	2	$H_xLi_{1-x}NbO_3$	$n_e = 2.33$ $n_0 = 2.25$	$S_z = 0.4$, $S_y = 0$
		$H:LiNbO_3 + LiNb_3O_8$	$n_e = 2.20$ to 2.05 $n_0 = 2.29$ to 2.05	not studied
(014)	6	α-$H_xLi_{1-x}NbO_3$	$n_e = 2.32$ $n_0 = 2.29$	$S_y = 7.5$, $S_z = -2.5$
		$H:LiNbO_3 + LiNb_3O_8$	$n_e = 2.23$ to 2.20 $n_0 = 2.23$ to 2.20	not studied
		β-$H_xLi_{1-x}NbO_3$	$n_e = 2.30$ $n_0 = 2.27$	$S_y = 6.7$, $S_z = -1.0$
		$H:LiNbO_3 + LiNb_3O_8$	$n_e = 2.20$ to 2.15 $n_0 = 2.20$ to 2.15	not studied
		γ-$H_xLi_{1-x}NbO_3$	$n_e = 2.32$ $n_0 = 2.26$	$S_y = 0.75$, $S_z = 0.25$
		$H:LiNbO_3 + LiNb_3O_8$	$n_e = 2.15$ to 2.05 $n_0 = 2.15$ to 2.05	not studied

*) RI's for this cut relate to waveguides produced with the multistage technique and thus, in fact, are identical to those for samples prepared with 1-melts. The warping relates to samples formed with 2-melts before surface destruction.

The mechanism of the PE doping process and features of annealing effects in H:LN are discussed in [15]. Now we shall note only that waveguides fabricated in 2-melts (excluding six-layer structures on (014)-cut) and subsequently treated in Li^+-containing melts, for example in pure $LiNO_3$, or annealed possess the same structure, optical properties, and parameter stability as those formed in 1-melts. It can be stated therefore with high degree of reliability that other properties, in particular electro-optical and electro-acoustical, are identical, too. So, the developed convenient and stable inorganic 2-melts can be used for quick waveguide manufacturing instead of benzoic acid.

At least, our studies disclosed that a phase-constitution diagram for the $LiNbO_3$–$HNbO_3$ system, drawn in [16] for the case of the PE process in LN powder, is inapplicable to PE in single crystals. Each plate orientation features its "own" constitution diagram.

3. PE Waveguides in $LiTaO_3$

Currently, the only way to obtain high-quality waveguides on LT at temperatures below the Curie point is through the PE process [17 to 19]. This technique, however, has been developed only with Z-cut LT. We found that it is possible also to dope X- and Y-cut plates. Here the most important features of processes are discussed.

3.1 Subacidic doping

In this case $H_xLi_{1-x}TaO_3$ phases, from low to high values of x, are formed at a very low PE intensity in 1-melts, for example in stearic or palmetic acids. On the surface

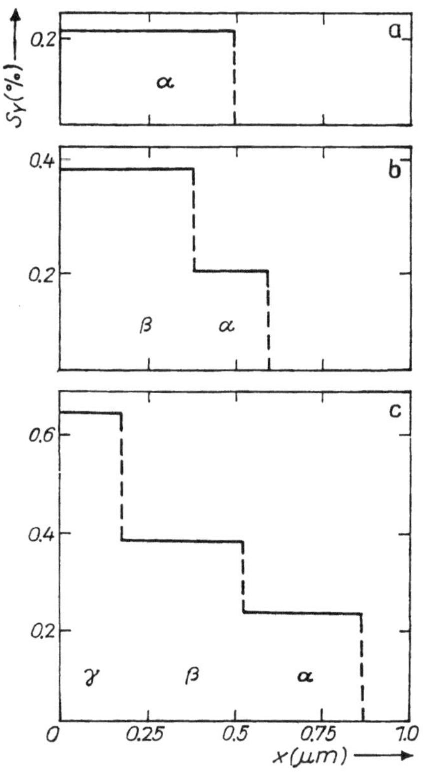

Fig. 1. Deformation patterns in $H:LiTaO_3$ X-cut structures formed by treatment in stearic acid melt at 300 °C during a) 1.5, b) 2, and c) 2.5 h

of X-LT plates α_x-, β_x, -γ_x-phases are sequentially formed with an increase of doping time (Fig. 1). These phases possess the $H_xLi_{1-x}TaO_3$ composition and the structure of LT but with other crystal lattice parameters. The warping of parameter a ($S_y = \Delta a/a$) for these phases is listed in Table 2 (only S_y-values are represented because S_z equals zero for X- and Y-cuts).

The longer treatment of X-plates in the melt also leads to surface destruction as well as for the case with Y-plates. The destructed surface of X-cut samples has typical cracks along the Y-axis or/and at an angle of 26° to this axis.

PE on Y-cut plates is much the same as on X-cut. The difference is that only two phases α_y and β_y of solid solutions are formed before destruction (Table 2). The difference is rather essential, and it is clear that Y-LT is prone to destruction more than X-LT.

According to the studies carried out, the RI of guides features a step-like distribution both in X- and Y-cut plates of LT. Therefore measurements of EI yield the guide thickness, h, even for single-mode layers, but in the case of e-modes only [15]. Fig. 2 shows the diffusion factor (effective coefficient), defined from $h^2 = 4Dt$, as a function of the doping time t, along with the results of IR spectroscopy. As can be noted, the ratio of absorption by OH groups around 3240 and 3480 cm^{-1} increases with doping time (curve I in Fig. 2 is the ratio of areas under the absorption intensity versus wave number curve as a function of doping time). This indicates a higher H^+ concentration in the surface layer. At a definite time the diffusion factor drops off

Table 2

LT cut	S parameter (%) in phase		
	α	β	γ
X	0.22	0.37	0.64
Y	0.18	2.10	surface destruction

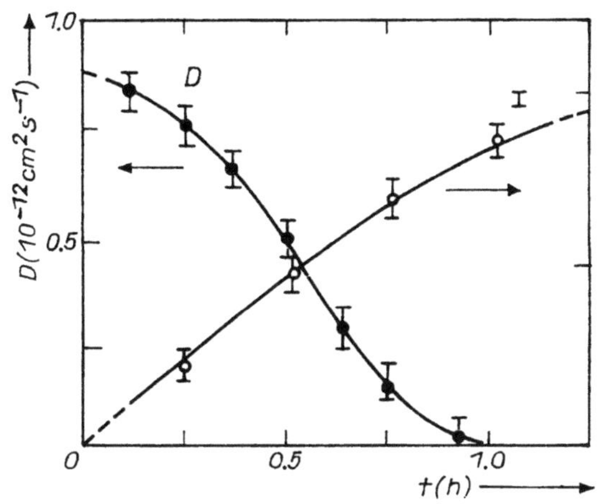

Fig. 2. Diffusion factor D and ratio of absorption intensity I by OH groups at 3240 to that at 3480 cm^{-1} as a function of doping time of an Y-cut $LiTaO_3$ crystal (treated in stearic acid melt at 270 °C)

to zero while the PE definite process goes on, thus leading to an increase in H^+ concentration. This continues till the strains developed do not relax by surface destruction.

In X-LT and Y-LT samples X-ray structure studies revealed maxima corresponding to a certain phase with a symmetry differing from that of LT. Probably it is a $LiTa_3O_8$ layer located at a definite depth of the PE front and capable to block Li^+ ion diffusion from the crystal depth to its surface. This leads to a self-completed PE process. Therefore only one-mode waveguides can be fabricated on X- and Y-cut plates in melts.

Short-term annealing at about 300 °C leads to only α-phase waveguides (with lowest deformations); their thickness h is in the interval 0.7 to 1.2μm and $\Delta n_e = 0.01$. Multistage treatment involving PE doping and annealing allows to produce double-mode waveguides with a stepwise RI pattern and RI increment $\Delta n_e = 0.03$. Further annealing results in triple-mode waveguide formation, but yet with a Gaussian-type RI pattern.

3.2 Strongly acidic doping

Doping of X-LT plates with strongly acidic melts results in the appearance of type γ_x-phases and a phase with $S_y = 2.15\%$ similar to the β_y-phase formed on Y-LT in subacidic melts (see Table 2).

Only the maximally warpable β_y-phase is formed on the Y-cut LT, just as in the case with subacidic doping the surface is destructed due to prolonged saturation.

The only way to fabricate waveguides in these melts is to use the above-described multistage treatment. We managed to get single- and double-mode waveguides with this technique.

According with [18], PE with Z-LT is also in all probability a self-completed process, since prolonged treatment in benzoic acid was found to slow down the doping process. The difference is that no surface destruction is observed in this case; this being probably due to the free nature of deformations on Z-cut LT crystals and thus to the absence of stresses, just as in the case of Z-LN. The results of [20] can be also satisfactorily explained with the self-completing mechanism.

All represented results prove that the PE kinetics of doping LN and LT in melts is much more complex than of usual ion exchange [21, 22]. Obviously, to solve this task and thus to describe and to control this process it is neccessary to take into consideration the observed effects of phase separation and blocking. That is the matter of our following studings.

References

[1] V. A. GANSHIN, YU. N. KORKISHKO, and V. Z. PETROVA, Obzory po elektronnoi tekhnike, Ser. Lasernaya tekhnika i optoelektronika 2, 1174 (1986).
[2] D. Y. ZANG, Optics Common. 47, 248 (1983).
[3] V. A. VOLKOV, V. P. VIRELKIN, V. A. GANSHIN, M. YU. KVASHA, YU. N. KORKISHKO, and S. M. FEDOTOV, Zh. tekh. Fiz. 57, 1221 (1987).
[4] D. Y. ZANG and C. S. TSAI, Appl. Phys. Letters 46, 703 (1985).
[5] V. A. GANSHIN and YU. N. KORKISHKO, Zh. tekh. Fiz. 57, 827 (1987).
[6] V. A. GANSHIN and YU. N. KORKISHKO, Zh. tekh. Fiz. 54, 383 (1984).
[7] V. A. GANSHIN, V. SH. IVANOV, YU. N. KORKISHKO, and V. Z. PETROVA, Zh. tekh. Fiz. 55, 2070 (1985).
[8] V. A. GANSHIN, V. SH. IVANOV, YU. N. KORKISHKO, and V. Z. PETROVA, Zh. tekh. Fiz. 56, 1354 (1986).
[9] J. L. JACKEL, Appl. Phys. Letters 41, 607 (1982).
[10] J. L. JACKEL and C. E. RICE, Proc. SPIE 460, 43 (1984).
[11] T. M. BOIKO, V. A. GANSHIN, and YU. N. KORKISHKO, Zh. tekh. Fiz. 55, 1441 (1985).
[12] J. L. JACKEL, C. E. RICE, and J. J. VESELKA, Electronics Letters 19, 387 (1983).
[13] J. L. JACKEL, C. E. RICE, and J. J. VESELKA, Ferroelectrics 50, 165 (1983).
[14] M. MIYAWAKI and S. OGURA, Appl. Phys. Letters 47, 918 (1985).
[15] YU. N. KORKISHKO and V. A. GANSHIN, Zh. tekh. Fiz. 58, 692 (1988).
[16] C. E. RICE and J. L. JACKEL, Mater. Res. Bull. 19, 591 (1984).
[17] V. V. ATCHUTIN and T. N. ZACHARIASH, Zh. tekh. Fiz. 54, 977 (1984).
[18] J. A. KOPILOV, V. B. KRAVCHENKO, E. N. MIRGORODSKAYA, and A. V. BOBILOV, Zh. tekh. Fiz. Pisma 9, 601 (1983).
[19] W. B. SPILLMAN, N. A. SANFORD, and R. A. SAREF, Optics Letters 8, 497 (1983).
[20] F. DUFFY, M. AL-SHUKRI, and M. DE LA RUE, Electronics Letters 23, 849 (1987).
[21] F. HELFFERICH and M. S. PLESSET, J. chem. Phys. 28, 418 (1958).
[22] C. D. W. WILKINSON and R. WALKER, Electronics Letters 14, 599 (1978).

(Received March 11, 1988; in revised form June 16, 1988)

phys. stat. sol. (a) 110, 403 (1988)

Subject classification: 61.80; 72.80; S5.11

A. N. Sevchenko Scientific Research Institute for Applied Physics Problems, V. I. Lenin Belorussian State University, Minsk[1])

Peculiarities of Radiation Defect Accumulation under High-Flux γ-Ray and Electron Irradiation of Silicon

By

P. F. LUGAKOV and T. A. LUKASHEVICH

The change of charge-carrier concentration is studied when irradiating floated zone n- and p-type Si with $\varrho = (1$ to $5) \times 10^2$ and (1 to 60) $\times 10^3\ \Omega$ cm by high fluxes of ^{60}Co γ-rays ($\Phi_\gamma = 1 \times 10^{17}$ to 2×10^{19} cm^{-2}) and electrons with energy $E_e = 4$ to 10 MeV ($\Phi_e = 1 \times 10^{15}$ to 5×10^{16} cm^{-2}). The temperature dependences of the Hall coefficient are measured. The analysis of the results on the accumulation under irradiation and annealing of radiation defects allows one to conclude that various trends of the dose dependences of hole concentration with the kinds of bombarding particles used are due to the difference in the production efficiencies of radiation defects under electron and γ-ray irradiation. The possibility of formation of phosphorus–two vacancy complexes by adding vacancies to E-centres during prolonged γ-irradiation is shown while electrons produce preferentially divacancies. Phosphorus–two vacancy complexes are stable up to $T = 400$ to 500 °C and introduce a deep acceptor level at $E_V + 0.35$ eV into the forbidden gap of Si. It is found that the annealing character of radiation defects in heavily irradiated Si does not depend on the kind of irradiation and is due to the rearrangements of intrinsic structural defects.

Изучено изменение концентрации носителей заряда при облучении высокими потоками γ-квантов Co^{60} ($\Phi_\gamma = 1 \times 10^{17}$ до 2×10^{19} cm^{-2}) и электронами с энергией $E_e = 4$ до 10 MeV ($\Phi_e = 1 \times 10^{15}$ до 5×10^{16} cm^{-2}), выращенного зонной плавкой n- и p-кремния с $\varrho = (1$ до $5) \times 10^2$ и (1 до 60) $\times 10^3$ Ω cm, соотвестсвенно. Измерялись температурные зависимости коэффициента Холла. Анализ результатов по накоплению при облучении и отжигу радиационных дефектов (РД) позволил сделать заключение, что неодинаковый ход дозовых зависимостей концентрации дырок при используемых видах бомбардирующих частиц обусловлен различием в эффективностях образования собственных и ,,примесных'' РД при электронном и γ-облучении. Показана возможность формирования при длительном γ-облучении комплексов фосфор-две вакансии путем присоединения вакансий к E-центру, тогда как электроны создают преимущественно дивакансии. Комплексы фосфор-две вакансии устойчивы до температур 400 до 500 °C и вносят в запрещенную зону кремния глубокий акцепторный уровень у $E_v + 0{,}35$ eV. Установлено, что характер отжига РД в сильнооблученном кремнии не зависит от вида облучения и обусловлен перестройками собственных дефектов структуры.

1. Introduction

When irradiating Si by electrons with $E_e < 10$ MeV and by ^{60}Co γ-rays the diffusion mechanism of defect production dominates, when point radiation defects uniformly distributed in the bulk of the crystal are produced, which are complexes of vacancies and interstitial Si atoms with each other or with chemical impurity atoms. The production efficiency of such complexes is determined both by the irradiation conditions

[1]) Ul. Kurchatova 7, SU-220080 Minsk 80, USSR.

(intensity, energy and kind of the bombarding particles, temperature of the crystal) and impurity content of the materials studied [1 to 4]. This paper studies the accumulation of radiation defects of various types at sufficiently high fluxes of irradiation by ^{60}Co γ-rays or electrons with $E_e = 4$ to 10 MeV.

2. Experimental Procedure

n- and p-Si grown by the float-zone technique with initial resistivity $\varrho = (1$ to $5) \times 10^2$ Ω cm (n-type) and $\varrho = (1$ to $60) \times 10^3$ Ω cm (p-type) was used. Concentrations of phosphorus in n-Si and boron in p-Si were (1 to 30) $\times 10^{12}$ cm^{-3} and oxygen and carbon concentrations were $\leqq 2 \times 10^{16}$ cm^{-3}. Irradiation was carried out by electrons with energy $E_e = 4$, 6, 10 MeV and ^{60}Co γ-rays at $T = 330$ K. The flux density (intensity) of the bombarding particles was $\approx 1 \times 10^{12}$ cm^{-2} s^{-1} and their flux Φ varies over the range $\Phi_e = (1 \times 10^{15}$ to $5 \times 10^{16})$ cm^{-2} and $\Phi_\gamma = (1 \times 10^{17}$ to $2 \times 10^{19})$ cm^{-2} for electrons and γ-rays, respectively. The experimental results were obtained from the analysis of the temperature dependences of the Hall coefficient measured at various stages of irradiation and 20 min isochronous annealing at $T = 100$ to 700 °C.

3. Experimental Results

At the initial stages of electron and ^{60}Co γ-ray irradiation of n-Si and high-resistivity typically heavily phosphorus-compensated p-Si E-centres are formed, whose accumulation results in a decrease of free electron concentration, n–p conversion of conductivity type in n-Si, and an increase of hole concentration due to the decompensation of the material when binding the compensating phosphorus impurity in neutral E-centres in p-Si [4 to 6]. Here it is characteristic that within some range of electron and γ-ray fluxes the hole concentration in crystals with initial n- (upon n–p conversion) and p-type conductivity remains constant, while the Fermi level (E_F) is stabilized at $E_v + (0.39 \mp 0.02)$ eV [4 to 6]. Fig. 1 shows the dose dependences ($T = 300$ K) of hole concentration at higher ^{60}Co γ-ray (curves 1 to 3) and 4 MeV electron irradiation fluxes (curves 4, 5). After a stabilization region ($\Phi_e \leqq 3 \times 10^{15}$ and $\Phi_\gamma \leqq 3 \times 10^{17}$ cm^{-2}) the hole concentration p is seen to decrease with Φ increase under

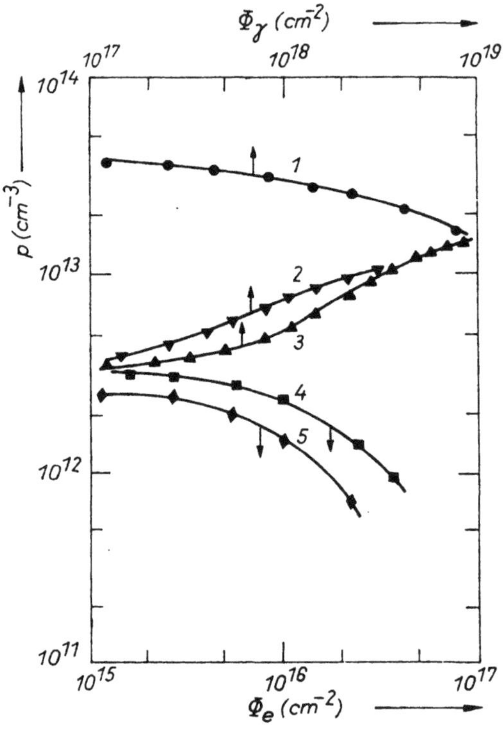

Fig. 1. Hole concentration as a function of (1) to (3) ^{60}Co γ-ray and (4), (5) 4 MeV electron flux; (3), (4) in n-Si and (1), (2), (5) p-Si at $T = 300$ K. n_0, p_0 = (1) 4×10^{13}, (2), (5) 8×10^{11}, (3), (4) 1×10^{13} cm^{-3}

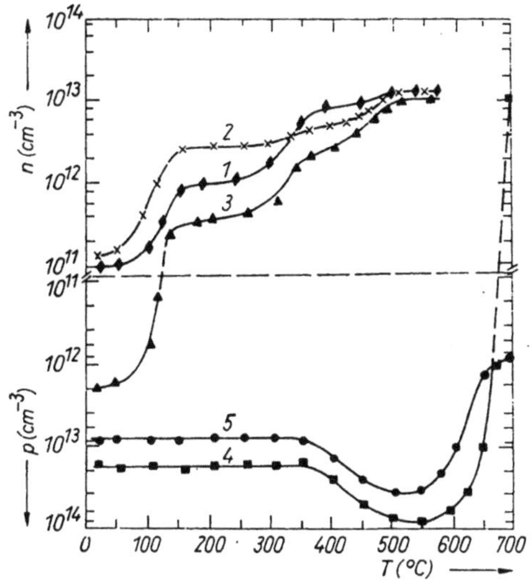

Fig. 2. Charge-carrier concentration restoration ($T_{meas} = 300$ K) during annealing of (1) to (4) n-Si and (5) p-Si irradiated by (2) to (5) ^{60}Co γ-rays and (1) 10 MeV electrons; Φ = (1) 1×10^{14}, (2) 4.1×10^{16}, (3) 1.4×10^{17}, (4) 1×10^{19}, (5) 3.2×10^{18} cm^{-2}; n_0, p_0 = (1) to (4) 1×10^{13}, (5) 8×10^{11} cm^{-3}

electron irradiation and vice versa it increases up to the values $p = (1 \text{ to } 2) \times 10^{13}$ cm^{-3} under ^{60}Co γ-ray irradiation. It should be noted that in p-Si of $p_0 \geqq 4 \times 10^{13}$ cm^{-3} one observes a decrease in p with γ-ray flux increase as with electron irradiation of crystals with higher resistivity (curves 1, 5).

Fig. 2 shows restoration of charge-carrier concentration at annealing of irradiated crystals. Samples of the initial n-Si (curves 1 to 4) and p-Si (curve 5) after electron (curve 1) and γ-ray irradiation (curves 2 to 5) were annealed. Annealing experiments were carried out on three groups of samples. The first group comprises n-Si crystals retaining the initial conductivity type (curves 1, 2) under both kinds of irradiation. The second group comprises the samples having experienced n–p conversion during irradiation by higher γ-ray fluxes when E_F is stabilized at $E_v + 0.39$ eV ($\Phi_\gamma < 2 \times 10^{17}$ cm^{-2}). The third group comprises the samples highly irradiated by ^{60}Co

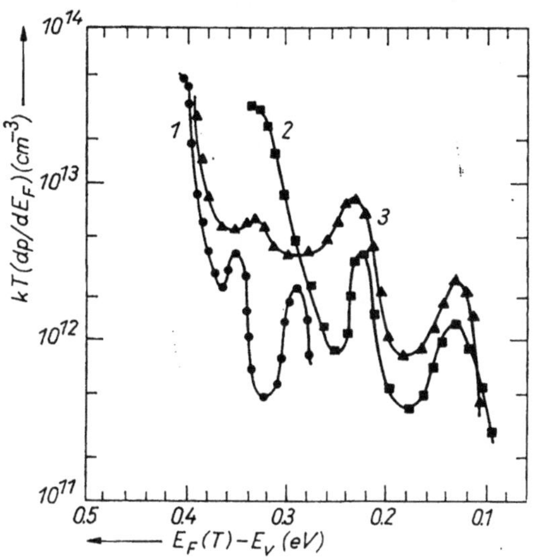

Fig. 3. ($kT\ \Delta p/\Delta E_F$) as a function of the Fermi level position in (1), (2) n-Si and (3) p-Si (1) after irradiation and (2), (3) isochronous annealing. n_0, p_0 = (1), (2) 1×10^{13}, (3) 8×10^{11} cm^{-3}; (1) $\Phi = 2.2 \times 10^{18}$, (2) 1×10^{19}, (3) 3.2×10^{18} cm^{-2}; (2) $T_{ann} = 450$, (3) 500 °C

γ-rays when p increases with rising Φ ($\Phi_\gamma > 1 \times 10^{18}$ cm^{-2}). Note that the defects in crystals of the first and second groups are annealed in three stages: 100 to 160 °C (first stage), 280 to 350 °C (second stage), 400 to 500 °C (third stage) while highly irradiated samples (third group) are characterized by the absence of these stages and the presence of a pronounced temperature range (350 to 550 °C) where the hole concentration increases but a complete restoration of the initial values of charge-carrier concentration occurs at the fourth stage at $T = 550$ to 700 °C. The energetic position of the defect levels observed after irradiation and at different stages of annealing is taken from the temperature dependences of the Hall coefficient processed by the differential procedure [7]. Fig. 3 shows some of the typical dependences of the function ($kT\ \Delta p/\Delta E_F$) on the Fermi level position E_F at various temperatures. From the analysis of these results it follows that after irradiation of Si by high ^{60}Co γ-ray fluxes ($\Phi_\gamma \geqq 1 \times 10^{18}$ cm^{-2}) the radiation defects with levels $E_v + 0.28$, $+0.35$, $+0.41$ eV are formed, while after their annealing at $T = 450$ to 500 °C the defects with levels $E_v + 0.12$, $+0.22$, $+0.33$ eV are produced. There are both donor- and acceptor-type defects among them.

4. Discussion

The experimentally observed (Fig. 1) various trends of dose dependences $p(\Phi)$ when irradiating by high fluxes of ^{60}Co γ-rays (curves 2, 3) and electrons (curves 4, 5) are, presumably, due to the difference in production efficiencies of intrinsic and impurity radiation defects for these kinds of bombarding particles. E-centres accumulate at the initial stages under both kinds of irradiation of n- and p-Si, but as Φ increases and phosphorus is exhausted (when binding in E-centres) the main channels of defect production under electron and γ-ray irradiation may appear to be different. Indeed, γ-ray-produced vacancies and interstitial Si atoms take part preferentially in the formation of secondary complexes with participation of impurity atoms, although they can also unite in divacancies, di-interstitials, and other intrinsic associates. In contrast, during electron irradiation with $E_e > 1$ MeV the probability of divacancy formation is high, and besides it increases strongly with increasing electron energy E_e from 1 to 10 MeV [8]. This is also confirmed by the results on the isochronous annealing (Fig. 2). As follows from the estimates the fraction of E-centres (first stage of annealing) and the defects being annealed at the third stage (curve 2) is high, while the fraction of divacancies being annealed at the second stage turns out to be negligible in this case as compared to (curve 2) electron irradiation (with the irradiation fluxes equivalent by the number of displacements). Therefore, the decrease of p (E_F shifts towards the middle of the forbidden gap) with increasing Φ_e (Fig. 1) can be accounted for by the preferential accumulation of divacancies introducing the donor level $E_v + 0.28$ eV into the lower half of the forbidden gap under electron irradiation.

At sufficiently high γ-ray fluxes when the whole phosphorus is bound in E-centres irradiation-generated vacancies can participate in accumulation of other more complicated vacancy-type defects. In particular, phosphorus–two vacancy complexes (PV_2) can be formed by adding vacancies to E-centres. Formation of such complexes contributing an acceptor level at $E_v + 0.34$ eV [9] to the lower half of the forbidden gap has been suggested to occur when heat treating irradiated Si and rearranging E-centres in more complicated defects being annealed at $T = 400$ to 500 °C which coincides with the range of the third stage of restoration of charge-carrier concentration in the crystals of the first and second groups studied. In [10] they observe a decrease in E-centre concentration when irradiating float-zone n-Si by high fluxes of electrons with $E_e = 1$ MeV which has been assumed to be due to the accumulation of PV_2 complexes. Besides, under electron (pulse) irradiation ($E_e = 900$ keV) radiation defects electrically inactive in float-zone n-Si and stable up to $T = 340$ to 420 °C involving phosphorus [11] are introduced.

In p-type Si when $E_F > E_v + 0.20$ eV, as with the crystals studied (Fig. 1) where $E_F \approx E_v + 0.39$ eV, E-centres and irradiation-generated vacancies are electrically neutral [3, 12]. Therefore, PV_2 complex formation should be accompanied by liberation of hole if E_F is located above the acceptor level of the PV_2 complex and the latter is negatively charged, i.e. $(PV)^0 + V^0 \rightleftarrows (PV_2)^- + p^+$. But if E_F is located below the PV_2 complex level, it is neutral and its formation occurs without changing the charge-carrier concentration according to the reaction $(PV)^0 + V^0 \rightleftarrows (PV_2)^0$. An experimentally observed (Fig. 1) increase in hole concentration when irradiating crystals with γ-rays, when E_F varies within the range from $E_v + 0.39$ eV to $E_v + 0.34$ eV, shows evidence for the fact that here the negatively charged complexes (acceptors) are really introduced and the holes are released. This suggests that the acceptor level $E_v + 0.35$ eV (Fig. 3, curve 1) determined by the differential procedure from the temperature dependences of the Hall coefficient [7] belongs to PV_2 complexes, all the more so, that its concentration does not exceed that of phosphorus atoms involved in the crystals studied. It should be noted that in p-Si crystals with $p_0 > 1 \times 10^{13}\,cm^{-3}$, when the boron atom concentration exceeds that of the compensating phosphorus impurity, the formation of E-centres and PV_2 complexes should not result in an appreciable growth of hole concentration ($T = 300$ K) as is observed in the experiment (Fig. 1, curve 1). This is also due to the fact that the change of hole concentration connected with the decompensation of the material during E-centre formation, turns out to be smaller than their initial concentration and PV_2 complexes are electrically neutral ($E_F \leqq E_v + 0.30$ eV) in such materials at $T = 300$ K, while accumulation of other radiation defects with donor levels in the lower half of the forbidden gap results in compensation of the material and decrease of p.

As for irradiation-produced interstitial silicon atoms, they participate in forcing interstitial carbon and boron into interstitial position or combine in di-interstitials. These radiation defects (interstitial carbon with donor level $E_v + 0.28$ eV and boron-containing complexes [13] are, on the whole, electrically neutral under experimental conditions ($E_F \approx E_v + 0.39$ eV). But at sufficiently high γ-ray fluxes ($\Phi_\gamma > 10^{18}\,cm^{-2}$), as with other kinds of irradiation, intrinsic di-interstitials (Si-P6 and Si-B3 centres [14, 15]) can accumulate. Si-B3 centres are stable up to $T = 400$ to 500 °C and the donor level $\geqq E_v + 0.40$ eV [14, 16] corresponds to them in the lower half of the forbidden gap. Ionization of a radiation defect with level $E_v + 0.41$ eV occurs also during high-flux γ-ray irradiation (Fig. 3). The annealing of such defects as well as simultaneous rearrangement-induced accumulation of less stable radiation defects of multivacancy complexes with acceptor levels $E_v + 0.12$, $+0.22$, $+0.33$ eV (Fig. 3) do result in the experimentally observed (Fig. 2) p increase under heat treatment of heavily irradiated Si over the temperature range 350 to 550 °C. The initial values of charge-carrier concentration are restored after annealing of acceptor-type multivacancy complexes at $T = 550$ to 700 °C (Fig. 2).

5. Conclusion

The results obtained on the accumulation under irradiation and annealing of radiation defects in n- and p-Si indicate possible formation of phosphorus-containing complexes phosphorus–two vacancies with acceptor level $E_v + 0.35$ eV, stable up to $T = 400$ to 500 °C, at sufficiently high γ-ray fluxes. Contrary to this, under identical conditions of irradiation by electrons with $E_e \geqq 4$ MeV divacancies are most effectively formed.

This explains the experimentally observed different trends of dose dependences of charge-carrier concentration in Si with initial n- and p-type conductivity for these kinds of bombarding particles. Note also that the annealing character of highly irradiated Si crystals does not practically depend on the kind of bombarding particles (γ-rays, electrons, neutrons, protons [16, 17]), since the change of electrophysical properties of such material is determined, mainly, by intrinsic structural defects.

References

[1] V. S. VAVILOV, Radiation Effects in Semiconductors, Moscow 1963 (in Russian).
[2] L. S. SMIRNOV (Ed.), Physical Processes in Irradiated Semiconductors, Nauka, Novosibirsk 1977 (in Russian).
[3] V. V. EMTSEV and T. V. MASHOVETS, Impurities and Point Defects in Semiconductors, Radio i Svyaz, Moscow 1981 (in Russian).
[4] I. D. KONOZENKO, A. K. SEMENYUK, and V. I. KHIVRICH, Radiation Effects in Silicon, Kiev 1974.
[5] P. F. LUGAKOV and T. A. LUKASHEVICH, Fiz. Tekh. Poluprov. **21**, 746 (1987).
[6] P. F. LUGAKOV, V. V. LUKYANITSA, and V. V. SHUSHA, Fiz. Tekh. Poluprov. **20**, 1894 (1986).
[7] H. J. HOFFMANN, Appl. Phys. **19**, 307 (1979).
[8] H. FLICKER and W. R. PATTERSON, Appl. Phys. **37**, 4998 (1966).
[9] L. A. KAZAKEVICH, P. F. LUGAKOV, V. V. LUKYANITSA, and I. M. FILIPPOV, Radiat. Eff. **105**, 181 (1988).
[10] A. V. DVURECHENSKII, B. P. KASHNIKOV, and L. S. SMIRNOV, Fiz. Tekh. Poluprov. **14**, 995 (1987).
[11] V. V. EMTSEV, T. V. MASHOVETS, and A. T. ABDUSATTAROV, Fiz. Tekh. Poluprov. **21**, 2106 (1987).
[12] G. D. WATKINS and J. R. TROXELL, Phys. Rev. Letters **44**, 593 (1980).
[13] L. S. KIMERLING, Inst. Phys. Conf. Ser. No. 31, 221 (1977).
[14] K. L. BROWER, Phys. Rev. B **14**, 872 (1976).
[15] Y. H. LEE, N. N. GERASIMENKO, and J. W. CORBETT, Phys. Rev. B **14**, 4506 (1976).
[16] V. I. KUZNETSOV and P. F. LUGAKOV, phys. stat. sol. (b) **112**, 457 (1982).
[17] P. F. LUGAKOV, T. A. LUKASHEVICH, and V. V. SHUSHA, phys. stat. sol. (a) **74**, 445 (1982).

(Received May 25, 1988)

phys. stat. sol. (a) **110**, 409 (1988)

Subject classification: 61.70; 78.70; S1.2

Departament de Matemàtica Aplicada III, E.T.S.E. Camins, Canals i Ports, Universitat Politècnica de Catalunya, Barcelona[1]) *(a)*
and Departament de Física de Materiales, Facultad de Ciencias Físicas, Universidad Complutense, Madrid[2]) *(b)*

Characterization of Defects in Deformed Titanium

By

A. SERRA (a) and N. DE DIEGO (b)

The nature of defects present in deformed titanium is studied by combining the positron annihilation technique and computer simulation. By performing isochronal annealing between room temperature and 1123 K it is shown that the lifetime decreases up to 700 and stays roughly constant up to 1000 K, achieving the bulk value at 1123 K. These results are interpreted as positron trapping at zonal dislocations present at twin boundaries, as inferred from computer simulation studies.

Die Art der Defekte, die in verformtem Titan vorhanden sind, werden mit Hilfe der Positronenannihilationstechnik und Computersimulation untersucht. Lebensdauermessungen im Temperaturbereich zwischen 293 und 1123 K werden durchgeführt und es wird beobachtet, daß die Lebensdauer mit zunehmender Temperatur abnimmt und zwischen 700 und 1000 K ungefähr konstant bleibt. Bei 1123 K werden die Werte für die defektefreie Probe erreicht. Aus den Lebensdauermessungen und den Computersimulationsergebnissen wird geschlossen, daß Zwillingsgrenzenversetzungen als Positroneneinfangzentren wirken.

1. Introduction

The positron annihilation technique has been widely used for studying the recovery of defects created by irradiation and deformation. In a deformed metal several kinds of positron traps are expected, namely dislocations, vacancies, and vacancy clusters. Perfect twins can contain dislocations with a structure differing from that of dislocations in the perfect lattice. Thus, twin boundaries should be also considered when elucidating the nature of defects capable of trapping positrons in deformed metals, especially in hexagonal structures. The problem is that it is not a simple matter to separate the single contributions in a sample containing several kinds of traps for positrons and a knowledge of the atomistic structure of the defects is needed.

In this work we present preliminary results on the characterization of defects in deformed titanium by positron annihilation, by making use of the present knowledge of such defects (Section 2) and their structure by computer simulation in h.c.p. metals (Section 3).

2. Plastic Deformation of Titanium

Titanium, like other h.c.p. metals, slips primarily in the $[11\bar{2}0]$ (or a) direction. However, $\boldsymbol{a}$ slip cannot accommodate changes in length parallel to the c-axis and $(\boldsymbol{a} + \boldsymbol{c})$ slip as well as twinning have been investigated in order to understand titanium

[1]) Jordi Girona Salgado 31, 08034 Barcelona, Spain.
[2]) Ciudad Universitaria, 28040 Madrid, Spain.

ductility. $(10\bar{1}2)$, $(11\bar{2}1)$, and $(11\bar{2}3)$ twins are expected to accommodate an extension along the c-axis, while $(11\bar{2}2)$, $(11\bar{2}4)$, and $(10\bar{1}1)$ twins allow a reduction along it.

Paton and Backofen [1] studied titanium single crystals under compression either parallel or normal to the c-axis from 300 to 1100 K. A reduction of the c-axis was accommodated almost entirely by $(11\bar{2}2)$ twinning up to 600 K and by $(10\bar{1}1)$ twinning in combination with $(\boldsymbol{a} + \boldsymbol{c})$ slip above 700 K. The only exception were three single $(11\bar{2}4)$ twins found in samples deformed at 453, 673, and 1073 K, respectively. Crystals reduced normal to the c-axis deformed by a combination of prism slip and $(10\bar{1}2)$ twinning at 300 K and by prism slip alone above 800 K. More recently, Minonishi et al. [2] have confirmed the results of Paton and Backofen and reported that a variety of accommodation slips can occur. In a following paper [3] the same authors report a transmission electron microscopy observation of an array of $1/3\ [11\bar{2}3]$ dislocations ahead of $(10\bar{1}1)$ twins by compressing single crystals parallel to the c-axis from 800 to 1100 K.

In a review of twinning Yoo [4] pointed out that the enhanced ductility observed after pre-straining of titanium and its alloys at low temperature is attributed to profuse fine $(11\bar{2}1)$ twins [5]. Polycrystalline titanium deformed in tension at room temperature was studied by Numakura et al. [6] and the importance of grain boundaries as sources of $1/3\ [11\bar{2}\bar{3}]\ (10\bar{1}1)$ dislocations was pointed out, a dislocations distributed uniformily in all grains were observed but no twins were found.

At low temperatures there is a preference for twinning over slip in titanium. This can be explained by the variation of critical shear stress (CSS) with temperature. For $(11\bar{2}2)$, $(11\bar{2}1)$, and $(10\bar{1}2)$ twins the CSS decreases with decreasing temperature [7] and for $1/3\ [11\bar{2}3]\ (10\bar{1}1)$ slip CSS increases rapidly with decreasing temperature [8].

3. Microstructure of Perfect and Twin Dislocations in H.C.P. Metals

Computer simulation has proved to be an important means of investigation for the microstructure of defects. Liang and Bacon [9], using pair potentials previously discussed [10, 11], found that the prism $\boldsymbol{b} = 1/3\ [2\bar{1}\bar{1}0]\ (01\bar{1}0)$ dislocation has a core spread over $1.7a$, $3.0a$, and $3.4a$ and confined to the slip plane for three of the models, one of them being selected by Johnson and Beeler [12] to represent titanium. The $(\boldsymbol{a} + \boldsymbol{c})$ dislocation with $\boldsymbol{b} = 1/3\ [11\bar{2}\bar{3}]$ Burgers vector has two stable core configurations for all potentials; in one of them the core decomposes into two partials with Burgers vectors half of $\boldsymbol{b}$, while the other core configuration is accommodated by both $(11\bar{2}1)$ and $(11\bar{2}2)$ microtwins bounding a high distortion.

The study of $(10\bar{1}2)$, $(11\bar{2}1)$, and $(11\bar{2}2)$ perfect twins showed that there is not an open volume in the twin plane [13]. However, for the $(11\bar{2}1)$ twin a stable boundary exists allowing a row of vacancies with a formation energy smaller than in the perfect crystal. Further investigations on vacancies in the twin plane are now in progress. Recently Serra et al. [14] simulated twin dislocations. Twin dislocations form a step in the twin plane which is responsible for its growth. For $(11\bar{2}1)$ and $(10\bar{1}2)$ twins the core of the dislocation, and hence the step, spreads over several lattice parameter distances in the boundary and free space is not allowed. For the $(11\bar{2}2)$ twin, however, the step is a zonal dislocation (see Fig. 1) where the atoms in the third atomic plane away from the original interface are displaced according to the twinning shear but the atoms in the intermediate planes must be shuffled to restore the h.c.p. structure. These shuffles are not unique and therefore the core of the dislocation can present different states, all being fairly localized with an open volume slightly smaller than in the $(\boldsymbol{a} + \boldsymbol{c})$ perfect dislocation.

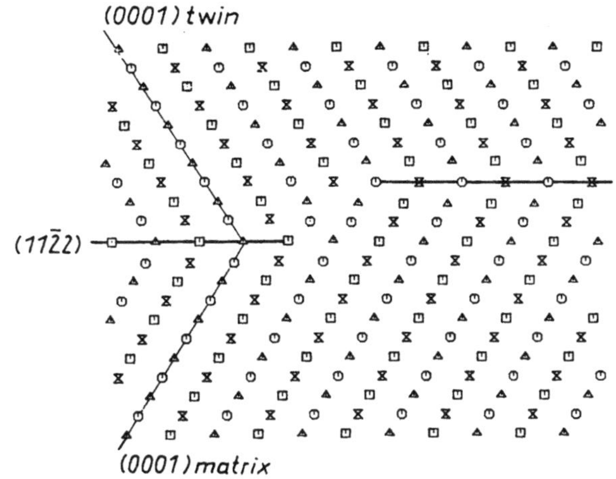

Fig. 1. (11$\bar{2}$2) twin stepped by a twin dislocation. The twin boundary and basal planes have been marked by solid lines

4. Experimental Technique and Results

The α-Ti samples were polycrystalline specimens having a purity of 99.95% supplied by Metal Crystals (Cambridge, UK). They were mechanically polished and rolled at 253 K to a thickness reduction of 40% and at 293 K to a thickness reduction of 28 and 67%. After deformation the samples were chemically polished. The recovery behaviour of the sample deformed at 293 K to a reduction of 28% was followed by annealing under an Ar atmosphere from 493 to 1123 K at temperature intervals of approximately 100 K during heating periods of one hour.

The positron lifetime spectra were recorded at room temperature by using a conventional fast system. As a positron source, about 10 μCi of an aqueous ^{22}NaCl solution were deposited onto a Ni foil (0.45 mg cm^{-2}). In order to extract the source contribution a well annealed Ti sample was measured. After subtracting the foil contribution (3.0% of 180 ps) the spectrum was analyzed with two components obtaining a satisfactory fit. The shorter component yielded a value of (152 ± 2) ps in agreement with previously reported values for bulk α-Ti [15, 16]; the long component, $\tau_2 = 500$ ps, of faint intensity, $I_2 = 2.7\%$, was ascribed to annihilations in the salt. This value was subtracted from the spectra in further analysis as the salt contribution. In all cases the spectra could be decomposed only in one component after subtracting the Ni foil and salt components.

In Table 1 the results for the as-deformed samples are shown; T_r denotes the temperature at which the rolling has been performed and d_r indicates the thickness reduction.

The measured lifetimes are higher than the bulk value, indicating that positron trapping takes place at defects formed during deformation. However, no differences are observed in the lifetime for the different rolling conditions. The results for samples labelled 1 and 2 are in agreement with the Doppler broadening measurements made by

Table 1

sample No.	T_r (K)	d_r (%)	τ (ps)
1	293	28	194 ± 2
2	293	67	193 ± 2
3	253	40	196 ± 2

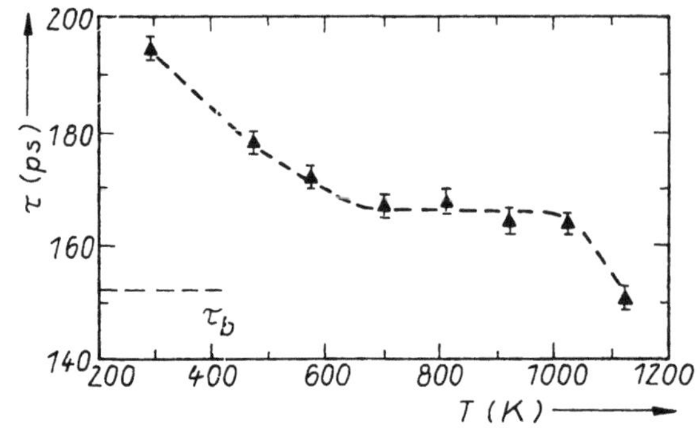

Fig. 2. The lifetime as a function of isochronal annealing for deformed α-Ti. The error bars correspond to the statistical uncertainty; the dashed line has been drawn as a guide for the eye

Hood and Schultz [17] who observe saturation trapping after a thickness reduction of 15%. By decreasing the rolling temperature no changes are observed in the lifetime, indicating that under these conditions the defect structure is basically the same.

The isochronal annealing curve for the sample labelled 1 is shown in Fig. 2, where the lifetime has been plotted as a function of the annealing temperature. The initial value of 194 ps drops to 167 ps at 700 K and remains roughly constant up to 1000 K, achieving the bulk value after further annealing at 1123 K.

5. Discussion

The results obtained for deformed α-Ti clearly reveal the trapping of positrons at defects. In the as-rolled sample several kinds of traps are expected to be present, namely, dislocations, vacancies, vacancy clusters, and twins.

Hood and Schultz [17] have measured the Doppler broadening S-parameter in α-Ti deformed by compression and they have interpreted the damage signal as positron trapping at point defects. The lifetime for vacancies in α-Ti has been reported by Yli-Kauppila et al. [15] to be 222 ps. Thus, the observed value, $\tau = 195$ ps, can be attributed neither to trapping at vacancies nor to vacancy clusters. Furthermores, Hood et al. [18] have found that vacancies migrate at 250 K giving rise to three-dimensional clusters that dissolve at 300 K. The observed trapping signal can be interpreted in terms of positron trapping at dislocations; however, the value $\tau = 195$ ps, seems still too low to be associated with lattice dislocations, for they usually exhibit a lifetime value very close to the one characteristic of vacancies.

The partial recovery that takes place at 673 K indicates the disappearance of defects created by the deformation process (see Fig. 2). This fact has been also reported by Hood and Schultz [18] who observe a strong decrease of the trapping signal after annealing at 648 K. The most remarkable feature of the above-mentioned figure is that the lifetime stays roughly constant up to temperatures close to the α to β transformation point. This behaviour suggests the presence of some kind of very stable defects that survive the annealing treatments above 650 K.

We conclude that the measured lifetime up to 673 K contains the contribution of two kinds of traps, i.e., dislocations and some kind of very stable trap that we tentatively ascribe to deformation twins containing dislocations. Since the spectra cannot be fitted by a two-component analysis, it is suggested that the lifetime associated to "lattice" and twin dislocations should be different and very close to each other.

As mentioned before, it is a well established fact that the lifetime associated with dislocations is close to the vacancy value, although the mechanisms leading to trapping at dislocations are still under active discussion. One of the currently proposed models simply assumes that positrons are trapped and annihilated at the dislocation line. An alternative interpretation has been proposed by Doyama and Cotterill [19] and developed by Smedskjaer et al. [20]; according to their model the dislocation line would be too "narrow" to be able to trap a positron, while the strong traps would be defects associated with dislocations such as vacancies or jogs. Some recent experiments [21, 22] have been interpreted by making use of this model; however, Park et al. [23] have found two different lifetimes for screw and edge dislocations in Fe, suggesting that positrons would be sensitive to the open volume associated with the dislocations.

6. Conclusions

Positron lifetime measurements performed in rolled polycrystalline Ti have shown the presence of two recovery stages that we tentatively ascribe to dislocations and twins. From the present results it is suggested that the positron lifetime associated to twin dislocations differs from the one characteristic of lattice dislocations. According to the above-mentioned computer simulation results, vacancies and dislocations tend to minimize their open volumes when they are localized in twin boundaries rather than in an otherwise perfect lattice. This effect could be responsible for the differences observed in the lifetimes associated to lattice and twin dislocations. Further experiments are now in progress to investigate this effect by deforming Ti under very carefully controlled conditions, in order to identify the precise nature of the positron traps created in Ti by plastic deformation.

Acknowledgements

The authors are grateful to Prof. D. J. Bacon for the critical review of the manuscript and helpful suggestions and to Dr. O. Ruano (C.E.N.I.M., Madrid) for deforming the samples. This work has been made under grant Acción Integrada 18/36.

References

[1] N. E. Paton and W. A. Backofen, Metallurg. Trans. **1A**, 2839 (1970).
[2] Y. Minonishi, S. Morozumi, and H. Yoshinaga, Scripta metall. **16**, 427 (1982).
[3] Y. Minonishi, S. Morozumi, and H. Yoshinaga, Scripta metall. **19**, 1241 (1985).
[4] M. H. Yoo, Metallurg. Trans. **12A**, 409 (1981).
[5] L. P. Odinokova, Russian Metallurgy **56**, 62 (1986).
[6] H. Numakura, Y. Minonishi, and M. Koiwa, Scripta metall. **20**, 1581 (1986).
[7] H. Conrad, Progr. Metal Sci. **26**, 123 (1981).
[8] N. E. Paton, J. C. Williams, and G. P. Rauscher, Titanium Science and Technology, Ed. R. Jaffee and H. M. Burte, Plenum Press, New York 1973 (p. 1049).
[9] H. Liang and D. J. Bacon, Phil. Mag. **A53**, 181 (1986).
[10] D. J. Bacon and J. W. Martin, Phil. Mag. **A43**, 883 (1981).
[11] D. J. Bacon and H. Liang, Phil. Mag. **A53**, 163 (1986).
[12] R. A. Johnson and J. R. Beeler, Interatomic Potentials and Crystalline Defects, Ed. J. K. Lee, AIME, New York 1981 (p. 165).
[13] A. Serra and D. J. Bacon, Phil. Mag. **A54**, 793 (1986).
[14] A. Serra, D. J. Bacon, and R. C. Pond, Acta metall. **36**, 3183 (1988).
[15] J. Yli-Kauppila, P. Moser, H. Künzi, and P. Hautojärvi, Appl. Phys. A **27**, 31 (1982).

[16] G. M. HOOD, M. ELDRUP, and N. J. PEDERSEN, Proc. V. Internat. Conf. Positron Annihilation, Lake Yamanaka, Ed. R. R. HASIGUTI and K. FUJIWARA, Japan Institute of Metals, Sendai 1979 (p. 751).
[17] G. M. HOOD and R. J. SCHULTZ, J. nuclear Mater. **120**, 94 (1984).
[18] G. M. HOOD, R. J. SCHULTZ, and J. A. JACKMAN, Radiat. Eff. **90**, 149 (1985).
[19] M. DOYAMA and R. M. J. COTTERILL, Proc. V. Internat. Conf. Positron Annihilation, Lake Yamanaka, Ed. R. R. HASIGUTI and K. FUJIWARA, Japan Institute of Metals, Sendai 1979 (p. 89).
[20] L. C. SMEDSJKAER, M. MANNINEN, and M. J. FLUSS, J. Phys. F **10**, 2237 (1980).
[21] C. HIDALGO, S. LINDEROTH, and N. DE DIEGO, Phil. Mag. **A54**, L61 (1986).
[22] C. HIDALGO, S. LINDEROTH, and N. DE DIEGO, Phys. Rev. B **36**, 6740 (1987).
[23] Y. K. PARK, J. T. WABER, and M. MESHII, Phys. Rev. B **34**, 823 (1986).

(Received July 21, 1988)

phys. stat. sol. (a) **110**, 415 (1988)

Subject classification: 66.30; S10

Shanghai Institute of Metallurgy, Academia Sinica[1])

Diffusivity of Oxygen in the Orthorhombic $YBa_2Cu_3O_y$ Phase

By

X. M. Xie, T. G. Chen, and J. Huang

The diffusion coefficient of oxygen in the orthorhombic $YBa_2Cu_3O_y$ phase is calculated based on internal friction data. From the calculation, the diffusivity can be expressed as $D = 1.830 \times 10^{-4} \exp(-1.23\ \text{eV}/kT)\ \text{cm}^2/\text{s}$ in the temperature range from 400 to 550 °C. The result show that the oxygen diffusion rate in this phase is very slow.

Коэффициент диффузии кислорода в орторомбической $YBa_2Cu_3O_y$ фазе был вычислен на основании данных внутреннего трения. Коэффициент диффузии может быть выражен как $D = 1,80 \times 10^{-4} \exp(-1,23\ \text{eV}/kT)\ \text{cm}^2/\text{s}$ в температурном интервале 400 до 550 °C. Эти результаты показывают, что диффузия кислорода в этой фазе очень медлена.

1. Introduction

Since the superconductivity of $YBa_2Cu_3O_y$ phase is closely related to the oxygen content [1], it is important to study the diffusion behavior of oxygen atoms in this phase. Morris et al. [2] and Routbort et al. [3] studied the oxygen diffusion process in the $YBa_2Cu_3O_y$ phase, and believed that the diffusion rate of oxygen is so fast in the temperature range from 400 to 500 °C that the oxygen content in the specimens can attain an equilibrium state just in a few minutes. Tu et al. [4] supported this result and suggested further that the rate of oxygen uptake is predominated by surface reactions. However, the above results seem to contradict the superconducting shell structure observed by Ginley et al. [5]. Their research showed that, even in single phase $YBa_2Cu_3O_y$, superconductivity occurs in only thin shells surrounding normal grains. The shells comprise no more than 25% of the material. This result implies that the oxygen diffusion in $YBa_2Cu_3O_y$ is very slow and oxygen cannot reach an equilibrium state in the specimens even by long time annealing. Huang et al. [6] investigated

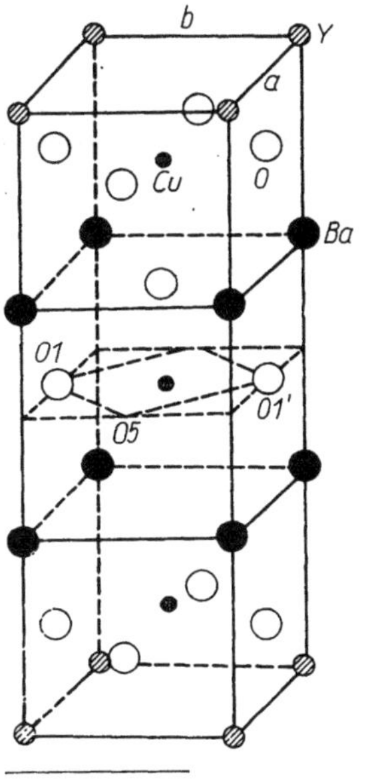

Fig. 1. Structure model of the orthorhombic $YBa_2Cu_3O_y$ phase

[1]) 865 Chang Ning Road, Shanghai 200050, People's Republic of China.

the oxygen diffusion behavior in $YBa_2Cu_3O_y$ phase by in-situ resistometry and gave a result similar to that reported by Ginley et al. [5].

Chen et al. [7] and Xie et al. [8] studied the $YBa_2Cu_3O_y$ compound by the internal friction (IF) method. They found an IF peak at about 204 °C which they thought to be induced by the jumping of oxygen atoms between O1 and O5 sites in the orthorhombic $YBa_2Cu_3O_y$ phase (see Fig. 1). From the IF peak temperature and the height of the peak, they calculated the potential energy, the vibration frequencies, and the occupancies of oxygen atoms on the O1 and O5 sites. As well known, all the oxygen sites except O1 and O5 are fully occupied by oxygen atoms [9], so that we come to the conclusion that oxygen diffusion can take place only between the two Ba layers and along the path O1–O5–O1′. Therefore, it is possible to calculate the diffusion coefficient based on Chen et al. and Xie et al. results.

2. Calculation of the Diffusion Coefficient

Fig. 1 shows the structure of the orthorhombic $YBa_2Cu_3O_y$ phase and the oxygen diffusion path. This diffusion process can be described in terms of a one-dimensional model, as shown in Fig. 2.

Now consider three adjacent lattice planes, designated as 1, 2, and 3 which are perpendicular to the b-axis and pass through O1, O5, and O1′ sites with a concentration gradient $\partial C/\partial x$ along the x-direction, that is, the b-axis in YBCO. Let n_1, n_2, and n_3 be the numbers of oxygen atoms per unit area on planes 1, 2, and 3; Γ_1 and Γ_2, the jump rates of oxygen atoms on planes 1(3) and 2; for $\delta t = 1/\Gamma_1 + 1/\Gamma_2$, the number of oxygen atoms through the G_1 plane per unit area along the x-direction is

$$J_{G_1}\,\delta t = (n_1P - n_1PP - n_2P + n_2PP - n_3PP)\,.$$

So the net flux is

$$J_{G_1} = (n_1P - n_1PP - n_2P + n_2PP - n_3PP)/\delta t\,. \qquad (1)$$

Similarly, for plane G_2

$$J_{G_2} = (n_1PP + n_2P - n_2PP - n_3P + n_3PP)/\delta t\,. \qquad (2)$$

Here, P is the possibility of forward or backward jumping of oxygen atoms, $P = 1/2$. For the case of steady-state diffusion, $J_{G_1} = J_{G_2} = J$, from (1) and (2), we obtain

$$J = \frac{1}{4}\,\frac{(n_1 - n_3)\Gamma_1\Gamma_2}{\Gamma_1 + \Gamma_2}\,. \qquad (3)$$

The quantity $(n_1 - n_3)$ can be expressed in terms of the volume concentration gradient,

$$n_1 - n_3 = (C_1 - C_3)\,b = -b^2\,\frac{\partial C}{\partial x}\,.$$

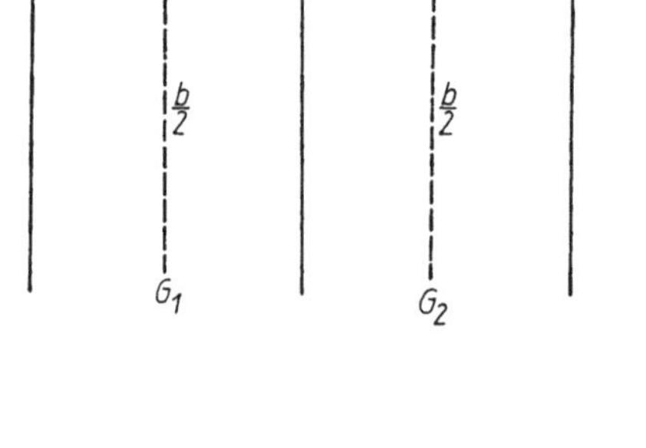

Fig. 2. One-dimensional model of oxygen diffusion in the orthorhombic $YBa_2Cu_3O_y$ phase

It is noteworthy that this equation can be obtained only if the concentration on plane 2 is negligible, and fortunately, this requirement can be satisfied in the temperature range from 400 to 550 °C [8, 10]. Therefore, we have

$$J = -\frac{1}{4}\frac{b^2\Gamma_1\Gamma_2}{\Gamma_1+\Gamma_2}\frac{\partial C}{\partial x}.$$

Comparing with Fick's first law $J = -D\,\partial C/\partial x$, we have

$$D = \frac{1}{4}\frac{b^2\Gamma_1\Gamma_2}{\Gamma_1+\Gamma_2}. \tag{4}$$

Now come back to the case of $YBa_2Cu_3O_y$, $b = 0.38864$ nm, $\Gamma_1 = \nu_B$, and $\Gamma_2 = \nu_A$ where ν_A and ν_B are the jump frequencies of oxygen atoms on O4 and O5 sites. At thermodynamic equilibrium the occupancies of oxygen atoms on O4 and O5 sites should obey

$$C_A\nu_A = C_B\nu_B\,,$$

so that

$$D = \frac{1}{4}\frac{b^2C_B}{C_A+C_B}\nu_B\,. \tag{5}$$

According to the Arrhenius equation

$$\nu_A = \nu_{A_0}\,e^{-H_A/kT}\,,$$

$$\nu_B = \nu_{B_0}\,e^{-H_B/kT}\,,$$

where ν_{A_0} and ν_{B_0} are the frequency factors, and H_A, H_B represent the potential energy of oxygen atoms on O5 and O4 sites, respectively (see Fig. 3), $H_B = H_A + \Delta E$. Assuming $\nu_{A_0} = \nu_{B_0} = \nu_0$, from (5), we obtain

$$D = \frac{1}{4}\frac{b^2C_B}{C_A+C_B}\nu_0\,e^{-H_B/kT} =$$

$$= D_0\exp\left(-\frac{H_B}{kT}\right); \tag{6}$$

$$D_0 = \frac{1}{4}\frac{b^2C_B}{C_A+C_B}\nu_0\,.$$

Here k is the Boltzmann constant and T the absolute temperature.

As reported by Xie et al. [8] and Chen et al. [10], the height of the 204 °C internal friction peak is proportional to the occupancy of oxygen atoms on O5 sites, that is C_A, and the activation energy of the peak is approximately equal to the potential

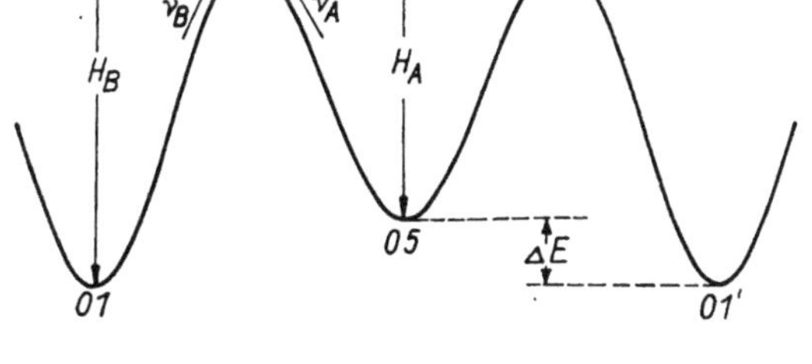

Fig. 3. Potential energy of oxygen atoms on O1(O1′) and O5 sites

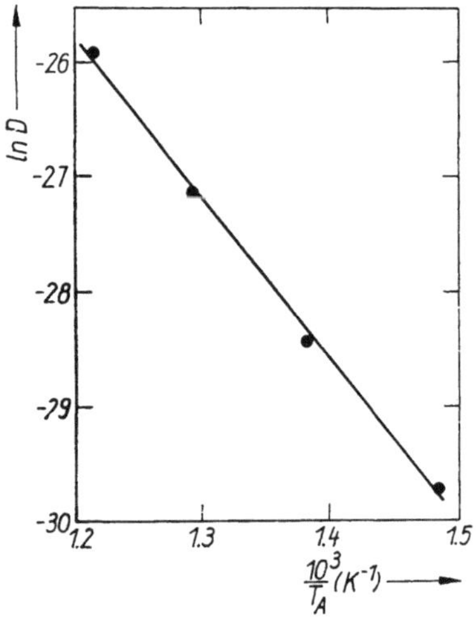

Fig. 4. Temperature dependence of the diffusion coefficient of oxygen atoms in the orthorhombic $YBa_2Cu_3O_y$ phase from 400 to 550 °C

energy of oxygen atoms on O5 sites. They have calculated ν_0, C_A, C_B, and H_A, H_B [7, 8, 10]. The calculated data are listed in Table 1. By substituting the calculated data into (6), the oxygen diffusion coefficients at different temperatures (T_A) are obtained. The results are also shown in Table 1.

Table 1

Diffusivity of oxygen atoms at different temperatures ($\nu_0 = 5 \times 10^{11}\ s^{-1}$ [7])

T_A (°C)	C_A*)	C_B*)	H_A**) (eV)	ΔE*) (eV)	H_B (eV)	D_0 (cm²/s)	D (cm²/s)
400	0.0155	0.804	1.030	0.197	1.227	1.83×10^{-4}	1.25×10^{-13}
450	0.0185	0.762	1.033	0.205	1.238	1.82×10^{-4}	4.5×10^{-13}
500	0.0255	0.615	1.036	0.201	1.236	1.79×10^{-4}	1.64×10^{-12}
550	0.0461	0.604	1.057	0.170	1.227	1.74×10^{-4}	5.56×10^{-12}

*) From [10].
**) From [8].

From Table 1, we can see that $H = 1.23$ eV and $D_0 = 1.80 \times 10^{-4}$ cm²/s in the temperature range from 400 to 550 °C. So that the diffusion coefficient can be expressed as

$$D = 1.80 \times 10^{-4} \exp(-1.23\ \text{eV}/kT)\,. \tag{7}$$

The diffusivity versus temperature curve is shown in Fig. 4.

3. Discussion

1. We can see from the calculation that the oxygen diffusion rate in the orthorhombic $YBa_2Cu_3O_y$ phase is very slow. The disagreement between our results and those of Morris et al. and Routbort et al. may result from the difference of the specimen state. Because the $YBa_2Cu_3O_y$ bulk material contains a great number of holes, unlike the traditional diffusion process in metals which takes place only through the outer surface, the diffusion in $YBa_2Cu_3O_y$ can occur through the intersurfaces of the holes

which are linked to the environment. This will greatly increase the surface area through which oxygen atoms can pass, and also remarkably decrease the actual diffusion distance, so that this may mislead people to wrong conclusions. Obviously, it is very difficult to study the oxygen diffusion process by means of general experimental methods. On the contrary, data obtained from the IF technique are irrelevant to the porosity of the specimens, hence, the diffusion coefficient obtained this way may be more reliable.

2. While describing the process of oxygen uptake, we must take into account surface reactions of oxygen atoms. This will make the oxygen diffusion rate slower. Our experimental work elsewhere has demonstrated that the rate of oxygen release is even much slower than the rate of oxygen uptake. Obviously, this is caused by the difference in activation energy of the reactions between oxygen and the surface of the $YBa_2Cu_3O_y$ phase. Tu et al. [4] have obtained an activation energy of oxygen decomposition on the surface of the specimen of 1.7 eV, but the activation energy of oxygen combination on the surface of the $YBa_2Cu_3O_y$ has not yet been reported.

3. The superconducting shell structure can be easily understood based upon the above calculation. Taking an infinitely flat specimen as an example, if the initial weight of the specimen is W_i and the equilibrium weight of the specimen W_f, then the time required for the specimen to gain half the wieght difference $[(W_f - W_i)/2]$ can be expressed as [11]

$$\sqrt{Dt} = 0.44d\ .$$

Here d is the thickness of the specimen.

Taking $T_A = 400$ °C and $d = 0.1$ mm, then from Table 1

$$t = 43022\ \text{h} = 4.9\ \text{year}\ .$$

Considering the porosity of the specimen, if $d = 10\ \mu\text{m}$, we have

$$t = 430\ \text{h}\ .$$

Therefore, it is obvious that oxygen cannot attain an equilibrium state in the normal annealing time. A superconducting shell is certain to appear. The superconducting shell structure will greatly decrease the critical current density of the specimen and thus bring great difficulties for its application.

References

[1] R. J. Cava, B. Batlogg, C. H. Chen, E. A. Rietman, S. M. Zahurk, and D. J. Werder, Phys. Rev. B **36**, 5719 (1987).

[2] D. Morris, U. M. Scheven, L. C. Bourne, M. L. Cohen, M. F. Crommie, and A. Zettl, to be published.

[3] J. L. Routbort, S. J. Rothman, B. K. Flandermeyer, L. J. Nowicki, to be published.

[4] K. N. Tu, S. I. Park, and C. C. Tsuei, Appl. Phys. Letters **51**, 2158 (1987).

[5] D. S. Ginley, E. L. Venturini, J. F. Kwak, R. J. Baughman, B. Morosin, and J. E. Schirber, Phys. Rev. B **36**, 829 (1987).

[6] Ji Huang, X. M. Xie, and T. W. Li, submitted to Solid State Commun.

[7] T. G. Chen, J. H. Zhang, J. Huang, Y. Chen, M. J. Yang, X. M. Xie, T. W. Li, and L. M. Xie, Chinese Phys. Letters **5**, 185 (1988).

[8] X. M. Xie, T. G. Chen, and Z. L. Wu, to be published.

[9] J. D. Jorgensen, M. A. Beno, D. G. Grace, I. K. Schuller, C. U. Segre, K. Zhang, and M. S. Kleefisch, Phys. Rev. B **36**, 3608 (1987).

[10] T. G. Chen, X. M. Xie, Y. Chen, and L. M. Xie, Proc. 2nd Nat. Conf. High-T_c Superconductors, April 1988, Bao Ji (China) (p. 196).

[11] Seminar on Atom Movements, Amer. Soc. Metals, Cleveland (Ohio) 1951.

(Received July 26, 1988)

phys. stat. sol. (a) **110**, 421 (1988)

Subject classification: 61.70; S1.1

Department of Mechanical Engineering and Materials Science, William Marsh Rice University, Houston[1])

Carbon–Vacancy Interactions in B.C.C. Iron

By

R. B. McLellan and Margot L. Wasz

Results have been taken from recent work on the thermodynamics of vacancy–interstitial interactions in metals in order to elucidate the effect of such interactions upon the thermodynamic functions and kinetic behavior of the C–b.c.c. iron system. It is shown that even with C–vacancy binding enthalpies (ε_{b_1}) as high as 1 eV, such interactions would not have an observable effect on the thermodynamic functions of the C atoms nor upon C diffusion or Fe-atom diffusion. However a consideration of positron annihilation data indicates that ε_{b_1} must be $\leqq 0.75$ eV.

Neue Ergebnisse bezüglich der Thermodynamik der Wechselwirkungen zwischen Leerstellen und interstitionellen Fremdatomen werden verwendet, um die Effekte solcher Wechselwirkungen auf die Thermodynamik und Kinetik der C–(k.r.z.)Fe-Lösungen zu untersuchen. Es wird gezeigt, daß selbst wenn die C–Leerstelle-Bindungsenthalpie (ε_{b_1}) 1 eV beträgt, solche Wechselwirkungen meßbare Effekte weder auf die thermodynamischen Funktionen der gelösten C-Atome, noch auf das Diffusionsverhalten der C- und Fe-Atome ausüben würden. Eine Berücksichtigung der vorhandenen Positronannihilationsdaten zeigt aber, daß ε_{b_1} unterhalb des Höchstwertes 0,75 eV liegt.

1. Introduction

Recently there has been considerable discussion concerning the binding enthalpies of carbon–vacancy complexes in α-iron. Walz and Blythe [1] have considered the annealing of dilute Fe–C solutions (e^-)-irradiated at low temperatures and have deduced the range of 0.1 eV $\leqq \varepsilon_{b_1} \leqq$ 0.65 eV for the C–monovacancy binding enthalpy ε_{b_1}. Other authors [2] have deduced higher values of ε_{b_1} in the range of 0.8 to 1.0 eV. Mandino and Seeger [3] have studied the 535 K stage of recovery of neutron-irradiated C-doped α-iron and deduced the relation

$$\nu = \nu_0 \exp\left[\frac{-H^{\text{diss}}_{1\text{v}-1}}{kT}\right] \tag{1}$$

for the detrapping rate of C from vacancies. In (1), $\nu_0 \approx 10^{13}\ \text{s}^{-1}$ and $H^{\text{diss}}_{1\text{v}-1}$, the C–monovacancy dissociation enthalpy is 1.6 to 1.7 eV. Combining this value of $H_{1\text{v}-1}$ with the interstitial migration enthalpy H^{m}_{i} and the simple model,

$$H^{\text{diss}}_{1\text{v}-1} = H^{\text{m}}_{\text{i}} + \varepsilon_{b_1} \tag{2}$$

yields the result $\varepsilon_{b_1} = 0.8$ eV. Similar reasoning was used by Vehanen et al. [4] who deduced $\varepsilon_{b_1} = 0.85$ eV.

[1]) Houston, Texas 77251, USA.

However Arndt and Damask [5], using stored energy techniques, concluded that the C–vacancy pair was less stable than these latter estimates and that $\varepsilon_{b_1} = 0.41$ eV. The paper of Walz and Blythe [1] contains a summary of the estimates of ε_{b_1} and the techniques used by the various authors.

The question to be answered in this report is a simple one. If the C–vacancy binding enthalpy is as high as some estimates suggest, such complexes may be present in sufficient concentrations and over sufficiently large ranges of temperature so as to be visible in the equilibrium properties of the thermodynamic and kinetic behavior of the Fe–C system. The present work will provide an answer to this question and deduce an upper bound for ε_{b_1}.

2. Thermodynamic Model

The equilibrium thermodynamic functions of interstitial solid solutions have recently been deduced by considering a model in which a fraction of the interstitial (i) atoms may form complexes with vacancies by "decorating" interstitial sites nearest neighbor to a vacancy [6]. If the i–vac (i.e. i–vacancy) binding energies are large, the total vacancy concentration present will be considerably larger than that present in the pure metal at a given temperature and may give rise to phenomena observable in the thermodynamic [6] and kinetic behavior of the system [7].

Complexes containing up to z decorating i-atoms (z is the number of interstitial sites of a given kind neighboring a vacancy) are considered, but only groups containing one vacancy. The general solution for the configurational chemical potential μ_i^c for a solid solution of i-concentration θ (i.e. ratio of i to metal atoms) is [6]

$$\mu_i^c = \overline{H}_i^\infty + kT\Psi \ln \theta - kT(1 - \Psi) \ln (\theta - \delta) , \tag{3}$$

where $\overline{H}_i^\infty$ is the enthalpy required to introduce an i-atom into an interstitial site distant from a vacancy, and Ψ is the summation²),

$$\Psi = C_{1v}^0 \sum_{i=0}^{z} \frac{(i)^2 g_i \theta^{(i-1)}}{(1 - Q)} \left(1 + \frac{Q'}{i(1 - Q)}\right) \exp \left(\frac{\varepsilon_{b_i}}{kT}\right), \tag{4}$$

where C_{1v}^0 is the vacancy concentration of the solute-free crystal, ε_{b_i} is the binding energy of an i-order complex, and $g_i = (z!/(z-i)!\, i!)$. The quantities Q and Q' are given by

$$Q = C_{1v}^0 \sum_{i=0}^{z} g_i \theta^i \exp \left(\frac{\varepsilon_{b_i}}{kT}\right), \tag{5}$$

$$Q' = C_{1v}^0 \sum_{i=0}^{z} i g_i \theta^i \exp \left(\frac{\varepsilon_{b_i}}{kT}\right), \tag{6}$$

and finally $\delta = Q'(1 - Q)^{-1}$. These formulas are derived in detail in [6] and the approximation $C_{1v}^0 = \exp(-E_{1v}^f/kT)$, where E_{1v}^f is the vacancy formation energy, is used (i.e. the vacancy formation entropy S_{1v}^f is neglected).

Now if we restrict our consideration to sufficiently low solute concentrations (i.e. $\theta_i \leqq 10^{-4}$) then complexes of order $i > 1$ may be neglected for any reasonable value of ε_{b_i} because of the factors θ^i in Q and Q'. In considering "trace" levels of interstitial impurities in b.c.c. iron, it is certainly permissible to ignore all clusters larger than

²) The symbol i denotes "interstitial", but is also a summation index counting the number of i-atoms decorating a vacancy.

$i = 1$. In this limit (3) assumes the simple form

$$\mu_i^c = \overline{H}_i^\infty + kT \ln \theta + kT(1 - zC_{1v}^0 B_1) \ln (1 - zC_{1v}^0 B_1)\,, \tag{7}$$

where we have used the abbreviation $B_1 = \exp(\varepsilon_{b_1}/kT)$.

If we use the value $E_{1v}^f = 1.5$ eV taken from the positron annihilation works of Schaefer et al. [8] (1.53 eV), Matter et al. [9] (1.60 eV), and Kim and Buyers [10] (1.4 eV), and take ε_{1v}^b at its largest suggested value of 0.85 eV, then even at 1200 K $1 \gg zC_{1v}^0 B_1$ so that (7) can be approximated to

$$\mu_i^c = \overline{H}_i^\infty + kT \ln \theta - kTzC_{1v}^0 B_1\,. \tag{8}$$

The partial molar enthalpy of the solute may be written immediately from (8) in the form

$$\overline{H}_i = \overline{H}_i^\infty + zC_{1v}^0 B_1 (E_{1v}^f - \varepsilon_{b_1})\,. \tag{9}$$

Note that in deriving (9) from (8) by using the identity $\overline{H}_i = \big(\partial(\mu_i/T)/\partial(1/T)\big)_p$, terms arising from the excess entropy, which was omitted from (3), do not appear. It is also clear that even for $\varepsilon_{b_1} = 0.85$ eV the difference between $\overline{H}_i$ and $\overline{H}_i^\infty$ is too small to be detected thermodynamically. The quantity $\overline{H}_i$ has been determined [11] in the range $\theta = 5 \times 10^{-5}$ to 5×10^{-4} and the T-range 846 to 1098 K and $\overline{H}_i$ found to be constant to within 0.4 kJ/mol (≈ 0.02 eV!). Thus the thermodynamic functions of Fe–C solutions offer no help in deciding upon the value of ε_{b_1}. It is however interesting to note that $\overline{H}_i$ remains constant through the magnetic disordering temperature range whereas the activation energy for C migration is strongly affected by the magnetic transformation [12].

3. Kinetic Considerations

Strong i–vac interactions may also be observable in the kinetic behavior of the Fe–C system provided ε_{b_1} is large enough and the measured diffusivities span a sufficiently large range of reciprocal temperature. Both the diffusivities of C and Fe should be considered.

3.1 Carbon diffusivity

It is, of course, well known that the diffusivity of C in b.c.c. iron exhibits a large non-linearity in Arrhenius behavior in the temperature region of the Curie transformation [12, 13]. This effect need not concern us here.

It is clear that i–vac complex formation will decrease the C diffusivity and that the extent of the decrease will be proportional to the fraction of total C atoms involved in complex formation. It is easy to show that the general expression for this fraction f_i^{1v} is [6]

$$f_i^{1v} = \frac{C_{1v}^0}{\theta(1 - Q)} \sum_{i=0}^{z} g_i \left(\frac{\theta}{\beta}\right)^i \exp\left(\frac{\varepsilon_{bi}}{kT}\right). \tag{10}$$

Thus for simple $i = 1$ complexes,

$$f_i^{1v} = \frac{zC_{1v}^0 B_1}{\beta(1 - Q)}\,. \tag{11}$$

The factor β appearing in (10) and (11) is included because C occupies the octahedra sites in the b.c.c. lattice and there are $\beta = 3$ such sites per lattice atom. The previous formulas were derived [6] for the case $\beta = 1$ (octahedral occupancy in f.c.c. lattices). The constant β was not included previously since θ does not appear in (9). It is clear from (11) that even for $\varepsilon_{b_1} = 1$ eV, $f_i^{1v} < 10^{-5}$ at all relevant temperatures and thus it is not possible to detect complex formation effects in the C-diffusivity data.

3.2 Iron atom diffusivity

There are many determinations of the diffusivity of Fe in C-doped and nominally pure Fe. The data are depicted in Fig. 1 and the author–symbol–reference code is given in Table 1.

Diffusion data taken in the range of the Curie transformation, e.g. those due to Hettich et al. [19] have not been included. All the lines in Fig. 1 correspond to a least-squares regression of all the α-Fe data sets. This fitting procedure yields the relation

$$D = (3.09 \pm 0.14) \times 10^{-4} \exp\left(-\frac{244.8 \pm 11.7}{RT}\right) \text{m}^2/\text{s}\,, \tag{12}$$

where R is equal to 8.3143×10^{-3} kJ/mol K. It is relevant to note that the C-contents of the "irons" used in the experiments listed in Table 1 range from "not stated" up

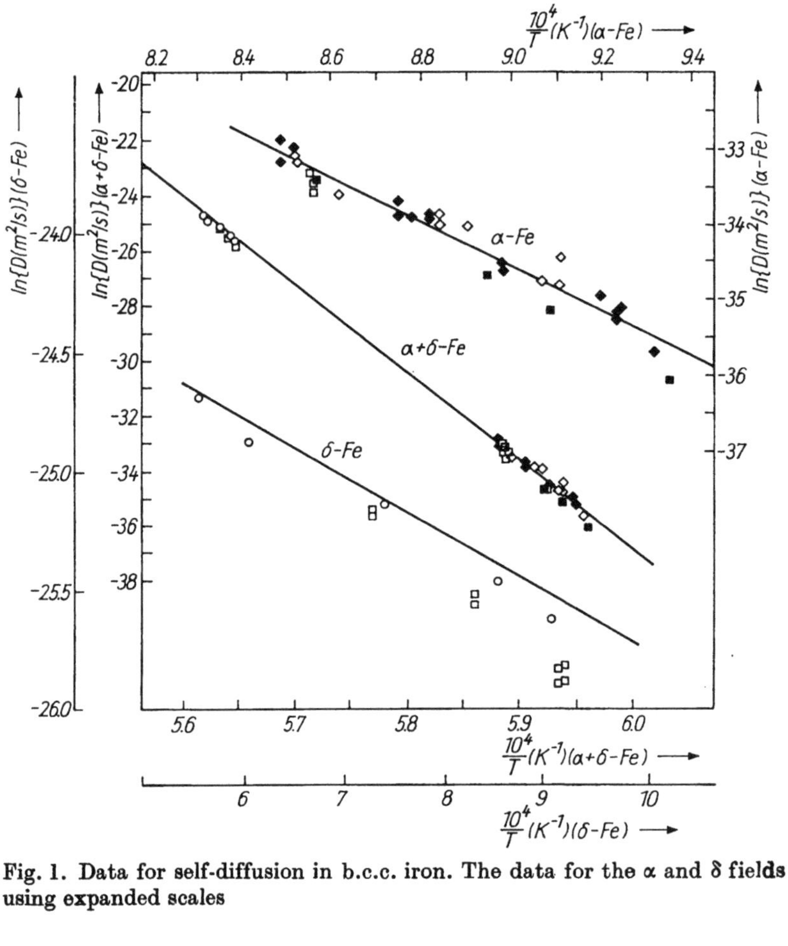

Fig. 1. Data for self-diffusion in b.c.c. iron. The data for the α and δ fields are shown separately using expanded scales

Table 1

Fe self-diffusion data

ref.	symbol	structure
[14]	○	δ
[15]	□	α, δ
[16]	◆	α
[17]	■	α
[18]	◇	α

to $\theta = 4.6 \times 10^{-3}$ for the (◇) data of Birchenall and Mehl [18]. It is clear from Fig. 1 that the Fe-diffusion data do not show any deviations in the entire span of $(1/T)$ and θ which would indicate an effect due to i–vac interactions.

Now it is easy to show [6] that the ratio λ of the total vacancy concentration to the concentration of vacancies formed in the limit $\theta = 0$ is given by

$$\lambda = \frac{Q}{C_{1v}^0} = 1 + z\left(\frac{\theta}{\beta}\right) B_1 . \tag{13}$$

If we take, as an extreme case $\varepsilon_{b_1} = 0.85$ eV, $T = 10^3$ K (i.e. α-iron), and $\theta = 4.6 \times \times 10^{-3}$, then $\lambda = 1.68 \times 10^2$. Thus on going from $\theta = 0$ to 4.6×10^{-3} the total vacancy concentration will increase by more than 10^2. However, this will not lead to a measurable increase in Fe self-diffusion because of the low dissociation rate of i–vac complexes. An approximate calculation of D^*, the Fe-diffusion rate in the C-doped solution, can be made if it is assumed that the predominant dissociation mode is the jump of a C atom from the complex (creating a "free" vacancy having its normal hopping frequency) rather than the vacancy jump itself. With this assumption,

$$\frac{D^*}{D} \approx 1 + z\left(\frac{\theta}{\beta}\right) B_1 \exp\left(\frac{-H_{1v-C}^{\text{diss}}}{kT}\right), \tag{14}$$

where D is the diffusivity for $\theta = 0$. Using the value of H_{1v-C}^{diss} from Mondino and Seeger [3] (1.6 eV) shows that $D^* \approx D$ even for the highest energy levels.

However, even though self-diffusion data will not shed light on the values of ε_{b_1} it is clear from the above that the total vacancy content, and thus the apparent vacancy formation energy, may very well be interesting in this context. In fact De Schepper et al. have indicated this [2].

4. Apparent Vacancy Formation Energy

The total vacancy concentration in the doped crystal is given by [6]

$$C_{1v}^{\mathrm{T}} = Q(1 - Q)^{-1} . \tag{15}$$

Defining that apparent vacancy formation energy as

$$E_{1v}^{\mathrm{f\,eff}} = -k\left(\frac{\partial \ln C_{1v}^{\mathrm{T}}}{\partial(1/T)}\right)_p$$

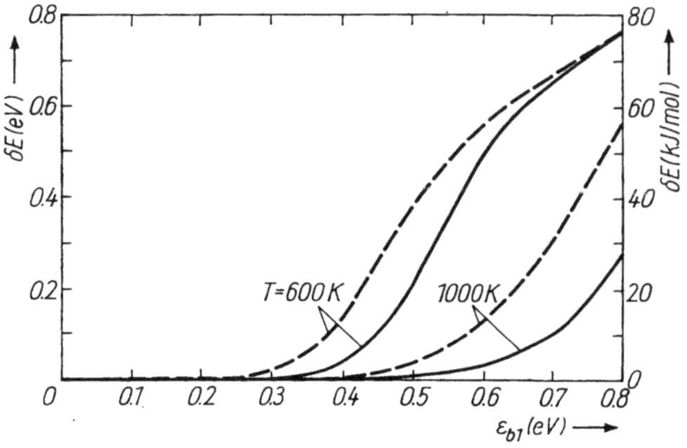

Fig. 2. Variation of the apparent vacancy formation energy in C-doped b.c.c. iron as a function of the C–vacancy binding energy, ε_{b_1}. ——— $\theta/3 = 1 \times 10^{-5}$, – – – 5×10^{-5}

leads, from (15), to the expansion ($\beta = 1$)

$$E_{1v}^{\text{feff}} = \frac{1}{Q(1-Q)} \sum_{i=0}^{z} g_i \theta \, \Delta\varepsilon_i \exp\left(-\frac{\Delta\varepsilon_i}{kT}\right), \tag{16}$$

where $\Delta\varepsilon_i = E_{1v}^{f} - \varepsilon_{b_1}$. Restricting the complexes to $i = 0, 1$ and introducing $\beta = 3$ for C in b.c.c. iron, gives

$$E_{1v}^{\text{feff}} = \frac{E_{1v}^{f} + z\left(\dfrac{\theta}{3}\right)(E_{1v}^{f} - \varepsilon_{b_1})\, B_1}{1 + z\left(\dfrac{\theta}{3}\right) B_1}. \tag{17}$$

The quantity $\delta E = E_{1v}^{f} - E_{1v}^{\text{feff}}$ is shown in Fig. 2 at 600 and 1000 K for two values of $\theta/3$ (i.e. 10^{-5} (———) and 5×10^{-5} (--------)) as a function of ε_{b_1}. At the lowest temperature δE begins to increase sharply at $\varepsilon_{b_1} = 0.4$ to 0.5 eV but at 1000 K the change in E_{1v}^{f} is not large for the lower doping level until $\varepsilon_{b_1} \approx 0.7$ eV.

Now the three values of E_{1v}^{f} quoted [8 to 10] are all in good mutual accord. The iron used by Schaefer et al. [8] had C and N contents of less than $\theta = 10^{-6}$, Matter et al. [9] gave no chemical analysis, but the iron used by Kim and Buyers [10] had a total interstitial content of $\theta = 1.32 \times 10^{-5}$. The good agreement between the sets of positron annihilation data and the results of Fig. 2 indicate that $\varepsilon_{b_1} \approx 0.75$ eV is an upper limit but much lower values are probably more appropriate.

5. Conclusions

Carbon–vacancy binding enthalpies as high as 1 eV (... kJ/mol) will not lead to observable effects in the thermodynamic functions nor in the C and Fe diffusion behavior in C–b.c.c. Fe solid solutions where the solute concentration is less than $\theta \approx 10^{-4}$. This is true even though the total vacancy concentrations produced would be $\geqq 10^2$ times larger than those present in undoped α-iron.

The present positron annihilation data suggests that $\varepsilon_{b_1} = 0.75$ eV (72 kJ/mol) is an upper limit for the C–vacancy binding enthalpy. It is clear that this technique can be a useful tool for the study of vacancy–interstitial interactions when positron annihilation measurements are undertaken on carefully doped metals. The effects of the interstitial additives upon the dilation of the parent lattice must of course be taken into account.

Finally it should be pointed out that there are thermodynamic systems (e.g. C–γ-Fe) where the solute concentrations can reach much higher levels ($\theta \approx 0.07$) so that C–vacancy interactions become an important consideration in understanding the diffusivities of both solute and host atoms [7, 20].

Acknowledgement

The authors are grateful for the support provided by the Robert A. Welch Foundation.

References

[1] F. Walz and H. J. Blythe, phys. stat. sol. (a) **104**, 343 (1987).
[2] L. De Schepper, G. Knuyt, and L. M. Stals, phys. stat. sol. (a) **67**, 153 (1981).
[3] M. Mondino and A. Seeger, Scripta metall. **11**, 817 (1977).
[4] A. Vehanen, P. Hautojärvi, J. Johansson, J. Yli-Kauppila, and P. Moser, Phys. Rev. B **25**, 762 (1982).
[5] R. A. Arndt and A. C. Damask, Acta metall. **12**, 341 (1964).
[6] R. B. McLellan, J. Phys. Chem. Solids **49**, 1213 (1988).
[7] R. B. McLellan, Acta metall. **36**, 1923 (1988).
[8] H. E. Schaefer, K. Maier, M. Weller, D. Herlach, A. Seeger, and J. Diehl, Scripta metall. **11**, 803 (1977).
[9] H. Matter, J. Winter, and W. Triftshäuser, Appl. Phys. A **20**, 135 (1979).
[10] S. M. Kim and W. J. L. Buyers, J. Phys F **8**, 103 (1978).
[11] W. W. Dunn and R. B. McLellan, Metallurg. Trans. **2A**, 1079 (1971).
[12] M. Wuttig, Scripta metall. **5**, 33 (1971).
[13] R. B. McLellan, M. L. Rudee, and T. Ishibachi, Trans. AIME **233**, 1938 (1965).
[14] R. J. Borg, D. Y. F. Lai, and O. H. Krikorian, Acta metall. **11**, 867 (1963).
[15] C. M. Walter and N. L. Peterson, Phys. Rev. **178**, 922 (1969).
[16] F. S. Buffington, K. Hirano, and M. Cohen, Acta metall. **9**, 434 (1961).
[17] R. J. Borg and C. E. Birchenall, Trans. AIME **218**, 980 (1960).
[18] C. E. Birchenall and R. F. Mehl, Trans. AIME **188**, 144 (1950).
[19] G. Hettich, H. Mehrer, and K. Maier, Scripta metall. **11**, 795 (1977).
[20] C. Köstler, F. Faupel, and T. Hehenkamp, Scripta metall. **20**, 1755 (1986).

(Received August 5, 1988)

phys. stat. sol. (a) **110**, **429** (1988)

Subject classification: 64.70; 62.65; 77.20; S11.1

Department of Physics, Vilnius University[1]) *(a)*
and Institute of Crystallography, Academy of Sciences of the USSR, Moscow (b)

Microwave and Ultrasonic Investigations of Superionic Phase Transitions in $CsDSO_4$ and $CsDSeO_4$

By

R. MIZERIS (a), J. GRIGAS (a), V. SAMULIONIS (a), V. SKRITSKI (a),
A. I. BARANOV (b), and L. A. SHUVALOV (b)

Results are given of the investigation of the temperature and frequency dependence of the complex dielectric permittivity and electrical conductivity of the $CsDSO_4$ and $CsDSeO_4$ single crystals in the frequency range 15 MHz to 52 GHz as well as velocity and attenuation of the longitudinal ultrasonic waves in the vicinity of a superionic phase transition temperature.

Es werden die Ergebnisse der Untersuchung der Temperaturabhängigkeit und der Frequenzabhängigkeit der komplexen Dielektrizitätskonstante und der elektrischen Leitfähigkeit im Frequenzbereich von 15 MHz bis 52 GHz sowie auch der Temperaturabhängigkeit der Geschwindigkeit und der Absorption der longitudinalen Ultraschallwellen, in der Umgebung des superionischen Phasenübergangs von $CsDSO_4$- und $CsDSeO_4$-Einkristallen dargestellt.

1. Introduction

During recent years much attention has been paid to the investigation of the superionic conductors $CsHSO_4$(CHS), $CsHSeO_4$(CHSe), and their deuterated isomorphs CDS and CDSe. At room temperature they belong to the monoclinic space group $P2_1/m$ (CDS) and $P2_1/c$ (CDSe) [1 to 4]. At $T_c = 414$ K (CHS), 412 K (CDS), 398 K (CHSe), and 399 K (CDSe) they undergo ferroelastic phase transitions of the first order to the superionic tetragonal phase. The ferroelastic space group of this phase is $I4_1/amd$ [5]. In the ferroelastic phase the crystals consist of chains of O–H(D)–O bonded $SO_4(SeO_4)$ groups running along the *b*-axis [6]. Only two of the four $SO_4(SeO_4)$ oxygens are H-bonded so that the number of protons is half the number of possible proton sites [7]. Therefore, protons move rather freely through the crystal and cause anomalously high ion conductivity above T_c. The phase transition of the order–disorder type is caused by protons disordering on hydrogen bonds [8].

Dielectric properties of CHS and CHSe crystals had been investigated earlier only at low frequencies. Komukae et al. [9] found that both real and imaginary parts of the complex dielectric permittivity ε_a^*, ε_b^*, and ε_c^* of CHS increased discontinuously at T_c, and at 10 kHz showed magnitudes of order of 10^3. Due to the large electric conductivity there is no dielectric data in the superionic phase. Wolak and Czapla [10] measured the dielectric permittivity ε_c of CHSe in the frequency range 1 kHz to 1 MHz. At T_c it shows an abrupt jump to about 350 and remains almost independent of temperature both at $T < T_c$ and $T > T_c$. According to Yokota et al. [11] both ε_b' and ε_b'' of CHSe at 1 MHz exhibit sharp peaks at T_c. Above T_c ε' is of the order of 10^2 ($\varepsilon_c' > \varepsilon_b'$) and increases with the increase of temperature.

[1]) Sauletekio al. 9, SU-232054 Vilnius, USSR.

The dielectric permittivity and conductivity at frequencies of the order of or below MHz are usually masked by contact blocking effects [12]. The values of the intrinsic dielectric permittivity of superionic conductors usually are much smaller [13] than mentioned above. Thus, there are no reliable data about the intrinsic values and the frequency and temperature dependences of the dynamic dielectric permittivity and conductivity in the superionic phase of the crystals of this family. The most interesting frequency range from the point of view of the dynamics of the conducting ions is the microwave and far-IR range.

In the present paper we report the results of a study of the temperature and frequency dependences of the real and imaginary parts of the complex dielectric permittivity $\varepsilon_a^* = \varepsilon_a' - i\varepsilon_a''$ and conductivity of $CsDSO_4$ and $CsDSeO_4$ single crystals in the frequency range 15 MHz to 52 GHz as well as velocity and attenuation of the longitudinal waves in the vicinity of the superionic phase transition temperature.

2. Experimental

The needle-shaped single crystals were grown by evaporation from an aqueous solution. The components of the complex dielectric permittivity ε_a^* or conductivity $\sigma_a^* = \sigma + i\omega\varepsilon_0\varepsilon'$ over the range 15 MHz to 4 GHz were obtained by a coaxial computer-controlled dielectric spectrometer set-up [14] from the results of the measurements of the reflection coefficient of the TEM wave in the line loaded with the cylindrical sample, placed at the end of the line between the inner conductor and the short piston. The sizes of the sample were about 0.2 mm thick and 3 mm^2 in area. In the mm-wave region a waveguide computer-controlled dielectric spectrometer set-up [15] was used. The permittivity and conductivity were computed from the results of the measurements of the modulus of the reflection and transmission coefficients of the H_{10}-wave produced by the needle-shaped crystal placed across the centre of a rectangular waveguide so that the a-axis of the crystal was parallel to the vector of the electric field. The crystal was fixed in good contact with the walls of top and bottom of the waveguide by Ag paste. In order to ensure the maximum precision of measurements the diameter of the crystal was chosen 0.25 mm.

Ultrasonic velocity and relative attenuation coefficient were measured by the pulse-echo method. The mechanical system contained the sample, the fused quartz delay line, and two $LiNbO_3$ piezoelectric transducers. Silicone oil was used as an acoustic bond. The longitudinal wave was propagating along the a-axis. The accuracy of thermal stabilization was better than 0.05 K.

3. Microwave Dielectric Permittivity

Fig. 1 shows the temperature dependence of ε_a' and ε_a'' along the a-axis of the $CsDSO_4$ crystal measured on heating at different frequencies. The value of ε_a' at room temperature is almost the same as in $CsHSO_4$ [9]. It increases slightly with the increase of temperature and at $T_c = 411.5$ K exhibits a peak. On cooling the phase transition with the similar peak of ε' is accompanied with a thermal hysteresis of 5 K. With the increase of frequency the magnitude of the peak value of ε' decreases and at frequencies $\nu > 240$ MHz it exhibits an upward jump to the superionic phase. The imaginary part of the complex permittivity exhibits an upward jump with increasing temperature at all frequencies.

Fig. 2 shows the similar behaviour of ε' and ε'' of $CsDSeO_4$. The dependence $\varepsilon'(T)$ at lower frequencies is similar to that obtained by Yokota et al. [11] for $CsHSeO_4$. The values of the dielectric permittivity of these crystals at microwaves are much

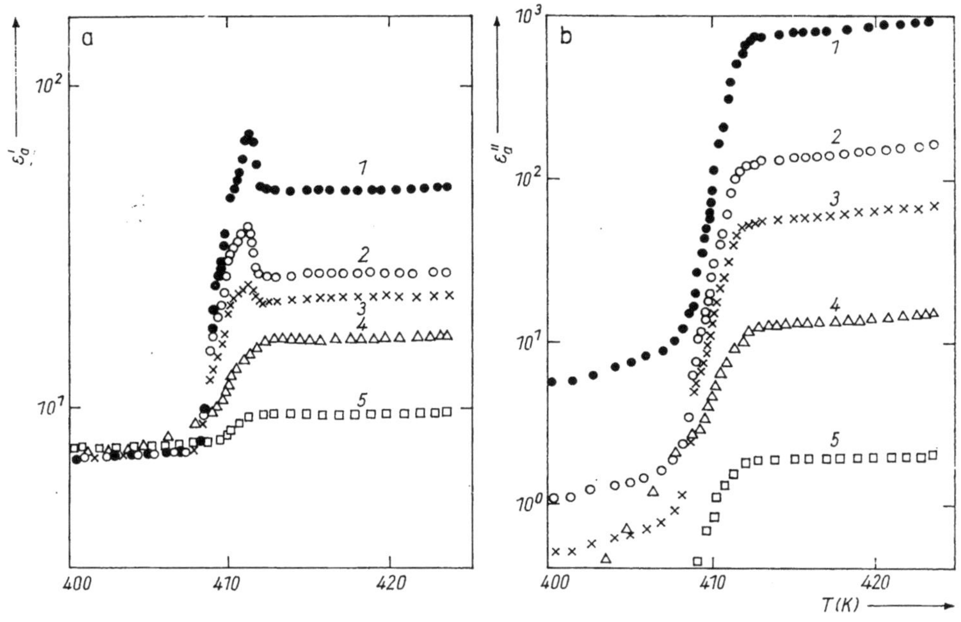

Fig. 1. Temperature dependence of a) ε'_a and b) ε''_a of $CsDSO_4$ crystal at different frequencies: (1) 15; (2) 95 MHz; (3) 240; (4) 985 MHz; (5) 32.5 GHz

smaller than were reported for CHS and CHSe [9 to 11]. The sharpness of the transition depends on the thermal history of the sample: only virgin samples show sharp jumps of ε' and ε'' at T_c.

The results indicate a critical behaviour of ε' near T_c. They lead to the suggestion that the peak of ε' is the result of the collective excitation of the vibrational motion of mobile ions.

Farther from T_c the temperature dependence of the dielectric permittivity can be understood on the basis of a modified q-component Pott model [16] within the framework of the molecular field approximation. The deuteron transfer process consists of two steps [7]: $SO_4(SeO_4)$ group rotation inducing cross-linking between previously isolated chains and the translation of deuterons between the two equilibrium states in the new O–D–O bond. These two steps contribute to the dielectric permittivity and its temperature dependence.

When a deuteron is in the lattice site its average dipole momentum is equal to zero. When it leaves the site the polarizability of the lattice increases and the spherical geometry of the moving interstitial ion changes. As a result the ion acquires a dipole moment. At the temperature T_c of the first-order phase transition of order–disorder type the temperature dependence of the dielectric permittivity is given by

$$\varepsilon = \varepsilon_\infty + \frac{eN}{T_c q - KN}, \tag{1}$$

where ε_∞ is the permittivity which is caused by the electronic polarization and phonons, N is the density of the mobile ions, e their charge, q the number of states for disordered ions, and $K = I_{11}^{(2)} - I_{11}^{(1)}$, $I_{11}^{(2)}$ and $I_{11}^{(1)}$ are interaction potentials of parallel and nonparallel oriented dipoles, respectively.

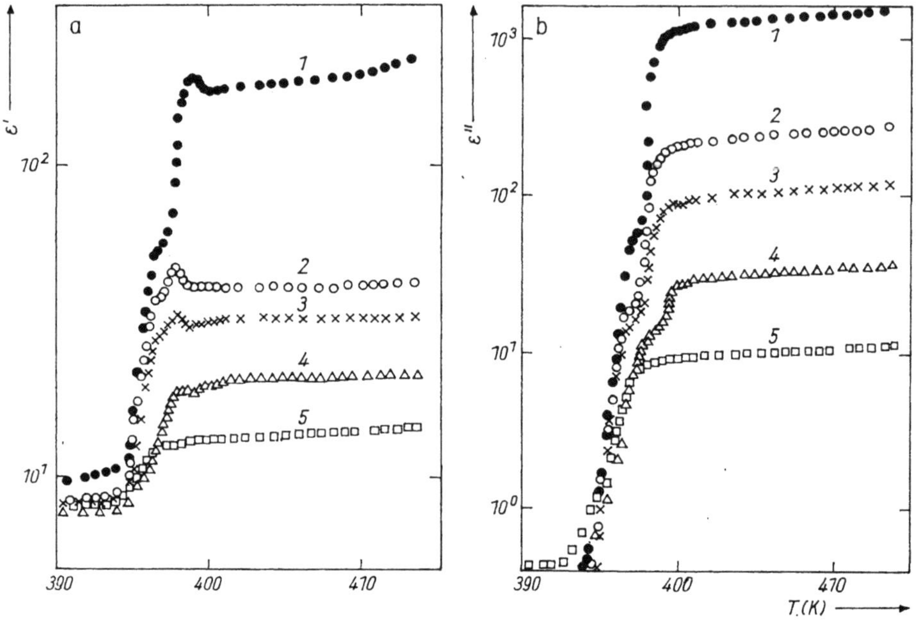

Fig. 2. Temperature dependence of a) ε' and b) ε'' of $CsDSeO_4$ crystal at different frequencies (see Fig. 1)

It follows from (1), that when $T_cq > NK$, the jump of N, i.e. the jump of conductivity at T_c causes an upward jump of dielectric permittivity. The magnitude of $\Delta\varepsilon = \varepsilon - \varepsilon_\infty$ may be different: $\Delta\varepsilon = 3$ in AgI [13], $\Delta\varepsilon = 1.5$ in Ag_2HgI_4, $\Delta\varepsilon = 1.2$ in Cu_2HgI_4 [17]. The larger K is, the larger is the magnitude of the $\Delta\varepsilon$ jump. With the increase of temperature the static permittivity in the superionic phase should follow the Curie-Weiss law [8], the dynamic one may decrease [13] or increase [17]. The dynamics of the ions in the lattice sites, i.e. the rotation of the $SO_4(SeO_4)$ ions in our case causes the increase of dielectric permittivity in the superionic phase with temperature. Fig. 1 and 2 show that the contribution of this mechanism prevails at frequencies $\nu < 100$ MHz. With the increase of frequency the contribution of $SO_4(SeO_4)$ rotations to ε decreases. At microwaves the contributions of $SO_4(SeO_4)$ and deuteron dynamics compensate each other so that the dielectric permittivity is small and weakly depends on temperature. At very high frequencies the dielectric permittivity may not exhibit any anomaly at T_c or even show a downward jump as in the case of AgI and $Na_3Se_2(PO_4)_3$ crystals [18].

The imaginary part of the complex dielectric permittivity $\varepsilon'' = \sigma/\omega\varepsilon_0$ reflects the behaviour of the dielectric losses caused by the temperature and frequency dependence of polarization.

4. Temperature Dependence of Conductivity

Fig. 3 a and b show the temperature dependence of ionic conductivity along the a-axis at various frequencies. The conductivity in both phases varies according to Arrhenius behaviour

$$\sigma T = \sigma_0 \exp(-E/kT) .$$

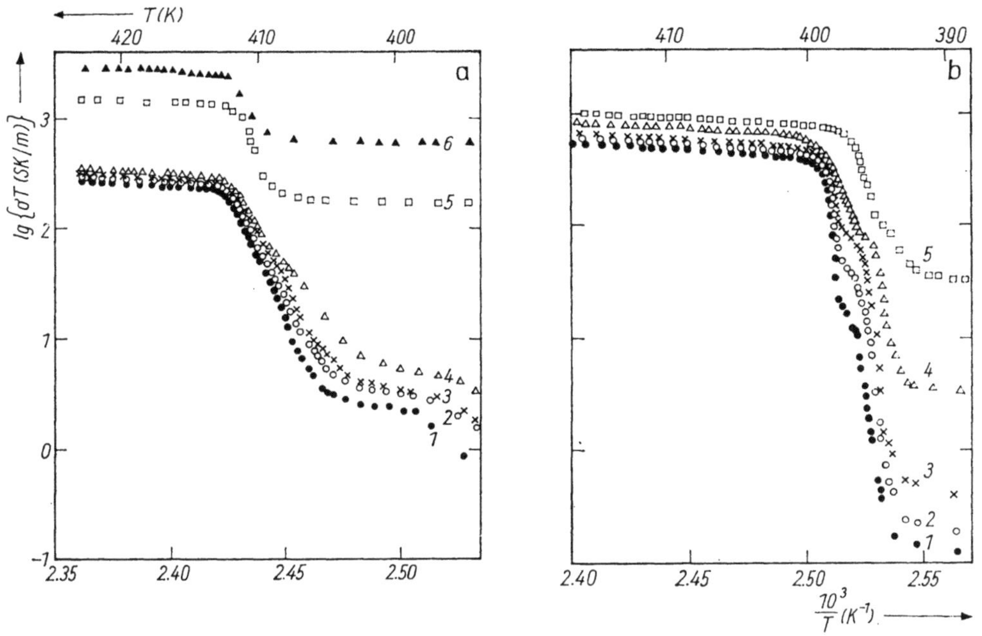

Fig. 3. The ionic conductivity as a function of $1/T$ of a) $CsDSO_4$ at frequencies: (1) 15; (2) 95; (3) 240; (4) 985 MHz; (5) 32.5, and (6) 52 GHz; b) $CsDSeO_4$ at frequencies: (1) 15; (2) 95; (3) 240; (4) 980 MHz; and (5) 3.9 GHz

In the superionic phase $E = 0.26$ eV for $CsDSO_4$ and 0.20 eV for $CsDSeO_4$. The activation energy is smaller than it was reported for CHS and CHSe at 10 kHz [1], it slightly decreases with the increase of frequency and is close to the value of the activation energy of the proton diffusion coefficient [7].

The relationship between the alternating current conductivity and dielectric losses with the dc conductivity, σ_0, being included is given by

$$\sigma(\omega) = \sigma_0 + \varepsilon_0\omega\varepsilon''(\omega) . \tag{2}$$

As the second term of (2) is prevailing the temperature dependence of $\sigma(\omega)$ below T_c and is mainly caused by $\varepsilon''(T)$, it is much weaker than that of the corresponding σ_0, therefore the values of the activation energy and conductivity depend on frequency.

5. Frequency Dependence of Permittivity and Conductivity

Frequency dependence of the dielectric permittivity and conductivity of the $CsDSO_4$ crystal (Fig. 4) in the high conducting phase show a pronounced relaxation-type dispersion with temperature independent relaxation time. A similar behaviour is for $CsDSeO_4$ crystals. Such a dispersion is a common feature for superionic conductors and had been found in many superionics [17, 18]. Thus, the relaxation dynamics of $CsDSO_4$ and $CsDSeO_4$ is entirely different from the Debye shape relaxation dynamics of the structurally similar ferroelectric CsD_2PO_4-type crystals [19]. The more sophisticated theories (e.g. [20]) can explain any behaviour $\sigma(\omega)$ of the microwave conductivity. However, our findings are very simple.

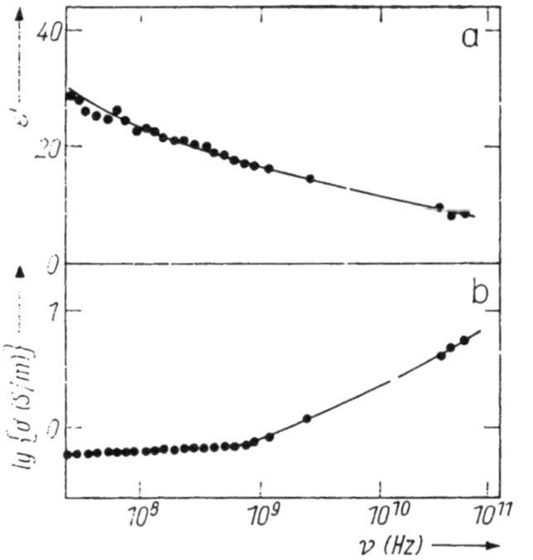

Fig. 4. Frequency dependence of a) permittivity and b) conductivity of $CsDSO_4$ at $T = 414$ K

At frequencies $\nu < 10^8$ Hz the dielectric permittivity ε' and losses ε'' decrease with the increase of frequency so that their ratio is independent of frequency, i.e.

$$\frac{\varepsilon''(\omega)}{\varepsilon'(\omega)} \approx \text{const}\,. \tag{3}$$

The ratio (3) implies the fact that the ratio of the energy lost per cycle to the energy stored in the crystal is constant and independent of frequency. This non-Debye relaxational dispersion law can be understood within the framework of a "screened hopping" model [21] involving many-body interactions. With reference to the model let us assume that an ion with a charge q occupies position i with a screening charge $(1 - p)\,q$ surrounding it due to the interaction with other ions. When the ion jumps to position j it leaves the screening charge at position i. The screening would gradually adjust to the position of the ion at j. The adjustment time is τ_s. When $\omega < 1/\tau_s$ the ratio (3) is determined by the value of the screening parameter p,

$$\frac{\varepsilon''(\omega)}{\varepsilon'(\omega)} = \frac{1 - p}{p}\,.$$

Heavy screening, i.e. $p \to 0$ gives large values of losses and its frequency dependence is $\varepsilon'' \sim 1/\omega$. In that case according to (2) conductivity $\sigma(\omega) > \sigma_0$ and up to 1 GHz it is almost independent of frequency. At microwaves with $\omega > 1/\tau_s$ and larger than the frequency with which the ions oscillate between i and j sites, the system cannot follow the microwave field, screening is weak ($p \to 1$), losses are low and ε'' may be characterized by the relation $\varepsilon''(\omega) \sim \omega^{n-1}$ with $n = 0.8$. In that case the conductivity $\sigma(\omega)$ increases with the increase of frequency. Thus, the frequency dependence of $\sigma(\omega)$ of $CsDSO_4$ and $CsDSeO_4$ crystals exactly like that of other fast-ion conductors, amorphous materials, and glasses [22], may be represented by (2). When ε'' weakly depends on frequency, the electrical conductivity is given by the expression of the universal response $\sigma(\omega) \sim \omega^n$.

6. Ultrasonic Velocity and Attenuation

Fig. 5a shows the temperature dependence of the ultrasonic velocity and attenuation of a $CsDSO_4$ crystal. The ultrasonic attenuation is so high that at higher frequencies the acoustic signal disappears near T_c. In the superionic phase the attenuation remains

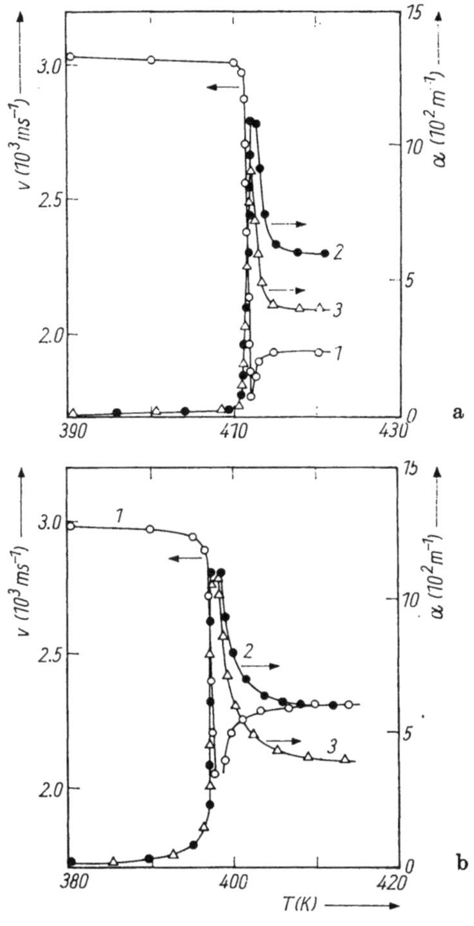

Fig. 5. Temperature dependence of (1) velocity and (2), (3) attenuation of the longitudinal ultrasonic waves in a) $CsDSO_4$ and b) in $CsDSeO_4$ at frequencies (MHz): (1) and (3) 4.5; (2) 12.5 MHz

relatively high and increases with the increase of frequency. Similar results were obtained for $CsDSeO_4$ crystal (Fig. 5b). Our measured value of ultrasonic velocity in $CsDSeO_4$ at room temperature is in good agreement with Brillouin data [23]. The change of velocity in $CsDSeO_4$ is smaller than in $CsDSO_4$, however.

The main feature is the large decrease of the ultrasonic velocity of these crystals in the superionic phase in which they become plastic. To our opinion such a decrease of velocity is caused by a strong acousto-ionic interaction which arises from the elastic modulation of the chemical potential [24]. In that case the step of velocity at T_c is given by

$$\Delta v = \frac{NA^2}{2\varrho vkT}, \tag{4}$$

where A is the acousto-ionic coupling constant, ϱ the mass density, N the number of mobile carriers per unit volume, k the Boltzmann constant, and v the ultrasonic velocity at $T < T_c$. The approximation (4) is valid for $\omega\tau \ll 1$ in the superionic phase (τ is the relaxation time for a mobile ion system). The peak of the attenuation near T_c is associated both with the increase of N and the change of the relaxation time ($\omega\tau > 1$ at $T < T_c$ and $\omega\tau < 1$ at $T > T_c$). In addition, the critical behaviour of the attenuation and velocity near T_c reveal themselves in a small dip of velocity and can be explained by the model [4] of the phase transition.

7. Conclusions

The dielectric permittivity and losses of $CsDSO_4$ and $CsDSeO_4$ single crystals in the superionic phase decrease with the increase of frequency. At microwaves ε' is of the same order as in other superionics [13, 17]. The contribution of $SO_4(SeO_4)$ and deuteron dynamics to ε' compensate each other so that it is independent of temperature. At T_c the permittivity shows an upward jump $\Delta\varepsilon \approx 2$. Frequency dependences of ε' and σ show a non-Debye relaxational dispersion. It contributes to ε' and ε'' and causes the frequency dependence of the microwave conductivity. Critical dynamics reveals itself near T_c in the dielectric and ultrasonic parameters.

References

[1] A. I. BARANOV, L. A. SHUVALOV, and N. M. SHCHAGINA, Zh. eksper. theor. Fiz., Pisma **36**, 381 (1982).
[2] A. I. BARANOV, L. A. SHUVALOV, and N. M. SHCHAGINA, Kristallografiya **29**, 1203 (1984).
[3] B. V. MERINOV, A. I. BARANOV, B. A. MAKSIMOV, and L. A. SHUVALOV, Kristallografiya **31**, 450 (1986).
[4] N. M. PLAKIDA, JINR-Rep. P17-84-760, Dubna 1984.
[5] A. M. BALAGUROV, A. I. BESKOROVNYI, I. D. DATT. L. A. SHUVALOV, and N. M. SHCHAGINA, JINR-Rep. P14-86-39, Dubna 1986.
[6] K. ITOCH, T. OZAKI, and E. NAKAMURA, Acta cryst. **B37**, 1908 (1981).
[7] R. BLINC, J. DOLINŠEK, G. LAHAJNAR, I. ZUPANČIČ, L. A. SHUVALOV, and A. I. BARANOV, phys. stat. sol. (b) **123**, K83 (1984).
[8] N. M. PLAKIDA, Zh. eksper. theor. Fiz., Pisma **41**, 95 (1985); JINR-Rep. P17-85-225, Dubna 1985.
[9] M. KOMUKAE, T. OSAKA, Y. MAKITA, T. OZAKI, K. ITOH, and E. NAKAMURA, J. Phys. Soc. Japan **50**, 3187 (1981).
[10] J. WOLAK and Z. CZAPLA, phys. stat. sol. (a) **67**, K171 (1981).
[11] S. YOKOTA, N. TAKANOHASHI, T. OSAKA, and Y. MAKITA, J. Phys. Soc. Japan **51**, 199 (1982).
[12] M. E. LINES, Phys. Rev. B **19**, 1189 (1979).
[13] H. ROEMER and G. LUTHER, Ferroelectrics **38**, 919 (1981).
[14] A. BRILINGAS, R. L. DAVIDOVICH, J. GRIGAS, S. LAPINSKAS, M. A. MEDKOV, V. SAMULIONIS, and V. SKRITSKII, phys. stat. sol. (a) **96**, 101 (1986).
[15] A. BRILINGAS, J. GRIGAS, and V. KALESINSKAS, Izv. Akad. Nauk SSSR, Ser. Fiz. **15**, 2196 (1987).
[16] W. URBONAVICHYUS, V. E. SCHNEIDER, J. GRIGAS, and R. L. DAVIDOVICH, Zh. eksper. teor. Fiz. **83**, 275 (1982).
[17] T. WONG, M. BRODWIN, and R. DUPON, Solid State Ionics **5**, 489 (1981).
[18] A. A. VOLKOV, G. V. KOZLOV, G. I. MIRZOYANTS, and J. PETZELT, Japan. J. appl. Phys. **24**, Suppl. 2, 531 (1985).
[19] R. R. LEVITSKY, I. R. ZACHEK, E. V. MITS, J. GRIGAS, and W. PAPROTNY, Ferroelectrics **67**, 109 (1986).
[20] L. PIETRONERO, S. STRÄSSLER, and H. R. ZELLER, Solid State Commun. **30**, 305 (1979).
[21] A. K. JONSCHER, phys. stat. sol. (b) **84**, 159 (1977).
[22] A. K. JONSCHER, Nature (London) **267**, 673 (1977).
[23] G. G. LUSHNIKOV, S. D. PROHOROVA, I. G. SINII, and G. A. SMOLENSKII, Fiz. tverd. Tela **29**, 496 (1987).
[24] J. B. BOYCE and B. A. HUBERMAN, Physics Rep. **51**, 189 (1979).

(Received May 25, 1988)

phys. stat. sol. (a) **110**, 437 (1988)

Subject classifications: 62.30 and 63.20; S11.1

Department of Physics, University of Windsor[1]) *(a)*
and Institute of Physics, Poznań Technical University[2]) *(b)*

Propagation of Plane Elastic Waves in $LiCsSO_4$ Single Crystals[3])

By

M. CZAJKOWSKI (a), M. DROZDOWSKI (b), and M. KOZIELSKI (b)

Using the values of elastic constants the phase velocities and energy flux vectors associated with propagating elastic waves in XY, YZ, and XZ planes in $LiCsSO_4$ single crystals are calculated. It is also pointed out, that for many directions in these planes the normals to the wavefronts of elastic waves are not collinear with the vectors representing either the energy flux or the particle displacement.

An Hand bekannter Werte der elastischen Konstanten von $LiCsSO_4$-Einkristallen werden die Phasengeschwindigkeiten und die Energieflußvektoren der elastischen Wellen in der XY-, YZ- und XZ-Ebene bestimmt. Diese Energie begleitet die sich in diesen Ebenen fortpflanzenden elastischen Wellen. Es wird gezeigt, daß die zur Wellenfront senkrechte Richtung sich häufig nicht mit der Ausbreitungsrichtung der elastischen Energie deckt, die durch den Vektor des Energieflusses dargestellt wird.

1. Introduction

In anisotropic single crystals the normals to the wavefronts of elastic waves very often are not collinear with the vectors representing either the energy flux or the particle displacement [1]. This problem can play a dominant role both in the description of the elastic properties of the crystals and in the interpretation of high-frequency pulsed ultrasound experiments [2].

In this paper the phase velocities and energy flux vectors associated with longitudinal (L) and two transverse (T_1 and T_2) elastic waves propagating in XY, YZ, and XZ planes in $LiCsSO_4$ single crystals have been calculated.

The $LiCsSO_4$ single crystals at room temperature (RT) belong to an orthorhombic space group Pcmn(D_{2h}^{16}) whose unit cell contains four formula units [3]. They have nine independent elastic constants [4]. Their values have been measured using different experimental methods [5 to 9]. The values of elastic constants of $LiCsSO_4$ determined at RT by the Brillouin method [5], which are used in subsequent calculations are listed in Table 1.

Table 1
Elastic constants (in 10^{10} N m^{-2}) of $LiCsSO_4$ single crystals at RT

C_{11}	C_{22}	C_{33}	C_{44}	C_{55}	C_{66}	C_{12}	C_{13}	C_{23}
5.64	6.20	3.51	1.09	1.56	1.67	3.21	1.36	2.70

[1]) Windsor, Ontario N9B3P4, Canada.

[2]) Piotrowo 3, PL-60-965 Poznań, Poland.

[3]) This work was supported in part by the Natural Sciences and Engineering Research Council of Canada and by Polish Research Project CPBP 01.12.8.15.

2. Theory

The method proposed by Cummins and Schoen [10] has been used to evaluate phase velocities of the elastic waves. If the elastic constants C_{ijkl} of the crystal are known, the phase velocities v of the three elastic waves propagating in the direction $\hat{q}$ (described by an angle α) can be carried out directly from

$$|C_{ijkl}l_jl_k - \varrho v^2\delta_{il}| = 0\,, \tag{1}$$

where l_j, l_k are direction cosines of $\hat{q}$, and ϱ is the density of the crystal.

The components of the energy flux vectors associated with each of the three elastic waves propagating in the crystal can be calculated from

$$S_i = \frac{A^2\omega^2}{2v} C_{ijkl}u_j(l_ku_l + l_lu_k)\,, \tag{2}$$

where A is the constant amplitude, ω the phase frequency, and u_j are direction cosines of the particle displacement (described by angle γ) associated with each of the three velocities v [11]. In the following section the elastic wave for which the displacement vector is more or less in the direction of the normal to the wavefront is referred to simply as a longitudinal wave. The transverse elastic wave with the displacement vector perpendicular to the propagating plane and the other one with its displacement vector lying in the plane is referred to as T_2 and T_1 transverse wave, respectively.

3. Results

Cross sections of the phase velocities for the longitudinal and two transverse elastic waves propagating in XY, YZ, and XZ planes for $LiCsSO_4$ crystal are shown in Fig. 1. As can be seen from Fig. 1 the X, Y, and Z axes are not degenerated for transverse waves as for these directions both velocities are different [12]. However, degenerated axes for transverse waves occur in the XZ plane at $\alpha = 30°$, 150°, 210°, and 330°, and in the XY plane (within calculation accuracy) in the directions with $\alpha = 25°$ to 40°, 140° to 155°, 205° to 220°, 320° to 335°.

The components of the energy flux vectors in XY, YZ, and XZ planes for corresponding elastic waves are shown in Fig. 2. Here, it can be seen that in XY and XZ planes for some α ranges the components of energy flux vectors, for both T_1 and T_2 transverse elastic waves are either equal or almost equal.

As noted previously, in anisotropic single crystals the normals to the wavefronts of the elastic waves very often are not collinear with the vectors representing the energy flux. In order to explain this feature we have evaluated the deviation $(\alpha - \beta)$ between the normal to the wavefront and the direction of the energy flux vector described by the angles α and β, respectively. The values of the deviation $(\alpha - \beta)$ versus the angle α for longitudinal and two transverse elastic waves propagating in XY, YZ, and XZ planes are shown in Fig. 3. Because of the crystal symmetry we have restrained the calculations to the α range between 0° and 90°. As can be seen from Fig. 3 for directions along X, Y, and Z axes the deviations equal zero for longitudinal and both transverse elastic waves. Moreover, for longitudinal elastic waves in the XY plane the deviation is equal to zero for the direction $\alpha = 58.2°$. Also for T_1 transverse elastic waves in each investigated plane there exists one direction at which the deviation equals zero. These directions are as follows: $\alpha = 44.1°$ in the XY plane, $\alpha = 50.1°$ in the YZ plane and $\alpha = 50.2°$ in the XZ plane.

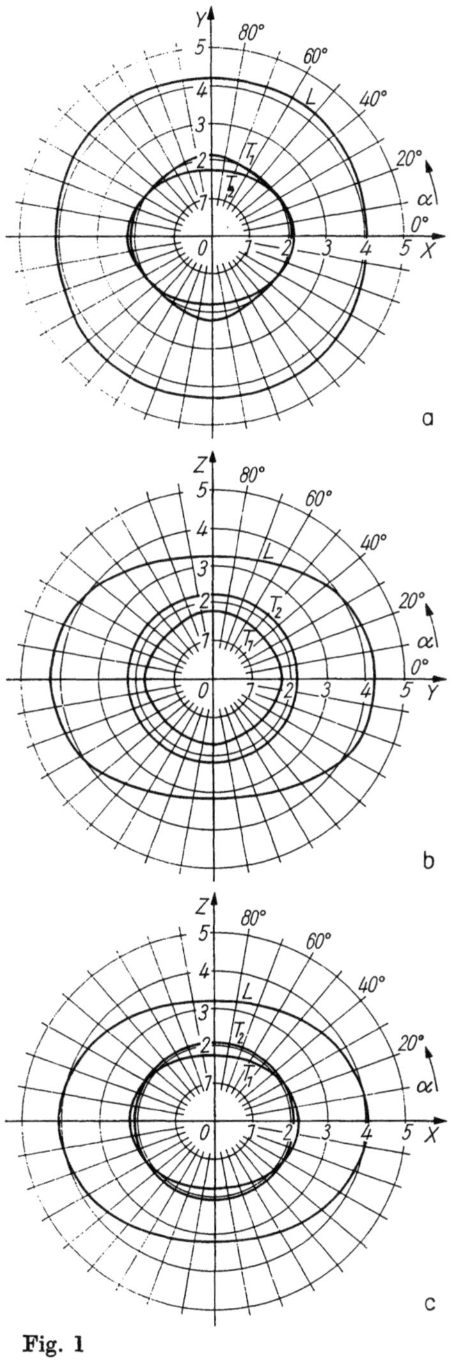

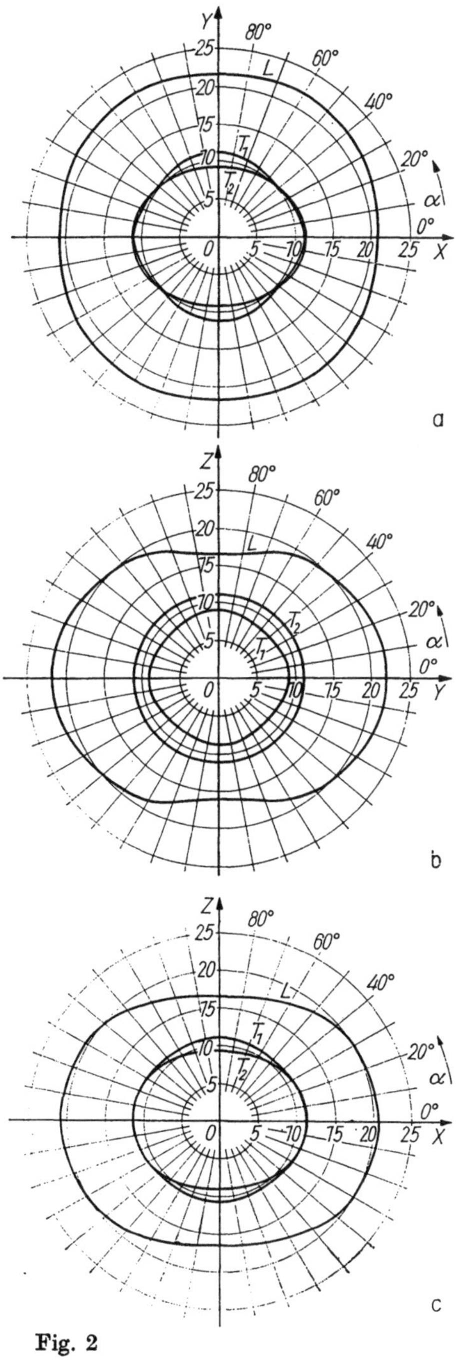

Fig. 1 Fig. 2

Fig. 1. Cross sections of the phase velocities v (km/s) for longitudinal (L) and transverse (T_1 and T_2) elastic waves in $LiCsSO_4$ single crystals. a) XY, b) YZ, c) XZ plane

Fig. 2. Cross sections of the energy-flux vectors S/A^2 (10^{37} kg m^{-2} s^{-3}) for longitudinal (L) and transverse (T_1 and T_2) elastic waves in $LiCsSO_4$ single crystals. a) XY, b) YZ, c) XZ plane

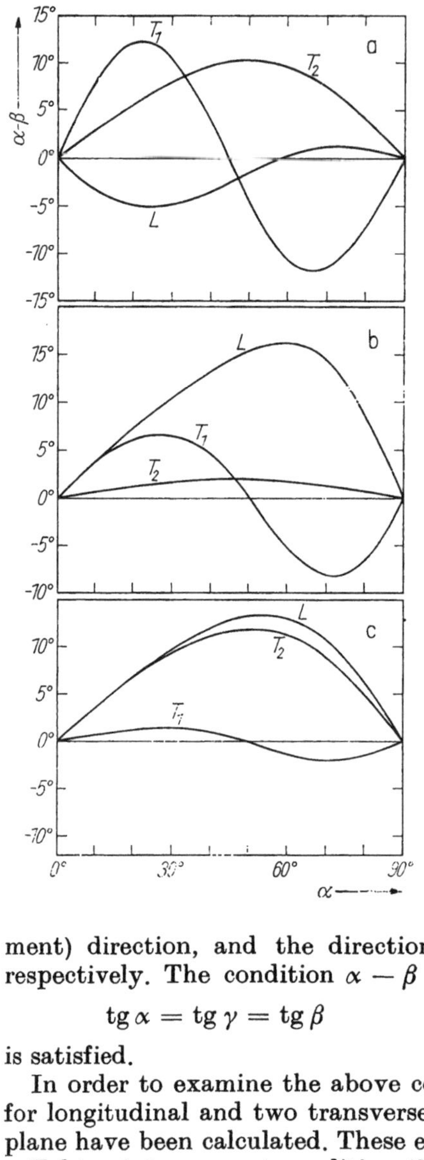

Fig. 3. Deviation angle (α–β) for longitudinal (L) and transverse (T_1 and T_2) elastic waves propagating in a) XY, b) YZ, and c) XZ planes in $LiCsSO_4$ single crystals

4. Discussion

At room temperature $LiCsSO_4$ belongs to a class of pseudohexagonal crystals (mmm orthorhombic point group). Crystals belonging to this class possess three twofold axes of symmetry which are taken as the X, Y, and Z axes [4]. They are pure-mode axes for longitudinal and both transverse elastic waves, which means that the appropriate l_i, u_i, and S_i are collinear for these directions and consequently the deviations $(\alpha - \beta)$ equal zero [12].

Let for a given elastic wave tg α, tg γ, and tg β describe the direction of the normal to the wavefront, polarization (particle displacement) direction, and the direction of the component of the energy flux vector, respectively. The condition $\alpha - \beta = 0$ requires that

$$\operatorname{tg} \alpha = \operatorname{tg} \gamma = \operatorname{tg} \beta \tag{3}$$

is satisfied.

In order to examine the above condition the expressions for tg α, tg 2γ, and tg β for longitudinal and two transverse elastic waves propagating in XY, YZ, and XZ plane have been calculated. These expressions are given in Table 2.

Taking into account condition (3), the deviation $(\alpha - \beta)$ is equal to zero for the longitudinal wave, if

$$\operatorname{tg} \alpha = \sqrt{\frac{C_{12} + 2C_{66} - C_{11}}{C_{12} + 2C_{66} - C_{22}}}, \tag{4a}$$

$$\operatorname{tg} \alpha = \sqrt{\frac{C_{23} + 2C_{44} - C_{22}}{C_{23} + 2C_{44} - C_{33}}}, \tag{4b}$$

$$\operatorname{tg} \alpha = \sqrt{\frac{C_{13} + 2C_{55} - C_{11}}{C_{13} + 2C_{55} - C_{33}}}, \tag{4c}$$

Table 2

The expressions for tg α, tg 2γ, and tg β for longitudinal and two transverse elastic waves propagating in XY, YZ, and XZ planes

	L, T_1	T_2
	XY plane	
tg α	l_2/l_1	l_2/l_1
tg 2γ	$\dfrac{2l_1l_2(C_{12}+C_{66})}{l_1^2(C_{11}-C_{66})+l_2^2(C_{66}-C_{22})}$	$u_1=u_2=0$ $u_3=1$
tg β	$\dfrac{C_{12}l_1u_1u_2+C_{22}l_2u_2^2+C_{66}u_1(l_1u_2+l_2u_1)}{C_{12}l_2u_1u_2+C_{11}l_1u_1^2+C_{66}u_2(l_1u_2+l_2u_1)}$	$\dfrac{C_{44}}{C_{55}}\dfrac{l_2}{l_1}$
	YZ plane	
tg α	l_3/l_2	l_3/l_2
tg 2γ	$\dfrac{2l_2l_3(C_{23}+C_{44})}{l_2^2(C_{22}-C_{44})+l_3^2(C_{44}-C_{33})}$	$u_2=u_3=0$ $u_1=0$
tg β	$\dfrac{C_{23}l_2u_2u_3+C_{33}l_3u_3^2+C_{44}u_2(l_2u_3+l_3u_2)}{C_{23}l_3u_2u_3+C_{22}l_2u_2^2+C_{44}u_3(l_2u_3+l_3u_2)}$	$\dfrac{C_{55}}{C_{66}}\dfrac{l_3}{l_2}$
	XZ plane	
tg α	l_3/l_1	l_3/l_1
tg 2γ	$\dfrac{2l_1l_3(C_{13}+C_{55})}{l_1^2(C_{11}-C_{55})+l_3^2(C_{55}-C_{33})}$	$u_1=u_3=0$ $u_2=1$
tg β	$\dfrac{C_{13}l_1u_1u_3+C_{33}l_3u_3^2+C_{55}u_1(l_1u_3+l_3u_1)}{C_{13}l_3u_1u_3+C_{11}l_1u_1^2+C_{55}u_3(l_1u_3+l_3u_1)}$	$\dfrac{C_{44}}{C_{66}}\dfrac{l_3}{l_1}$

for XY, YZ, and XZ planes, respectively. Using the C_{ij} values listed in Table 1 we obtained for longitudinal waves propagating in the XY plane $\alpha = \pm 58.2°$, that confirms pseudohexagonal structure of $LiCsSO_4$ crystals. However, there is no real solution of (4b) and (4c) for the given elastic constant values. It means that condition (3) cannot be satisfied and in consequence there exists no direction (but X, Y, and Z axes) in these two planes for which pure-mode axes for longitudinal waves occur.

A detailed analysis of the direction of the displacements for T_1 and T_2 transverse waves propagating in XY, YZ, and XZ planes has been carried out. From this analysis we have pointed out that for the T_1 elastic wave (with its displacement vector lying in the propagating plane) the energy flux vector is collinear with the wave normal, if

$$\text{tg}\,\gamma = -\,\text{tg}\,\alpha\,. \tag{5}$$

Taking into account condition (5) the deviation $(\alpha - \beta)$ is equal to zero for the T_1 elastic wave, if

$$\operatorname{tg} \alpha = \sqrt{\frac{C_{12} + C_{11}}{C_{12} + C_{22}}}, \tag{6a}$$

$$\operatorname{tg} \alpha = \sqrt{\frac{C_{23} + C_{22}}{C_{23} + C_{33}}}, \tag{6b}$$

$$\operatorname{tg} \alpha = \sqrt{\frac{C_{13} + C_{11}}{C_{13} + C_{33}}}, \tag{6c}$$

for XY, YZ, and XZ planes, respectively. Again, substituting the values of elastic constants for $LiCsSO_4$ crystals given in Table 1 we obtained for the XY plane $\alpha = \pm 44.1°$, for the YZ plane $\alpha = \pm 50.1°$, and for the XZ plane $\alpha = \pm 50.2°$. The values of α calculated from (6a) to (6c) are in good agreement with those given in Fig. 3. However, these directions are not pure-mode axes for T_1 elastic waves.

A similar analysis has been performed for the T_2 elastic wave (with its displacement vector perpendicular to the propagating plane) in the XY, YZ, and XZ planes. As can be seen from Fig. 3, there is no direction (excluding X, Y, and Z axes), for which the deviation $(\alpha - \beta)$ is equal to zero. According to the theory, the polarization of the transverse elastic wave (T_2) for the crystals belonging to the mmm orthorhombic point group is perpendicular to the XY, YZ, and XZ planes, which are planes of mirror symmetry [12].

Acknowledgement

The authors are indebted to Prof. T. Krajewski from A. Mickiewicz University in Poznań for helpful discussions.

References

[1] G. W. Farnell, Canad. J. Phys. **39**, 65 (1961).
[2] P. C. Waterman, Phys. Rev. **113**, 1240 (1959).
[3] K. S. Aleksandrov, L. I. Zherebtsova, I. M. Skornev, A. I. Kruglik, O. V. Rozanov, and I. N. Flerov, Soviet Phys. — Solid State **22**, 2150 (1980).
[4] J. F. Nye, Physical Properties of Crystals, Clarendon Press, Oxford, 1957.
[5] M. Drozdowski and F. Holuj, Ferroelectrics **76**, 113 (1987).
[6] K. S. Aleksandrov, M. P. Zaitseva, L. A. Shabanova, and O. V. Shimanskaya, Soviet Phys. — Solid State **27**, 1426 (1981).
[7] A. T. Anistratov, A. V. Zamkov, L. A. Kot, I. N. Stolovitskaya, and L. A. Shabanova, Ferroelectrics **48**, 103 (1983).
[8] G. Pakulski, B. Mróz, and T. Krajewski, Ferroelectrics **48**, 259 (1983).
[9] B. Mróz, H. Kiefte, M. J. Clouter, and J. A. Tuszyński, Phys. Rev. B **36**, 3745 (1987).
[10] H. Z. Cummins and P. E. Schoen, in: Laser Handbook, Ed. F. T. Arecchi and E. D. Schultz-Du Bois, North-Holland Publ. Co., 1972.
[11] A. I. Kruglik, M. A. Simonov, E. P. Zhelezin, and N. V. Belov, Dokl. Akad. Nauk SSSR **247**, 1384 (1979).
[12] E. Dieulesaint and D. Royer, Ondes Elastiques dans les Solides, Masson et Cie, Paris 1974.

(Received July 25, 1988)

phys. stat. sol. (a) **110**, 443 (1988)

Subject classification: 64.75; 71.25; S1.3

Department of Materials Science and Engineering, Tsinghua University (a) and Center of Condensed Matter and Radiation Physics, CCAST (World Laboratory) (b), Beijing[1])

The Band Structure and the Anomalous Alloying Behavior of Noble Metal Based Systems[2])

By

L. J. HUANG (a), E. MA (a), and B. X. LIU (a, b)

Alloying behavior is investigated by considering the change of the original electronic structures of the constituent metals and the interactions initiated in between. A co-resonance model is proposed to depict the physical picture of alloy phase formation. It is shown that the model can explain the intrigue alloying behaviors of noble–transition metal systems.

Les comportements alliants sont examiné par la considération sur le changement des structures électroniques originals des méteaux constituants et sur les interactions entre les atoms des méteaux constituants. Un modèle de la co-résonance est proposé pour peindre la peindre physique de la formation des phases alliants. Il est montré que le modèle sait expliquer les comportements alliants intrigants des systèmes des noble–transition méteaux.

1. Introduction

It is a well known phenomenon that the alloying behavior of silver is considerably different from that of copper and gold, especially when alloyed with transition metals such as iron, cobalt, and nickel. This can be readily found in the documented equilibrium phase diagrams. In these systems, silver can only form limited solid solutions and is sometimes even completely insoluble in the liquid state. On the contrary, quite wide solid solution regions as well as liquid intersolubility can be found in the copper- and gold-based systems [1, 2]. The above intricate alloying behavior of silver has long been concerned by the metallurgists as well as solid state physicists, but yet no proposed criterion has so far provided a satisfactory answer. For example, in the Darken-Gurry diagram, which is constructed by electronegativity versus atomic radius to explain the alloy behavior of binary systems, it was found for the silver-based alloy systems that the location of silver was even not in the center of the solubility ellipse [3]. This result indicates the failure of applying electronegativity for the silver-based systems [4]. Most recently Gschneidner [5] proposed rules to predict whether or not two elements will form an extensive terminal solid solution. However, the total reliability to predict the alloying behavior of silver is low. The lattice rigidity [6] argument applies well for the anomaly of Ag–Cu and Au–Cu alloying behaviors when silver and gold act as solvents, but the reverse makes it a dilemma. In the seventies, a semi-empirical model was proposed by Miedema and co-workers [7] to calculate heats of formation and heats of solution for binary systems. This model has been successfully applied to many topics of alloy theory. However, despite its consideration of some electronic parameters of the constituent metals, it is a "macroscopic atom" model

[1]) Beijing, People's Republic of China.

[2]) Project supported by the National Natural Science Foundation of China.

which is established in the normal space and also fails to interpret the alloying behavior of silver-based alloy systems [8].

The above-mentioned issue is the motivation of the present study. The authors attempt in this study to consider the alloying behavior from the original electronic structures of the elemental metals involved and consider the alloy phase formation through the process of the corresponding new electronic structure being formed. In addition, this study is also triggered by the evidence that a great number of conventionally inaccessible alloy phases have been formed with some newly developed methods. A typical example is the Ag–Ni system prepared by ion beam mixing. Silver and nickel are almost immiscible in the solid as well as in the liquid states, but ion beam mixing has succeeded in forming metastable phases, indicating that the high energy excitation process can bring silver and nickel atoms together and form some new alloy phases [9]. This progress also requires reasonable explanations from the basic physics underlying the alloy phase formation.

2 Model and Discussion

When a new alloy phase is to form by two constituent metals which have their own stable electronic structures, the original structures should be first changed severely and the initiated interactions between the different atoms should also provide a favorable condition for the new structure. As for the transition metals, the well established tight binding method shows that their outmost d-electrons play a predominant role in determining the properties of the solids [10]. The alloying behaviors of the binary transition metals may therefore depend mainly on their d-electrons. For transition metals, following the atomic sphere approximation [11], it is rigorously correct to describe the d-electrons as free between the spheres and to consider the effect of the spheres as being that of scattering the electrons. According to the scattering theory, energy levels of transition metals with and without scattering match far from resonance, but are shifted near the resonance state [12]. Furthermore, the total cross section, which stands for the probability of the interaction, also approaches a maximum at resonance, i.e. the electronic structures of the metals will change considerably during the resonance process. Considering the case when a solute metal with a foreign electronic structure is added to its solvent and forms a new alloy structure, the necessary condition is that both the solute and the solvent metal change their individual electronic structures and adjust into a tolerable one which should be thermodynamically stable for the newly formed phase. We thus propose that the formation of an alloy phase be realized through a coresonance process, i.e. when it happens, atoms of both constituent metals could be initiated to or approach their respective resonance states and the individual scattering cross section thus tends to a maximum at approximately the same time, resulting in severe changes of the original electronic structures. Consequently, the new electronic structure is to form, if the interactions between atoms initiated during the co-resonance process also provide a favorable condition for its formation. The interactions in the scattering region are complex and the resonance energies, E_d, and the corresponding width of resonances, Γ, are the characteristic parameters of the whole complex structure [13]. The resonance energy, E_d, can be calculated empirically by combining the energy levels of the band structure of the metal with respect to the crystalline symmetry,

$$E_d = (\tfrac{2}{5}\,\Gamma_{12} + \tfrac{3}{5}\,\Gamma'_{25}) - \Gamma_1 \tag{1}$$

and the width of resonance, Γ, can be obtained by the following equation:

$$\Gamma = 0.1214\hbar^2 k_d^5 r_d^3/m\,, \tag{2}$$

where k_d is the wave number, r_d a length characterizing the transition metals, and m the electron mass. These two parameters, E_d and Γ, can be readily found in the literature [12].

Following the above ideas, the probability of initiation of co-resonance would mainly depend on ΔE_d, the difference of the resonance energies between the constituent metals. The smaller ΔE_d is, the greater is the probability for co-resonance to occur. The other characteristic parameter governing the resonance process is the resonance lifetime τ, which is defined as $\tau = \hbar/\Gamma$. By simply combining linearly the resonance lifetime $\bar{\tau} = (\tau_1 + \tau_2)/2$, a map of ΔE_d versus $\bar{\tau}$ can be constructed. Fig. 1 shows the map and the three curves in the map are for the noble metals silver, gold, and copper combining with 3d transition metals, respectively. In order to be more explicit, ln (ΔE_d) is used instead of ΔE_d in constructing the map. It is clearly shown that these curves are very much alike in shape and each curve is divided into two branches having positive and negative differentials, respectively. By least square fitting, the left branches of these curves can be expressed by

$$\Delta E_d \approx A_k \exp\left(\frac{4ke_0\bar{\tau}}{\hbar}\right), \tag{3}$$

where $k = 1, 2, 3$ for Ag, Au, and Cu, respectively. While the right branches of these curves can be expressed by

$$\Delta E_d \approx B_k \exp\left(\frac{-\pi ke_0\bar{\tau}}{4\hbar}\right), \tag{4}$$

where $k = 1, 2, 3$ for Ag, Au, and Cu, respectively, and $\hbar$ is Planck's constant and e_0 the electronic charge. All the data used in the above calculation were taken from [12] and are listed in Table 1. For comparison, Miedema's heats of formation [14] are also listed in Table 1. The constants A_k and B_k obtained by least square fitting are listed in Table 2.

As we suggested above the alloying process is realized through a co-resonance process, which should include the initiation of co-resonance and the interactions between the constituent metal atoms. It could be seen that the differential of the curve is a quantity with the dimension of power, we therefore propose to take the differential

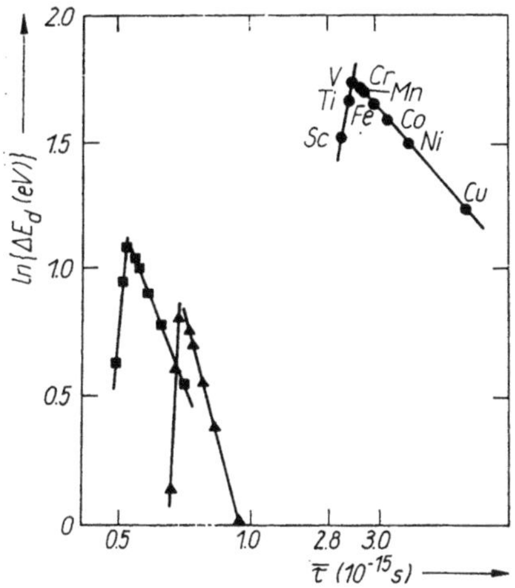

Fig. 1. ln ΔE_d vs. $\bar{\tau}$ for the noble–transition metal systems (3d period). Ag-based systems (●), Au-based systems (■), and Cu-based systems (▲)

Table 1

Heat of formation*) and combined electronic structure parameters of noble–3d transition metals

system (A–B)	ΔE_d (eV)	$\bar{\tau}$ (10^{-15} s)	Miedema's model H (kJ/g-atom)
Ag–Sc	4.56	2.86	− 50
Ag–Ti	5.27	2.87	− 21
Ag–V	5.64	2.89	+ 19
Ag–Cr	5.52	2.93	+ 36
Ag–Mn	5.42	2.94	+ 15
Ag–Fe	5.15	2.97	+ 39
Ag–Co	4.87	3.02	+ 26
Ag–Ni	4.42	3.11	+ 22
Au–Sc	1.87	0.49	−111
Au–Ti	2.58	0.51	− 84
Au–V	2.95	0.53	− 29
Au–Cr	2.83	0.56	0
Au–Mn	2.73	0.57	− 17
Au–Fe	2.46	0.61	+ 12
Au–Co	2.18	0.66	+ 11
Au–Ni	1.73	0.75	+ 11
Cu–Sc	1.15	0.70	− 43
Cu–Ti	1.86	0.72	− 27
Cu–V	2.23	0.74	+ 6
Cu–Cr	2.11	0.77	+ 20
Cu–Mn	2.01	0.79	+ 5
Cu–Fe	1.74	0.82	+ 22
Cu–Co	1.46	0.87	+ 13
Cu–Ni	1.01	0.95	+ 10

*) The data are from [14].

as the parameter presenting the positive or negative interactions for forming a new phase under the condition that the co-resonance is inititated. This is reasonable, as the four resonance characteristic parameters of the constituent metals are all involved, and the importance of this differential that divides the curves into two branches is shown clearly in Fig. 1.

Under this assumption, the alloying behavior of the noble metal-3d transition metals alloys and the intricate silver problem can then get an explanation. From Fig. 1, one sees that ΔE_d are small for both Au-based and Cu-based systems (also cf. Table 1). The co-resonance is therefore easy to be initiated and the mutual solubility can thus be large. Nevertheless, for the right branches of these curves, the differentials are both

Table 2

The constants A_k and B_k obtained by least square fitting

element	k	A_k (eV)	B_k (eV)
Ag	1	7.18×10^{-8}	1.71×10^{2}
Au	2	3.29×10^{-3}	1.22×10^{1}
Cu	3	2.14×10^{-6}	4.40×10^{1}

negative, which is assumed to be unfavorable for the alloying process, the solubilities should be less than those lying in the left branches. By examining the corresponding equilibrium phase diagrams, this is indeed the case. However, for the silver-based systems, the curve is clearly separated from the other two, because the ΔE_d for these systems are much greater. This means that it is difficult to initiate the co-resonance and is therefore unfavorable for alloying. In the meantime, for the systems located on the left branch of the silver-based systems, their differentials are positive and hence these systems can still be partially intersoluble, like the Ag–Sc, Ag–Ti, and Ag–V systems, while for the systems on the right branch with the negative differentials, alloying should be very difficult. In fact, most of the systems on the right branch are insoluble, some are completely immiscible even in the liquid state, e.g. Ag–Fe and Ag–Ni systems. It turns out that the distinctive alloying behavior of silver with the transition metals can be interpreted by the large differences of ΔE_d, which are almost two times bigger than those of gold and copper, together with the negative differentials.

It should be pointed out that some similarities exist between our approach and Miedema's heat of formation model. For the binary systems of noble and transition metals, the heats of formation change from negative to positive with increasing atomic numbers in one period (Table 1) and the change of the sign of the heats of formation always appears at the maximum value of ΔE_d for the individual silver-, gold-, and copper-based systems. This also happens in correspondence with the change of the differentials of each curve from positive to negative and the turning point is at the top of the curve or nearby. However, according to Miedema's model, the heats of formation of Au–Ni, Cu–Ni, and Ag–Ni systems are approximately the same. But the actual alloying behaviors of these three systems are quite different. This failure of Miedema's model could be understood by following the proposed co-resonance model. The drop of ΔE_d is in coincidence with the alloying behaviors of these three systems from almost insoluble to completely intersoluble. One could thus see that the possible mechanism which the silver-based alloys follow, is specialized. Moreover, the data presented in Table 1 could also tell something of the subtile discrepancies of the alloying behavior in one series, e.g. Miedema's heats of formation give the same value for Au–Co and Au–Ni systems and approximately the same for the Au–Fe system, but the solubilities of these three systems increase in the sequence of iron, cobalt, and nickel, which is exactly reflected by the decreasing sequence of ΔE_d. The different alloying behaviors of Ag–Ni and Fe–Cu systems with the same value of the heat of formation (22 kJ/g-atom) could be told from the different data of ΔE_d in Table 1 as well. Further studies following this approach indicate that this model could also be applied for the binary systems of noble and transition metals in the other two periods.

With the viewpoint of the co-resonance, if a process involving high energy deposition rate and varying energy distribution instead of the conventional steady-state method is applied, some new alloy phases should be produced. Some recent reports have already presented the fact. The extended solubilities of silver-based alloys have been obtained by the scheme of ion beam mixing, e.g. Ag–Ni system [9]. By employing this method, the authors also succeeded in forming a metastable Ag–Cr alloy [15] and extended the solubilities of Fe–Cu alloys [16]. These results provided additional evidence supporting our above viewpoints.

3. Concluding Remarks

In conclusion, by considering the change of the original electronic structures of the constituent metals, and the interactions between them, the alloying behavior can be explained, especially when other methods are inconvenient to apply. If the alloy phase formation is conceived as a co-resonance process, the large ΔE_d and hence the difficulty of initiating the co-resonance in the silver-based systems may be the reason for the insolubility of these systems.

Acknowledgement

The partial financial aid from the International Atomic Energy Agency is gratefully acknowledged.

References

[1] C. J. Smithells (Ed.), Metal Reference Book, Butterworths, New York 1976.
[2] L. H. Bennett, T. B. Massalski, and B. C. Giessen (Ed.), Alloy Phase Diagrams, North-Holland Publ. Co., New York 1983.
[3] D. Feng, Y. N. Wang, and D. R. Qiu, Metal Physics, Science Press, Beijing 1964.
[4] W. Hume-Rothery, J. less-common Metals **4**, 390 (1962).
[5] K. A. Gschneidner, Jr., in: Theory of Alloy Phase Formation, Ed. L. H. Bennett, AIME, Warrendale 1980 (p. 1).
[6] K. A. Gschneidner, O. D. McMasters, D. G. Alexander, and R. F. Venteicher, Metallurg. Trans. **1**, 1961 (1970).
[7] A. R. Miedema, R. F. de Boer, and R. Boom, CALPHAD **4**, 341 (1977).
[8] A. R. Williams, C. D. Gelatt Jr., and V. L. Moruzzi, Phys. Rev. Letters **44**, 429 (1980).
[9] B. Y. Tsaur and J. Mayer, Appl. Phys. Letters **37**, 389 (1980).
[10] D. G. Pettifor, Phys. Rev. Letters **42**, 846 (1979).
[11] O. K. Anderson, Solid State Commun. **13**, 501 (1973).
[12] W. A. Harrison, Electronic Structure and Properties of Solids, W. H. Freeman, San Francisco 1980.
[13] P. Roman, Advanced Quantum Theory, Addison-Wesley, Massachusetts 1965.
[14] A. R. Miedema, Philips tech. Rev. **36-8**, 217 (1976).
[15] L. J. Huang, E. Ma, and B. X. Liu, Chin. Phys. Letters **3**, 197 (1986).
[16] L. J. Huang and B. X. Liu, Nuclear Instrum. and Methods **B18**, 256 (1987).

(Received March 14, 1988; in revised form September 5, 1988)

phys. stat. sol. (a) **110**, **449** (1188)

Subject classification: 62.20; S1.2

Lehrstuhl I, Institut für Werkstoffwissenschaften der Universität Erlangen-Nürnberg[1])

Comparison between Experimental Determination and Calculation of Elastic Properties of Nickel-Base Superalloys between 25 and 1200 °C

By

H.-A. KUHN and H. G. SOCKEL

The elastic coefficients S_{ij} of the single crystalline nickel base superalloy MAR-M002 mod. are determined as a function of temperature by the resonance method of Förster. The orientation — independent S_{ij} are evaluated by the linear regression method. The application of averaging concepts allows the calculation of the elastic properties of the quasi-isotropic polycrystalline superalloys MAR-M247 and MAR-M200 at elevated service temperatures. The computed polycrystalline moduli of elasticity are compared with measured values.

Die elastischen Nachgiebigkeiten S_{ij} der einkristallinen Nickel-Basis-Superlegierung MAR-M002 mod. werden in Abhängigkeit von der Temperatur mit der dynamischen Resonanzmethode nach Förster bestimmt. Mittels linearer Regression werden die drei unabhängigen Einkristallkonstanten berechnet. Mit Grenzwertkonzepten werden die elastischen Eigenschaften quasiisotroper Vielkristalle MAR-M247 und MAR-M200 im Einsatzbereich hoher Temperaturen ermittelt und mit gemessenen Werten verglichen.

1. Introduction

For the design of turbine disks and blades the knowledge of elastic properties such as Young's modulus E, shear modulus G, and Poisson's ratio ν at elevated temperatures is required. Commercially employed superalloys are available in polycrystalline, directionally solidified, and single-crystalline modifications. The latter two exhibit improved high-temperature creep resistance and fatigue behaviour. In addition, they differ from conventionally cast and wrought alloys in the anisotropy of elasticity. Due to the cubic crystal system, the complex elastic response of a single crystalline turbine blade is described by three elastic coefficients S_{11}, S_{12}, and S_{44}. In the case of polycrystalline material the designing engineer has to take into account the dependence of E, G, and ν on microstructure and anisotropy.

Using the single-crystal elastic coefficients S_{ij} or constants C_{ij} in suitable structure-correlated boundary concepts the influence of grain orientation and distribution on the polycrystalline moduli can be calculated.

If S_{ij} or C_{ij} are known, the time-consuming measurements of E and G become avoidable. The aim of this work is to apply boundary and averaging methods of Voigt [1], Reuss [2], Hill [3], Kröner and Gairola [4], and Hütt [5] to the single-crystal superalloy MAR-M002 mod. and the polycrystalline alloys MAR-M247 and MAR-M200 of similar composition.

The treated superalloys consist of a f.c.c. solid solution strengthened matrix and of approximately 60 Vol% precipitated γ'-phase with an ordered f.c.c.-type structure.

---

[1]) Martensstr. 5, D-8520 Erlangen, FRG.

Table 1
Alloy compositions (in wt%) (balance: nickel)

	C	Cr	Co	Ti	Al	W	Zr	B	Fe
MAR-M247 polycr.	0.15	8.8	10.0	1.2	5.5	10.0	0.06	0.015	<0.5
MAR-M002 mod. single cryst.	0.15	9.0	10.0	1.2	5.5	10.0	0.00	0.000	<1.0
MAR-M200 polycr.	0.14	8.8	10.2	2.2	5.1	12.8	0.05	0.020	0.2

	Mo	Ta	Hf	Nb	Cu
MAR-M247 polycr.	0.6	3.1	1.6	0.00	0.0
MAR-M002 mod. single cryst.	0.0	2.5	1.5	0.00	0.0
MAR-M200 polycr.	0.0	0.0	0.0	0.97	0.1

The alloy compositions of the single-crystalline material MAR-M002 mod. and the polycrystalline modifications of MAR-M247 and MAR-M200 are listed in Table 1.

2. Fundamental Considerations. Textureless Aggregates

The different models [1 to 5] describing the influence of the grain arrangement lead to various upper and lower bounds of elastic properties of a random aggregate of crystals. With increasing information about the microstructural state the range between the upper and the lower limits for a certain model becomes smaller.

Neither Voigt [1] nor Reuss [2] considered a statistic orientation distribution of crystals, grain interaction, and surface strain in their averaging of the elastic stiffnesses C_{ij} and the elastic compliances S_{ij}. In analogy to the Kirchhoff-rules for the electrical resistivity the lower Reuss bound corresponds to the parallel connection of grains with minimal and maximal E, whereas the upper bound of Voigt agrees with a serial connection of the same grains,

$$E^{-1}_{\text{Reuss}} = 0.2(3S_{11} + 2S_{12} + S_{44})\,, \tag{1a}$$

$$E_{\text{Voigt}} = \frac{(C_{11} + C_{12}) + (C_{11} - C_{12} - 0.6C)}{(C_{11} + C_{12} - 0.2C)}\,, \tag{1b}$$

$$C = (C_{11} - C_{12} - 2C_{44})\,, \tag{1c}$$

where S_{ij} are elastic complicances of the crystallographic ⟨001⟩-orientation, C_{ij} elastic constants of the crystallographic ⟨001⟩-orientation.

The Young's modulus in an arbitrary direction defined by the polar angle θ and the azimuthal angle φ between the specimen axes and the ⟨001⟩-orientation of the crystal is given by

$$E = \frac{1}{S'_{11}} = \frac{1}{S_{11} - 2SJ} \tag{2a}$$

with

$$S = S_{11} - S_{12} - 0.5S_{44} \tag{2b}$$

and

$$J = \sin^2\theta\cos^2\theta + 0.125\sin^4\theta\,(1 - \cos 4\varphi)\,. \tag{2c}$$

The shear modulus is given by

$$G = \frac{1}{S'_{44}} = \frac{1}{S_{44} + 4SJ} . \tag{2d}$$

Poisson's ratio ν is a function of S_{11} and S_{12},

$$\nu = \frac{-(S_{12} + SJ)}{S_{11} - 2SJ} . \tag{2e}$$

Hill [3] was more successful in approximating real values by developing the arithmetic mean of Reuss and Voigt averages. In addition to this treatment Hütt [5] gave a consistent iteration procedure for constructing mean values by combining Hill bounds (VRH) and parallel connections of the same order. Upper and lower bounds of the property M usually converge after the fourth iteration step,

$$M_{\mathrm{R}_n} = \frac{M_{\mathrm{R}_{n-1}} + M_{\mathrm{P}_{n-1}}}{2} , \tag{3a}$$

$$\frac{1}{M_{\mathrm{P}_n}} = \frac{1}{2}\left(\frac{1}{M_{\mathrm{R}_{n-1}}} + \frac{1}{M_{\mathrm{P}_{n-1}}}\right) \tag{3b}$$

($M_{\mathrm{R}_1} = E, G, \nu$ of Voigt, $M_{\mathrm{P}_1} = E, G, \nu$ of Reuss, n step of iteration). The second-order limits of Hashin and Shtrikman (HS) [6] describe a homogeneous and elastically isotropic microstructural state. Kröner's [7, 8] third-order bounds as well as the self-consistent value of the shear modulus take into account the grain interaction in a macroscopically isotropic polycrystalline aggregate of cubic crystals. The self-consistent moduli are obtained for a state of perfect disorder, which means a grain shape and size independent distribution of G. In this work polycrystalline elastic properties are obtained by the formulation of third-order bounds (see Appendix) by Gairola and Kröner [4].

The range between upper and lower bounds of a special averaging concept depends on the ratio of anisotropy A, which is defined for cubic crystals by

$$A = \frac{2(S_{11} - S_{12})}{S_{44}} . \tag{4}$$

The self-consistent procedure is an iterative scheme, too. But the self-consistent value differs from the lower mean value of the third-order average by less than 1.5% in the case of an anisotropy ratio of 3.21 [4]. For VRH, HS, and Kröner bounds Fig. 1

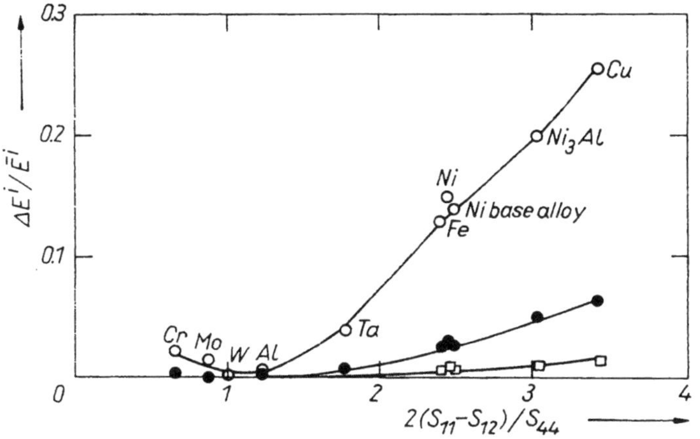

Fig. 1. Differences between limit and mean value of Young's modulus calculated by various averaging concepts (i: ○ 1st order: VRH bounds; ● 2nd order: HS bounds, □ 3rd order: Kröner bounds) as a function of the ratio of anisotropy. Elastic coefficients S_{ij} are taken from [9]

shows the difference between the mean values and the limits of the polycrystalline Young's moduli for several pure metals and nickel alloys with differently strong anisotropy. This difference increases with the anisotropy ratio. The theory of Kröner and Gairola is recommended for nickelbase superalloys with an A of 2.5 to 3.

3. Experimental Procedure

The orientation-dependent reciprocal compliances $1/S'_{11}$ and $1/S'_{44}$ (see (2)) of the single-crystalline alloy MAR-M002 mod. and the elastic parameters E and G of the corresponding polycrystalline alloys MAR-M247 [10] and MAR-M200 [11, 12] were determined by measuring the resonance frequencies of rod-shaped single-crystalline and prismatic polycrystalline specimens using a Förster Elastomat equipment, type 1.024 [13]. From these resonance frequencies of bending and torsion the Young's modulus and the shear modulus were calculated by means of theoretical relations [14]. A high-temperature equipment [15] made it possible to measure resonance frequencies in the temperature range from 25 to 1200 °C. Above 25 °C the dimensions of the specimens had to be corrected by the temperature-dependent thermal expansion coefficient [16]. The factor of shear correction was given by Spinner and Tefft [14]. Due to the uncertainties in determining density, specimen dimensions, and resonance frequencies the experimental error of E ($= 1/S'_{11}$) and G ($= 1/S'_{44}$) varied between 0.6% at 25 °C and 1% at 1200 °C.

Coupling of different vibration modes in single crystals with the specimen axis deviating from the ⟨001⟩ crystal axis [17] were avoided by specially formed samples. For the transverse vibration slender cylindrical rods (diameter to length ratio < 0.1) and for the torsional vibrations short and thick cylindrical samples (diameter to length ratio < 0.5) were prepared. Young's moduli and shear moduli of such specimens are called free moduli.

Fig. 2 shows the temperature dependence of the reciprocal S'_{11} of several single-crystalline specimens. They differ in the orientation of the sample axes to the crystallographic ⟨001⟩ axis.

For the calculation of elastic properties of polycrystalline material it is indispen-

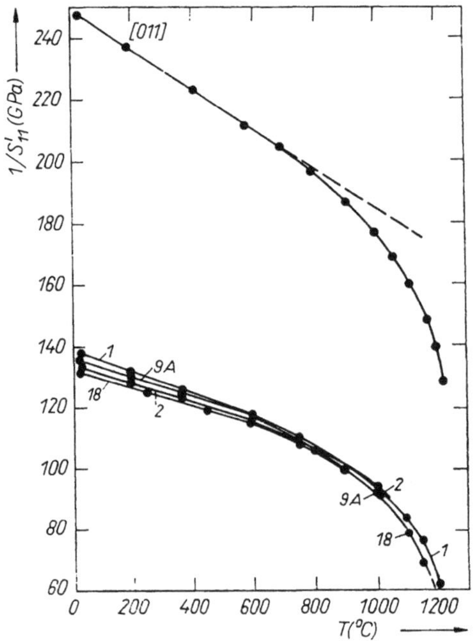

Fig. 2. Temperature dependence of $1/S'_{11}$ of a single-crystalline nickel-base superalloy MAR-M002 mod. with different orientations. The quantity J describing the orientation deviation from ⟨001⟩ is given by (2c):

specimen	J
1	0.048
2	0.0285
9A	0.035
18	0.021
[011]	0.255

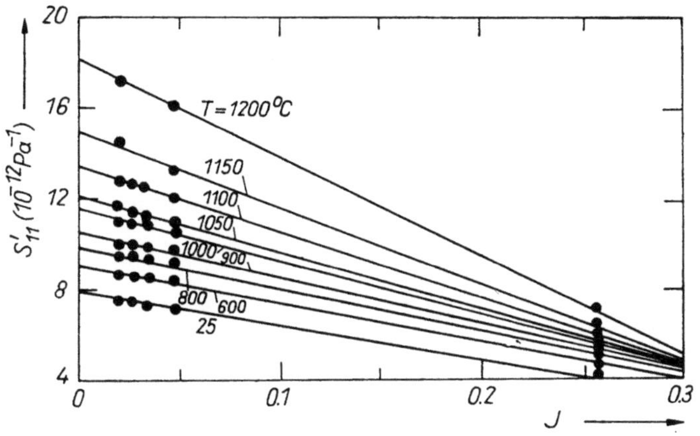

Fig. 3. Determination of S_{11} ($S_{11} = S'_{11}$ for $J = 0$) by linear regression (MAR-M002 mod.)

sable to obtain the orientation-independent compliances of the corresponding single crystal MAR-M002 mod. By means of linear regression as shown for S'_{11} in Fig. 3 the S_{11} and S_{44} were calculated from 25 up to 1200 °C. $E = 1/S'_{11}$ and $G = 1/S'_{44}$ were evaluated from the measured resonance frequencies of differently orientated single crystals. The orientation and the corresponding factor J were determined by the Laue back-reflection X-ray technique. The obtained most important elastic properties of MAR-M002 mod. are listed in Table 2.

Table 2
Elastic constants and coefficients of MAR-M002 mod.
(S_{ij} in 10^{-12} Pa^{-1}, C_{ij}, E, G, K in GPa)

T (°C)	S_{11}	S_{12}	$S_{44} = 1/C_{44}$	C_{11}	C_{12}	K	$E_{\langle 100 \rangle}$	$G_{(100)[001]}$	ν	A
25	7.843	−3.078	8.000	258.6	167.0	197.5	127.5	125.0	0.392	2.73
600	8.960	−3.524	9.348	227.8	147.0	174.3	111.6	107.0	0.393	2.67
800	9.671	−3.830	10.101	215.1	141.0	165.8	103.2	99.0	0.396	2.67
900	10.288	−4.096	10.615	205.4	135.9	159.0	97.2	94.2	0.398	2.71
1000	11.198	−4.495	11.299	193.4	129.7	151.0	89.3	88.5	0.401	2.78
1100	13.071	−5.395	12.145	182.2	128.0	146.1	76.5	82.2	0.413	3.04
1200	18.115	−7.843	13.661	162.9	124.4	137.2	55.2	73.2	0.433	3.80

The magnitude of the slope of the S'_{11}–J curve is equal to $2S$. Therefore the third independent, but not measurable coefficient is given by

$$S_{12} = S_{11} - S - 0.5 S_{44} \,. \tag{5}$$

The bulk modulus K is defined by

$$K = \left(3(S'_{11} + 2S'_{12})\right)^{-1} \,. \tag{6}$$

Due to the counteracting J-dependences of S'_{11} and $2S'_{12}$, the bulk modulus is an orientation-independent property.

4. Results and Discussion. Quasi-Isotropic Aggregates

The unprimed S_{ij} allow the calculation of all interesting elastic properties of MAR M002 mod. as a function of temperature and for given orientation of the sample. For instance at 25 °C the Young's modulus varies for different orientations between 127.5 GPa in ⟨001⟩ and 309.7 GPa in ⟨111⟩. Fig. 4 a and b represent sectional views of the E- and A-bodies. A is defined by (4), whereas the A-body is computed by considering the orientation dependence of S_{ij} as given by (2). A single-crystalline specimen

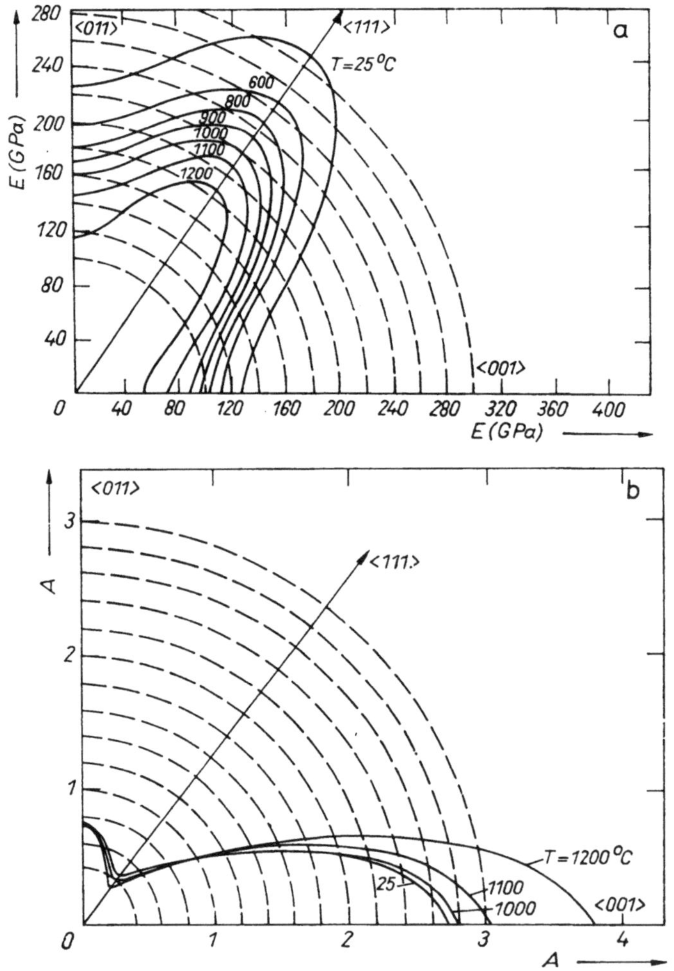

Fig. 4. Sectional views of a) the E- and b) the A-bodies for several temperatures

▶

Fig. 5. Application of several theoretical concepts of the calculation of elastic properties of a polycrystalline superalloy with same composition as the single crystal MAR-M002 mod.: a) Young's modulus, b) shear modulus, c) Poisson's ratio vs. temperature. a) triangles and squares MAR-M200 [11] and [12] respectively), circles MAR-M247 [10]; b) solid lines MAR-M002 poly.

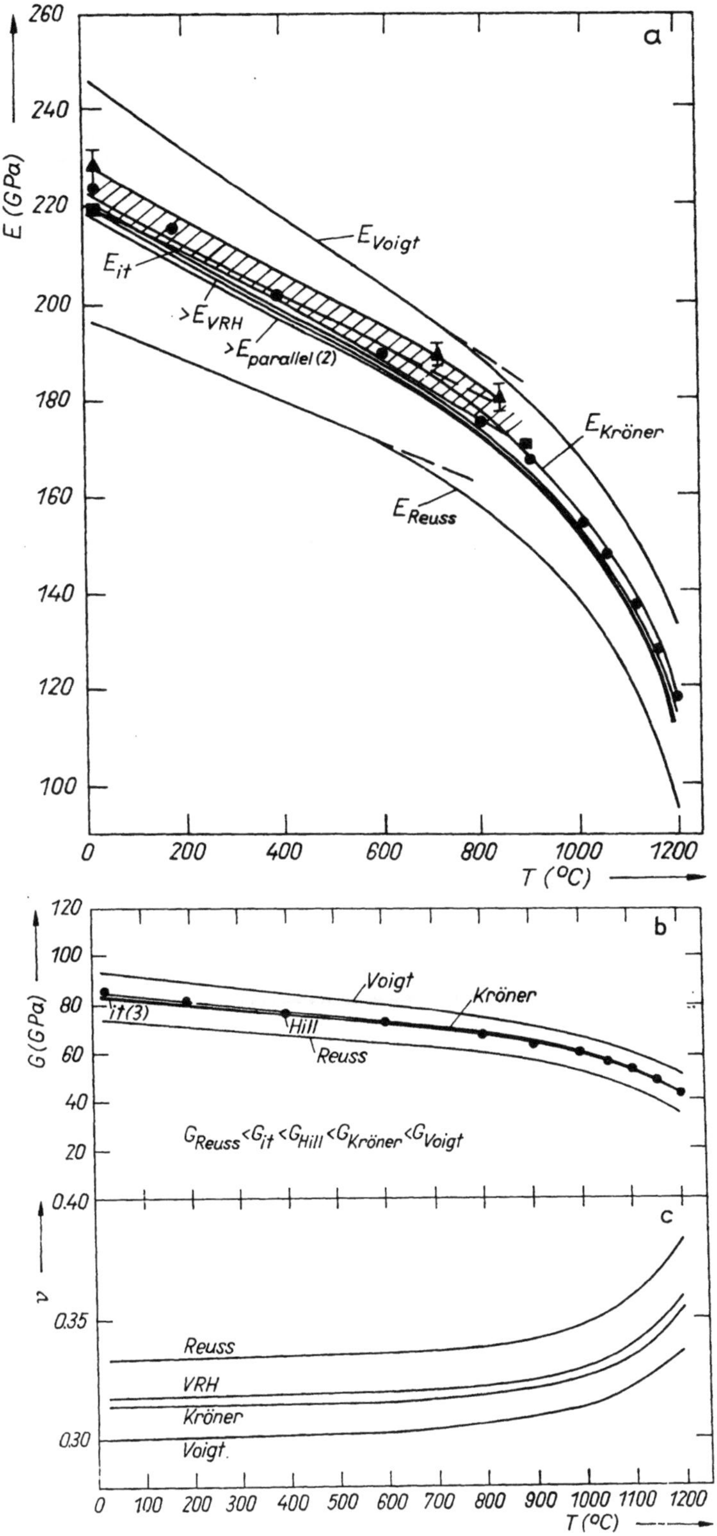
a
E (GPa)
260
240
220
200
180
160
140
120
100
E_{Voigt}
E_{it}
$>E_{VRH}$
$>E_{parallel (2)}$
$E_{Kröner}$
E_{Reuss}
0
200
400
600
800
1000
1200
T (°C)
b
G (GPa)
120
100
80
60
40
20
Voigt
Kröner
it(3)
Hill
Reuss
$G_{Reuss} < G_{it} < G_{Hill} < G_{Kröner} < G_{Voigt}$
c
ν
0.40
0.35
0.30
Reuss
VRH
Kröner
Voigt
0
100
200
300
400
500
600
700
800
900
1000
1100
1200
T (°C)

Fig 5.

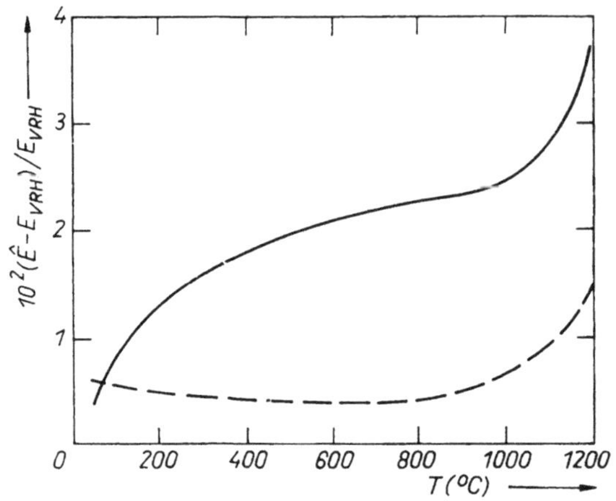

Fig. 6. Temperature dependence of the Young's modulus of polycrystalline MAR-M002 mod. according to the iteration model of [5] (1st order bounds (dashed line)) and the Kröner bound (3rd order bounds (solid line)) relative to the VRH mean value

with an $A = 1$ behaves as an isotropic polycrystalline specimen. After a nearly constant regime up to 900 °C the anisotropy defined by the unprimed S_{ij} increases with temperature from 2.7 to 3.8 at 1200 °C.

The application of the theoretical models of [1 to 5] allows the calculation of E, G, and ν of an isotropic polycrystalline MAR-M002 mod. (Fig. 5). Because of nearly the same composition of alloying elements this fictitious polycrystal MAR-M002 mod. stands a comparison with the measured datas of MAR-M200 and MAR-M247. Discussing the averaging concepts of first order [1, 2, 3], we note that the difference between Reuss and Voigt bounds decreases with increasing temperature from 48 GPa at 25 °C to 30 GPa at 1000 °C and extends to 38 GPa at 1200 °C. The serial connection for the simulation of real grain arrangement in the Hill model creates higher mean values than the fourth iteration step of [5] leading to convergence. Only three steps are required to obtain the quasi-isotropic shear modulus and Poisson's ratio. The Young's moduli and shear moduli averaged by Kröner's model lie above the Hill and iterative mixture model values, while the Poisson's ratios vary in the opposite sense. As Fig. 6 shows for the case of E how the differences between Kröner's bound of third order and the VRH mean value and between iterative mixture model and VRH limit depend on temperature.

Other authors [11, 12] have reported only few $E(T)$ curves of the comparable isotropic cast alloy MAR-M200. Above 900 °C values on MAR-M247 [10] but no data on MAR-M200 are available. Up to 800 °C the VRH limit and the four-step iterative scheme of MAR-M002 mod. agree with measurements of [12]. The theory of Kröner shows the same temperature dependence as that mentioned and lies between the values of [11] and [12]. A better proof of the usefulness of Kröner's concept is the fact that the measured curve of the temperature behaviour of E and G of MAR-M247 agree with computed properties of MAR-M002 mod. Especially at high temperature the iterative mixture scheme is a competitive averaging method too. Moreover in the case of the shear modulus Hill's mean value is a sufficiently exact boundary concept to calculate polycrystalline properties of nickel base superalloys with an anisotropy ratio A of nearly 3.

5. Summary and Conclusions

1. The elastic compliances S_{ij} of the single-crystalline superalloy MAR-M002 mod. were determined in the temperature range from 25 to 1200 °C.

2. Using this orientation-independent S_{ij} in the averaging concepts of Hill, Hütt, and Kröner, these theories were proved to be useful to confirm measured elastic properties of quasi-isotropic aggregates of grains.

3. Further investigations have to take into account the influence of fibre-texture as postulated by [18 to 20].

Acknowledgements

We are grateful to the Forschungsgemeinschaft Verbrennungskraftmaschinen (FVV) and the Arbeitsgemeinschaft Industrieller Forschungsvereinigungen (AIF) for the financial support of this project. The research project was accompanied and supported by a working group with members from ABB, AEG, KHD, Siemens KWU, and MTU under the leadership of Dr. V. Thien, Siemens KWU Mülheim. Concerning the publication of the results the authors are indebted to Prof. Dr. H. Mughrabi for the critical reading of the manuscript.

Appendix

Kröner averages of third order:

$$G^{(3)} = \frac{G_1G_2(258K + 396G_2 + 264G_1) + 81KG_2^2 + 108G_2^3 + 36KG_1^2 + 32G_1^3}{252G_1G_2 + 195KG_2 + 144G_2^2 + 180KG_1 + 104G_1^2}, \tag{A1}$$

$$G_{(3)} = \frac{G_1G_2(540KG_1^2 + 945KG_1G_2 + 1300G_2G_1^2 + 390G_2^2K + 1200G_2^2G_1)}{G_1G_2(999KG_1 + 540G_1^2 + 666KG_2 + 1720G_1G_2 + 240G_2^2) + 48KG_2^3 + 162KG_1^3} \tag{A2}$$

with

$$G_1 = \frac{1}{2(S_{11} - S_{12})}, \qquad G_2 = \frac{1}{S_{44}}, \qquad K = \frac{1}{3(S_{11} + 2S_{12})},$$

$$E = \frac{4.5\mathrm{K}(G_{(3)} + G^{(3)})}{3K + 0.5(G_{(3)} + G^{(3)})}. \tag{A3}$$

References

[1] W. Voigt, Lehrbuch der Kristallphysik, Teubner-Verlag, Leipzig 1928.
[2] A. Reuss, Z. angew. Mech. **9**, 49 (1929).
[3] R. Hill, Proc. Phys. Soc. (London) **65**, 349 (1952).
[4] B. K. D. Gairola and E. Kröner, Internat. J. Engng. Sci. **19**, 865 (1981).
[5] R. Hütt, Mater. Sci. Engng. **29**, 55 (1977).
[6] Z. Hashin and S. Shtrikman, J. Mech. Phys. Solids **10**, 335 (1962).
[7] E. Kröner, Z. Phys. **151**, 504 (1958).
[8] E. Kröner, J. Phys. Solids **25**, 137 (1977).
[9] G. Simmons and H. Wang, Single Crystal Elastic Constants and Calculated Aggregate Properties (Handbook), 2nd ed., M.I.T. Press, Cambridge (Ma.) 1971.
[10] H.-A. Kuhn, Doctorate Thesis, Universität Erlangen-Nürnberg, 1987.
[11] D. P. Dandekar, A. G. Martin, and J. V. Kelley, Met. Trans **12A**, 801 (1981).
[12] High Strength Nickel Base Alloys, The International Nickel Company Inc., New York 1964.

[13] Institut Dr. Förster, Bedienungsanleitung Elastomat 1.024, Fa. Förster, Reutlingen.
[14] J. SPINNER and W. E. TEFFT, Proc. ASTM **61**, 1221 (1961).
[15] W. GRELLNER, Doctorate Thesis, Universität Erlangen-Nürnberg, 1980.
[16] P. L. M. KLOSE, Doctorate Thesis, Universität Erlangen-Nürnberg, 1987.
[17] R. F. S. HEARMON, Proc. Phys. Soc. (London) **55**, 67 (1943).
[18] H. PURSEY and H. L. COX, Phil. Mag. **45**, 295 (1954).
[19] G. KNEER, Doctorate Thesis, Bergakademie Clausthal, 1964.
[20] R. A. ADAMESKU, E. A. MITJUSHOV, L. L. MITJUSHOVA, and K. S. JUSHKOV, Z. Metallkde. **76**, 747 (1985).

(Received August 8, 1988)

phys. stat. sol. (a) **110**, 459 (1988)

Subject classification: 64.70 and 77.80; S11

Department of Physics, Martin-Luther-University Halle/Wittenberg, Halle (Saale)[1])

Ferroelastic Phase Transition in $LiRb_4H(SO_4)_3 \cdot H_2SO_4$

Landau-Theory and Experimental Results

By

H. HEMPEL, H. MAACK, and G. SORGE

Elastic properties of $LiRb_4H(SO_4)_3 \cdot H_2SO_4$, determined by measuring the velocities of ultrasonic waves, and dielectric permittivities are shown to be in good agreement with the theoretical predictions based on Landau-theory for a proper ferroelastic phase transition of the Aizu-species 4F2.

Es wird gezeigt, daß die elastischen Eigenschaften von $LiRb_4H(SO_4)_3 \cdot H_2SO_4$, die durch Messungen von Ultraschallwellen bestimmt wurden, und die dielektrischen Permeabilitäten in guter Übereinstimmung mit den theoretischen Vorhersagen der Landau-Theorie für einen eigentlich ferroelastischen Phasenübergang der Aizu-Spezies 4F2 sind.

1. Introduction

Recently Wolejko et al. [1, 2] reported the pyroelectric, dielectric, elastic, and thermal properties of $LiRb_4H(SO_4)_3 \cdot H_2SO_4$ connected with the ferroelastic phase transition at about 137 K. Our experimental results confirm their dielectric and thermal data, but we found a different behaviour of the pyroelectric coefficient and of some elastic stiffness coefficients.

2. Experimental

All the samples used in our experiments had been cut from the same crystal, which was grown isothermally from aqueous solution.

The ultrasonic measurements were performed with the MATEC 6600 system using the pulse-echo-overlap technique. We measured the velocities of transverse and longitudinal waves of 20 MHz propagating along the principal crystallographic axes, thus determining the main diagonal elements of the elastic stiffness tensor. The results are shown in Fig. 1, 2 in terms of the sound velocities $v_{11} \dots v_{66}$. Below T_c the results of the longitudinal waves, propagating in (100) and (010) directions, as well as those for v_{55} were not well reproducible, what may be caused by the ferroelastic domain structure [2].

The ultrasonic data did not reveal as large anomalies as the elastic data from pendulum and resonance methods, reported in [2]. This is valid for the longitudinal modes in x- and y-directions and was confirmed by our own measurements of s_{11} by the piezoelectric resonance method. The reason for this different temperature dependence of the elastic stiffness and the elastic compliance will be explained below.

The change of polarization P_3 with temperature (Fig. 3) was determined by integrating the sample current when changing the temperature. Contrary to the results of Wolejko et al. [1] we did not find any maximum of the pyroelectric coefficient in the vicinity of the phase transition temperature.

[1]) Friedemann-Bach-Platz 6, DDR-4020 Halle (Saale), GDR.

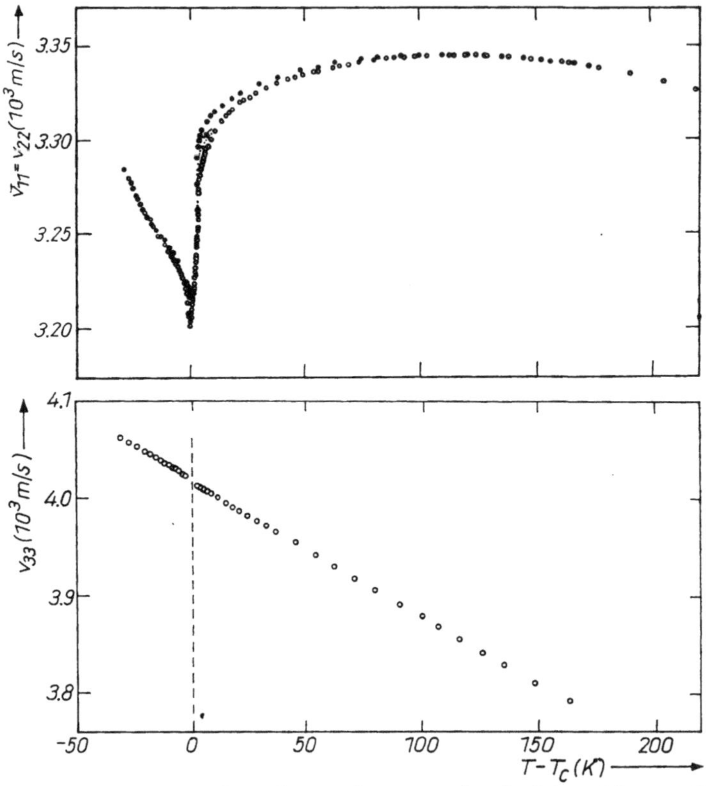

Fig. 1. Temperature dependence of the sound velocities of longitudinal ultrasonic waves

The measurement of the dielectric permittivity along the z-axis (Fig. 4) was carried out with the automatic capacitance bridge 4270A of Hewlett & Packard. The small steps in the ε_{33} versus temperature curve are due to the error of digitalization of the data. At the phase transition temperature no step was observed, we found only a change in the slope of the temperature dependence at T_c.

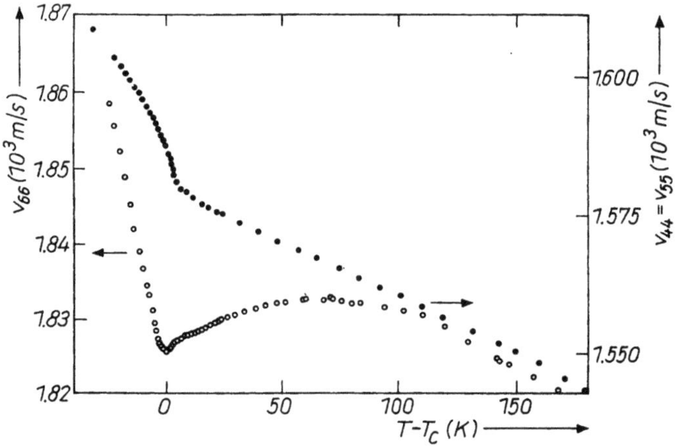

Fig. 2. Temperature dependence of the sound velocities of transverse ultrasonic waves

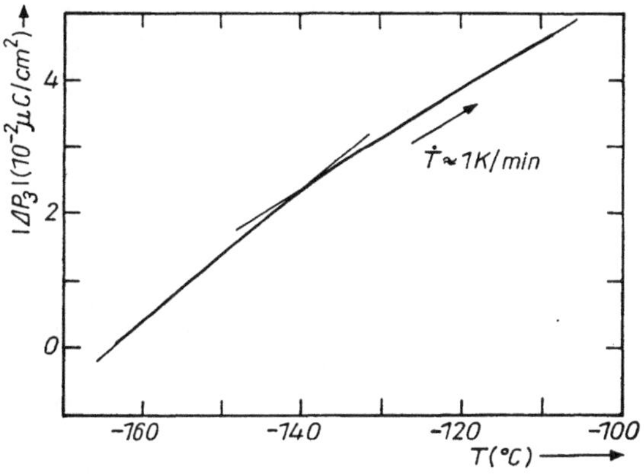

Fig. 3

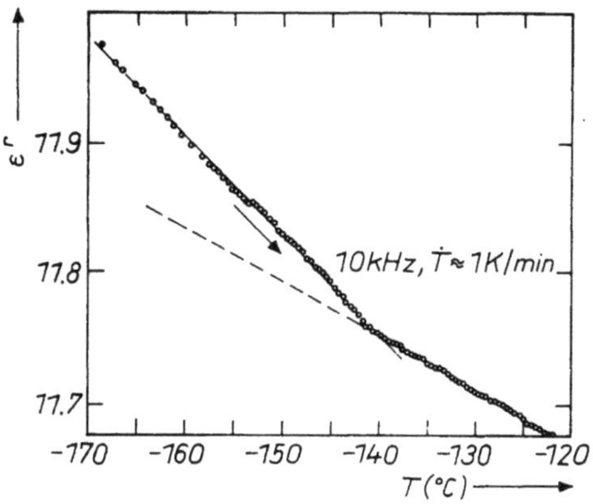

Fig. 4

Fig. 3. Pyroelectric effect

Fig. 4. Temperature dependence of the free dielectric permittivity ε_{33}

Specific heat was measured by adiabatic calorimetry. The observed anomaly (Fig. 5) at T_c, although very small, shows the lambda-form typical for phase transitions of second order. This confirms the fact that no thermal hysteresis could be detected from all our measurements.

3 Thermodynamic Theory

From the habit of the as-grown crystal as well as from the absence of optical birefringence in c-direction, checked by polarization microscopy, one can conclude tetragonal symmetry of the room temperature phase of $LiRb_4H(SO_4)_3 \cdot H_2SO_4$. On the other hand, the crystal is polar in both phases with the same direction of the polarization vector. At first, the symmetry change at the phase transition was assumed to be 4mmFmm2 [1, 2], while the structural investigation of the crystal yielded the

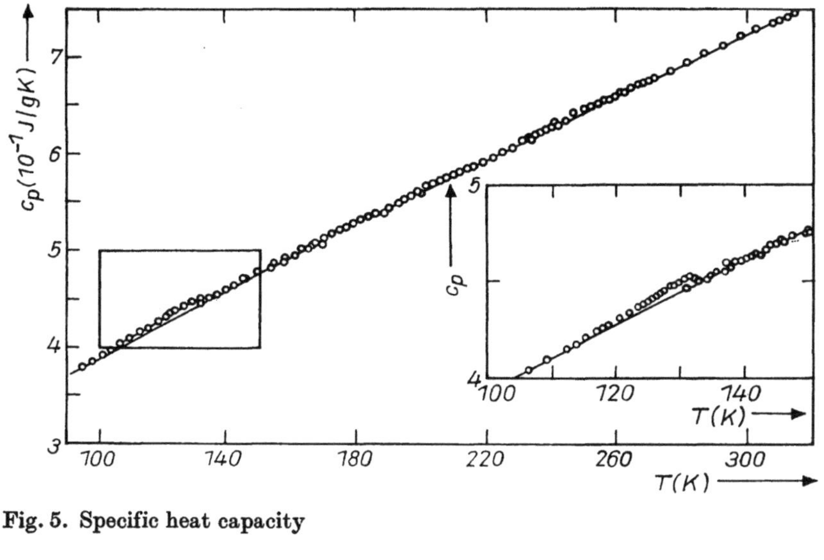

Fig. 5. Specific heat capacity

point group 4 for the room-temperature phase [3]. So, in accordance with the experimental results, one has to look for a ferroelastic, but non-ferroelectric transition from the point group 4. Furthermore, because of the strong temperature dependence of the elastic compliance s_{11} [2] the transition seems to be a proper one.

3.1. Aizu species 4F2

According to Janovec [4] there are two possible proper ferroelastic transitions from the point group 4, they result in the point groups 2 and 1, respectively. The species 4F1 must be excluded, since the low-temperature phase should be ferroelectric in this case, what was not observed. The order parameter for the transition 4F2 is $(S_1 - S_2)/2$ or S_6. In the latter case, however, the order parameter susceptibility is c_{66}, so that c_{66} should go to zero at the phase transition temperature. This could not be confirmed by the ultrasonic data (Fig. 2).

Using standard group theoretical methods (e.g. [4, 5]), for the first case one gets the following thermodynamic potential:

$$F = F_0 + F_{\text{elast}} + F_{\text{el-mech}} + F_{\text{electr}} \tag{1}$$

with

$$\begin{aligned} F_{\text{elast}} = {} & A(S_1 - S_2)^2 + B(S_1 - S_2)^4 + C(S_1 + S_2)^2 + DS_3^2 + \\ & + E(S_4^2 + S_5^2) + NS_6^2 + WS_3S_6^2 + G(S_1 + S_2)\, S_3 + \\ & + H(S_1 + S_2)\,(S_1 - S_2)^2 + I(S_1 - S_2)^2\, S_3 + J(S_1 + S_2)\,(S_4^2 + S_5^2) + \\ & + K(S_1 - S_2)\,(S_4^2 - S_5^2) + L(S_1 + S_2)\, S_6^2 + M(S_1 - S_2)^2\,(S_4^2 + S_5^2) + \\ & + nS_6(S_1 - S_2) + o(S_1 + S_2)\,(S_1 - S_2)\, S_6 + p(S_1 - S_2)\, S_3S_6 + \\ & + r(S_1 - S_2)\, S_4S_5 + vS_4S_5S_6\,, \\ F_{\text{el-mech}} = {} & OP_3S_6^2 + P(S_1 - S_2)^2\, P_3 + Q(S_1 + S_2)\, P_3 + RS_3P_3 + \\ & + T(S_1 - S_2)\,(P_1^2 - P_2^2) + s(S_1 - S_2)\, S_6P_3 + tS_6(P_1^2 - P_2^2) + \\ & + u(S_1 - S_2)\, P_1P_2\,, \\ F_{\text{electr}} = {} & UP_3^2 + V(P_1^2 + P_2^2)\,. \end{aligned}$$

The coefficients in (1) stand for the appropriate combinations of linear and nonlinear elastic stiffness coefficients, piezoelectric and dielectric moduli. For example, $D = c_{33}^0/2$, $R = h_{33}^0$, and so on. As usually done in Landau theory, we consider A to be temperature dependent according to

$$A = \frac{c_{11}^0 - c_{12}^0}{4} = a(T - T_0)\,, \tag{2}$$

while the other coefficients are temperature independent.

From the conditions of thermodynamical equilibrium in the absence of external stresses and fields one obtains the dependence of the strain and the polarization components on the order parameter $(S_1 - S_2)/2$,

$$\begin{aligned} (S_1 + S_2) &= X^1(S_1 - S_2)^2\,, & S_3 &= X^2(S_1 - S_2)^2\,, \\ S_4 &= S_5 = 0\,, & S_6 &= y(S_1 - S_2)\,, \\ P_1 &= P_2 = 0\,, & P_3 &= X^3(S_1 - S_2)^2\,, \end{aligned} \tag{3}$$

y and $X^1 - X^3$ (like the $X^4 - X^{17}$ below) being temperature independent combinations of the coefficients from (1). The introduction of (3) into (1) leads to

$$(S_1 - S_2)^2 = \begin{cases} 0 & \text{for} \quad T > T_c\,, \\ a(T_c - T)/2X^4 & \text{for} \quad T < T_c\,. \end{cases} \tag{4}$$

Due to the pseudoproper coupling between $(S_1 - S_2)$ and S_6 the phase transition temperature T_c in (4) is different from T_0 according to

$$T_c = T_0 + \frac{y(n + Ny)}{a}. \tag{5}$$

The second derivatives of F with respect to the irreducible components of the strain tensor give the isothermal elastic moduli, which depend on temperature via A (cf. (2)), the order parameter $(S_1 - S_2)$ and χ_{op}, standing for the order parameter susceptibility ($\chi_{op} = \partial^2 F/\partial(S_1 - S_2)^2$),

$$\begin{aligned}
c_{11} &= 2A + 2C + X^5(S_1 - S_2) + X^6(S_1 - S_2)^2 + Y^1\,,\\
c_{22} &= 2A + 2C - X^5(S_1 - S_2) + X^6(S_1 - S_2)^2 + Y^1\,,\\
c_{33} &= 2D + Y^2\,,\\
c_{44} &= 2E + X^7(S_1 - S_2) + X^8(S_1 - S_2)^2\,,\\
c_{55} &= 2E - X^7(S_1 - S_2) + X^8(S_1 - S_2)^2\,,\\
c_{66} &= 2F + X^9(S_1 - S_2)^2 + Y^3 - n^2/\chi_{op}\,,\\
c_{12} &= -2A + 2C - X^6(S_1 - S_2)^2 + Y^1\,,\\
c_{13} &= G + X^{11}(S_1 - S_2) + Y^4\,,\\
c_{23} &= G - X^{11}(S_1 - S_2) + Y^4\,,\\
c_{16} &= n + X^{12}(S_1 - S_2) + X^{13}(S_1 - S_2)^2 + Y^5\,,\\
c_{26} &= -n + X^{12}(S_1 - S_2) - X^{13}(S_1 - S_2)^2 + Y^5\,,\\
c_{36} &= X^{14}(S_1 - S_2) + Y^6\,,\\
c_{45} &= X^{15}(S_1 - S_2)\,.
\end{aligned} \tag{6}$$

Following the same procedure, one gets the isothermal dielectric impermeabilities of the stress-free crystal,

$$\begin{aligned}
\beta_{11} &= 2V + X^{16}(S_1 - S_2)\,,\\
\beta_{22} &= 2V - X^{16}(S_1 - S_2)\,,\\
\beta_{33} &= 2U + Y^7\,,\\
\beta_{12} &= X^{17}(S_1 - S_2)\,.
\end{aligned} \tag{7}$$

(The relation $\varepsilon_{ii} = 1/\beta_{ii}$ is valid only for the paraelastic phase!) Together with (2) and (4), (6) and (7) describe the temperature dependence above as well as below the phase transition temperature T_c. The terms $Y^1 - Y^7$ are of the common form

$$Y^1 = -\left(C_1 + C_2(S_1 - S_2)\right) \frac{S_1 - S_2}{\chi_{op}}\,, \tag{8}$$

so that they are zero in the paraelastic phase.

3.2 Aizu species 4mmFmm2

Since some experimental results, e.g. divergence of s_{11} at T_c and no, or at least very weak, optical activity at room temperature, indicated the possibility that the high-temperature phase could have the symmetry 4mm, we investigated the case 4mmFmm2, too. The order parameter is also $(S_1 - S_2)$ (S_6 is excluded as in Section 3.1). The resulting thermodynamic potential is that of (1), when all coefficients written in small letters are set zero.

The discussion of this potential results in the following changes of (3) to (8):

(i) $S_6 = 0$; $T_c = T_0$ (since $n = y = 0$),

(ii) $c_{16} = c_{26} = c_{36} = c_{45} = \beta_{12} = 0$,

(iii) in c_{66}: $Y^3 = 0$, $n = 0$,

(iv) in c_{13}, c_{23}, c_{33}, and β_{33} one has to include a term $\sim (S_1 - S_2)^2$,

(v) the Y^i $(i \neq 3)$ are no longer temperature dependent; they represent a step of the respective coefficient at T_c.

4. Discussion

4.1 Ultrasonic data

Since the components of the elastic stiffness tensor are connected with the respective ultrasonic velocities via $c_{11} = \varrho v_{11}^2$, for small changes we have $dc_{11}/c_{11} = 2\varrho\, dv_{11}/v_{11}$, i.e. the same temperature dependence.

Above T_c the temperature dependence of v_{33}, v_{44}, and v_{55} was found to be linear up to room temperature. The other modes show a nonlinearity of the temperature dependence in the range $T_c + 120\text{ K} > T > T_c$. For v_{11} and v_{22} a temperature dependence above T_c is expected from the term $2A$ in (6). It should be linear in the temperature range, where the Curie-Weiss law is valid. While Wolejko [2] found this linearity within $T_c < T < T_c + 15\text{ K}$ from pendulum data, we found an interval of 5 K above T_c only (cf. Fig. 1a).

The elastic compliance coefficients s_{ij} follow from the inversion of the c_{ij} tensor. Assuming the coefficient $n = c_{16}^0$ to be zero at first, one gets for the temperature dependence of s_{11} in the paraelastic phase

$$s_{11} = 1/8C + 1/8A\,, \tag{9}$$

(when $(c_{13}^0)^2 \ll c_{33}^0(c_{11}^0 + c_{12}^0)$, which is usually fulfilled) which leads to a divergence of s_{11} at T_0. (In this case, which is true for the species 4mmFmm2, one gets no linear coupling between S_6 and $(S_1 - S_2)$, resulting in $y = 0$ and $T_0 = T_c$.) The measured value of $1/s_{11}$ [2] decreases strongly when the temperature goes to T_c from above, thus indicating that n has a small value only. To be concrete, it should be $(c_{16}^0)^2 \ll c_{66}^0(c_{11}^0 + c_{12}^0)$.

For the v_{44} and v_{55} modes the anomaly of the sound velocity is somewhat smeared although the damping constant shows a sharp maximum at T_c. According to the proposed model, one gets two different parts in the temperature dependence of these modes below T_c, one of them being proportional to $(T_c - T)$, and the other one proportional to $(T_c - T)^{1/2}$.

Except for v_{33}, below T_c the quantitative comparison of the experimental data with the theory is difficult, however, since (5) and (6) refer to the single-domain state of the crystal and in the measurements the crystal was split into domains [2].

4.2 Dielectric and pyroelectric data

In accordance with the results from Section 3 the dielectric permittivity ε_{33} shows a weak linear temperature dependence below as well as above T_c, the slope changing at T_c. Furthermore, we found $\varepsilon_{11} = \varepsilon_{22}$ in the whole temperature range under investigation (not shown in Fig. 3). As in the case of v_{44} and v_{55} this could result from the domain splitting, too.

The temperature dependence of the polarization P_3 also shows the predicted change in the slope of the curve at T_c (cf. Fig. 3 and (3) and (4)).

The proposed thermodynamic model of a proper ferroelastic transition gives an adequate description of the whole experimental data, while any attempt to explain the elastic and dielectric properties of $LiRb_4H(SO_4)_3 \cdot H_2SO_4$ with an improper ferroelastic phase transition failed. Especially the qualitatively different behaviour of c_{11} and $1/s_{11}$, the temperature dependence of c_{11} and c_{22} above T_c, and the absence of step-like changes of c_{33} and β_{33} at T_c remained unclear in this case.

Acknowledgements

The crystal was kindly supplied by Prof. T. Krajewski (Poznan). The measurements of heat capacity were performed with the support of Prof. B. A. Strukov (Moscow).

References

[1] T. Wolejko, P. Piskunowicz, T. Breczewski, and T. Krajewski, to be published.
[2] T. Wolejko, G. Pakulski, and Z. Tylczynski, to be published.
[3] Z. Czapla, private communication.
[4] V. Janovec, V. Dvorak, and J. Petzelt, Czech. J. Phys. **B25**, 1362 (1975).
[5] P. Toledano, M. M. Fejer, and B. A. Auld, Phys. Rev. B **27**, 5717 (1983).
[6] K. Aizu, J. Phys. Soc. Japan **28**, 706 (1970).

(Received July 7, 1988)

phys. stat. sol. (a) **110**, 467 (1988)

Subject classification: 64.70; 62.50; 78.30; S10

Department of Physics, N. A. S. College, Meerut[1])

High-Pressure Raman Spectroscopic Study of the Ice Ih → Ice IX Phase Transition

By

A. K. Garg

The phase boundaries of the ice phase diagram are not clear in the low-temperature region even under low pressures. A detailed study of the Raman spectra of ice is reported at 88 K and in the pressure range 0 to 0.6 GPa. It is found that upon pressurising at 150 K ice Ih transformed into ice IX, although a transition ice Ih → ice II is expected from the existing phase diagram. No signs of recovering ice Ih can be detected after depressurising ice IX at 150 K and leaving for few hours. The ice Ih and ice IX data are analysed to determine various temperature and pressure coefficients.

Die Phasengrenzen des Phasendiagramms von Eis sind im Tieftemperaturbereich schon bei niedrigen Drücken unsicher. Es wird eine ausführliche Untersuchung der Ramanspektren von Eis bei 88 K und im Druckbereich von 0 bis 0,6 GPa durchgeführt. Es wird gefunden, daß nach Druckbelastung bei 150 K Ih-Eis in IX-Eis übergeht, obwohl aus dem vorliegenden Phasendiagramm ein Übergang Ih-Eis → II-Eis erwartet wird. Nach Entlastung des IX-Eis bei 150 K und einigen Stunden Wartezeit kann kein Anzeichen einer Ih-Eis-Rückbedeckung beobachtet werden. Die Ih-Eis- und IX-Eis-Daten werden analysiert, um verschiedene Temperatur- und Druckkoeffizienten zu bestimmen.

1. Introduction

Besides the ordinary hexagonal low-pressure form of ice naturally occurring, there are at least eight high-pressure forms and one unstable cubic low-pressure form of ice. The study of the transformation from one solid form to another is bound to be very interesting for this important simple molecule. The pressure–temperature phase diagram of H_2O is presented in Fig. 1 for pressures up to 1 GPa. The solid and dash-dotted lines are directly measured stable and metastable lines, respectively, whereas dashed lines represent boundaries that are extrapolated or estimated from the available data. The phase boundaries in the low-temperature region are still very vague even under low pressure. This may be because of the difficulties in carrying out high-pressure experiments at low temperatures. The large heat capacity, optical windows, high-pressure seals, and handling of the cell create numerous problems under these conditions. Most of the reported spectroscopic work on the high-pressure ice forms is done on metastable "reclaimed" samples at 77 K and atmospheric pressure mainly by Whalley and his group at National Research Council of Canada. The high-pressure forms of ice do not revert to a low-pressure form when the pressure is released provided that it is quenched to liquid nitrogen temperature and maintained at this temperature. This made it possible to examine high-pressure forms at atmospheric pressure and low temperature. However, in order to study the dynamic behaviour near a phase

[1]) Meerut-250001, India.

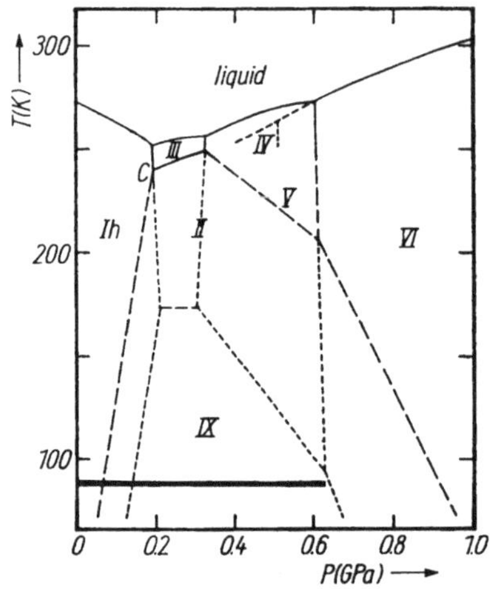

Fig. 1. Phase diagram of H_2O

transition, to understand phase boundaries, and to determine the important quantities like $\partial v/\partial p$ and $\partial v/\partial T$, one is bound to work in the region of true thermodynamic stability.

The purpose of the present study was to explore spectroscopically the ice Ih → ice IX (II) phase transition at a temperature as low as possible, close to that of liquid nitrogen. The bold line in Fig. 1 represents the region of interest. No high-pressure spectroscopic data were available on ice in this region. Most of the studies of ice phase transitions reported in the literature are based either on dielectric measurements [1] or on the study of the accompanying volume change or its effects [2].

The Raman spectrum of ice Ih (space group $P6_3/mmc$) has been thoroughly studied at atmospheric pressure [3 to 5] but assignments are only partial and some are still disputed in the literature, mainly because water molecules in ice Ih are orientationally disordered. Some high-pressure Raman data of ice Ih are available at 255 K [6 to 8]. The half-widths at this temperature are quite large and therefore, an investigation of high-pressure shifts of ice Ih Raman bands at a lower temperature is desirable.

Both ice II (space group, $R\bar{3}(C_{3i}^2)$) and ice IX (space group, $P4_12_12(D_4^4)$) are orientationally ordered forms of ice. Ice IX can be obtained [1] by cooling its disordered form (ice III) at a rate not less than (1 to 2) K/min to a temperature below around 170 K keeping the pressure constant. But ice II can be produced [9] under pressure only from ice I at a temperature much below the upper limit of the stability region of ice II. The Raman spectra of reclaimed ice II and IX have been reported [10, 11] at temperatures between 100 and 25 K but only a partial assignment is possible to date. The ice III → ice IX phase transition was first studied [1] by measuring the dielectric properties. A spectroscopic study of this transition has only recently been reported [12].

Upon pressurising ice Ih at a temperature below 170 K one may expect its transition to ice II from the existing ice phase diagram. If it does not happen, the next possibility may be a transition to ice IX. However, before pressurising ice Ih at such a low temperature it should be noted that

(i) even below 100 K temperature, ice Ih retains very large proton disorder, because the short-range proton ordering process due to cooling is defeated by an exponential increase in the relaxation time below 140 K (extrapolated value greater than 10^{16} s at 77 K);

(ii) the heat capacity of ice Ih is anomalous near 100 K [13] and the thermal equilibrium is only slowly established between 85 and 100 K. The super-pressure phenomenon begins to take place;

(iii) since most of the high-pressure phases of ice are quenchable under ordinary pressure at 77 K, it can be suggested that the transformation at such low temperatures, if not unlikely, becomes extremely difficult to occur.

2. Experimental Method

2.1 High-pressure low-temperature ice Raman cell

The Raman cell used in the present observations was originally constructed to work with a Cary 81 Raman spectrometer [14]. The cell was modified to use with a Spex Ramalog 5M system and was cooled by flowing cold nitrogen gas [15]. To go down to liquid nitrogen temperature for the present work the cell was further modified in such a way [16] that it allowed liquid nitrogen to flow continuously through it. The active volume of the cell is about 2×10^{-6} m^3 and it was designed to go up to 1 GPa. Fig. 2 illustrates the details of the cell. It is a piston-type cell made of maraging steel. The piston (A) is driven by a relatively low-pressure hydraulic ram (B) and a flexible copper coupling (C) is used to connect it with a simple oil handpump. The optical part of the cell consists of two relatively small in line sapphire windows (D, E) for laser in and laser out and one larger sapphire window (F) for collecting Raman scattering at 90° angle. The hole between laser in and laser out windows is filled with two matching sapphire rods (G, H), leaving unoccupied space in the central portion of the cell which is filled with the specimen and pressed by the piston. The outer vacuum jacket (I) has three big glass windows (J, K, L) to allow optical measurements and a lid (M) which incorporates the low pressure ram. A small channel (N) was drilled along the whole length of the piston to allow the trapped air to come out when the piston is pushed into the waterfilled bore (O). This was necessary to avoid air bubbles in the cell. After loading, the top of the channel is sealed using an annealed copper cone (P) pressed by a screw (Q). A tiny ruby crystal (R) was placed in the bottom part of the bore away from the focal point of the laser beam for pressure measurements.

The cell was cooled by a controlled continuous flow of liquid nitrogen through copper tubing (S) soldered on a brass jacket which grips the inner cell very tightly at low temperature. The temperature was measured within ±0.1 K by a calibrated copper–constantan thermocouple (T) tactfully sealed (U) to the cell as shown. To overcome small vacuum leaks in the cell the vacuum pump connected at (V) required a continuous operation. With the maximum permissible pumping rate of liquid nitrogen the inner cell's temperature stabilised at 88 K. The cell cannot hold pressure inside it unless the seals of all sapphire windows and piston are perfect. The piston seal was made by using a wedge ring of phosphor–bronze (W) and a 0-ring. The wedge ring was coated with soft alloy tin solder. This ring does not seal until the pressure becomes sufficiently high and so a supplementary seal is provided by the 0-ring. The seal between sapphire window support plug (X) and the cell body was made essentially by a silicon rubber 0-ring held between a window sealing soft solder coated ring (Y) and a phosphor–bronze ring (Z). These windows are Poulter type with window diameter and diameter of circular aperture of its plug critically chosen to satisfy the standard unsupported area expression. To reduce the stress the edges of the sapphire windows and rods were rounded and the window support plugs were properly blended in after their edges were rounded.

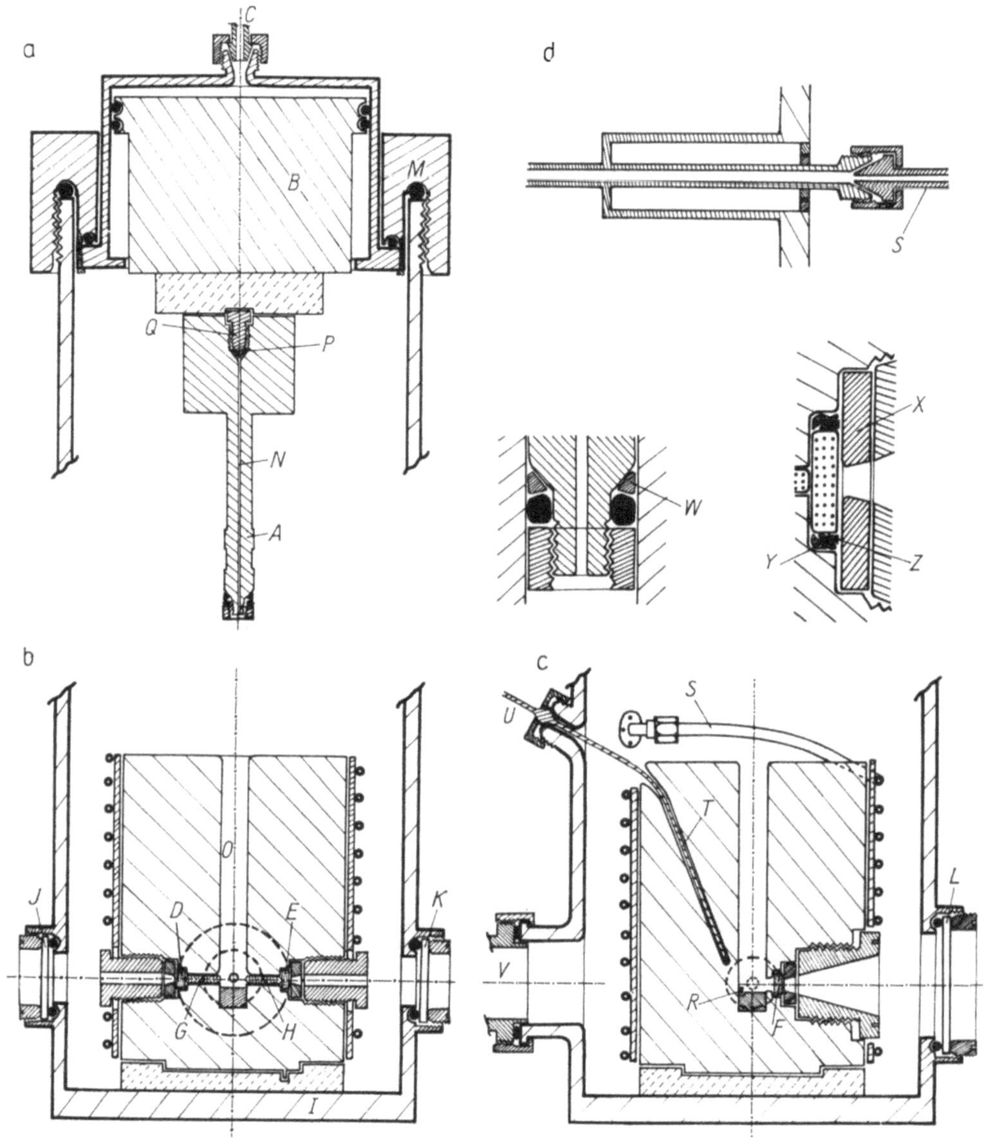

Fig. 2. High-pressure Raman cell for ice measurements: hatched areas metal; pointed areas sapphire; dash — hatched areas tuflon; ● rubber ○ ring. a) Cross section of upper part of the cell; b) cross section of lower part across the laser window; c) cross section of lower part across Raman window; d) details of piston, sapphire window, and nitrogen carrying tube seals

2.2 Pressure measurement

The ruby fluorescence line shift with pressure was used for pressure calibration (± 0.01 GPa). The reproducibility of the wave number scale was checked in each run by simultaneous measurement of the sharp atomic emission line at 14431.1 cm^{-1} of neon gas lamp. Both ruby lines, R_1 and R_2, were kept under observation. These lines shift at the same rate (separation ΔR is constant) under hydrostatic pressure, but in

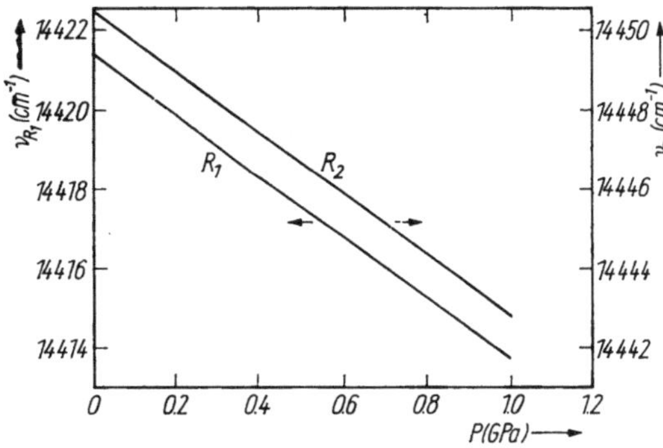

Fig. 3. Pressure calibration curves for ruby at 88 K

case the pressure becomes non-hydrostatic R_1 and R_2 move at different rates and consequently ΔR changes. The development of non-hydrostatic component may also increase their line-widths. At room temperature (293 K) the value of $(\partial\nu_R/\partial p)_T$ for ruby lines is (-7.57 ± 0.04) cm^{-1}/GPa. At low temperature a small decrease was reported earlier [17] and a value of (-7.3 ± 0.2) cm^{-1}/GPa was obtained at 166.8 K taking a constant pressure shift upon cooling at a constant applied force. This is not confirmed by a more accurate and direct calibration [18] which gave a temperature independent value of (-7.6 ± 0.2) cm^{-1}/GPa between 0 and 300 K. Fig. 3 shows the calibration curves of ruby lines at 88 K having this value of the slope, $(\partial\nu_R/\partial p)_T$. In some cases each of the two ruby lines appeared as a doublet with a splitting of about 2 cm^{-1}. The splitting of the ground state into a Kramer doublet may produce a similar effect but the resulting splitting is only 0.38 cm^{-1} The presence of ruby 1 and ruby 2 in the ruby chip seems to be a more acceptable explanation [18].

2.3 Recording of Raman spectra

The Raman spectra were excited with 514.5 nm radiation of a Spectra Physics argon ion laser of 1 W output power and recorded on a Spex Ramalog 5 M system. Redistilled and deionized water was used and before filling, it was boiled to remove dissolved gases from it. No air bubble was left inside the cell because laser light is very badly scattered from it. The filled cell was tested on a hydraulic press for any leaking sapphire window or piston seal before mounting the cell inside its outer vacuum jacket. The application of a small pressuıe of about 0.6 MPa to solidifying water produced spectroscopically better quality ice Ih.

The aim of the present study was to observe the behaviour of ice Ih as its pressure is increased up to about 0.7 GPa at a temperature as close to liquid nitrogen as possible. Unfortunately, it did not prove possible to pressurise ice at the lowest 88 K temperature of the cell, so the following procedure was adopted. For each pressure reading the ice was heated to $\approx$ 150 K, pressurised cyclically for several minutes within $\pm$ 0.05 GPa of the final stable pressure and then cooled to 88 K. The temperature was not allowed to rise above 150 K to avoid any possible complication. The Raman spectrum was recorded with an accuracy of $\pm$0.1 cm^{-1} in the lattice vibrational region and $\pm$1 cm^{-1} in the O–H stretching region.

3. Results and Discussion

3.1 Raman spectra of ice phase transition

The observed pressure dependence of the Raman spectrum of H_2O ice at 88 K in the lattice vibrational and O–H stretching regions are presented in Fig. 4 a and b, re-

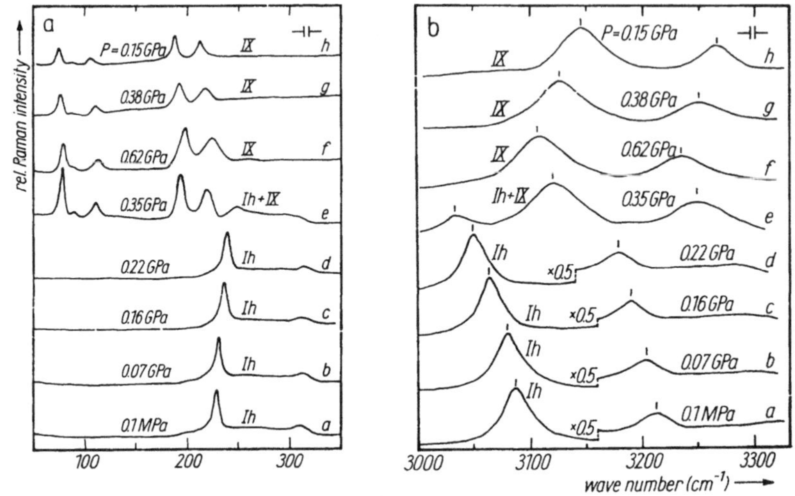

Fig. 4. Pressure dependence of Raman spectra of H_2O ice establishing the ice Ih → ice IX transition a) in the lattice vibrational region and b) in the O–H stretching region at 88 K

spectively. The pressure was increased from 0.1 MPa to 0.62 GPa and then decreased from 0.62 to 0.15 GPa. Thus the sequence of recording of the spectrum was from a to h.

The Raman spectra of the reclaimed samples (at atmospheric pressure) of ice Ih (100 K), ice II (73 K), and ice IX (38 K) in the lattice vibrational region as reported by earlier workers [3, 10] are reproduced in Fig. 5a for comparison. In this region the Raman spectrum of ice Ih [3] is dominated by a strong peak at about 227 cm^{-1} and there is no intense band below 200 cm^{-1}. The Raman spectra of both ice IX and ice II have [10] four sharp intense bands roughly at (70, 100, 185, and 205 $\pm$ 5) cm^{-1} but

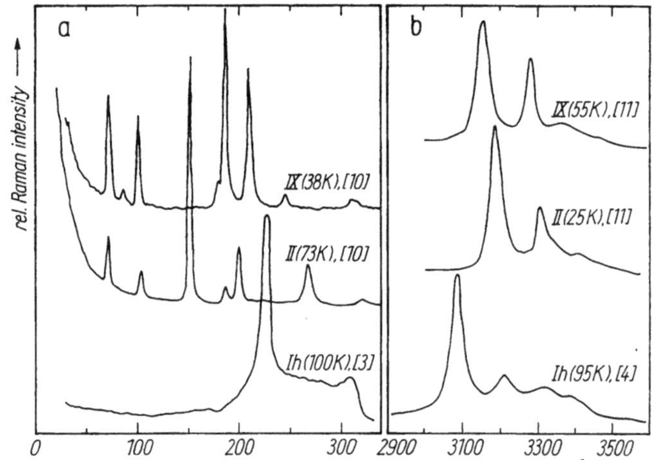

Fig. 5. Comparison of Raman spectra of H_2O ices Ih, II, and IX a) in the lattice vibrational region and b) in the O–H stretching region at atmospheric pressure (reclaimed sample) and low temperature as indicated

ice II gives an additional prominent peak at $\approx$ 150 cm^{-1} which is the most intense low-frequency peak of the Raman spectrum of ice II. The total absence of any feature around 150 cm^{-1} in the Raman spectrum of ice IX may be used to distinguish ice IX from ice II. Fig. 5b compares the reported [4, 11] Raman spectra of the reclaimed samples of ice Ih (95 K), ice II (25 K), and ice IX (55 K) in O–H stretching region at atmospheric pressure. The strongest band always has the lowest frequency in this region, and for ice Ih it probably corresponds to coupled in-phase symmetric stretching (ν_1) vibrations of water molecules. The next band on the higher-frequency side is the second intense band. The frequencies of these two main peaks are (3083 and 3209 $\pm$ $\pm$ 2) cm^{-1} in ice Ih at 95 K; 3193.4 and 3311 cm^{-1} in ice II at 90 K, and 3157.8 and 3278.9 cm^{-1} in ice IX at 90 K.

The spectrum of ice at 0.1 MPa and 88 K (curves a of Fig. 4 a and b) corresponds to ice Ih and upon pressurising it up to 0.22 GPa (curves b to d of Fig. 4 a and b) there is no indication of any phase change. The bands in the lattice vibrational region (Fig. 4a) shift towards the higher-frequency side whereas those in the O–H stretching region towards the lower-frequency side continuously as the pressure increases. At 0.35 GPa pressure (curve e of Fig. 4a) four sharp additional bands appeared at 77.2, 110.4, 193.6, and 218.8 cm^{-1} and the ice Ih peak intensity decreased markedly, indicating the onset of a phase transition. Although both ices, IX and II, have four peaks near these positions, the total absence of the most intense peak of ice II around 150 cm^{-1} in the observed spectrum unambiguously identifies the new phase as ice IX. In the O–H stretching region also (curve e of Fig. 4b) new bands appeared at 0.35 GPa near 3120 and 3249 cm^{-1}. These positions are close to O–H stretching bands of ice IX and not to those of ice II, confirming the above identification. The spectrum at 0.35 GPa (curves e of Fig 4 a and b) represents a mixture of ice Ih and ice IX. As the pressure is further increased to 0.62 GPa (curves f of Fig. 4 a and b) the ice Ih peaks and its feature vanished from Raman spectra, establishing that the transformation of ice Ih into ice IX has completed. There are no signs of recovering ice Ih after depressurising ice IX to 0.15 GPa (curves g and h of Fig. 4 a and b). Ice IX peaks shifted regularly to lower frequency side in lattice vibrational region and to higher frequency side in O–H stretching region as the pressure is released and no new bands appeared.

Ice Ih thus transformed into ice IX when a pressure of at least 0.35 GPa was applied to it at a temperature below 150 K. Ice II was not present, even in traces, in the observed spectrum and transformation of ice II into ice IX is an improbable transformation [12]. Therefore, ice Ih transformed into ice IX without converting into ice II. It has been reported [1, 12] that ice IX could be obtained from ice III at around 0.3 GPa by cooling it at a rate not less than (1 to 2) K/min through the region of stability of ice II without converting [9] into ice II. At present not much is known about the phase diagram of ice (Fig. 1) in the region of interest and there are serious uncertainties in it. At a temperature below 150 K application of pressure transformed ice Ih into ice IX either directly or through some intermediate ice phase. In the latter case ice III may be suggested as a possible candidate. It has been reported earlier [12] that ice III appeared as an intermediate phase in the ice Ih → ice II transition at a temperature as low as 210 K. One may consider the possibility of extension of this down-falling of the ice III cusp (C in Fig. 1) into the region of present interest. This requires actual observation of ice III during the pressurisation process. Further work is necessary.

3.2 Analysis of pressure and temperature data

The observed frequencies of the Raman bands in ice Ih and ice IX at 88 K and different pressures are listed in Table 1. The values at atmospheric pressure reported by

Table 1

Frequencies of Raman modes of ices Ih and IX at different pressures and 88 K

H_2O ice Ih at 88 K				H_2O ice IX at 88 K						
pressure (GPa)	frequency of Raman bands (cm^{-1})			pressure (GPa)	frequency of Raman bands (cm^{-1})					
	lattice vibrational	O–H stretching			lattice vibrational				O–H stretching	
0.0001	228.3	3087.7	3213	0.0001	73.1***)	101.7***)	185.6***)	208.4***)	3157.8****)	3278.9****)
	227.6*)	3083**)	3209**)		76.2***)					
0.059	231.3	3077.2	3205	0.145	75.4	106.1	188.1	212.1	3145.3	3266
0.066	230.6	3080	3204	0.353	77.2	110.4	193.6	218.8	3120	3249
					89.2					
0.099	231.7	3075	3201	0.381	76.9	109.1	192.3	217.6	3126	3250
0.158	235.4	3063.8	3191.6	0.62	79.1	112.1	198.1	223.6	3106	3234
0.161	236.8		3184.5							
0.178		3058								
0.218	238.8	3050.2	3178.5							
0.263	243.5	3046.7								
0.353	248.4	3034								

*) [3] at 100 K.
**) [4] at 95 K.
***) [10] at 80 K.
****) [11] at 90 K.

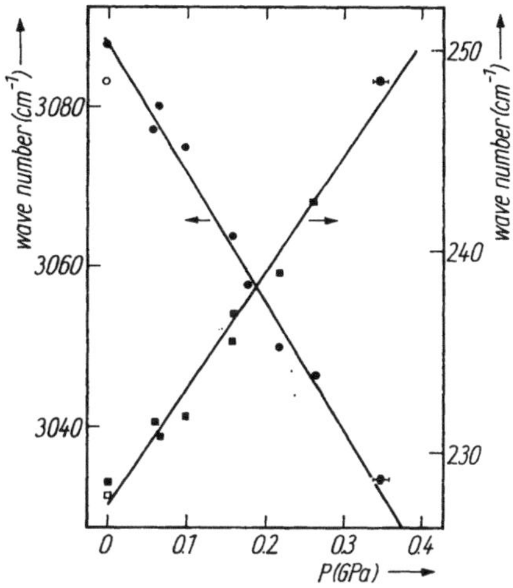

Fig. 6. Variation of frequencies of Raman modes with pressure for ice Ih in lattice vibrational (■) and O–H stretching (●) regions at 88 K (earlier reported data at atmospheric pressure, □ [3] at 100 K and ○ [4] at 95 K)

earlier workers [3, 4, 10, 11] are also included but the temperatures of lattice vibrational and O–H stretching modes are 100 and 95 K, respectively, for ice Ih, whereas for ice IX these temperatures are 80 and 90 K, respectively. Ice IX data at 0.1 MPa are for the reclaimed sample [10, 11].

It can be noted that for ice Ih (Table 1) the frequency of the main peak in the lattice vibrational region increases from 228.3 to 248.4 cm^{-1} as the pressure is increased from 0.1 MPa to 0.353 GPa. A least-square linear fit to the data is shown in Fig. 6 and the obtained value of $\partial \nu/\partial p$ is 57.8 cm^{-1}/GPa for this peak. In the O–H stretching region, upon pressurising the frequency of the peaks decreases. The lowest-lying most intense peak in this region shifts linearly from 3087.7 to 3034 cm^{-1} as the pressure is changed from 0.1 MPa to 0.353 GPa. The fitting of data (Fig. 6) gave a slope ($\partial \nu/\partial p$) of -159.8 cm^{-1}/GPa.

The ice IX crystal has theoretically twenty-one Raman active fundamental lattice vibrations of $4A_1$, $5B_1$, $4B_2$, and 8E symmetry. However, only eleven modes could be reported [10] so far at ambient pressure (Fig. 5a). Other modes may have very low intensity and/or the modes could be grouped together. To date no information is available about the symmetry of the vibration which causes a particular Raman peak. In our Raman spectra (Fig. 4a) at high pressures all the four intense bands of ice IX and in addition a fifth weak feature at around 90 cm^{-1} were present.

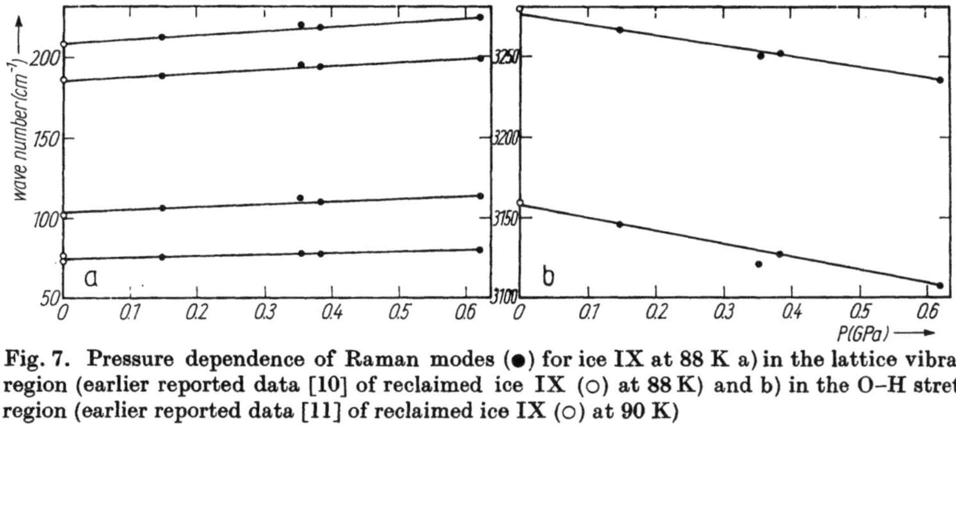

Fig. 7. Pressure dependence of Raman modes (●) for ice IX at 88 K a) in the lattice vibrational region (earlier reported data [10] of reclaimed ice IX (○) at 88 K) and b) in the O–H stretching region (earlier reported data [11] of reclaimed ice IX (○) at 90 K)

Table 2
Various coefficients of ices Ih and IX at 88 K

coefficient	units	H_2O ice Ih		H_2O ice IX					
		lattice vibrational	O–H stretching	lattice vibrational				O–H stretching	
$\nu_{88\text{ K}, 0.1\text{ MPa}}$	cm^{-1}	227.2	3089	74.5	103.5	185	208.5	3158	3276
$(\partial\nu/\partial p)_{T = 88\text{ K}}$	cm^{-1}/GPa	57.8	−159.8	7	14	20.5	24	−83	−68
$(\partial\nu/\partial T)_{p = 0.1\text{ MPa}}$	cm^{-1}/K	−0.051	0.24	−0.01*)	−0.035*)	−0.035*)	−0.04*)	0.2**)	0.14**)
$(\partial\nu/\partial T)_V$	cm^{-1}/K	−0.041	0.21						
$(\partial\nu/\partial R_{O-O})_P$	cm^{-1}/nm	−29310	137950						
$(\partial\nu/\partial R_{O-O})_T$	cm^{-1}/nm	−5730	15870						
Γ		2.302	−0.47						

*) See [10].
**) See [11].

The frequencies of the four peaks of ice IX in the lattice vibrational region at different pressures are plotted in Fig. 7a. It can be noted that the frequency increases linearly with pressure. The values of the slopes ($\partial\nu/\partial p$) obtained from the least-square fits [19] are listed in Table 2. Roughly these are around $+20$ cm^{-1}/GPa except for the lowest-lying peak. Extrapolation of the straight lines of Fig. 7a to atmospheric pressure gave predicted frequencies of the four peaks of ice IX as 74.5, 103.5, 185, and 208.5 cm^{-1}, at 88 K and 0.1 MPa. These frequencies are close to the reported observed values 73.1 and 76.2, 101.7, 185.6, and 208.4 cm^{-1}, for reclaimed ice IX at 80 K [10]. In the O–H stretching region the two prominent bands of ice IX lie above 3100 cm^{-1} (Table 1) and their pressure dependence between 0.145 and 0.62 GPa (at 88 K) is plotted in Fig. 7b. There is a negative pressure shift and from a least-square linear fit, the obtained values of $\partial\nu/\partial p$ are -83 and -68 cm^{-1}/GPa for the two bands. Extrapolating the two straight lines of Fig. 7b to atmospheric pressure one gets predicted frequencies of the two O–H stretching bands of ice IX at 88 K and 0.1 MPa as 3158 and 3276 cm^{-1}. These values remarkably agree with the reported observed values [11] of 3157.8 and 3278.9 cm^{-1} for reclaimed ice IX at 90 K. It may be noted that all points corresponding to 0.353 GPa pressure at which the phase transition occurred, lie relatively from the straight lines of Fig. 7 a and b. The frequencies of the lattice vibrational modes are higher and of the O–H stretching modes are lower than those calculated at 0.353 GPa from the pressure dependence of these modes. At this pressure both ice IX and ice Ih were present and probably the equilibrium was not attained completely. The ice Ih shrinks upon transformation into ice IX (ice Ih, four molecules per unit cell, density 0.931; and ice IX, 12 molecules per unit cell, density 1.16) and therefore it affects the pressure. Very small non-hydrostatic component was actually noticed at this pressure.

Although transformation from ice Ih to ice IX is accompanied by more efficient packing, the optic mode frequency is found to decrease at the phase transition (Table 1). It is so because at the phase transition, in spite of a marked decrease in the volume, there is an increase in the O–H ... O distance from 274.7 pm (ice Ih) to 279.3 pm (ice IX). In one phase, the O–H ... O distance decreases with increasing pressure due to compression and therefore, the lattice vibrational mode frequency increases continuously with pressure. The O–H stretching modes showed just the opposite behaviour. As a given ice structure (ice Ih or IX) is pressurised, the O–H stretching frequency drops, which indicates that the hydrogen bond gets stronger with increasing pressure in the same structure. On passing through the phase transition (ice Ih → → ice IX) there is a sudden rise in O–H stretching frequency due to the appearance of a new ice phase (ice IX) of weaker hydrogen bond.

The shift of ice Ih Raman bands with temperature was also followed in our experiments at atmospheric pressure and the observed values of $(\partial\nu/\partial T)_p$ at 0.1 MPa and around 88 K are -0.051 and 0.24 cm^{-1}/K for the Raman bands at 228.3 and 3087.7 cm^{-1}, respectively. No change in the relative intensities of Raman features could be detected with change in temperature. The reported low-temperature data [10, 11] for reclaimed ice IX have been analysed both in lattice vibrational and O–H stretching regions and the obtained values of $(\partial\nu/\partial T)_p$ at atmospheric pressure are listed in Table 2. These values are around -0.04 and 0.2 cm^{-1}/K for lattice vibrational (except for the lowest-frequency mode) and O–H stretching modes, respectively. It may be noted that the temperature coefficients for ice IX modes at ambient pressure are only slightly smaller than the corresponding coefficients for ice Ih, whereas the pressure coefficients for ice IX at 88 K were not even half of the corresponding coefficients for ice Ih. The temperature and pressure coefficients of the lowest-frequency peak of ice IX are remarkably lower than those of other peaks of ice IX. It is too early to draw an inference.

3.3 Change with intermolecular distance of ice Ih

The coefficients $(\partial\nu/\partial T)_p$ and $(\partial\nu/\partial p)_T$ may be used to derive $(\partial\nu/\partial T)_V$ using the following relation:

$$\left(\frac{\partial\nu}{\partial T}\right)_V = \left(\frac{\partial\nu}{\partial T}\right)_p + \frac{\alpha}{\beta}\left(\frac{\partial\nu}{\partial p}\right)_T, \tag{1}$$

where α and β are volume thermal expansivity and compressibility, respectively. At 88 K and atmospheric pressure, $\alpha = 19\ \mathrm{MK^{-1}}$ [20] and $\beta = (0.11 \pm 0.01)\ \mathrm{GPa^{-1}}$ [21] for ice Ih. The obtained values of $(\partial\nu/\partial T)_V$ are $-0.041\ \mathrm{cm^{-1}/K}$ at $227.2\ \mathrm{cm^{-1}}$ and $0.21\ \mathrm{cm^{-1}/K}$ at $3089\ \mathrm{cm^{-1}}$.

The magnitudes of temperature coefficients at constant pressure and at constant volume suggest that the change in the lattice vibrational frequency of ice Ih with temperature near 88 K is about 80% due to the increase in temperature alone and the remaining 20% is due to the accompanied increase in volume. Again, the change in the O–H stretching frequency with temperature near 88 K is found to be about 90% due to the change in temperature alone and 10% due to the change in volume with temperature. Johari et al. [6, 7] reported that for ice Ih at 255 K, the change in lattice vibrational frequency with temperature is as much as 45% due to change in volume with temperature and the change in O–H stretching frequency with temperature is 30% due to change in volume with temperature. This, comparatively larger share of change in volume with temperature at 255 K, may be attributed mainly to the high value of α at 255 K ($\alpha = 155\ \mathrm{MK^{-1}}$ at 255 K and $\alpha = 19\ \mathrm{MK^{-1}}$ at 88 K).

In ice Ih, at 88 K and atmospheric pressure, the intermolecular distance $R_{\mathrm{O-O}}$ is 0.274 nm ([22] data adjusted using [20]). Therefore, under these conditions $(\partial R_{\mathrm{O-O}}/\partial T)_p = 1.7398\ \mathrm{nm/MK}$ and $(\partial R_{\mathrm{O-O}}/\partial p)_T = -10.072\ \mathrm{nm/TPa}$ from the values of α and β. Now writing

$$\left(\frac{\partial\nu}{\partial R_{\mathrm{O-O}}}\right)_p = \frac{(\partial\nu/\partial T)_p}{(\partial R_{\mathrm{O-O}}/\partial T)_p}, \tag{2}$$

and

$$\left(\frac{\partial\nu}{\partial R_{\mathrm{O-O}}}\right)_T = \frac{(\partial\nu/\partial p)_T}{(\partial R_{\mathrm{O-O}}/\partial p)_T}, \tag{3}$$

these coefficients were determined and listed in Table 2. It is found that for the same decrease in $R_{\mathrm{O-O}}$, the isobaric cooling shift in frequency is 5.1 times (8.7 times) as large as the isothermal compression shift in the lattice vibrational region (in the O–H stretching region) at 88 K in ice Ih. Johari et al. data [6, 7] at 255 K temperature showed that for the same decrease in $R_{\mathrm{O-O}}$ isobaric cooling shifts ν 2.2 times (3.6 times) more than does the isothermal compression in the lattice vibrational region (in the O–H stretching region) of ice Ih. The handling of $(\partial\nu/\partial R_{\mathrm{O-O}})$ in the model calculations of ice Ih needs extra care because it is not independent of the manner by which the change is affected. The value of the mode Grüneisen parameter Γ, which represents the change in frequency with volume was calculated at 88 K using the following relation:

$$\Gamma \equiv -\left(\frac{\partial \ln \nu}{\partial \ln V}\right)_T = \frac{1}{\beta\nu}\left(\frac{\partial\nu}{\partial p}\right)_T. \tag{4}$$

In the lattice vibrational and O–H stretching regions the obtained values are 2.302 and −0.47, respectively.

Comparing the present results at 88 K with those of Johari et al. at 255 K, it is clear that as the temperature is lowered from 255 to 88 K in ice Ih, the magnitudes of $(\partial \nu/\partial T)_p$ and $(\partial \nu/\partial T)_V$ decrease by nearly a factor 2.6 and 2, respectively, whereas the magnitudes of $(\partial \nu/\partial p)_T$, $(\partial \nu/\partial R_{O-O})_T$, and Γ increase by roughly 25% and that of $(\partial \nu/\partial R_{O-O})_p$ increases as much as 2.5 times. The rate at which the mode frequency shifts with temperature is a measure of the effect of temperature on the intermolecular coupling, intermolecular distance, and anharmonicity of the vibration. The observed remarkable decreases in this rate and in thermal expansivity of ice Ih as the temperature is lowered from 255 to 88 K, are interconnected. The slight increase, on the other hand, in the pressure coefficient upon cooling implies that the application of pressure changes the shape of the intermolecular potential of ice Ih more effectively at a lower temperature. Further experimental work is needed to verify it. Thermal expansivity and compressibility data of ice IX are not available at 88 K and therefore, a similar discussion of the ice IX pressure data may only be possible in future.

3.4 Effect of isotopic substitution

When the experiments were conducted with D_2O ice Ih, its transformation into phase IX was confirmed at 0.23 GPa and 88 K. All general features of H_2O ice were present in D_2O ice but the phase transition appeared to be more sluggish. Raman bands at 220.2 and 2287.4 cm^{-1} of D_2O ice Ih at atmospheric pressure and 88 K shifted at a rate of 40.2 and −112.2 cm^{-1}/GPa, respectively with increasing pressure and at a rate of −0.045 and 0.16 cm^{-1}/K, respectively with temperature. The values of the rate of change of frequency with pressure for O–H stretching and for O–D stretching modes have a ratio of 1.424:1 which is in agreement with the expected value $\sqrt{2}:1$. The effect of change in pressure and temperature, when converted in terms of the resulting change in O–D ... O separation, showed that for 1 pm decrease in O–D ... O separation, the Raman band of D_2O ice Ih in the lattice vibrational region at 220.2 cm^{-1} and in the O–D stretching region at 2287.4 cm^{-1} shifted by 25.87 and −91.97 cm^{-1}, respectively, for isobaric change, and by 3.99 and −11.14 cm^{-1}, respectively, for isothermal change. Thus at 88 K and atmospheric pressure in D_2O ice Ih, the isobaric cooling shift in the lattice vibrational region (O–D stretching region) is 6.5 times (8.25 times) larger than the corresponding isothermal compressional shift for the same decrease in O–D ... O separation.

Acknowledgements

The present work was carried out during author's stay at King's college, London. The collaboration with Mr. G. E. Slark, Dr. W. F. Sherman, and Prof. G. R. Wilkinson is gratefully acknowledged.

References

[1] E. Whalley, J. B. R. Heath, and D. W. Davidson, J. chem. Phys. **18**, 2362 (1968).
[2] O. Mishima and S. Endo, J. chem. Phys. **73**, 2454 (1980).
[3] P. T. T. Wong and E. Whalley, J. chem. Phys. **65**, 829 (1976).
[4] P. T. T. Wong and E. Whalley, J. chem. Phys. **62**, 2418 (1975).
[5] T. C. Shivkumar, S. A. Rice, and M. G. Sceats, J. chem. Phys. **69**, 3468 (1978).
[6] G. P. Johari and T. C. Shivkumar, J. chem. Phys. **69**, 5557 (1978).
[7] T. C. Shivkumar, H. A. M. Chew, and G. P. Johari, Nature **275**, 524 (1978).

[8] G. P. Johari and H. A. M. Chew, Phil. Mag. **B49**, 647 (1984).
[9] J. E. Bertie, L. D. Calvert, and E. Whalley, J. chem. Phys. **38**, 840 (1963).
[10] J. E. Bertie and B. F. Francis, J. chem. Phys. **77**, 1 (1982).
[11] J. E. Bertie and B. F. Francis, J. chem. Phys. **72**, 2213 (1980).
[12] B. M. Sukarova, W. F. Sherman, and G. R. Wilkinson, J. mol. Structures **115**, 137 (1984).
[13] O. Haida, T. Matsuo, H. Suga, and S. Seki, Proc. Japan. Acad. **48**, 489 (1972).
[14] J. M. Kennedy, W. F. Sherman, N. Treloar, and G. R. Wilkinson, Proc. V Internat. Conf. Raman Spectroscopy, Schulz Verlag, Freiburg 1976 (p. 76).
[15] B. Sukarova, W. F. Sherman, and G. R. Wilkinson, J. Phys. C **17**, 5833 (1984).
[16] G. E. Slark, A. K. Garg, W. F. Sherman, and G. R. Wilkinson, J. mol. Structures **115**, 161 (1984).
[17] D. M. Adams, R. Appleby, and S. K. Sharma, J. Phys. E **9**, 1140 (1976).
[18] R. A. Noack and W. B. Holzapfel, High Pressure Science and Technology, Plenum Press, New York 1979 (p. 748).
[19] A. K. Garg, Solid State Phys. (India) **27C**, 283 (1984).
[20] V. R. Brill and A. Tippe, Acta cryst. **23**, 343 (1967).
[21] A. J. le Adbetter, Proc. Roy. Soc. **A287**, 403 (1965).
[22] S. W. Peterson and H. A. Levy, Acta cryst. **10**, 344 (1957).

(Received October 1, 1987; in revised form August 25, 1988)

phys. stat. sol. (a) **110**, 481 (1988)

Subject classification: 62.40; S4

Physics Department, Faculty of Education, Ain Shams University, Cairo[1])

Effect of Loading on the Recovery of Internal Friction of an Al–1.0 wt%Mn–0.28 wt%Fe Alloy

By

G. GRAISS, G. SAAD, and A. FAWZY

The phenomenon of internal friction recovery of Al–1.0 wt%Mn–0.28 wt%Fe alloy is studied after cold-working the specimens by twist using the free decay method. The wire specimens are investigated for internal friction recovery while they were loaded — within the elastic limit — by different loads at different working temperatures. It is found that temporarily loading enhances the rate of recovery. The results are explained on the basis of dislocation damping adopting the Granato-Lücke model for recovery of internal friction. The activation energy for the process is found to be ≈0.1 eV, such a value leads to the assumption that recovery occurs through binding between the dislocation network and solute atoms of Mn and Fe acting as active pinners for dislocation motion.

Das Phänomen der Erholung der inneren Reibung einer Al–1,0 Gew%Mn–0,28 Gew%Fe-Legierung wird nach der Kaltbehandlung der Proben durch Verdrillung mit der Methode des freien Abklingens untersucht. Die Drahtproben werden für die Erholung der inneren Reibung untersucht, während sie — innerhalb des elastischen Grenzfalls — bei verschiedenen Arbeitstemperaturen belastet werden. Die Ergebnisse werden erklärt auf der Grundlage von Versetzungsdämpfung, wobei das Granato-Lücke-Modell für die Erholung der inneren Reibung angenommen wird. Die Aktivierungsenergie für den Prozeß wird zu ≈0,1 eV gefunden. Ein derartiger Wert führt zu der Annahme, daß die Erholung durch Bindung zwischen dem Versetzungsnetzwerk und gelösten Atomen von Mn und Fe erfolgt, die als aktive Verankerungen für Versetzungsbewegung wirken.

1. Introduction

The Köster effect [1] is known to be the transient increase in damping observed in metals after plastic deformation, and which was found to decrease more or less rapidly as a function of time after straining. Such a recovery of the effect is ascribed either to the pinning of dislocations by point defects or to rearrangement of the dislocation network resulting in an overall decrease in dislocation damping.

Reviewing the literature of previous work on the recovery of internal friction by a torsion pendulum one can recognize — as a matter of experimental procedure — that the stress acting on the test specimen is virtually zero. Now a question arises concerning the effect of temporarily loading the specimen. In other words if the test specimen is loaded — within the elastic limit — for certain periods of time just before measuring its internal friction, how such loading will affect the recovery behaviour. Thus, the present work is devoted to investigate under different working temperatures the effect of loading on the recovery of cold-worked internal friction in the as-quenched Al–1.0 wt%Mn–0.28 wt%Fe alloy.

[1]) Heliopolis, Cairo, Egypt.

2. Experimental Procedure and Technique

The preparation of the alloy in the form of a wire of 0.35 mm diameter was described elsewhere [2, 3]. All the investigated specimens were solution heat treated at 620 °C for 4 h, then quenched in iced water. Internal friction was determined by the free decay method by means of a low-frequency ($\approx$ 1 Hz) torsion pendulum with electromagnetic excitation and optical read-out. Pretorsional deformation [4] was introduced only at RT (20 °C). The degree of torsional deformation ϑ is measured by the dimensionless quantity ND/L where N is the number of twist turns, D and L are the diameter and length of the test specimen, respectively. A value of $\vartheta = 0.02$ was given to all test specimens. The recovery behaviour under the different loads 0.45, 0.95, 1.45, and 1.95 N was studied at working temperatures 20, 60, 90, and 120 °C. The internal friction Q^{-1} was first measured before deformation, this is denoted by Q_u^{-1}. The wire specimen then deformed with $\vartheta = 0.02$ and the internal friction is immediately measured, and Q_0^{-1} corresponding to a time $t = 0$ is determined. Setting the specimen at a certain working temperature, a certain load was applied for a certain time ranging between 1 min at the beginning of the recovery process to 3 min at its end. The load is then removed and Q^{-1} is measured. The above step is repeated by loading the specimen for an additional period of time, then the load is removed, and Q^{-1} is again measured. In every step Q_t^{-1} is determined where t is the time measured from the beginning of the recovery process. The short intervals during which the specimen was unloaded for internal friction measurement are actually involved in t, thus a negligibly small error in the value of Q_t^{-1} exists. Calculations which took such an error into consideration proved that it has no appreciable effect on the value of Q_t^{-1} as a function of t.

3. Results and Observations

The present results showed that the recovery of internal friction is affected by the amount of load applied and the working temperature. The recovery behaviour is presented as a relation between the normalized internal friction $S = (Q_t^{-1} - Q_u^{-1})/(Q_0^{-1} - Q_u^{-1})$ and the time t (as will be seen later, S indicates — on an atomic scale — a damping effect). Fig. 1 shows the recovery behaviour of the alloy specimens under the effect of elastic stressing by different loads at different temperatures as a relation between the normalized internal friction S and the time t. It is seen from Fig. 1 that the decay in S with time can be divided into three stages, namely a rapid initial stage, a nearly $t^{0.6}$ stage (this will be proved later from calculations based on Fig. 2), and finally a slow saturation stage. These stages involved in the recovery behaviour have also been noticed by the present authors in a previous work [5] where the recovery behaviour of the same alloy was investigated under an excitation-induced fatigue effect. We will be only interested in the first and second stages.

4. Discussion

The results of the present work will be analysed in view of the Granato-Lücke [6] break-away mechanism for the recovery of internal friction. In the Granato-Lücke model, attention is restricted to impurity atoms occupying dislocation sites, and hence acting as effective pinning points on a straight dislocation line.

In our solid-solution treated specimens dislocation lines — created by deformation — are surrounded by a Cottrell atmosphere of impurity atoms (Mn and Fe). Immediately after deformation the freshly created dislocations are vibrating with the full network loop length L_n, causing damping with the highest value. This corresponds to

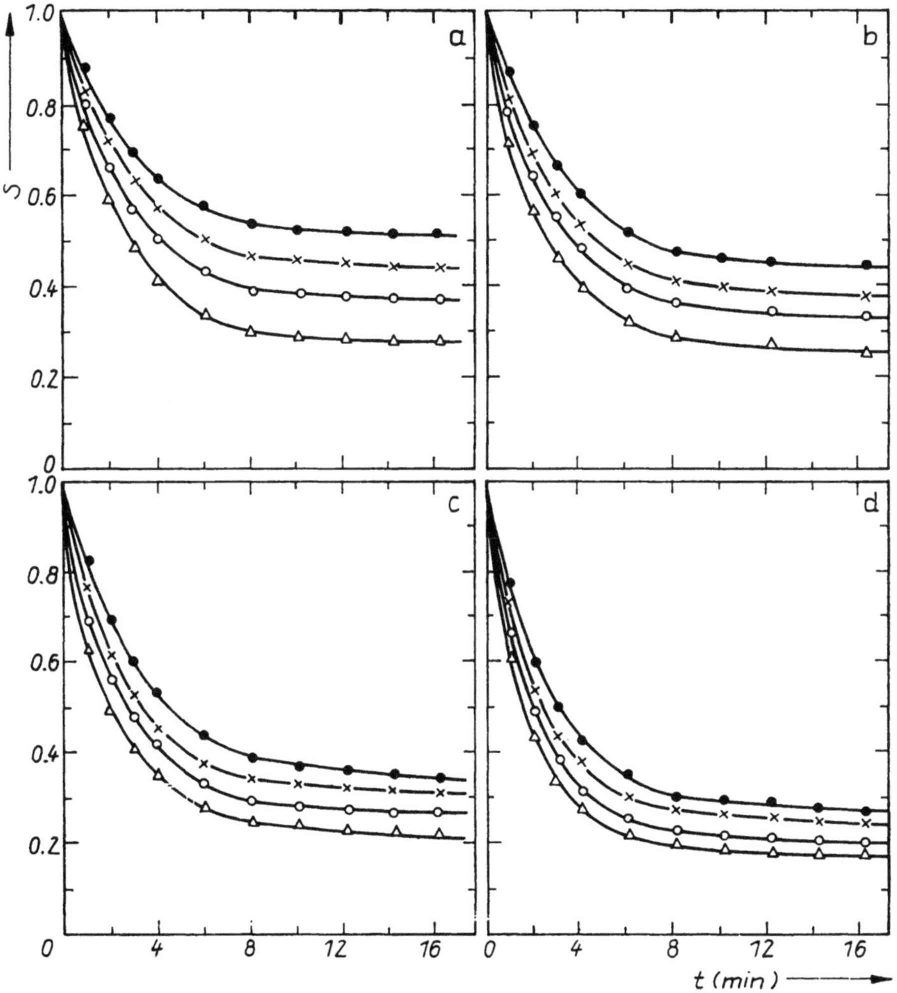

Fig. 1. Normalized internal friction S of predeformed and then loaded specimens as a function of time t. Different loads applied are (●) 0.45, (×) 0.95, (○) 1.45, (△) 1.95 N. The working temperatures are a) 20, b) 60, c) 90, and d) 120 °C

the value of S at $t = 0$ as seen from Fig. 1. On aging, the loop length distribution on each network length will be altered, and different network lengths must be considered to have different concentrations of pinning points. The most important is the decrease in the number of active dislocation sites resulting from an increase of the number of effective pinning points. In this case dislocation damping is determined by the dynamic loss corresponding to an average dislocation loop length L_p (L_p is the average distance between impurity pins which is less than L_n and decreases progressively during aging).

4.1 Rapid stage of recovery

This stage extended up to a time of ≈ 3 min in which the internal friction at a certain working temperature was found to be more drastically decreased the larger the applied load is.

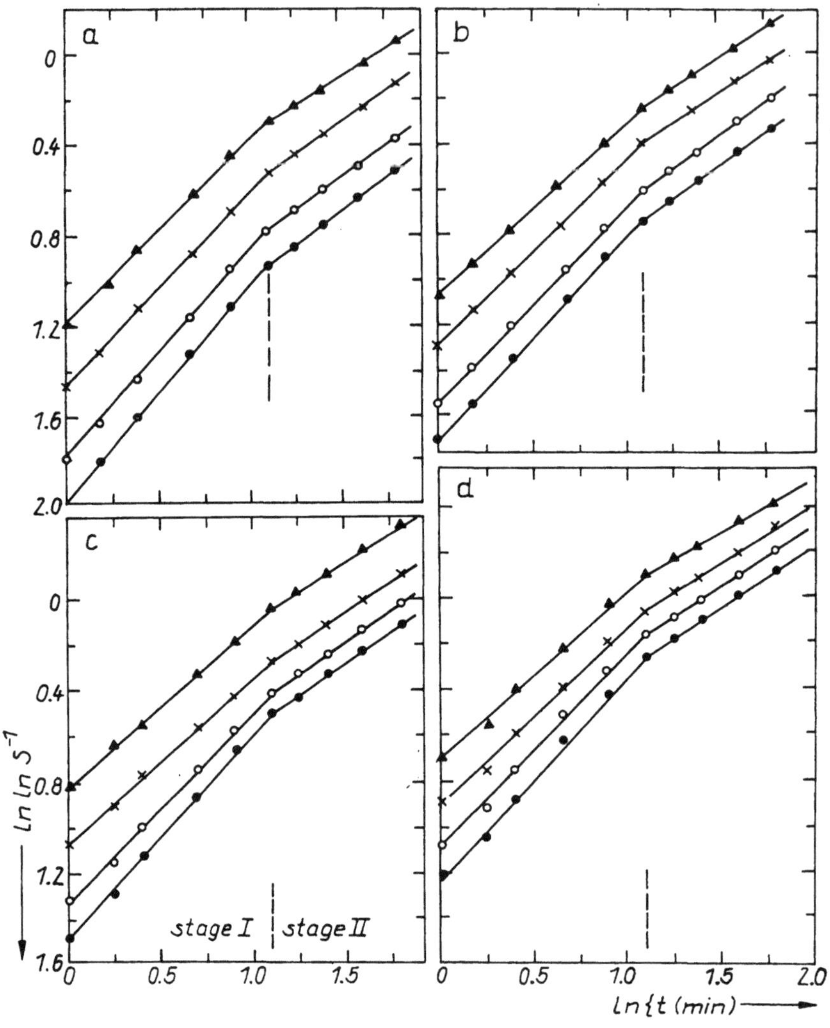

Fig. 2. Plot of ln ln S^{-1} vs. ln t for the differently loaded specimens: a) 0.45, b) 0.95, c) 1.45, and d) 1.95 N. The working temperatures are (●) 20, (○) 60, (×) 90, and (▲) 120 °C

The recovery behaviour as illustrated by the curves shown in Fig. 1 could be described by an Avrami-type relation of the form [5] $S = \exp(-Kt^n)$, where K is the decay rate and n the power of the time dependence of damping. S is a fractional quantity which represents the freely moving dislocations, i.e., those dislocations contributing to the damping effect. Hence the quantity $P = (1 - S)$ represents the fraction of dislocations which have been transferred during a time t to stable equilibrium sites, i.e., representing therefore the decay behaviour. To calculate both n and K the same procedure used by Kenawy and coworkers [5] is followed. Fig. 2 shows a plot of $\ln \ln (S^{-1})$ versus $\ln t$ for specimens differently loaded at different working temperatures. The power n of the time dependence of damping as calculated from the slopes of the parallel straight lines in Fig. 2 has an average value 0.9, indicating that the operating mechanism for the decay of internal friction in this stage

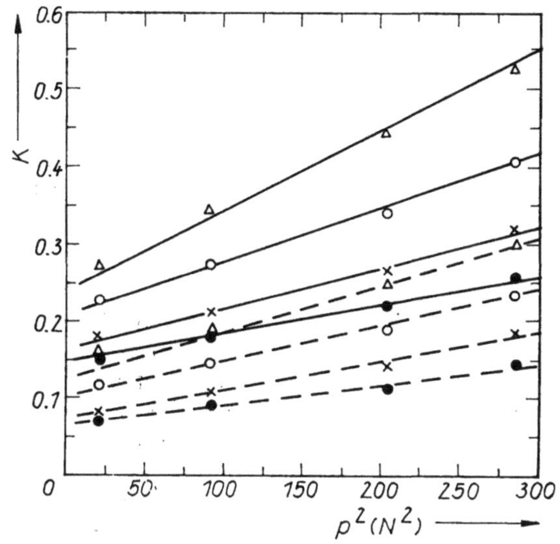

Fig. 3. Relation between the decay rate K and the square of the applied load p^2 at different working temperatures: (●) 20, (×) 60, (○) 90, (△) 120 °C. Solid lines refer to stage I and dashed lines to stage II

is the same irrespective of the conditions of loading and working temperature applied. However this value is high as compared with the ordinarily known limits [7] which range between $\frac{1}{3}$ and $\frac{2}{3}$. Since the value of n characterizes a certain stage of recovery, it is obvious that the relatively high value as obtained for the present alloy indicates that the relevant mechanism is highly time-dependent. In view of the Granato-Lücke model, as mentioned above, this may be attributed generally to the relatively high concentration of Mn and Fe impurities acting as effective pinners in the as-quenched alloy under solid-solution conditions.

Under the effect of loading, the metal crystal is elastically strained due to tensional forces acting upon it. The elastic energy U stored in the crystal during loading with a load p is proportional to p^2 [8]. In view of this the dependence of the decay rate K on the applied load p is shown in Fig. 3 (solid lines) as a relation between K and p^2. It is clear from Fig. 3 that K at a certain working temperature increases linearly with the elastic stored energy U. The elastic energy stored in a crystal lattice during loading is assumed to enhance the rate of pinning exerted by impurities on dislocations. Thus redistribution of dislocation loop lengths will be in favour of a reduction of the effective average loop length L_p. Such a reduction proceeds with a higher rate when a larger load is applied, accordingly, the value of K is increased.

The increased value of K with working temperature is referred to a higher rate with which the redistribution process just mentioned takes place. The effect of temperature change will be discussed later in more detail when the activation energy calculation is to be presented.

4.2 *Nearly $t^{0.6}$ stage of recovery*

This stage starts after the first initial stage and extends for nearly 10 min (see Fig. 1). The power n of the time dependence in this stage was found to have an average value of 0.6, indicating that the recovery process in stage II follows a nearly $t^{2/3}$ law in good agreement with Gordon and Nowick [9] and Kenawy and coworkers [5]. The rate of decay of K in this stage as a function of applied load p is shown in Fig. 3 (dashed lines). Straight lines are obtained as in the case of stage I, with slopes which are less as compared with those corresponding to the same working temperatures in stage I. The similarity of the two groups of curves in the two stages indicates that the operating mechanism for recovery behaviour in the second stage is the same as that in the first stage, namely the redistribution of dislocation loop lengths. The difference is that the rate with which the mechanism proceeds in the second stage is lower. This is conceivable in view of the fact that recovery is basically a time-dependent process, i.e., during stage II the "residual" fraction of dislocations that contribute to damping being less should correspond to a lower rate of recovery as compared to that in stage I.

4.3 Activation energy calculation

The activation energy for the internal friction recovery process may be obtained from a relation between the slopes $\partial K/\partial p^2$ of the straight lines of Fig. 3 at different working temperatures and $1/T$. This is shown in Fig. 4, from which a value of ≈ 0.1 eV for the activation energy is obtained. In view of such a low value for the activation energy we assume the recovery mechanism not to be diffusion. The same conclusion was reported by Chambers and Smoluchowski [10] as a result of their early work on Al and other metals. The value obtained (≈ 0.1 eV) is of the order of magnitude of the binding energy between point defects [11, 12]. Adopting the Granato-Lücke model for the recovery of internal friction as being the loop length redistribution due to the change in concentration of pinning points, we may assume the pinning process to be a binding between an impurity atom (Mn or Fe) and a moving dislocation. A particular atom can jump between the dislocation line and the surrounding atmosphere when it acquires an energy equal to its binding energy. Such an atom being bound to the dislocation site acts as an effective pinning point for the dislocation. Depending on the amount of energy supplied to the specimen either thermally or by loading, we may expect different concentrations of pinning points on a dislocation line leading to different network lengths L_p. At higher temperatures and/or by applying larger loads smaller values of L_p are progressively obtained corresponding to higher values of decay rate K. In view of this, the values of the slopes $\partial K/\partial p^2$ of the straight lines of Fig. 3 at different working temperatures are considered to reflect the pinning interaction of the dislocation segment with its atmosphere. Based upon this, it is conceivable to obtain an activation energy for pinning interaction of nearly the same order of magnitude as that of the binding energy. This may be supported by the finding of Chambers and Smoluchowski who reported [10] that the pinning energies are generally only several times kT. The value of activation energy as ob-

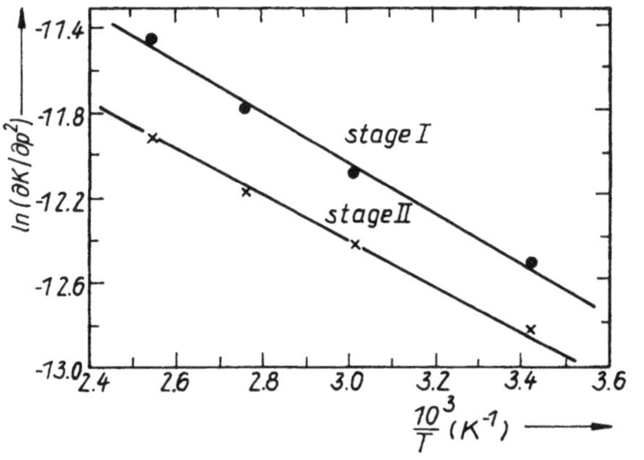

Fig. 4. Arrhenius plot of ln ($\partial K/\partial p^2$) as a function of $10^3/T$ for stages I and II

tained in the present work is of the same order of magnitude as that obtained by the same authors [5] on the same alloy studied using the effect of excitation-induced fatigue.

References

[1] W. Köster, Arch. Eisenhüttenwesen **14**, 271 (1940).
[2] G. Graiss and G. Saad, phys. stat. sol. (a) **92**, 489 (1985).
[3] M. A. Kenawy, G. Graiss, G. Saad, and A. Fawzy, J. Phys. D **20**, 125 (1987).
[4] T. H. Youssef and G. Saad, Phil. Mag. A **42**, 217 (1980).
[5] M. A. Kenawy, G. Saad, G. Graiss, and A. Fawzy, phys. stat. sol. (a) **107**, 171 (1988).
[6] A. V. Granato and Lücke, J. appl. Phys. **27**, 583, 789 (1956).
[7] R. De Batist, Internal Friction of Structural Defects in Crystalline Solids, North-Holland Publ. Co., 1972 (p. 376).
[8] I. H. Hall, Deformation of Solids, Thomas Nelson & Sons Ltd., Great Britain 1968 (p. 119).
[9] R. B. Gordon and A. S. Nowick, Acta metall. **4**, 514 (1956).
[10] R. H. Chambers and R. Smoluchowski, Phys. Rev. **117**, 725 (1960).
[11] O. Takai, R. Yamamoto, M. Doyama, and Y. Hisamatsu, Phys. Rev. B **10**, 3113 (1974).
[12] M. Doyama, J. nuclear Mater. **69/70**, 350 (1978).

(Received June 6, 1988)

phys. stat. sol. (a) 110, 489 (1988)

Subject classification: 68.45; 68.55; S7.12

Institute of Semiconductor Physics, Academy of Sciences of the USSR, Siberian Branch, Novosibirsk[1])

Relaxation Kinetics of a Non-Equilibrium Interface of a Multicomponent Liquid Phase–Binary Substrate on the Example of In–Ga–As/InAs System

By

Yu. B. Bolkhovityanov, R. I. Bolkhovityanova, T. A. Gavrilova, and A. E. Dolbak

It is shown that upon isothermal contact of a substrate and undersaturated multicomponent liquid whose solid equilibrium phase has a smaller lattice parameter then the substrate, relaxation of this non-equilibrium interface proceeds by the mechanism "dissolution–regrowth" with an abnormally high (for LPE) rate of growth of the new phase centres which is accompained by the capture of liquid drops.

Показано, что в случае изотермического контакта подложки и недосыщенной многокомпонентной жидкой фазы, равновесная которой твердая фаза имеет параметр решетки меньше, чем у подложки, релаксация такой неравновесной границы раздела происходит по механизму „растворение–рост", отличительной особенностью которого является чрезвычайно большая для жидкофазной эпитаксии скорость разрастания центров новой фазы, сопровождающаяся захватом капель растворителя.

1. Introduction

At isothermal contact of an undersaturated multicomponent liquid phase (say A–B–C) with a binary AB substrate a layer of ABC solid solution with an imperfect structure, in particular, with inclusion of some drops of the solvent, is formed in some cases. As shown in [1, 2] this form of relaxation of a non-equilibrium interface is observed if $\Delta a/a = (a_0 - a_s)/a_s < 0$ (here a_s is the substrate lattice constant and a_0 the lattice parameter of the solid phase whose equilibrium liquid phase is used for the contact). This imperfect start of heteroepitaxy in the case of $\Delta a/a < 0$ has been the cause for the practically total lack of investigations into this problem. However, for obtaining a complete pattern of the relaxation processes occurring at a liquid/solid interface before and during the LPE of III–V compounds, research of this problem is necessary.

The present work is devoted to studying the initial stages of solid solution formation for the case $a_0 - a_s < 0$, which occurs during contact of an undersaturated In–Ga–As solution with an InAs substrate.

2. Experimental

The contact temperature was chosen to be 600 °C, the InAs substrates were (111) A oriented and the composition of the In–Ga–As solution was calculated according to [3]. The quantity of gallium which was added to the In–As liquid phase was changed

[1]) Prospekt Akad. Lavrenteva 13, SU-630090 Novosibirsk 90, USSR.

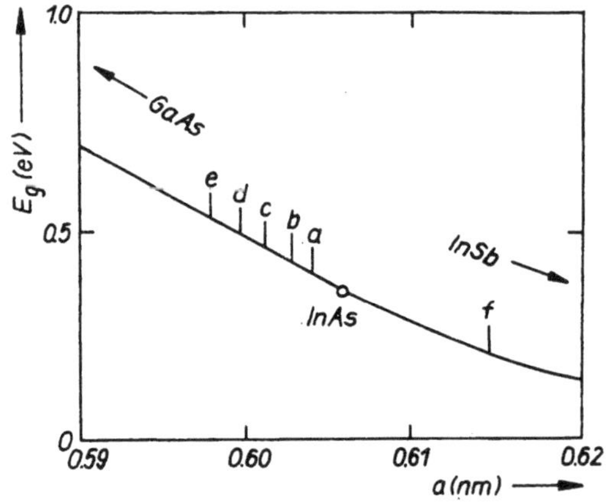

Fig. 1. The dependence of the width of the forbidden gap on the lattice parameter of the solid solution in the vicinity of the InAs composition. a to f are the solid phases whose equilibrium liquids were used in the experiments of isothermal contact with InAs

in such a way that the composition of the equilibrium $In_{1-x}Ga_xAs$ solid varied from $x = 0$ to 0.2. Fig. 1 shows the dependence of the width of the forbidden gap of the solid solution (or the binary compound), E_g, on the lattice parameter of this solid. This dependence is widely used in the planning of film growth on III–V compounds. Marks a to e denote the points which corresponds to those compositions of the $In_{1-x}Ga_xAs$ solution whose equilibrium liquids were used for the contact with the InAs substrate. The undersaturation of the In–Ga–As liquid was achieved by introducing a few per cent less arsenic than needed for the equilibrium liquid.

The contact of the In–Ga–As liquid and the InAs (III) A substrate was maintained for a time interval from 1 s to 3 min. Then the surface and the cross-sections of the samples were studied in detail with an optical microscope MBI-15; certain characteristic specimens were examined by Auger profiling in a "Riber" LAS-2000 instrument and a "Tesla" BS-350 scanning electron microscope.

3. Experimental Results

It was found that a new phase appears on the substrate surface in the form of isolated three-dimensional centres whose density and lateral growth rate increase with increasing x (points b to e in Fig. 1). The In–Ga–As liquid characterized by point a in Fig. 1 did not lead to the formation of the new phase, and the substrate surface was dissolved planarly. Then (points b to d) on the dissolving surface of InAs some triangular centres appeared which transformed to a spherical shape with increasing contact time (Fig. 2b). The density of the latter was in the range (10^1 to 10^2) cm^{-2}.

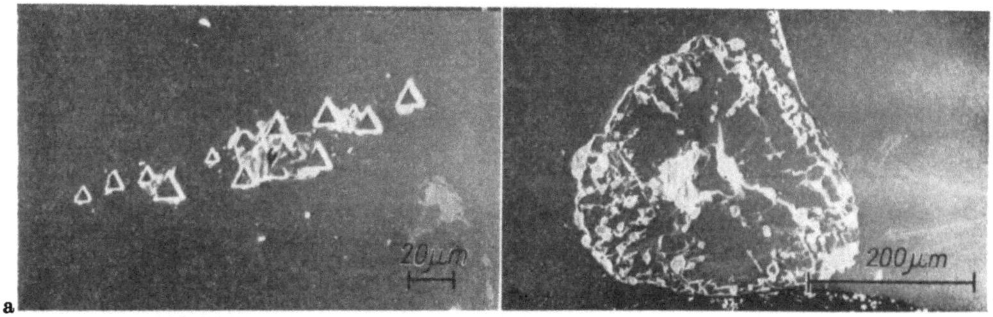

a b

Fig. 2. The shapes of centres of the new phase on the surface after contact. Scanning electron microscope "Tesla" BS-350

Fig. 3 presents the photomicrographs of a part of the surface of a spherical centre (Fig. 3a) and the cross-section of the same centre (Fig. 3b) on the same scale. The centre "dips" into the volume of InAs and a depression (marked by the arrow) is observed in the substrate at the centre boundary. If the centre front spreads across the substrate surface some drops of the solution are captured which appear in the cross-section (Fig. 3b) as dark spots. Centres seem to develop by diffusion of the dissolved substrate material through the liquid phase. The substrate is mainly dissolved at the centre front because of the high stresses at the edge of the centre.

Fig. 4 presents the density of the centres (Fig. 4a) and their lateral growth rate (Fig. 4b) as functions of the planned $In_xGa_{1-x}As$ composition. If the dependence $v(x)$ in Fig. 4b is extrapolated to $x = 0.2$, the lateral growth rate of this centre will exceed 10^{-2} cm/s. At such a rate and with the density of centres taken from Fig. 4a for $x = 0.2$, a continuous layer of a new phase is expected to form in less than 1 s. Fig. 3c shows the cross-section of a specimen in which a continuous, thick layer of $In_xGa_{1-x}As$ solid solution is seen. This layer appeared after 5 s contact of the InAs with the In–Ga–As liquid characterized by point e in Fig. 1. This is the result usually observed after isothermal contact of a multicomponent liquid and a binary solid, provided $\Delta a/a < 0$ [1, 2].

Fig. 5 presents Auger spectra of the basic elements of the specimen surface free from centres of the new phase (Fig. 5a) and of the centre (Fig. 5b). As is evident from the figure, the peak-to-peak signal of Ga is observed not only on the centre of the new phase, but also on the dissolving surface of the substrate; however, in the latter case it is appreciably smaller. During depth profiling of the sample by ion bombardment the Ga signal from the dissolving surface free from centres was decreasing simultaneously with decreasing oxygen peak-to-peak value, i.e., the layer of InGaAs discovered by Auger spectroscopy is situated within the natural oxide layer. This indicates, according to our data [4 to 6], that this layer is separating the solid from

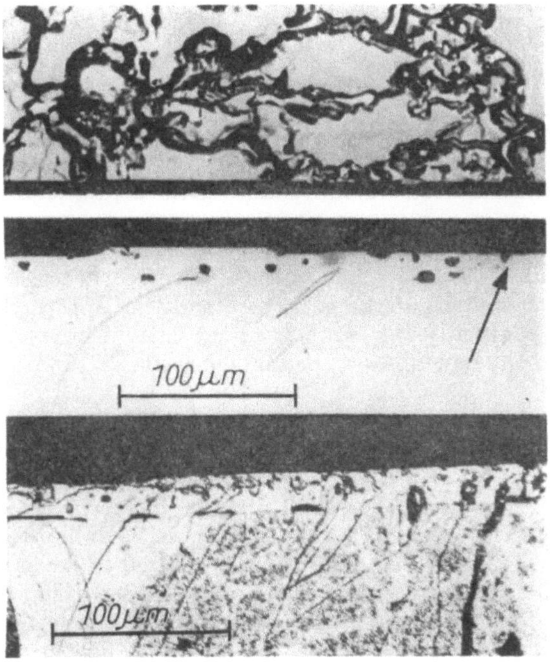

a

b

c

Fig. 3. Photomicrographs of a) the surface and b) the cross-sections of InAs samples after contact with an undersaturated In–Ga–As liquid phase. a), b) correspond to point b in Fig. 1, c) corresponds to point e in Fig. 1 and shows the cross-section of the specimen in which a continuos thick layer of InGaAs soulution is seen

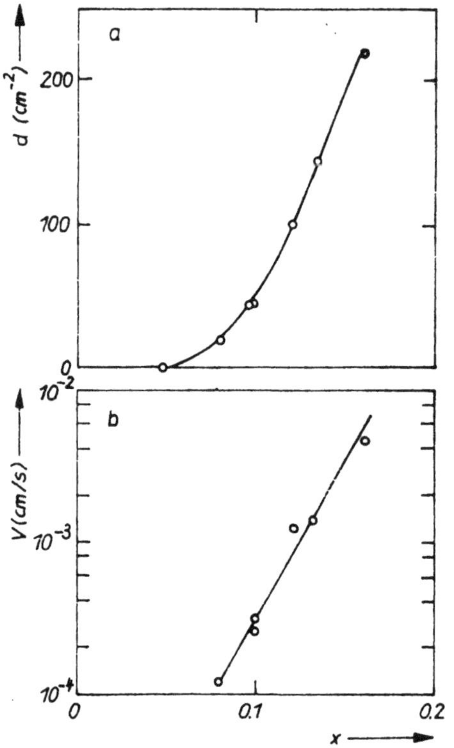

Fig. 4

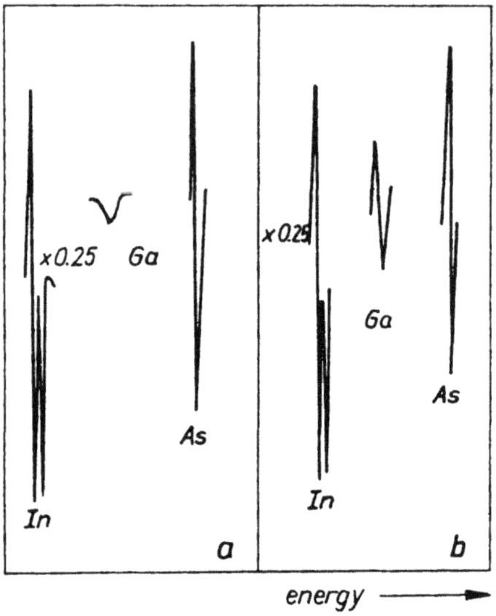

Fig. 5

Fig. 4. The dependences of a) the density of the centres and b) their lateral growth rate as functions of the composition of the solid planned to be grown

Fig. 5. Auger peaks from the InAs surface a) free from any centres and b) from the centre of the new phase. The peaks are: In 410 eV, Ga 1070 eV, As 1228 eV

the multicomponent liquid. Its thickness is extremely thin (0.5 to 3 nm) and is of non-epitaxial origin: it is formed by diffusion of the components into the dissolving crystal.

An undersaturated In–As–Sb liquid phase whose solid equilibrium composition is $InAs_{1-y}Sb_y$ with $y = 0.2$, was also used for the contact. Despite the considerable difference between the lattice parameters of the substrate and the $InAs_{1-y}Sb_y$ solid planned to be grown, which is practically the same as in the case of point e, Fig. 1, but of opposite sign, no centres of the new phase appeared even after a 10 min contact of the two phases.

4. Discussion

A substantial difference is observed when a new solid phase is formed at isothermal contact of an undersaturated multicomponent liquid and an InAs substrate in the cases $\Delta a/a < 0$ and $\Delta a/a > 0$. The latter case has been considered in several works [4, 7, 8] and it has been found that isolated centres of a new phase are formed for large differences of the lattice parameters of the substrate and the solid phase planned to be grown, and these centres are developing slowly. For example, in the case Ga–

As/GaP ($\Delta a/a = 3.6\%$) at 700 °C over 80% of the GaP substrate are covered with a new phase after several hours [4]. The same may be said for the In–Ga–As/GaAs and Ga–As–Sb/GaAs systems [7]. Consequently, the basic difference between the case In–Ga–As/InAs, for which $\Delta a/a < 0$, and the cases characterized by $\Delta a/a > 0$ consists in the fact that in the former case we observe a high lateral growth rate of centres of the new phase. Therefore, in the case $\Delta a/a < 0$, in the literature there is no evidence of isolated centres on the substrates which are commonly used as sources for saturating the liquid phase.

The following may be suggested as the probable cause of this difference: in the case $\Delta a/a > 0$, dissolution of the substrate surface covered with a diffusion layer in a state of compression is slow and limits the whole process of dissolution of the solid phase in the non-equilibrium multicomponent liquid phase. This is confirmed by the circumstance that in this case the dissolution of the solid surface is selective: on the substrate surface different forms of etch-pits are discovered [9, 10]. The selective dissolution of the substrate is a morphological feature of the limiting role of the dissolution reactions. For the opposite sign of the inequality, $\Delta a/a < 0$, dissolution of the substrate appears to be limited by the diffusion of the dissolved atoms in the liquid phase and the kinetics of the surface reactions of dissolution is fast. Smooth surfaces of the substrate after dissolution (phase a, Fig. 1) and the observation that part of the surface is free from centres (Fig. 2b) are an indirect confirmation of such a mechanism of solid phase dissolution.

As a result of the fast dissolution of the substrate (case $\Delta a/a < 0$) in the volume of the liquid phase high supersaturations are created which lead first of all to the rapid appearance of centres of the new phase and to lateral growth rates of the centres of the order of 10^{-2} cm/s extremely high for LPE, and, as it seems, even larger ones for larger Δa differences (for example, Ga–As/InAs). We believe that the main cause of such a difference in the investigated cases is the different sign of the stresses in the diffusion layer separating the substrate and the multicomponent liquid: for $\Delta a/a < 0$ this layer is subjected to tensile stress.

5. Conclusion

It is shown that in the case of isothermal contact of a substrate and an undersaturated multicomponent liquid phase whose equilibrium solid has a smaller lattice parameter than the substrate ($\Delta a/a < 0$), relaxation of this non-equilibrium interface proceeds by the mechanism of "etch-back and regrowth", and the processes of dissolution and growth take place in different parts of the substrate. The distinguishing feature of this mechanism is a lateral growth rate of centres of the new phase an abnormally high for LPE, which is accompanied by the capture of drops of the solution. It is supposed that the sign of the stresses that occur in a thin diffusion layer forming on the substrate surface at the initial moment of the contact is a factor which determines character and speed of dissolution of the substrate in a non-equilibrium liquid phase.

References

[1] Yu. B. Bolkhovityanov, Crystal Res. Technol. **18**, 679 (1983).
[2] Yu. B. Bolkhovityanov and S. I. Chikichev, in: Rost poluprovodnikovykh kristallov i plenok, Izd. Nauka, Novosibirsk 1984 (p. 54).
[3] T. U. Wu and G. L. Pearson, J. Phys. Chem. Solids **33**, 409 (1972).

[4] Yu. B. Bolkhovityanov, Yu. D. Vaulin, B. Z. Olshanetskii, and S. I. Stenin, Poverkh. Fiz. Khim. Mekh. No. 12, 68 (1984).
[5] Yu. B. Bolkhovityanov, Yu. D. Vaulin, B. Z. Olshanetskii, S. I. Stenin, and S. A. Teys, phys. stat. sol. (a) **84**, K13 (1984).
[6] Yu. B. Bolkhovityanov, Yu. D. Vaulin, B. Z. Olshanetskii, and S. I. Stenin, Poverkh. Fiz. Khim. Mekh. **9**, 47 (1985).
[7] Yu. B. Bolkhovityanov, J. Crystal Growth **55**, 591 (1981).
[8] Yu. B. Bolkhovityanov and Yu. D. Vaulin, Thin Solid Films **98**, 86 (1982).
[9] Yu. B. Bolkhovityanov, Electronnaya Tekh., Ser. Mater. No. 6, 167 (1982).
[10] Yu. B. Bolkhovityanov, R. I. Bolkhovityanova, and E. M. Trukhanov, Crystal. Res. Technol. **20**, 467 (1985).

(Received April 19, 1988)

phys. stat. sol. (a) **110**, 495 (1988)

Subject classification: 68.20; 79.20; S1.1; S1.2

Institute of Physical Chemistry, Polish Academy of Sciences, Warsaw[1]) *(a),*
Institute of Technical Physics, Hungarian Academy of Sciences, Budapest[2]) *(b),*
and Institute of General and Inorganic Chemistry,
Bulgarian Academy of Sciences, Sofia[3]) *(c)*

Surface Composition of the Ordered Fe–Co Alloys

By

P. Mrozek (a), M. Menyhárd (b), A. Jabłoński (a), and G. Tyuliev (c)

The procedure of quantitative AES analysis proposed for ordered alloys is applied to surface studies of the Fe–Co alloys. Surface composition of alloys after extensive sputtering is practically the same as bulk composition with the exception of $Fe_{50}Co_{50}$ alloy for which slight Co enrichment is found. The results agree quantitatively with the simple theoretical model of preferential sputtering. High temperature annealing at temperatures exceeding 973 K causes moderate Fe surface enrichment in agreement with the theory of Moran-Lopez and Falicov. Quantitative analysis by XPS provided results consistent with AES.

Die für geordnete Legierungen' vorgeschlagene Methode der Augerelektronen-Spektroskopie für quantitative Analysen wird zur Bestimmung der Oberflächenzusammensetzung von geordneten Fe–Co Legierungen angewandt. Die Oberflächenzusammensetzung ist nach extensivem Zerstäuben mit Argon-Ionen praktisch die gleiche wie im Volumen mit Ausnahme der $Fe_{50}Co_{50}$-Legierung, wo eine Kobaltanreicherung gefunden wird. Diese Resultate entsprechen quantitativ den Voraussagen des einfachen Modells des bevorzugten Zerstäubens der Atome in den Legierungen. Ausheizen bei Temperaturen über 973 K veranlaßt in der Oberfläche der Legierungen eine mäßige Eisenanreicherung. Diese Meßergebnisse stimmen mit der Theorie von Moran-Lopez und Falicov überein. Die quantitative Analyse mit der Photoelektronen-Spektroskopie ergibt mit Ergebnissen der Augerelektronen-Spektroskopie vergleichbare Ergebnisse.

1. Introduction

The surface composition of binary alloys may be in general different from the bulk composition due to the surface segregation phenomenon [1 to 5]. This fact has important consequences in the field of applied physics, materials science, and catalysis [6]. Numerous experimental and theoretical studies on surface composition were published in the past. However, the surface composition was determined only for a few ordered binary alloys [1, 7 to 10]. The $Pt_{50}Sn_{50}$ [7], $Pt_{75}Sn_{25}$ [8], and $Ni_{75}Sn_{25}$ [8] alloys exhibited Sn segregation to the surface. The surface of Au–Cu alloys at 25, 50, 75 at% Au was found to be enriched with Au [9] while in the $Au_{50}Sn_{50}$ alloy none of the components prevails at the surface [10]. The surface composition of $Al_{50}Co_{50}$ alloy can be affected only by ion sputtering [1]. Among the existing theories of surface segregation in the binary ordered alloys [5, 11, to 13] the most advanced approach was presented by Moran-Lopez et al. [12, 13]. This approach is based on the quasichemical approximation, where both, long-range and short-range order parameters were taken into account. The Moran-Lopez theory was applied in the case of the $Fe_{50}Co_{50}$ alloy to explain

[1]) Ul. Kasprzaka 44/52, PL-01-224 Warszawa, Poland.
[2]) P.O. Box 76, H-1325 Budapest, Hungary.
[3]) BG-1040 Sofia, Bulgaria.

the Fe enrichment [13] and also to the $Al_{50}Co_{50}$ alloy [1]. Both systems have the ordered B2 (CsCl)-type crystal structure in the vicinity of the equiatomic composition. The phase diagram is presented by Hansen et al. [14]. In the case of Fe–Co alloys the superstructure was confirmed for Co contents from 30 to 70 at% [15]. The order-disorder (O–D) transition temperature (T_c) versus composition shows a maximum equal to $T_c = 998.5$ K at 47.9 ± 0.3 at% Co [15]. The ordering process in the bulk of the $Fe_{50}Co_{50}$ alloy has been extensively studied [16 to 19]. The enthalpy of formation ΔH_f at 923 K as a function of composition has a minimum $\Delta H_f = -8.24$ kJ/mol at 45at% Co [20].

The catalytic properties of Fe–Co alloys were investigated first by Long et al. [21], who studied the hydrogenation of benzene to cyclohexane. An important contribution towards understanding the role of alloy catalysts in the hydrogenation reactions in Fischer-Tropsch synthesis (FTS) is due to Ponec and van Barneveld [22], who stated that the FTS cannot be performed on a monoatomic homogeneous catalyst. The Boudouard reaction on the surface of Fe–Co alloys was studied by Rienaecker and Toursel [23] and Temkin et al. [24] and more recently by Audier et al. [25 to 28], who dealt extensively with the problem of catalyst poisoning. The FTS reaction on Fe–Co unsupported catalysts was studied also by Nakamura et al. [29] who found that the $Fe_{50}Co_{50}$ catalyst was the most active one in $H_2 + CO$ reactions. The selectivity of this catalyst increased towards production of low molecular weight hydrocarbons $C_2 + C_3$ as compared with Fe. The last study encouraged other authors to investigate $Fe_xCo_{1-x}SiO_2$ catalysts in the low pressure FTS reaction ($x = 0.80$ [30], $x = 0.40$ to 0.95 at. fraction [31]) and the water gas shift reaction [30]. The high pressure FTS reaction was carried out over $Fe_{80}Co_{20}/SiO_2$ by Arcuri et al. [32] and Butt et al. [33]. It turned out that the alloy catalyst did not deactivate during a period of 200 h while Fe/SiO_2 deactivated much faster. The ammonia synthesis and nitrogen isotope exchange may also be performed on Fe–Co catalysts [34]. Several surface studies of Fe–Co alloys were performed under ultra-high vacuum conditions. Dittrich et al. [35] investigated the interaction of CO and hydrogen with Fe–Co films by means of the thermal desorption mass spectrometry. Auger electron spectroscopy (AES) was applied in studies of the surface composition at the (100) crystal plane of the ordered $Fe_{50}Co_{50}$ alloy [36, 37]. Bevolo [38] found slight Co surface enrichment in the AES analysis of the Ar^+ sputtered Fe–Co samples at room temperature. The Fe surface enrichment was determined recently by means of X-ray photoelectron spectroscopy (XPS) on Fe–Co alloys after mechanical polishing and annealing at 650 K [39].

The present work is a continuation of our experimental and theoretical studies of surface composition of ordered binary alloys [1]. Furthermore, there are relatively limited data on the surface composition of the Fe–Co system published in the literature, which is rather surprising considering the importance of this alloy in catalysis.

2. Experimental

The alloy samples and the standards were prepared by arc-melting weighted amounts of 4N iron and 4N cobalt in argon atmosphere. The details of preparation are given elsewhere [1]. The bulk atomic compositions (x^b_{Fe}) were determined by electron probe microanalysis (EPMA) and are listed in Table 1. The structure was examined using the X-ray diffractometer with Mo and Co radiation. The samples exhibited A2 structure in agreement with literature data [40]. Unfortunately, due to the similarity of atomic numbers, no superstructure of B2 type was found. The alloy buttons of $Fe_{50}Co_{50}$ and $Fe_{60}Co_{40}$ were then submitted to neutronographic experiments at room temperature on the TKSN-420 neutron spectrometer at the "Ewa" nuclear reactor

Table 1

The correcting factors for quantitative AES analysis of Fe–Co alloys

Auger transition	Auger electron energy (eV)	x^b_{Fe} (at. fraction)	r	d (nm)	M ($10^{28}/m^3$)	$F_{Fe,Co}$
Fe LMM	598	1.000	1.58 ± 0.01	1.089	8.48	—
Fe LMM	703		1.58 ± 0.01	1.181		
Fe LMM	598	0.700 ± 0.002	1.57 ± 0.01	1.080	8.63	1.0235
Co LMM	775		1.55 ± 0.01	1.229		
Fe LMM	598	0.604 ± 0.005	1.58 ± 0.01	1.077	8.68	1.0300
Co LMM	775		1.55 ± 0.01	1.226		
Fe LMM	598	0.513 ± 0.004	1.58 ± 0.01	1.074	8.73	1.0300
Co LMM	775		1.55 ± 0.01	1.223		
Fe LMM	703		1.58 ± 0.01	1.165		
Fe LMM	598	0.309 ± 0.002	1.59 ± 0.01	1.067	8.83	1.0299
Co LMM	775		1.56 ± 0.01	1.215		
Co LMM	775	0.000	1.55 ± 0.01	1.204	8.99	—

in Swierk. The superstructure lines (100), (111) were easily distinguished, and in this way the presence of B2 superstructure in all studied alloys was proved.

AES experiments were carried out using a RIBER single pass cylindrical mirror analyzer (CMA, model OPC103). The samples were bombarded at normal incidence with a 3 keV, 3 μA electron beam, coaxial with the CMA. The diameter of the beam was 50 μm. The base pressure in the chamber was in the range of 10^{-7} Pa. The samples were cleaned at elevated temperature by 2 keV Ar^+ ion sputtering at the incidence angle 77° with respect to the surface normal. The major contaminants of the samples were carbon, oxygen, and sulphur. After several cycles of Ar^+ bombardment and annealing the Auger peaks associated with contaminants were reduced to the background level. The spectra recorded for Fe and Co compared very well with the standard spectra published by Palmberg et al. [41]. The ultimate cleanliness was also comparable.

3. Quantitative Analysis

For a quantitative analysis two Auger transitions were monitored, the $L_{23}M_{23}M_{23}$ transition of Fe (598 eV) and the $L_{23}M_{45}M_{45}$ transition of Co (775 eV). In the case of $Fe_{50}Co_{50}$ at 1073 K the $L_{23}M_{45}M_{45}$ transition of Fe (703 eV) was applied. A 10 kHz modulation frequency with 4.82 eV amplitude was used. The Auger electron intensity (H_{Fe} or H_{Co}) was measured as peak-to-peak height in the differentiated spectra. The spectra were recorded either by an X–Y recorder or an on-line computer system [42].

The method based on the relative sensitivity factors was used for the quantitative analysis of the studied alloys [43]. The Auger electron intensity ratios were determined for the alloys (H_{Fe}/H_{Co}) and for pure metals (H^0_{Fe}/H^0_{Co}). The measurements were carried out for alloys in the temperature range from 683 to 1073 K and for pure metals at 683 K. For one point on the alloy sample several (five to eight) measurements of the ratio were made and the average value was calculated with the error bar being the 0.95 confidence interval. Ten or fifteen scans in the selected energy range were made during the measurements using the computer system in order to improve the Auger intensity to background ratio. The final errors for the ratios H_{Fe}/H_{Co} were determined assuming ± 1 count error for the analog-to-digital converter.

The surface composition of iron (in atom fraction) x^s_{Fe} was calculated under the assumption of uniform composition of the surface region within the sampling depth of Auger electrons. In this case the following nonlinear equation was solved [43]:

$$x^s_{Fe} = 1 - x^s_{Co} ,$$

$$x^s_{Co} = \left[\frac{H_{Fe}/H_{Co}}{(H^0_{Fe}/H^0_{Co}) \, F_{Fe, Co}} + 1 \right]^{-1} , \tag{1}$$

where x^s_{Co} is the surface concentration of Co (in atom fraction). The total correction factor $F_{Fe, Co}$ [1, 43] depends on x^s_{Co} and has the following form:

$$F_{Fe, Co} = \frac{M^0_{Co} r^0_{Co} d^0_{Co}}{M^0_{Fe} r^0_{Fe} d^0_{Fe}} \frac{r_{Fe} d_{Fe}(x^s_{Co})}{r_{Co} d_{Co}(x^s_{Co})} , \tag{2}$$

where M is the total number of atoms per unit volume, r the backscattering factor, and d the inelastic mean free path (IMFP). The superscript "0" refers to pure metals.

The methods for determining the correction factors, i.e. the backscattering factor, r, the inelastic mean free path, d, and the atomic density, M, are briefly described below.

3.1 The backscattering factor

As in the previous work [1] the Monte Carlo algorithm for the calculation of the backscattering factor was used [44 to 46]. The following definition was applied in the calculations [45, 46]:

$$r = 1 + [I_0 \sigma(E_0)]^{-1} \int_{E_c}^{E_0} \int_0^{90^\circ} I_B(E, \alpha) \, \sigma(E) \sec \alpha \, d\alpha \, dE , \tag{3}$$

where I_0 is the primary current, E_c the ionization energy corresponding to the Auger transition, E_0 the primary energy, $\sigma(E_0)$ the ionization cross-section at the primary energy E_0, and $I_B(E, \alpha)$ the angular and energy distribution of backscattered electrons. The calculated values of the backscattering factors are shown in Table 1.

3.2 The total number of atoms per unit volume and the monolayer thickness

For pure elements the number of atoms in the unit volume can be calculated from the usual expression [47]

$$M^0 = \frac{N_A \varrho 1000}{A} = (t^0)^{-3} , \tag{4}$$

where N_A is the Avogadro number, ϱ the density per unit volume, A atomic mass, and t^0 the monolayer thickness. In the present work the densities of 7.86×10^3 kg/m^3 and 8.80×10^3 kg/m^3 for pure Fe and Co, respectively, were used. These values were obtained from X-ray (Fe) and pycnometric measurements (Co). The alloy monolayer thickness is calculated assuming that Vegard's law is valid in the bulk and in the surface region of Fe–Co alloys [43]. All the values are listed in Table 1.

3.3 The inelastic mean free path

The IMFP for the standards and alloy samples was estimated using the published equation of Seah and Dench for pure elements [47]. Details of the calculations for alloys are given elsewhere [43].

4. The XPS Analysis

The ESCALAB II (Vacuum Generators) spectrometer has been used for X-ray photoelectron analysis of the $Fe_{30}Co_{70}$, $Fe_{50}Co_{50}$, and $Fe_{70}Co_{30}$ alloys. The experimental method of Seah [48] was applied to determine the surface composition. The alloy samples and standards were analysed using MgK_α radiation after Ar^+ ion bombardment ($E_{Ar^+} = 2$ keV) at room temperature and after annealing at 873 and 1123 K (1173 K in the case of $Fe_{70}Co_{30}$ alloy). The temperature was controlled using an optical pyrometer. The Co $2p_{3/2}$ ($E_B = 777.9$ eV) and Fe $2p_{3/2}$ ($E_B = 706.7$ eV) lines were recorded at two take-off angles 0° and 60° with respect to the surface normal. Due to the similarity of kinetic energies the depth of analysis was similar as in the case of AES measurements.

5. Results

Fig. 1a, b show the dependence of the Fe surface concentration on the sputtering time at lowest and highest $Fe_{50}Co_{50}$ sample temperatures during AES measurements. The results at other temperatures were similar. Figs. 2, 3 show the dependence of Fe surface concentration on the annealing time and $Fe_{50}Co_{50}$ sample temperature while Fig. 4 includes the dependence of Fe surface concentration on bulk iron content after extensive sputtering at 973 K. Fig. 5 shows the dependence of Fe surface concentration on the bulk Fe content after annealing. Fig. 1, 2, 4, 5a, b contain the results of AES measurements, Fig. 3, 5c the results of XPS measurements.

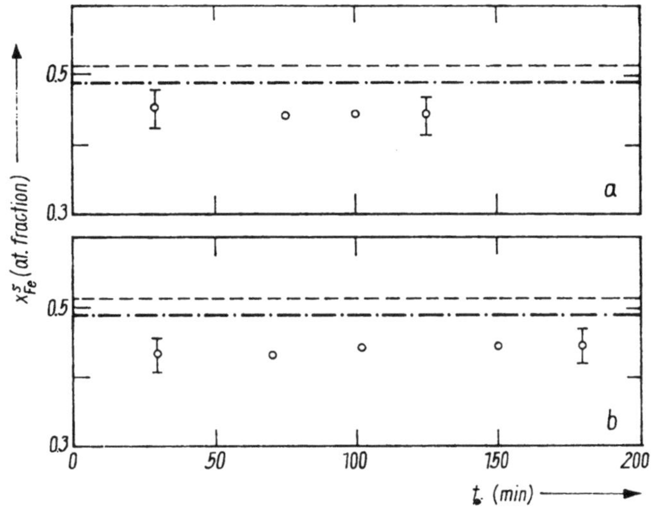

Fig. 1. Surface composition of the ordered $Fe_{50}Co_{50}$ alloy as a function of the sputtering time. Dashed line bulk Fe concentration; dash-dotted line surface Fe concentration in the steady state of sputtering according to theory of Ho et al. [54]; a) $T = 683$, b) 1073 K

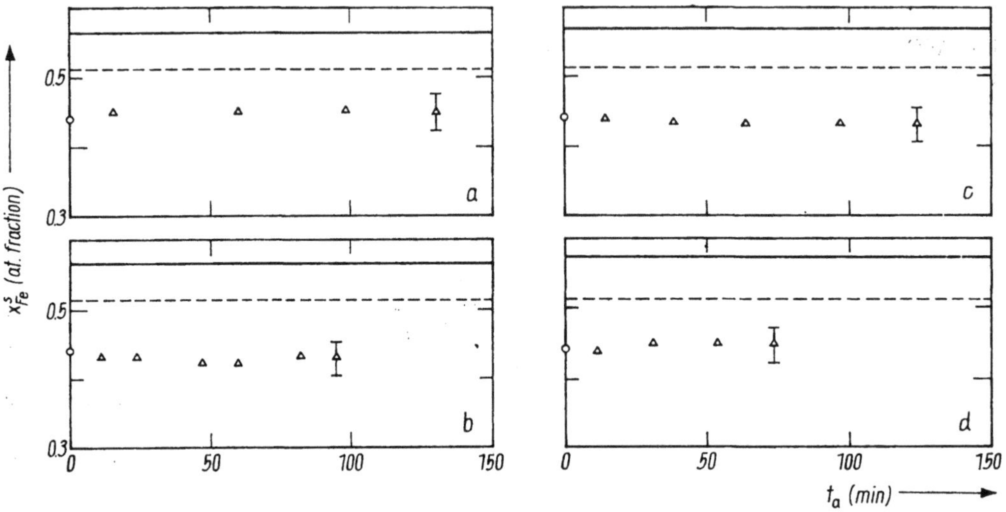

Fig. 2. Surface composition of the ordered $Fe_{50}Co_{50}$ alloy as a function of the annealing time (△). Circle denotes the surface composition at the last measurement during sputtering. Dashed line bulk Fe concentration; solid line surface Fe concentration calculated from the theory of Moran-Lopez et al. [12, 13]. a) $T = 683$, b) 873, c) 973, d) 1073 K

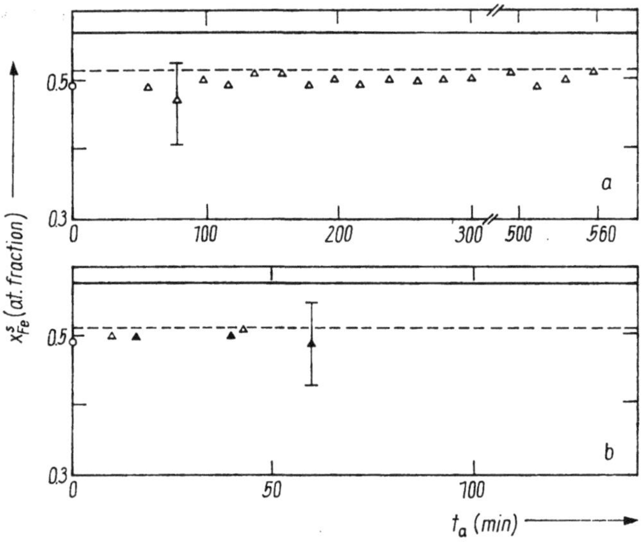

Fig. 3. Surface composition of the ordered $Fe_{50}Co_{50}$ alloy as a function of the annealing time (XPS measurements, open triangle take-off angle 0°, filled triangle take-off angle 60°). Dashed line bulk Fe concentration; solid line surface Fe concentration calculated from the theory of Moran-Lopez et al. [12, 13]. Circle denotes the surface composition after sputtering at room temperature: a) $T = 873$, b) 1123 K

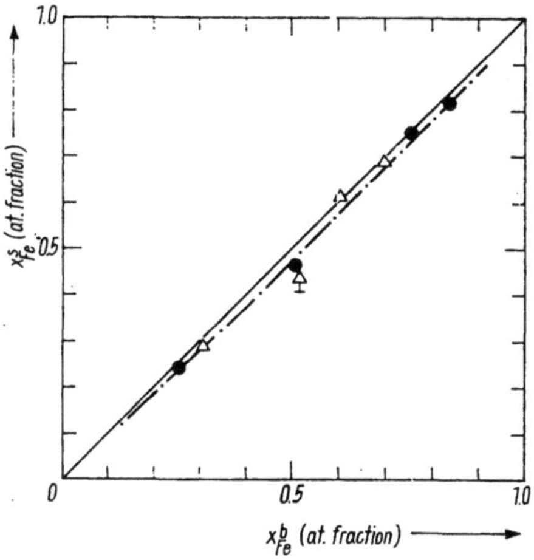

Fig. 4. Surface composition of the ordered Fe–Co alloys after sputtering at $T = 973$ K as a function of bulk composition. The error bars are of the size of the symbols. Solid straight line surface Fe concentration equal to bulk concentration; dash-dotted curve Fe surface concentration in the steady state of sputtering according to the theory of Ho et al. [54]; triangles this work, circles results taken from [38]

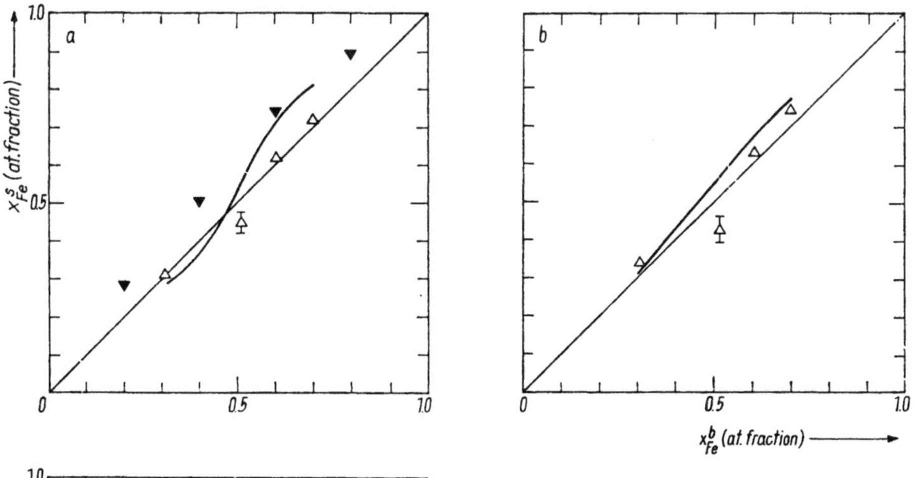

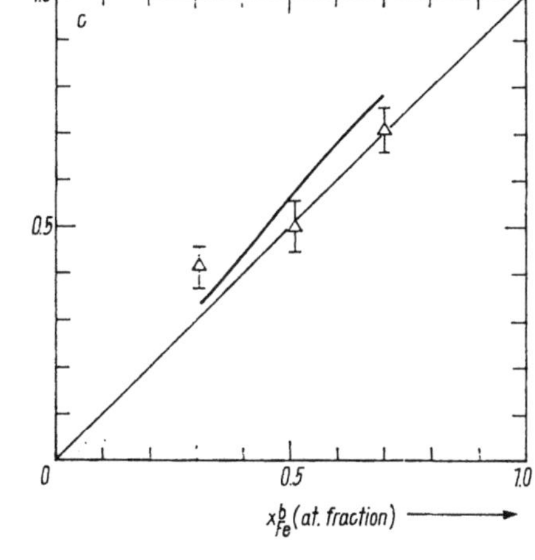

Fig. 5. Surface composition of the ordered Fe–Co alloys after annealing as a function of bulk composition. The error bars are of the size of the symbols. Solid line theory of Moran-Lopez et al. [12, 13]; ▽ this work, △ results taken from [39]; a) $T = 683$, b) 973, c) 1123 K (XPS measurements)

5.1 *The surface composition after extensive sputtering*

Sputtering of multicomponent alloys is accompanied by the formation of the so-called "altered layer" due to preferential sputtering [49]. Van Oostrom [50] proposed the following criterion for the steady state of sputtering:

$$\lim_{n \to \infty} (1 - Y_{Co}/Y_{Fe})^n = 0 , \tag{5}$$

where Y_{Fe} and Y_{Co} are the component sputtering yields defined by Betz and Wehner [49] and n is the number of removed surface layers. The sputtering yields for Fe and Co were estimated using the semiempirical formulas proposed by Matsunami et al. [51] and Oechsner [52, 53] (to account for the ion incidence angle). The estimated sputtering yield ratio was $Y_{Co}/Y_{Fe} = 0.94$. In our experimental conditions (ion flux in the range of 10^5 μA/m^2 and time of sputtering exceeding 1500 s) it was possible to remove several hundred layers and the steady state of sputtering should be achieved. The final surface composition of the altered layer in this state is determined by the ratio Y_{Co}/Y_{Fe} and the bulk composition [49]. The agreement between the theory and the experimental data on surface composition seems to be better [1], if the sputtering yield ratio is additionally corrected for the atomic densities of the pure elements according to Ho et al. [54]. We have found that the surface Fe concentration in Fe–Co alloys should be smaller by 2 to 3 at% than the bulk concentration, when the corrected ratio Y_{Co}/Y_{Fe} equal to 0.91 was used in the calculation. This is within our experimental error and practically all experimental results of the present study follow this rule (Fig. 4). Only in the case of $Fe_{50}Co_{50}$ alloy the theoretical predictions are slightly higher than the measured surface iron concentration.

5.2 *The surface composition after annealing*

Fig. 5a to c summarize the final surface concentration after prolonged annealing. The maximum time of annealing was chosen so that the level of impurities was still comparable with the background noise. For this reason the state of thermodynamical equilibrium of alloys was not attained even at the temperature 1123 K. The annealing time reached typically (except for $Fe_{50}Co_{50}$) 45 to 120 min at 683 K, 40 min at 973 K, and 50 to 60 min at 1173 K. XPS quantitative analysis of the $Fe_{70}Co_{30}$ alloy (Fig. 5c) at 1123 K indicates that the surface equilibrium Fe concentration is attained after 2 h of annealing and does not change significantly during the next 5 h of annealing. According to literature data [15 to 18] the bulk diffusion in Fe–Co alloys is slow below 723 K. For this reason the lowest annealing temperature chosen in the present work was 683 K. Furthermore, one can expect that above 683 K the rate of CO adsorption from the residual gases is considerably reduced [29, 35] .

The surface composition of $Fe_{30}Co_{70}$, $Fe_{60}Co_{40}$, and $Fe_{70}Co_{30}$ during annealing (Fig. 5) shows a different behaviour than the $Fe_{50}Co_{50}$ alloy. No change in Fe surface concentration took place at the surface of the $Fe_{50}Co_{50}$ alloy sample in the temperature range of 683 to 1123 K. Values of x^s_{Fe} resulting from AES and XPS are in agreement within experimental error (Fig. 2, 3). The surface of the other alloys during annealing at 973 and 1123 K is enriched in Fe by less than 0.10 at. fraction (Fig. 5b, c).

The theory published by Moran-Lopez et al. [12, 13] enables to calculate the composition of the first atomic layer of ordered alloys. The predictions of this theory can be compared with the experimental data obtained in the present work (Fig. 2, 3, 5). The surface composition is expressed in terms of nearest neighbour pair interaction bond energies, atomic coordination numbers, and temperature. The energies for

Table 2
The thermodynamical parameters used in the calculations of surface composition of Fe–Co alloys within the theory of Moran-Lopez et al. [12, 13]

T (K)	$U_{Fe-Fe} - U_{Co-Co}$ (kJ/mol)	Δ
683	2.536	0.52
873	2.758	0.56
973	3.109	0.64
1073	3.569	0.73
1123	3.749	0.77

Fe–Fe and Co–Co bonds in Fe–Co alloys usually are approximated using the expression [5]

$$U_{A-A} = -\frac{2\,\Delta H_T^A}{Z_A}, \tag{6}$$

where ΔH_T^A is the sublimation enthalpy of A [55] and Z_A the coordination number in the pure element A (A = Fe, Co). A small correction to U_{A-A} was necessary to include the difference in coordination numbers [56]. The heat of mixing can be expressed by [12, 13]

$$W = U_{Fe-Fe} + U_{Co-Co} - 2U_{Fe-Co}\,. \tag{7}$$

This parameter may be determined from the experimental values of the critical order–disorder temperature, T_C [12, 13, 57]. The dimensionless parameter Δ [12, 13] was calculated from

$$\Delta = (U_{Fe-Fe} - U_{Co-Co})/W\,. \tag{8}$$

The heat of mixing, W, may be in general composition and temperature dependent. It was shown by Inden [58] that in the Fe–Co system the temperature of the order–disorder transition and the enthalpy of ordering can be successfully described by means of the Bragg-Williams-Gorsky approximation. The value of W was estimated [58] as (4157 ± 1330) J/mol independent of composition and critical temperature. The same assumption was applied in the present theoretical approach. The set of values of W was calculated for the Fe–Co system using the critical temperatures measured by Oyedele and Collins [15] at twelve various compositions $0.305 \leqq x_{Fe}^b \leqq 0.704$. The mean value of W was established as (4880 ± 121) J/mol (the error at 0.95 confidence level). The resulting Δ values are listed in Table 2 and the theoretical results for x_{Fe}^s are shown in Fig. 2, 3, 5.

6. Discussion

Experimental studies of the surface composition of Fe–Co alloys are compiled in Table 3. The present results on iron surface concentration after extensive sputtering compare well with the results of Bevolo [38]. Noticeable discrepancies are observed in the case of other authors.

The iron surface concentration of unsupported Fe–Co catalysts was determined by means of AES after reduction in H_2 at 725 K [13, 29], calcination in O_2 followed by reduction in H_2 at 725 K and Ar^+ etching [29]. The normalized ratio for the high energy LMM Auger transitions in alloys defined as $(H_{Fe}/H_{Co})/(H_{Fe}^0/H_{Co}^0)$ was believed to be a measure of Fe surface concentration [13, 29]. This value was not corrected

Table 3
The surface composition of Fe–Co alloys published in the literature

ref.	Auger transition used	surface treatment	x^b_{Fe} (at. fraction)	x^s_{Fe} (at. fraction)
[13]	$Fe(L_{23}M_{45}M_{45})$ $Co(L_{23}M_{45}M_{45})$	H_2 reduction at 725 K	0.50	0.75 ± 0.05
[29]	$Fe(L_{23}M_{23}M_{23})$ $Co(L_{23}M_{45}M_{45})$	Ar^+ bombardment	0.25	0.41 ± 0.05
			0.50	0.77 ± 0.07
			0.75	0.91 ± 0.02
[37]	$Fe(M_{23}VV)$ $Co(M_{23}VV)$	Ar^+ bombardment and annealing		
		at 626 K	0.50	0.396 ± 0.03
		713 K	0.50	0.48 ± 0.03
		870 K	0.50	0.55 ± 0.03
		980 K	0.50	0.48 ± 0.03
		1100 K	0.50	0.49 ± 0.03
[38]	$Fe(L_{23}M_{23}M_{23})$	Ar^+ bombardment	0.253	0.240 ± 0.008
			0.504	0.467 ± 0.006
			0.751	0.752 ± 0.004
			0.832	0.820 ± 0.004
[39]	$Fe(2p_{3/2})$, $Co(2p_{3/2})$	polishing in cyclo-hexane and annealing at 650 K	0.198	0.28
			0.398	0.50
			0.602	0.74
			0.796	0.89
this work	$Fe(2p_{3/2})$, $Co(2p_{3/2})$	Ar^+ bombardment and annealing		
		at 1123 K	0.309	0.41 ± 0.04
		1123 K	0.513	0.50 ± 0.06
		1173 K	0.700	0.71 ± 0.05
	$Fe(L_{23}M_{45}M_{45})$, $Co(L_{23}M_{45}M_{45})$	1073 K	0.513	0.45 ± 0.03
	$Fe(L_{23}M_{23}M_{23})$, $Co(L_{23}M_{45}M_{45})$	973 K	0.309	0.34 ± 0.01
			0.513	0.43 ± 0.03
			0.604	0.63 ± 0.01
			0.700	0.74 ± 0.01

and it may explain the different results as compared with the present work. The values of the surface concentration given by Nakamura et al. [29] were recalculated applying the method proposed in the present work. The Fe–Co alloys after Ar^+ bombardment [29] exhibit then the following surface concentrations: 0.28 ± 0.03, 0.43 ± ± 0.02, 0.47 ± 0.01 at. fraction for $Fe_{25}Co_{75}$, $Fe_{50}Co_{50}$, and $Fe_{75}Co_{25}$, respectively. The above values indicate that the simplified procedure of quantitative analysis applied by Moran-Lopez et al. [13] and Nakamura et al. [29] leads to a considerable overestimation of the Fe concentration.

Allie et al. [36, 37] investigated the surface composition of a (100) $Fe_{50}Co_{50}$ single crystal, but the authors did not quantify the spectra. These results based on the low energy $Fe/M_{23}VV$(46 eV) and $Co/M_{23}VV$ (52 eV) Auger transitions were converted into the Fe surface concentrations using the present quantitative formalism. The

x^s_{Fe} values (Table 3) as a function of temperature are very close to 0.50 at.fraction. Fe surface enrichment is observed between 860 and 880 K reaching a maximum at 870 K. The x^s_{Fe} maximum occurs at a temperature fairly below the equilibrium state temperature of the surface, reported as 973 K [36] for such a state, where the low energy Fe and Co Auger spectra were independent of temperature and annealing time. Our results are somewhat lower than the data of Asami et al. [39] at 650 K (Fig. 5a), but the samples of the latter study were subjected to polishing in cyclohexane prior to annealing. The Fe enrichment on the surface might result from the interaction with the solvent or from an improper cleaning procedure.

The application of the theoretical model to calculate the surface composition of Fe–Co alloys requires also a comment. Moran-Lopez and Falicov [12, 13] used the values of pure element sublimation enthalpies to calculate the bond energies. The published value of $\Delta = 2.3$ [12, 13] for $Fe_{50}Co_{50}$ alloy at 996 K suggested that the pure element ΔH^A_T (A = Fe, Co) values were assumed by the authors to be equal to the bond energies. This is equivalent, as seen from (6), to the assumption that the iron and cobalt coordination numbers are the same and equal to the unrealistic value of 2 valid for a linear chain of atoms. Application of the surprisingly high value of the parameter $\Delta = 2.3$ in calculations provided the corresponding high value of iron surface concentration, i.e. 0.68 at.fraction at 725 K (temperature of H_2 reduction), which is quite close to their experimental results (0.75 ± 0.05 at.fraction). Our calculations were performed using the same theoretical model but the difference of bond energies $U_{Fe-Fe} - U_{Co-Co}$ in the alloy was determined by taking the real coordination of the pure metals (Section 5.2). The application of $\Delta = 2.3$ at 725 K [12, 13] causes a considerable overestimation of x^s_{Fe}.

Thus, we conclude that the Fe–Co ordered alloys after extensive Ar^+ ion sputtering exhibit a surface composition close to the bulk composition (except for the $Fe_{50}Co_{50}$ alloy, where a slight Co enrichment at the surface is observed). This result agrees well with the published experimental studies on sputtering of Fe–Co alloys and the theoretical model of the steady state of sputtering proposed by Ho et al. On the other hand, moderate Fe surface enrichment after annealing is established, which agrees well with the experimental observations of other authors using our quantitative formalism. The careful analysis of the existing data seems to exclude the possibility of a considerable iron segregation effect at the surface of the ordered Fe–Co alloys. The theoretical model of surface segregation in the ordered binary alloys was successfully applied to Fe–Co alloys. This model describes well the dependence of the surface composition on the bulk composition at high temperatures.

Acknowledgements

The AES measurements have been carried out in the Institute of Technical Physics of the Hungarian Academy of Sciences. The XPS measurements were performed in the Institute of General and Inorganic Chemistry of the Bulgarian Academy of Sciences in Sofia. Two of the authors (P.M. and A.J.) acknowledge the support of the Research Project 12.2.

References

[1] P. Mrozek, M. Menyhárd, J. Wernisch, and A. Jabłoński, phys. stat. sol. (a) **84**, 39 (1984).
[2] V. Kumar, Phys. Rev. B **23**, 3756 (1981).
[3] L. Z. Mezey and J. Giber, Surface Sci. **117**, 220 (1982).
[4] W. M. H. Sachtler, Appl. Surface Sci. **19**, 167 (1984).

[5] M. J. SPARNAAY, Surface Sci. Rep. 4, 101 (1985).
[6] P. H. HOLLOWAY, Adv. Electronics and Electron Phys. 54, 241 (1980).
[7] R. BOUWMAN, L. H. TONEMAN, and A. A. HOLSCHER, Surface Sci. 35, 8 (1973).
[8] P. BILOEN, R. BOUWMAN, R. A. VAN SANTEN, and H. H. BRONGERSMA, Appl. Surface Sci. 2, 532 (1979).
[9] J. M. MCDAVID and S. C. FAIN, JR., Surface Sci. 52, 161 (1975).
[10] S. H. OVERBURY and G. A. SOMORJAI, J. chem. Phys. 66, 3181 (1977).
[11] R. A. VAN SANTEN and W. M. H. SACHTLER, J. Catalysis 33, 202 (1974).
[12] J. L. MORAN-LOPEZ and L. M. FALICOV, Phys. Rev. B 18, 2542 (1978).
[13] J. L. MORAN-LOPEZ and H. WISE, Appl. Surface Sci. 4, 93 (1980).
[14] M. HANSEN and K. ANDERKO, Constitution of Binary Alloys, 2nd ed., McGraw-Hill Publ. Co., New York 1958.
[15] J. A. OYEDELE and M. F. COLLINS, Phys. Rev. B 16, 3208 (1977).
[16] A. CHAMBEROD, P. VILLEMAIN, and J. PAULEVE, J. Phys. Chem. Solids 33, 593 (1972).
[17] P. VILLEMAIN and A. CHAMBEROD, J. Phys. Chem. Solids 33, 1547 (1972).
[18] O. W. CLEGG and R. A. BUCKLEY, Metals Sci. J. 7, 48 (1973).
[19] B. DE MAYO, Phys. Rev. B 24, 6503 (1981).
[20] F. ARSLAN, H. B. BELL, and D. B. DOWNIE, Metals Sci. J. 12, 198 (1978).
[21] J. LONG, J. FRAZER, and E. OTT, J. Amer. Chem. Soc. 56, 1101 (1934).
[22] V. PONEC and W. A. VAN BARNEVELD, Ind. engng. Chem., Prod. Res. Dev. 18, 268 (1979).
[23] G. RIENAECKER and W. TOURSEL, Monatsber. Dtsch. Akad. Wiss. Berlin 8, 22 (1966).
[24] M. I. TEMKIN, L. O. APELBAUM, and K. A. LEZNEVA, USSR P. 312.831/M.K.P. C016, 31/18.
[25] M. AUDIER, M. COULON, and A. OBERLIN, Carbon 18, 73 (1980).
[26] M. AUDIER, A. OBERLIN, and M. COULON, J. Crystal Growth 55, 549 (1981).
[27] M. AUDIER, A. OBERLIN, M. OBERLIN, M. COULON, and L. BONNETAIN, Carbon 19, 217 (1981).
[28] M. AUDIER, M. COULON, and L. BONNETAIN, Carbon 21, 93, 99, 105 (1983).
[29] M. NAKAMURA, B. J. WOOD, P. Y. HOU, and H. WISE, Prepr. 7th Internat. Congr. Catalysis, Paper A29 (1980).
[30] J. A. AMELSE, L. H. SCHWARTZ, and J. B. BUTT, J. Catalysis 72, 95 (1981).
[31] R. M. STANFIELD and W. N. DELGASS, J. Catalysis 72, 37 (1981).
[32] K. B. ARCURI, L. H. SCHWARTZ, R. D. PIOTROWSKI, and J. B. BUTT, J. Catalysis 85, 349, (1984).
[33] J. B. BUTT, L. H. SCHWARTZ, M. BAERNS, and R. MALESSA, Ind. engng. Chem., Prod. Res. Dev. 23, 51 (1984).
[34] P. J. SMITH, D. W. TAYLOR, D. A. DOWDEN, C. KEMBALL, and D. TAYLOR, Appl. Catal. 3, 303 (1982).
[35] TH. DITTRICH, H. W. BUSCHMANN, and W. VIELSTICH, Ber. Bunsenges. phys. Chem. 85, 289 (1981).
[36] G. ALLIE, C. LAUROZ, and P. VILLEMAIN, Surface Sci. 104, 583 (1981).
[37] G. ALLIE and C. LAUROZ, Surface Sci. 116, L179 (1982).
[38] A. I. BEVOLO, J. Vacuum Sci. Technol. A 1, 1004 (1983).
[39] K. ASAMI and K. HASHIMOTO, Corros. Sci. 24, 83 (1984).
[40] T. NISHIZAWA and K. ISHIDA, Bull. Alloy Phase Diagrams 5, 251 (1984).
[41] P. W. PALMBERG, G. E. RIACH, R. E. WEBER, and N. C. MACDONALD, Handbook of Auger Electron Spectroscopy, Physical Electronics Industries, Inc., Minnessota 1972 (p. 51, 53).
[42] M. MENYHARD and A. SULYOK, in: Proc. 5th Seminar Electron Spectroscopy of Socialist Countries, Dresden, August 26 to 29, 1984, Technische Universität Dresden, 1984 (p. 85).
[43] A. JABŁONSKI, Surface Interface Anal. 4, 135 (1982).
[44] A. JABŁONSKI, Surface Sci. 87, 539 (1979).
[45] A. JABŁONSKI, Surface Interface Anal. 2, 39 (1980).
[46] A. JABŁONSKI, Surface Sci. 124, 39 (1983).
[47] M. P. SEAH and W. A. DENCH, Surface Interface Anal. 1, 2 (1979).
[48] M. P. SEAH, Surface Interface Anal. 2, 222 (1980).
[49] G. BETZ and G. K. WEHNER, in: Sputtering by Particle Bombardment II, Ed. R. BEHRISCH, Springer-Verlag, Berlin/Heidelberg/New York/Tokyo 1983, (p. 16).
[50] A. VAN OOSTROM, J. Vacuum Sci. Technol. 13, 224 (1976).

[51] N. Matsunami, Y. Yamamura, Y. Itizawa, W. Itoh, Y. Kazamata, S. Miyagawa, K. Morita, R. Shimizu, and H. Tawara, Atomic Data nuclear Data Tables **31**, 1 (1984).
[52] H. Oechsner, Z. Phys. **261**, 37 (1973).
[53] H. Oechsner, Appl. Phys. 8, 185 (1975).
[54] P. S. Ho, J. E. Lewis, and W. K. Chu, Surface Sci. **85**, 19 (1979).
[55] R. Hultgren, P. D. Desai, D. T. Hawkins, M. Gleiser, and K. K. Kelley, Selected Values of the Thermodynamic Properties of the Elements, American Society for Metals, Metals Park (Ohio) 1973 (p. 126, 180).
[56] Shu Zhen and G. J. Davies, phys. stat. sol. (a) **86**, 155 (1984).
[57] M. A. Krivoglaz and A. A. Smirnov, Theory of Ordering Alloys, Gos. Izd. Fiz.-Mat. Lit., Moscow 1958 (p. 170) (in Russian).
[58] G. Inden, Z. Metallk. **68**, 529 (1977).

(Received January 25, 1988; in revised form July 29, 1988)

phys. stat. sol. (a) **110**, 509 (1988)

Subject classification: 68.48 and 68.55; S4

California Institute of Technology, Pasadena[1])

Pt_2Al_3 Formation on Evaporated and Large-Grained Al Substrates

By

E. MA and M-A. NICOLET

The growth of Pt_2Al_3 during thermally induced Pt–Al reaction is studied with two types of samples: bilayered thin films (Pt/Al) and Pt thin film on large-grained Al substrate (Pt/《Al》). The same behavior upon vacuum furnace annealing is observed for both types of sample configurations: Pt_2Al_3 forms as the initial growing phase in a laterally uniform fashion, following a diffusion-limited process. A parabolic growth law of the form $x^2 = kt$ is obtained in the temperature range 200 to 300 °C, where x is the thickness of the growing Pt_2Al_3 layer, t is the annealing duration, and the growth constant k is given by $k = 1.02 \times 10^{-2}$ (cm²/s) exp $((-1.15 \pm 0.1)$ eV$/k_BT)$. The morphological rearrangements in the Al thin film during reaction is observed and discussed. The common Pt_2Al_3 growth kinetics on the two types of Al substrates suggests that neither the microstructure of nor the morphological rearrangements in the Al layer affect the Pt_2Al_3 growth significantly.

Das Wachstum von Pt_2Al_3 während der thermisch induzierten Reaktion von Pt–Al wird an zwei Probenarten untersucht. Doppelschichten von dünnen Filmen (Pt/Al) und dünne Pt-Filme auf einer großkörnigen Al-Unterlage (Pt/《Al》). Beide Probenarten zeigen dasselbe Verhalten: Pt_2Al_3 erscheint als erste Phase, die in lateral gleichförmiger Weise nach einem diffusionsbeschränkten Vorgang wächst. Das Wachstum folgt dem Gesetz $x^2 = kt$ im Temperaturbereich von 200 bis 300 °C, wo x die Dicke der wachsenden Pt_2Al_3-Schicht ist, t die Dauer der Erwärmung, und die Wachstumskonstante k den Wert $1{,}02 \times 10^{-2}$ (cm²/s) exp $((-1{,}15 \pm 0{,}1)$ eV$/k_BT)$ hat. Die morphologischen Umwandlungen im Al-Film während der Reaktion werden diskutiert. Die gemeinsame Wachstumskinetik für beide Al-Unterlagen weist darauf hin, daß weder die Mikrostruktur noch die morphologischen Veränderungen des Al das Wachstum von Pt_2Al_3 wesentlich beeinflussen.

1. Introduction

Aluminum and its alloys are frequently used as metallizations for Si integrated circuits [1]. To prevent spiking, an interlayer usually involving transition metals is required to separate the Al from the Si substrate. The knowledge of Al–metal thin-film reactions, in parallel with those well-documented for Si–metal systems [2], is therefore of interest [3].

The Pt–Al thin-film reaction has been investigated by a number of authors [4 to 11]. The concensus is that Pt_2Al_3 forms first during furnace annealing at temperatures ≥ 225 °C. The only exception to this is an early work by Howard et al. who identified $PtAl_4$ as the initial phase [4]. Recent work by Colgan [7] and Zhao et al. [11] show that Pt_2Al_3 grows in a layer-by-layer fashion. Al has been determined to be the dominant moving species during Pt_2Al_3 growth [7, 10]. Diffusion-limited kinetics has been reported for Pt_2Al_3 grown on large-grained Al substrates (《Al》) in the temperature range 255 to 305 °C [11]. The complication encountered so far is a significant mor-

[1]) 116-81, Pasadena, CA 91125, USA.

phological degradation in the Al film associated with the Pt–Al reaction in thin-film bilayers (Pt/Al). Severe void formation accompanied by a thickening of the surrounding Al has been observed by Colgan et al. and described in detail in [8, 9]. Information regarding subsequent aluminide formation and dissociation in Pt/Al thin-film diffusion couples can be found in [7].

The present work provides information on the following two aspects. The Pt_2Al_3 growth kinetics is measured in Pt/Al bilayers over an extended temperature range and compared with that for Pt/《Al》. In addition, we examine and discuss the microscopic features in reacted Pt/《Al》 couple and contrast them with those of Pt/Al.

2. Experimental Procedures

The thin-film Pt/Al bilayers were prepared by sequential e-beam evaporation of Al and Pt onto oxidized Si wafers with the configuration: Si(111)/SiO_2/Al($\approx$ 400 nm)/Pt ($\approx$ 180 nm). The oil-free evaporator was evacuated to a base pressure of $<5 \times 10^{-8}$ Torr. The vacuum during evaporation increased to $\approx 3 \times 10^{-7}$ Torr for Al and close to 1×10^{-6} Torr for Pt. Pt was also evaporated onto in-situ sputter-cleaned 《Al》 substates (grain size a few mm to 1 cm in diameter) in a similar evaporation unit. Detailed procedures to prepare the Pt/《Al》 samples has been described in [11]. Annealings were performed in a vacuum furnace at around 7×10^{-7} Torr. The furnace temperatures were calibrated to an accuracy of ± 2 °C. 2 MeV $^4He^+$ backscattering spectrometry (BS) was used to monitor the reaction in every reacted diffusion couples. Glancing angle X-ray diffraction (Read camera) was employed for phase identification. One reacted Pt/《Al》 sample was examined with cross-sectional transmission microscopy (XTEM) in a Philips 430 microscope.

3. Results and Discussions

Fig. 1 presents overlapped BS spectra of Pt/Al thin-film bilayered samples as-deposited and annealed at 265 °C for 40 min. Distinct steps in both Pt and Al signals

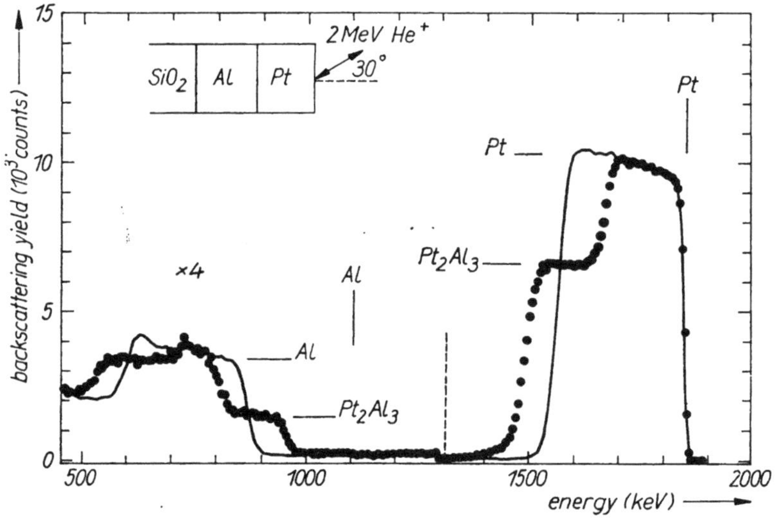

Fig. 1. 2 MeV $^4He^+$ backscattering spectra of Pt/Al samples as-deposited (solid line) and annealed at 265 °C for 40 min (circles), showing the formation of Pt_2Al_3 interlayer after annealing. The $^4He^+$ beam is incident 30° from the sample normal, and the detection angle is 170°. The configuration of the as-deposited sample is shown in the insert

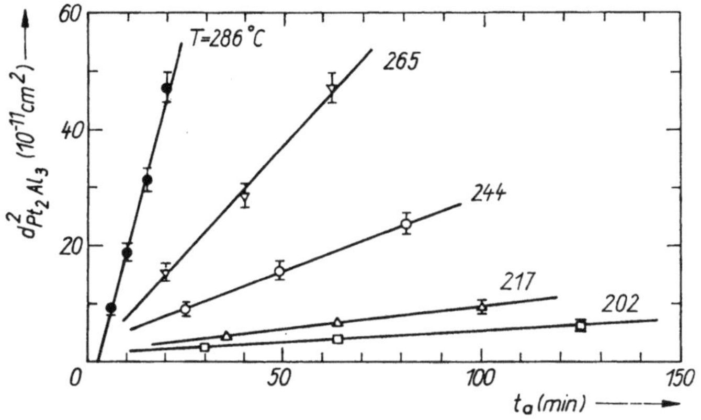

Fig. 2. Plot of squared Pt_2Al_3 thickness vs. annealing duration. A bulk density of 1.262×10^{22} Pt_2Al_3/cm^3 [12] is assumed for the Pt_2Al_3 interfacial layer

are evident in the spectrum for the annealed sample. A composition ratio of Pt : Al = = 3:2 ± 5% is calculated from these steps. The phase actually formed was identified to be Pt_2Al_3 by X-ray diffraction. The interfacial Pt_2Al_3 layer expands in width along with annealing duration. Fig. 2 is a plot of the squared thickness of Pt_2Al_3 versus annealing duration. The thickness values were obtained by assuming a bulk density of 1.262×10^{22} Pt_2Al_3 units/cm^3 for the Pt_2Al_3 interlayer ($D_m = 9.87$ g/cm^3) [12]. Least-squares fits to the data for each temperature exhibit good linearity, indicating a diffusion-limited growth law of the form $x^2 = kt$, where x is the thickness of the Pt_2Al_3 compound, t the annealing duration, and k the growth constant. Figures corresponding to Fig. 1 and 2 for large-grained Al substrates (Pt/《Al》) have been presented in [11].

We compare in the Arrhenius plot of Fig. 3 the growth kinetics for Pt_2Al_3 growth on the two different types of Al substrates. The growth constants at various temperatures, k, are obtained from the slopes of the linear fits shown in Fig. 2 and [11]. The solid line displayed in Fig. 3 is a linear fit to the data for Pt/Al samples (open squares). The temperature dependence of the growth constant k obtained from this linear fit

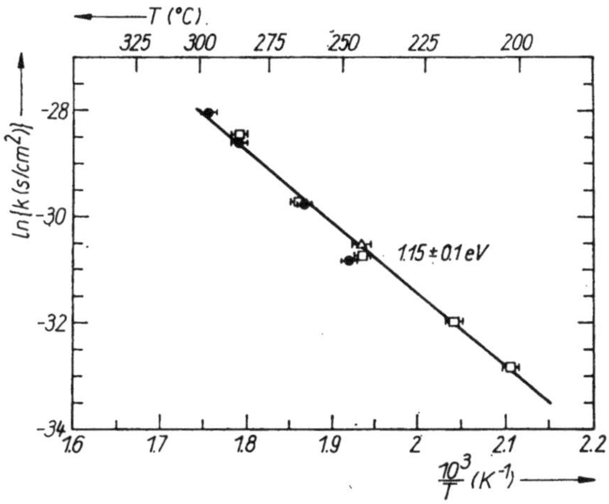

Fig. 3. Arrhenius plot of Pt_2Al_3 growth constant k for Pt/Al (open squares are from Fig. 2 and open triangle from another set of samples with reduced layer thicknesses annealed at 244 °C) and for Pt/《Al》 (full circles, from [11]). The solid line is a linear fit to the data points for Pt/Al

is $k = 1.02 \times 10^{-2}$ (cm²/s) exp $((-1.15 \pm 0.1)$ eV$/k_BT)$. Some additional Pt/Al samples with reduced thickness of Pt and Al layers (120 nm Pt on 270 nm Al) were prepared independently to verify the kinetics. Annealing of these samples at 244 °C yields another kinetics point embodied in Fig. 3 with an open triangle. The growth constant concurs with that of the previous samples to within 10%. The data for Pt/《Al》 (full circles) (250 to 300 °C) also fall closely to the solid line, although a linear fit of these four points gives a slightly different activation energy of 1.45 ± 0.1 eV [11]. Comparison of kinetics in the temperature range 200 to 250 °C is not feasible since the Pt/《Al》 samples do not react below 250 °C. This delayed initiation of reaction, as well as an apparent incubation time, has been attributed to the inevitable presence of residual impurities at the sputter-cleaned 《Al》 surface [11, 13].

The reaction of thin-film Pt/Al bilayers is known to generate voids in the Al film. This effect is detectable in Fig. 1. The original overlap of the signals of Si in SiO_2 and of Al at around 620 keV is replaced after reaction by overlaps ranging from 560 to 720 keV. Such a change is consistent with the picture of void formation and the concurrent increase of the amount of the surrounding Al described by Colgan et al. [8, 9]. When voids form in the Al, the overlap between the signals of Si and Al displaces to higher energies. On the other hand, the Al coming from the voids adds to the surrounding Al or reacts with Pt to thicken the Pt_2Al_3 layer, both displacing the overlap between Si and Al signals to lower energies. Colgan et al. propose that the controlling mechanism is Al diffusion along the Pt/Al interface driven by surface tension. According to this model, the interfacial diffusion is made possible by compound formation at the Pt/Al interface which breaks down the original oxide layer there. Al is then transported vertically by Al grain boundaries and laterally by the Al/Pt_2Al_3 interface. As a result, local voids nucleate and grow at Al grain boundaries. The thickness of the surrounding Al film is increased by the Al displaced from the voids while being reduced by the Pt_2Al_3 growth.

We propose a modified version of this model based on the fact that Al is the dominant moving species in the Pt_2Al_3 layer. If the Al transport in Pt_2Al_3 is governed by a vacancy mechanism, there must be a large number of vacancies that propagate through the compound layer in a direction opposite to that of the Al flux. This vacancy flux will produce excess vacancies in Al and promote the morphological rearrangements there. If the Al transport in Pt_2Al_3 is via interstitial mechanism, vacancies will also be created in Al when Al atoms move across the Al/Pt_2Al_3 interface into the aluminide. The condensation of excess vacancies at sinks, such as grain boundaries, to nucleate voids in the layer of the moving species has been observed in a number of thin-film reactions, e.g., Mo–Si [15] and Ni–Zr [16]. That the interface where the reaction takes place is a source of vacancies has recently been shown by radioactive tracer experiments in the Si–Pd system [14]. We therefore propose that the observed morphological instabilities in Al, i.e., the void formation and Al interfacial diffusion, is triggered by the excess vacancies, probably generated mainly by compound formation at the Pt_2Al_3/Pt interface. This idea is consistent with the observation [8, 9] that void growth halts when all the Pt layer is consumed, thereby cutting off the supply of additional vacancies. It also follows that the morphological rearrangements in Al will be pronounced only if the transportation mechanism involves a sufficient number of vacancies, meaning that this phenomenon is system-specific. It is indeed true that such morphological changes do not occur in all the Al–metal reactions [8].

The use of large-grained Al substrate should minimize the void formation in Al by virtually eliminating the nucleation sites. It is not plausible to use backscattering technique to detect morphological rearrangements in the bulk 《Al》 as is possible with

Al films on SiO_2. A XTEM examination of one of our reacted Pt/《Al》 samples was performed. It revealed that the 《Al》 underneath a uniform Pt_2Al_3 layer is laterally quite smooth, with no signs of voids. We cannot conclude from this fact, however, that voids are definitely absent because the technique only samples a minute section of the sample. The present result, nevertheless, is consistent with the idea that the void formation is associated with Al grain boundaries. The fact that Pt_2Al_3 grows uniformly via similar kinetics on both types of Al substrates suggests that the morphological changes in the Al film influence the Pt_2Al_3 growth insignificantly. The structural rearrangement in the Al appears to arise as a consequence of the reaction, but without affecting it.

4. Concluding Remarks

Similar aluminide growth kinetics on two different types of Al substrates, namely, evaporated and large-grained Al substrates, has been observed so far for the initial phases in three Al–metal systems: Al–Pt [11], Al–Ni [13, 17], and Al–Ti [18]. Al moves predominantly during the growth of all three compounds, Pt_2Al_3, $NiAl_3$, and $TiAl_3$ [19]. Evidently, the growth kinetics of these aluminides are insensitive to the microstructure of the moving species layer. The use of high purity large-grained Al substrate in our work does show advantages: in Ni–Al reaction, it avoids the Ni penetration into Al associated with the presence of impurities and Al grain boundaries [17]; in Pt–Al reaction, it minimizes the morphological degradation in Al film associated with reaction. Presumably, the use of large-grained Al substrates would also help elucidate the irregular reaction behaviors observed in other Al–metal systems [4, 20] where Al grain boundaries allegedly play an important role.

Acknowledgements

The authors wish to thank X.-A. Zhao for his contributions in this work, and C. W. Nieh and W. Flick for help in making XTEM observations. We are grateful to Prof. W. L. Johnson for his interest and helpful discussions on thin-film interfacial reactions. This work was supported in part by National Science Foundation-MRG Grant DMR-8811795 and by a grant from Intel Corporation.

References

[1] D. B. Fraser, in: VLSI Technology, Ed. S. M. Sze, McGraw-Hill Publ. Co., New York 1983, (p. 347).
[2] M-A. Nicolet and S. S. Lau, in: VLSI Electronics: Microstructure Science, Vol. 6, Chap. 6, Ed. N. G. Einspruch and G. B. Larrabee, Academic Press, New York 1983.
[3] E. G. Colgan and J. W. Mayer, Mater. Res. Soc. Symp. Proc., in the press.
[4] J. K. Howard, R. F. Lever, P. J. Smith, and P. S. Ho, J. Vacuum Sci. Technol. **13**, 68 (1976).
[5] S. P. Murarka, I. A. Blech, and H. J. Levinstein, J. appl. Phys. **47**, 5175 (1976).
[6] M. Nastasi, L. S. Hung, and J. W. Mayer, Appl. Phys. Letters **43**, 831 (1983).
[7] E. G. Colgan, J. appl. Phys. **62**, 1224 (1987).
[8] E. G. Colgan, C.-Y. Li, and J. W. Mayer, Appl. Phys. Letters **51**, 424 (1987).
[9] E. G. Colgan, C.-Y. Li, and J. W. Mayer, J. Mater. Res. **2**, 557 (1987).
[10] J. M. Legresy, B. Blanpain, and J. W. Mayer, J. Mater. Res., in the press.
[11] X.-A. Zhao, E. Ma, H.-Y. Yang, and M-A. Nicolet, Thin Solid Films **153**, 379 (1987).
[12] W. B. Pearson, in: A Handbook of Lattice Spacings and Structures of Metals and Alloys, Vol. 2, Pergamon Press. Oxford 1967.
P. Villars and L. D. Calvert, in: Pearson's Handbook of Crystallographic Data for Intermetallic Phases, Vol. 2, Amer. Soc. Metals, Metals Park (OH) 1985.

[13] X.-A. ZHAO, H.-Y. YANG, E. MA, and M-A. NICOLET, J. appl. Phys. **62**, 1821 (1987).
[14] C. M. COMRIE and J. M. EGAN. J. appl. Phys. **64**, 1173 (1988).
[15] M. J. RICE, JR., and K. R. SARMA, J. Electrochem. Soc. **128**, 1368 (1981).
[16] S. B. NEWCOMB and K. N. TU, Appl. Phys. Letters **48**, 143 (1986).
[17] E. MA, M-A. NICOLET, and M. NATHAN, J. appl. Phys., in the press.
[18] X.-A. ZHAO, F. C. T. So, and M-A. NICOLET, J. appl. Phys. **63**, 2800 (1988).
[19] E. G. COLGAN and J. W. MAYER, Nuclear Instrum. and Methods **B17**, 242 (1986).
[20] R. F. LEVER, J. K. HOWARD, W. K. CHU, and P. J. SMITH, J. Vacuum Sci. Technol. **14**, 158 (1977).
E. G. COLGAN, Thesis, Cornell University, 1987.

(Received August 25, 1988)

phys. stat. sol. (a) **110**, 515 (1988)

Subject classification: 73.40; 78.30; S5.11; S7.12

Shanghai Institute of Metallurgy (a)
and Laboratory for Infrared Physics, Shanghai Institute of Technical Physics (b), Academia Sinica[1])

The Influence of Strain and Dislocations on Transport Properties of GaAs/Si Strained-Layer Heterojunctions

By

ER-WANG MAO (a), WEN-QIN ZHAO (b), HONG-RUI ZHANG (a), AI-ZHEN LI (a), JIAN-MIN CHEN (a), and GUO-PING FANG (a)

The strain and dislocation effects in various thickness GaAs/Si samples are studied by means of Raman scattering and Hall measurement at different temperatures. It is observed that the strain effect is insensitive to GaAs layer thickness above 100 nm. Dislocation defects play an important role on the Hall mobility of the samples. The effect decreases as the thickness of GaAs layer increases.

Die Dehnungs- und Versetzungseffekte in GaAs/Si-Proben mit unterschiedlicher Dicke werden mittels Ramanstreuung und Halleffektmessungen bei verschiedenen Temperaturen untersucht. Es wird beobachtet, daß der Dehnungseffekt gegenüber GaAs-Schichtdicken oberhalb 100 nm unempfindlich ist. Versetzungsdefekte spielen eine wesentliche Rolle für die Hallbeweglichkeit der Proben. Der Effekt nimmt mit zunehmender Dicke der GaAs-Schicht ab.

1. Introduction

GaAs/Si has received more and more attention because of the low cost of Si and its perfect planar device technology, as well as the high mobility and saturation velocity of GaAs. It is a promising material for future optoelectronic and optical communication devices. However, the lattices of GaAs and Si are mismatched. The strain and dislocation defects exist in the interface of GaAs/Si heterojunction. Obviously it is important to study the effect of strain and dislocations on transport properties for the application of this kind of material. Several authors [1, 2] investigated the interface properties of GaAs/Si but did not link it to the mobility of the materials. In this paper, we investigated the strain in GaAs layers of various thickness by means of Raman scattering spectra. We also measured the Hall mobility to reveal the effects of strain and dislocations on the transport properties of the material.

2. Experiments

Three different thicknesses of GaAs layers (30, 60, 90 nm) were grown on n-Si substrate in the MBE system. The Si substrate was 4° off (100) surface to (011). After chemical etching, the following multiple step growth procedure for GaAs directly deposited on Si was used. Prior to epitaxial growth, the Si substrate was heated to near 850 °C for about 30 min for desorption of oxide in situ, then the substrate temperature was reduced to 400 to 450 °C and a few monolayers of As were deposited as

[1]) Shanghai, People's Republic of China.

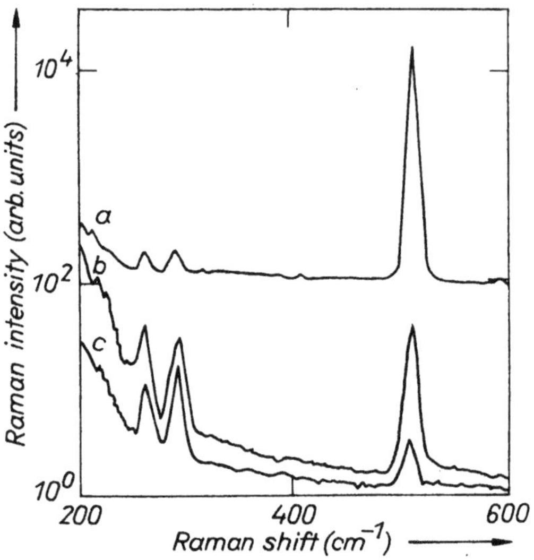

Fig. 1. Raman spectra of GaAs/Si: (a) sample A, $d = 30$, (b) sample B, 60, (c) sample C, 90 nm

a prelayer on the Si substrate. Then a 10 nm metastable GaAs grain initial nucleation layer was grown at a growth rate of 200 nm/h. The Ga oven shutter was then closed and the substrate temperature increased to 570 °C under As flux conditions for 30 min to allow the metastable GaAs grains to rearrange to become single crystalline [3].

The Raman scattering spectra were measured in backscattering configuration at room temperature, using SPEX Raman spectrometer. The Ar^+ ion laser power was 600 mW. Fig. 1, curve a shows the Raman spectrum of a GaAs/Si strained-layer heterojunction with GaAs layer of 30 nm thickness. The peak at 520 cm^{-1} is the LO phonon mode of Si and two small peaks are LO, TO phonon modes of GaAs, respectively. The Raman spectra of GaAs/Si samples of other thicknesses are shown in Fig. 1, curve b and c.

The Hall measurements were performed using the Van de Pauw method at various temperatures. The temperature range was 10 to 300 K by using a CSA-202 displex closed-cycle refrigeration system. The magnetic field was in the range from 0 to 0.75 T. The samples were cut to rectangular shape. A constant current of 10 μA was passed through samples during the measurement. The results are presented in Fig. 2.

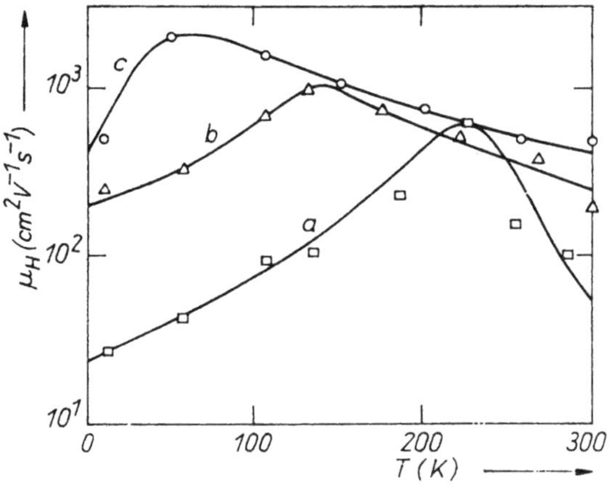

Fig. 2. Temperature dependence of GaAs Hall mobility of (a) $d = 30$, (b) 60, (c) 90 nm

3. Results and Discussion

Comparing the results of Fig. 1, curves a and b, we can see that the peaks in the spectra of GaAs shifted but those of Si did not. The reason is that the lattices of GaAs and Si are mismatched, when GaAs grows on Si substrate, the lattices of both materials are disturbed. The Si substrate is much thicker than the GaAs epitaxial layer, the deformation of Si can be neglected. Thus, we use the Raman scattering technique to detect the lattice deformation by observing the frequency shift of the GaAs optical phonon mode. In the backscattering configuration, according to the selection rule, for (100) GaAs the TO mode is forbidden, but we do observe the TO mode in our measurements. We argue that this TO mode does not come from the polycrystallization of the material, because the X-ray diffraction of our samples showed the single crystal patterns. These features have also been observed in a number of (100) GaAs samples [2, 4]. We guess that selection rules in the interface GaAs/Si might be broken by the deformation.

The strain in the interface can be deduced from the frequency shift of the optical phonon mode of the material. The biaxial stress in the interface can be divided into uniaxial stress and hydrostatic component of the applied stress, the uniaxial stress causes polarization dependent splitting or shifts which are proportional to the stress. According to Cerdeiras et al.'s theory [5], the threefold degeneracy of the $k = 0$ optical phonon in a diamond-type material is lifted by the strain. The frequencies of the optical modes in this case obey the following secular equation:

$$\begin{vmatrix} p\varepsilon_{xx} + q(\varepsilon_{yy} + \varepsilon_{zz}) - \lambda & 2r\varepsilon_{xy} & 2r\varepsilon_{xz} \\ 2r\varepsilon_{xy} & p\varepsilon_{yy} + q(\varepsilon_{xx} + \varepsilon_{zz}) - \lambda & 2r\varepsilon_{yz} \\ 2r\varepsilon_{yz} & 2r\varepsilon_{yz} & p\varepsilon_{zz} + q(\varepsilon_{xx} + \varepsilon_{yy}) \end{vmatrix} = 0 .$$

Here $\lambda = \Omega^2 - \omega_0^2$, $\Omega \approx \omega_0 + \lambda/2\omega_0$ is the strain dependent frequency, ω_0 the zero strain frequency, p, q, r are three constants describing the changes in effective elastical constants. There are long-range Coulomb forces existing in materials induced from the polarity of III–V compound semiconductors and the TO and LO modes for small values of k are split. This small splitting is considered to be independent of strain. Therefore, we can obtain the strain dependent optical phonon frequency by adding the splitting to the frequency obtained by secular equation,

$$\Omega^{\mathrm{TO}} = \omega_0^{\mathrm{TO}} + 2\,\Delta\Omega_{\mathrm{H}} + \tfrac{1}{3}\,\Delta\Omega ,$$

$$\Omega^{\mathrm{LO}} = \omega_0^{\mathrm{LO}} + 2\Delta\Omega_{\mathrm{H}} - \tfrac{2}{3}\,\Delta\Omega .$$

Here

$$\Delta\Omega_{\mathrm{H}} = \frac{x}{6\omega_0}\,(p + 2q)\,(s_{11} + 2s_{12}) ,$$

$$\Delta\Omega = \frac{x}{2\omega_0}\,(p - q)\,(s_{11} - s_{12}) .$$

$\Delta\Omega_{\mathrm{H}}$ is a shift due to hydrostatic component of the applied stress, i.e. hydrostatic stress. x is the applied stress and s_{ij} are elastic compliance constants referred to the cubic axes. With the equation presented above and the parameters listed in Table 1, we get strain and stress from experimental results. The results are listed in Table 2. It can be seen that the elastic strain of the GaAs and Si is very small. Assuming that the misfit of the heterojunction is totally comensated by elastic deformation, following

Table 1
Parameters of GaAs and Si used from [5]

	ω_0 (10^{14} s^{-1})	s_{11} (10^{-7} cm^2/N)	s_{12} (10^{-7} cm^2/N)	$-\frac{p+2q}{6\omega_0}$	$\frac{p-q}{2\omega_0}$
GaAs	0.51	1.175	−0.365	0.9	0.10
Si	0.98	0.768	−0.214	0.9	0.31

Table 2
Strain of GaAs and Si obtained from experimental results

	GaAs			Si		
GaAs thickness (nm)	30	60	90	30	60	90
ε_{xx}	0.0013	0.0012	0.0012	0.0011	0.0011	0.0011
ε_{zz}	−0.0012	−0.0011	−0.0011	−0.0009	−0.0009	−0.0009

the equations

$$a^{\perp}(d_1 + d_2) = a_1 d_1 + a_2 d_2 ,$$

$$\overset{i}{\varepsilon}_{xx} = \overset{i}{\varepsilon}_{yy} = \frac{a^{\perp} - a_i}{a_i} ,$$

$$\varepsilon_{zz} = \frac{a_2^i - a_i}{a_i} .$$

Here a_1, d_1, and a_2, d_2 are the lattice constants and thicknesses of the substrate and epitaxial layer, respectively. We can get that, as $d_1 \gg d_2$ and $a^{\perp} \approx a_{Si}$, the strain is

$$\varepsilon_{xx} = \varepsilon_{yy} = -3.89\% .$$

This is much larger in absolute value than our result. Therefore, the contribution of elastic deformation to the misfit of the heterojunction is small. We can also see that, as the thickness of the GaAs epitaxial layer increased, the strain remains constant. This is understandable. The interface structure in heterojunction and superlattices can be predicted from the forces acting on dislocation lines [6]. Two kinds of important forces determine the interface structure. One is the elastic stress F_ε and the other the dislocation tension F_λ,

$$F_\varepsilon = \frac{2G(1+\gamma)}{(1-\gamma)} bh\varepsilon \cos\lambda ,$$

$$F_\lambda = \frac{Gb^2}{4\pi(1-\gamma)} (1 - \gamma \cos^2 \alpha) \left(\ln \frac{h}{b} + 1\right).$$

One can get critical layer thickness,

$$h_c = \frac{b}{2\pi f} \frac{(1 - \gamma \cos^2 \alpha)}{(1+\gamma)\cos\lambda} \left(\ln \frac{hc}{b} + 1\right).$$

This is about 10 nm for GaAs/Si [7]. In our experiments the GaAs layer is about 30 nm, thus the strain has reached its maximum value and remained constant. Thus we conclude that the misfit of the thick epitaxial layer material is mainly compensated by dislocation defects. This can also be recognized from the broadening of the Raman scattering spectra.

From the results of Raman scattering experiments we expect that the effect of strain on mobility will be small and the influence on mobility is due to the dislocation defects. We measured the Hall mobility of those samples. There are two conductive regions in GaAs/Si, the GaAs layer and the Si substrate, so we have to separate out the GaAs Hall mobility from the experimental data carefully. We chose the Si with high resistivity and calculated the GaAs Hall mobility from Petriz as following [8, 9]:

$$\bar{\mu} = \frac{n_1\mu_1^2 + n_2\mu_2^2}{n_1\mu_1 + n_2\mu_2},$$

$$\bar{n} = \frac{n_1\mu_1 + n_2\mu_2}{n_1\mu_1^2 + n_2\mu_2^2}.$$

n_1, n_2, and μ_1, μ_2 are the sheet concentrations and Hall mobilities of the GaAs and Si, respectively. So we can get the Hall mobility of GaAs from well-known Si data,

$$\mu_1 = \frac{\bar{n}\bar{\mu}^2 - n_2\mu_2^2}{\bar{n}\bar{\mu} - n_2\mu_2}.$$

The Hall mobility of the sample A presented a peak at 225 K. The peak at higher temperature implies that the localization effect induced from the interface dislocation defects play an important role. The decrease of the mobility at low temperature agrees with a Coulomb scattering by dislocation defects. When the thickness of the GaAs layer increased, the peak of the Hall mobility moved to lower temperature. When the GaAs layer was thicker than 90 nm, the peak shifted to about 55 K. This result is coincident with the Raman scattering spectrum, we can see that the peak of the GaAs optical phonon mode becomes narrower as the thickness of the GaAs layer increased. Thus we come to a point that when we are going to fabricate a device on this kind of material, a thick buffer layer is necessary.

4. Conclusion

Three different thicknesses of GaAs on Si substrate were measured by means of Raman scattering technique and Hall measurement. The elastic deformation theory has been used to analyse the stress and strain in the interface. The results show that for a GaAs/Si strained heterojunction, the elastic deformation of GaAs is not important for the lattice misfit. The Hall measurement reveals that the dislocation defects play an important role in compensating the mismatch of the lattices. There is a peak in the $\mu_H - T$ curve, which is shifted to lower temperature as the thickness of the GaAs layer increased. The mobility decrease at low temperature is in agreement with Coulomb scattering by interface dislocation defects. The effects of dislocation defects on mobility can be eliminated when the GaAs layer is thick enough.

Acknowledgements

We are grateful to Mr. J. X. Wang and M. S. Qiu for MBE sample preparation. This work was supported by the Fundation of Academia Sinica.

References

[1] W. T. Masselink, T. Henderson, J. Klem, R. Fischer, P. Pearah, H. Morkoc, M. Hafich, P. D. Wang, and G. Y. Robinson, Appl. Phys. Letters **45**, 1309 (1984).
[2] W. I. Wang, Appl. Phys. Letters **44**, 1149 (1984).
[3] A. Z. Li, J. X. Wang, J. H. Qiu, B. W. Liang, Y. L. Cheng, and C. C. Li, J. Vacuum Sci. Technol. B **6**, 713 (1988).
[4] See, for example: C. S. Rama Rao, S. Sundaram, R. L. Schmidt, and J. Comas, J. appl. Phys. **54**, 1808 (1984).
[5] F. Cerdeira, C. J. Buchenauer, F. H. Pollak, and M. Cardona, Phil. Mag. B **5**, 580 (1972).
[6] W. A. Jesser, Phil. Mag. **15**, 1097 (1967).
[7] R. Hull and A. Fischer-Colbrie, Appl. Phys. Letters **50**, 851 (1987).
[8] R. L. Petriz, Phys. Rev. **110**, 1254 (1958).
[9] R. D. Larrabee and W. R. Thurber, IEEE Trans. Electron. Devices **27**, 32 (1980).

(Received February 9, 1988; in revised form July 14, 1988)

phys. stat. sol. (a) **110**, 521 (1988)

Subject classification: 73.60; S5; S8

Laboratoire de Physique des Matériaux et Composants de l'Electronique, Faculté des Sciences et des Techniques, Universite de Nantes[1])

Experimental Study of Oxygen Effect on the Variation of Resistivity with Temperature for Polycrystalline Selenium Thin Films

By

J. C. BERNÈDE, G. SAFOULA, A. AMEZIANE, and P. BURGAUD

A detailed correlation between the temperature dependence of the resistivity of polycrystalline selenium thin films and the presence of oxygen at the grain boundaries is established. The temperature dependence of the dc and ac resistivity and the thermoelectric power of polycrystalline selenium thin film structures are studied as functions of thermal history and ambient gas exposure (argon, air atmosphere, dry oxygen, helium). Annealing at 413 K for 4 h under oxygen results in a decrease of the room temperature resistivity and an increase of the resistivity with increasing temperature from 290 to 350 K. Evidence that these effects are due to oxygen diffusion at the grain boundaries is deduced from ac conductivity dependence of planar structures, thermoelectric dower measurements, and ESCA studies.

Eine ausführliche Korrelation zwischen der Temperaturabhängigkeit des Widerstand von polykristallinen dünnen Selenschichten und der Anwesenheit von Sauerstoff an den Korngrenzen wird aufgestellt. Die Temperaturabhängigkeit des Gleich- und Wechselstromwiderstands und der Thermospannung von polykristallinen Selendünnschichtstrukturen werden als Funktionen der thermischen Vorgeschichte und Umgebungsgaseinflusses (Argon, Luftatmosphäre, trockener Sauerstoff, Helium) untersucht. Temperung bei 413 K für 4 h unter Sauerstoff führt zu einer Abnahme des Zimmertemperaturwiderstands und einem Anwachsen des Widerstands mit steigender Temperatur von 290 bis 350 K. Hinweise, daß diese Einflüsse durch Sauerstoffdiffusion an den Korngrenzen hervorgerufen werden, werden aus der Wechselstromleitfähigkeitsabhängigkeit der planaren Strukturen, Messungen der Thermospannung und ESCA-Untersuchungen gezogen.

1. Introduction

The electrical conductivity σ of trigonal selenium has unusual properties: plastic deformation of the crystals significantly changes the voltage and temperature dependence of σ indicating that lattice defects play an important role in these phenomena [1, 2]. Abkowitz and Enck [3] have shown that heat treatment under oxygen of polycrystalline trigonal selenium compaction converts the original simply activated conductivity to a complicated function of temperature, characterized by a local conductivity maximum near room temperature.

Even in single crystals the dependence of the conductivity on temperature and electric field needs the introduction of the hypothesis of a strongly inhomogeneous distribution of hole density in the single crystals to explain the σ variations [1].

In this paper the results of a study of the effect of annealing under oxygen of thin polycrystalline selenium films, on the temperature variation of the resistivity and the thermoelectric power are described. The results are interpreted in terms of grain boundary modification by oxygen diffusion.

[1]) 2, Rue de la Houssinière, F-44072 Nantes Cédex 03, France.

2. Experimental Techniques

The substrates were polished glass or mica chemically cleaned and heated at 400 K during 1 h prior to deposition.

The substrate temperature, T_s, which was controlled by a copper–constantan thermocouple attached with silver past to the surface of the sample, was $T_s = 300$ or 400 K. The selenium films thermally evaporated in a high vacuum (base pressure 10^{-5} Pa) were amorphous in the former case and crystallized in the trigonal structure in the latter case. The selenium evaporation rate ($\approx$ 4 nm/s) and the film thickness (500 to 15000 nm) were measured in situ by the vibrating quartz method. The thickness was controlled by a stylus. Amorphous thin films were crystallized by heating at 340 K for 24 h and then at 403 K for 24 h under 10^5 Pa argon pressure to avoid any sublimation of the material. The structure of the films was checked using an X-ray goniometer. In order to verify the influence of halogen on selenium thin film properties, chlorine-doped selenium films were also deposited [4 to 6].

For planar samples metal electrodes were evaporated before selenium thin films were (Fig. 1b). For sandwich structures (Fig. 6b) samples were cooled, by a flow of liquid nitrogene, to 175 K during the deposition of the upper electrode.

The Cr electrodes were thermally evaporated by an electron beam.

The dc conductance was measured by a conventional method with an electrometer between 80 and 450 K. The ac conductance was measured by a multi-frequency LCR meter (10 kHz to 10 MHz). All electrical measurements were carried out in the dark.

The thermoelectric power (TEP) measurement method was described elsewhere [7, 8].

Electron spectroscopy for chemical analysis (ESCA) measurements (ESCA analysis has been carried out in Nantes with a Leybold spectrometer — University of Nantes, C.N.R.S.) were performed with a magnesium X-ray source (1253.6 eV) operating at 10 kV and 10 mA. The energy resolution was 1 eV at a pass energy of 50 eV. For insulating substrates, thin films were grounded with silver paste. The change in composition with depth through the film was traced by recording successive XPS spectra obtained after argon ion etching for short periods. As estimated from standard selenium thin films the rate of sample erosion was approximately 50 nm $\min^{-1}$, sputtering was accomplished at pressures of less than 5×10^{-4} Pa, a 10 mA emission current, and a 5 kV beam energy using an ion gun.

Observations of the surface topography have been performed by a scanning electron microscope (SEM) in order to estimate the grain size and the compaction of the films and have been reported elsewhere [4 to 6].

3. Experimental Results

X-ray patterns show that, whatever the deposition process is, the selenium thin film is crystallized in the hexagonal form [9, 10], and there is no new peak diffraction appearing after annealing whatever the ambient gas is (Ar, O, air, He). From the SEM photographs [11] we have determined the grain size distribution. We can estimate the grain size to be about 0.5 to 2 μm for the films with and without chlorine, respectively [4, 6]. The thicknesses of the films ranged from 0.5 to 15 μm. It turns out that the thickness variations did not affect notably the experimental results. Qualitatively identical results have been obtained with chlorine-doped selenium films or pure selenium films. As shown in a previous paper [4] chromium was the metal providing the more suitable ohmic contacts.

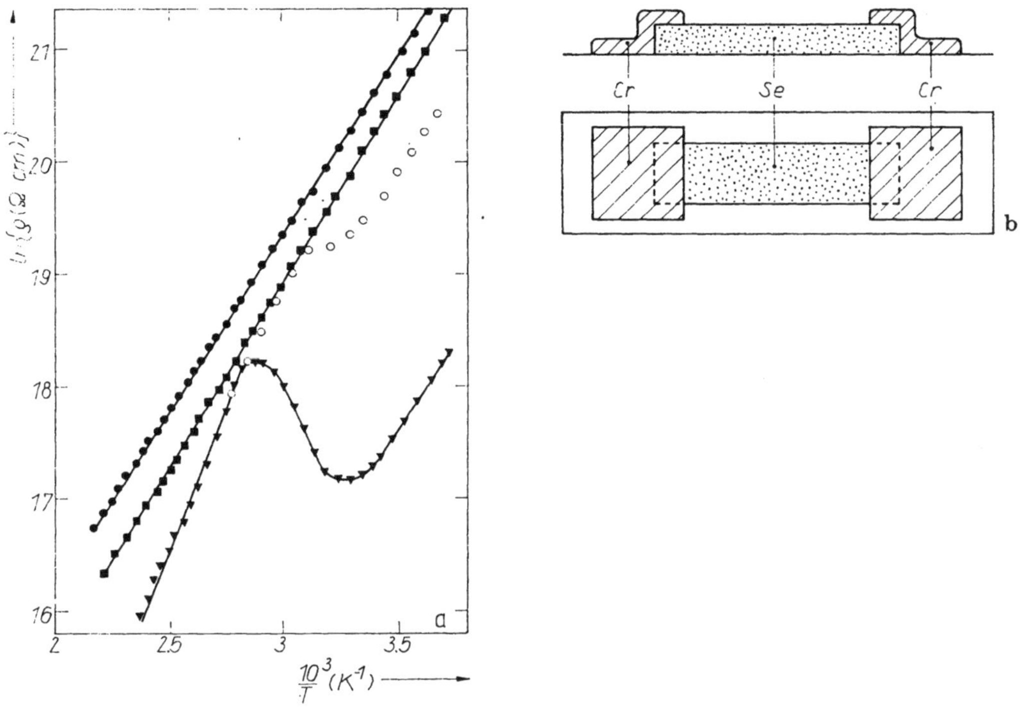

Fig. 1. a) ln ϱ vs. $1/T$ for Cr–Se(Cl)–Cr planar sample (Se being chlorine-doped selenium) ■ before annealing, ● after annealing 4 h at 415 K under argon atmosphere, ○ after annealing 4 h at 385 K under room atmosphere, ▼ after annealing 4 h at 415 K under room atmosphere. b) Planar structure

3.1 Planar Cr–Se–Cr structures

The experimental results for the dark resistivity and its temperature variation in the case of pure selenium thin films before and after annealing are shown in Fig. 1a. There is no striking aging effect before or after annealing under argon atmosphere. After annealing under room atmosphere the value of the resistivity decreases while the resistivity increases with the temperature between 290 to 350 K and a local resistivity maximum occurs. Fig. 1a shows that this effect increases with the temperature and/or the duration of annealing. The same effect is present after annealing under dry oxygen atmosphere which shows that oxygen and not water is the atmospheric component required during heat treatment in order to generate the new temperature variation of the resistivity. The ac conductance versus frequency is shown in Fig. 2. There is a transition with increasing frequency from high resistivity to low resistivity. Fig. 3 shows the conductance versus temperature behaviour between 290 to 410 K at 10 MHz. The modification of the resistivity with temperature after heating the sample in oxygen is not apparent at 1 or 10 MHz.

The thermoelectric power of Cr–Se–Cr planar samples has been measured before and after annealing under room atmosphere and oxygen atmosphere (Fig. 4). We see no modification of the TEP by annealing. In Fig. 5 we can see that after annealing under helium the initial value of the room temperature resistivity is recovered and that the local resistivity maximum near room temperature has nearly disappeared.

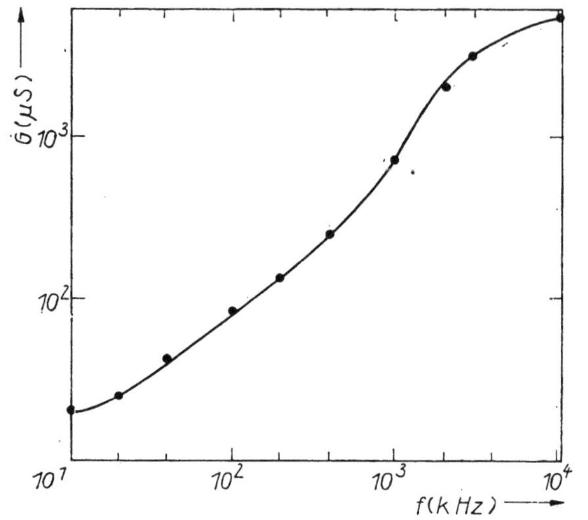

Fig. 2. ac conductance G vs. frequency f for Cr–Se(Cl)–Cr planar sample

3.2 Cr–Se–Cr sandwich structures

In all the sandwich structures selenium thin films have been deposited at $T_s = 400$ K. The experimental results for the dark resistivity and its temperature variation for pure selenium thin films before and after annealing are shown in Fig. 6. We see the same decrease of the resistivity and increase of the resistivity with temperature between 290 to 350 K as for planar samples.

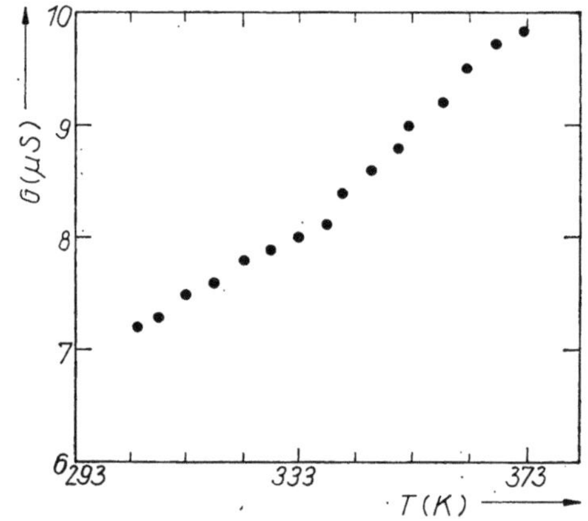

Fig. 3

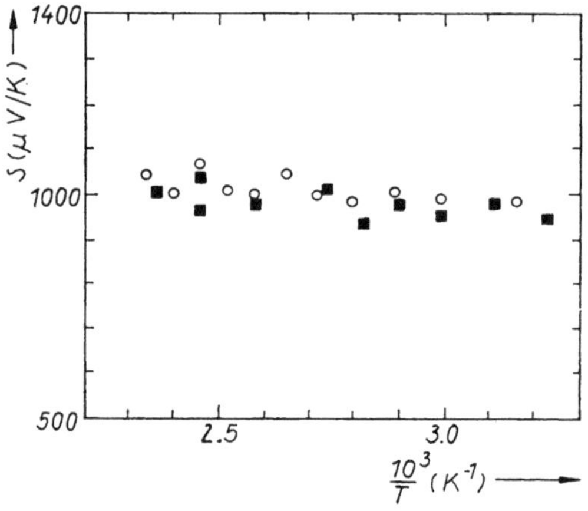

Fig. 4

Fig. 3. G vs. T at 10 MHz for Cr–Se(Cl)–Cr planar sample after annealing under room atmosphere (4 h, 385 K)

Fig. 4. Thermoelectric power variations with temperature ○ before annealing, ■ after annealing under room atmosphere

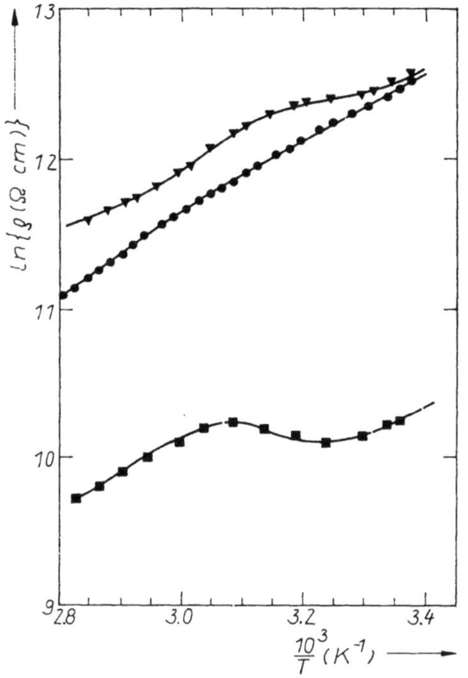

Fig. 5. ln ϱ vs. $1/T$ for Cr–Se–Cr planar sample ● before annealing, ■ after annealing under dry oxygen (4 h, 415 K), ▼ after annealing under helium pressure (4 h, 415 K)

3.3 Electron spectroscopy studies

XPS measurements have been performed on Cr–Se–Cr planar samples before and after annealing under oxygen and helium. Depth oxygen content was obtained by ion sputtering. Fig. 7 shows that the oxygen content in annealed selenium films decreases very slowly with sputering time. Even after 20 min of sputtering, oxygen is present, while before annealing there is no oxygen in the film after 6 min of sputtering. We can see also that after annealing under helium pressure no oxygen can be detected in the film.

Spectra of the Se 3d region are also shown in Fig. 7b, we can see that there is no SeO_2 formation.

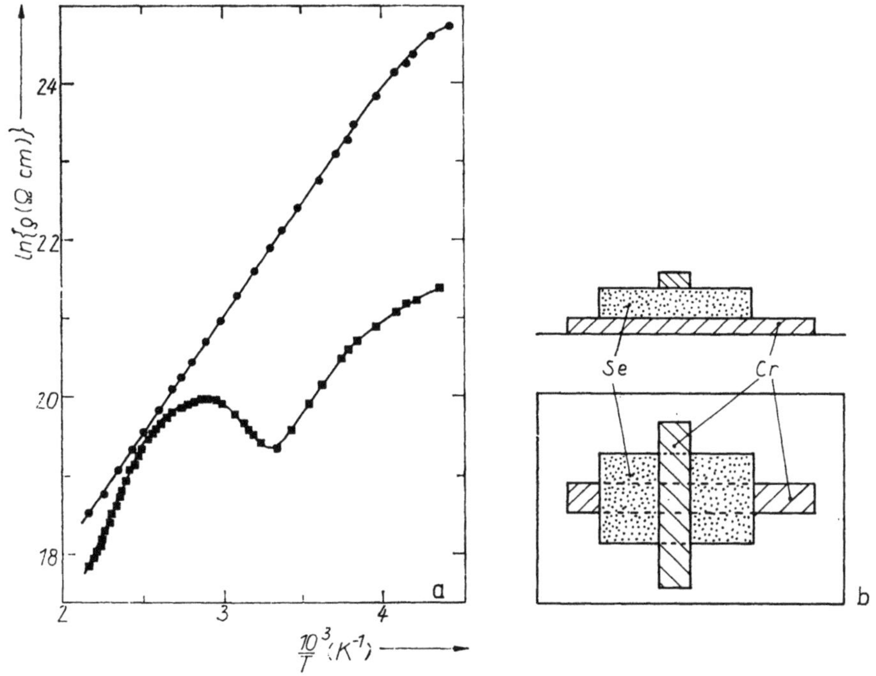

Fig. 6. a) ln ϱ vs. $1/T$ for Cr–Se–Cr sandwich sample ● before annealing, ■ after annealing 4 h at 415 K under room atmosphere. b) Schematic drawing of the structure

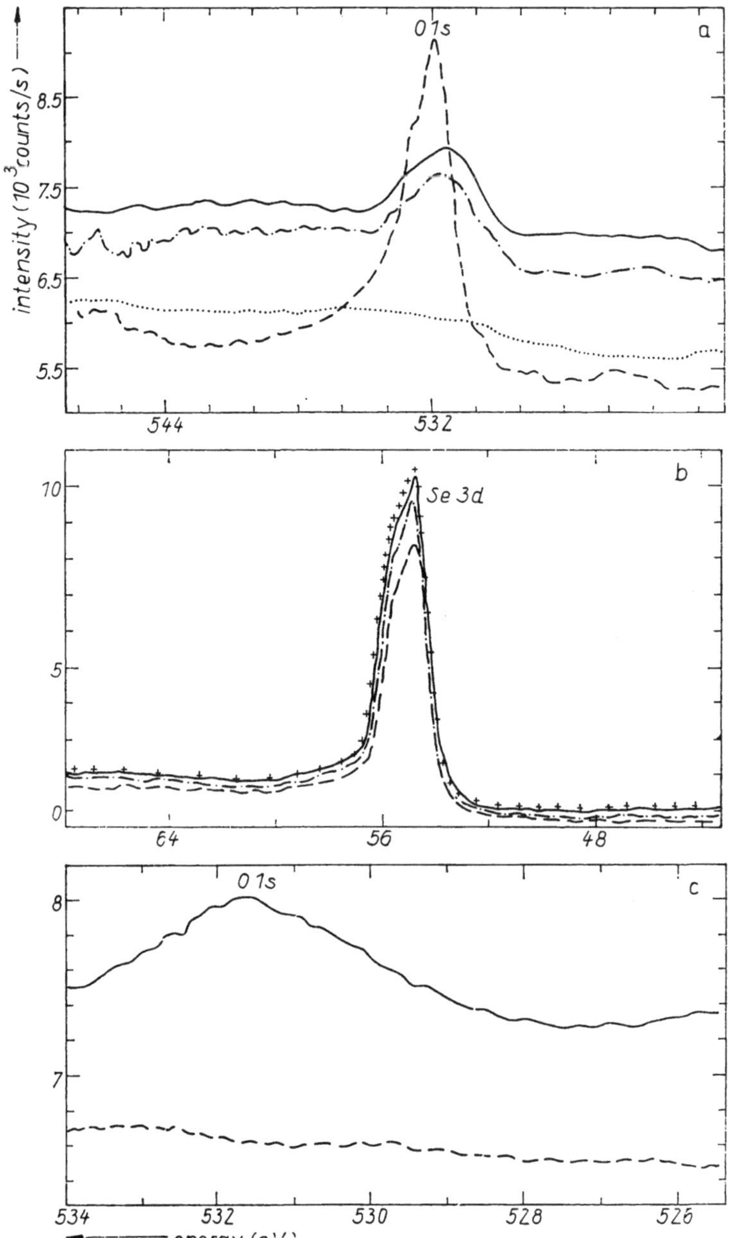

Fig. 7. ESCA study of a) O 1s peak of selenium thin film after annealing under room atmosphere (4 h, 415 K) ---- before sputtering, ——— after 8 min of sputtering, —·— after 20 min of sputtering; ····· selenium thin film not annealed after 8 min of sputtering. b) Se 3d peak of selenium thin film after annealing under room atmosphere (4 h, 415 K) ---- before sputtering, ——— after 8 min of sputtering, —·— after 20 min of sputtering; + + + + selenium thin film not annealed after 8 min of sputtering. c) O 1s peak of ——— selenium thin film after annealing under oxygen atmosphere (4 h, 415 K), ---- the same selenium thin film after annealing under helium atmosphere (4 h, 415 K)

4. Discussion

As shown by annealing under dry oxygen and the ESCA study, oxygen is the atmospheric component required during heat treatment in order to generate a decrease of the room temperature dc resistivity and an increase in dc resistivity at temperatures between 290 to 350 K. The effect of oxygen on ϱ indicates that oxygen is probably incorporated into the bulk and/or the grain boundaries of the polycrystalline films. The lateral resolution of the chemical analysis technique used in this work (ESCA) was not good enough to determine to what extent the oxygen segregated at the grain boundaries of the polycrystallites of the selenium films.

Frequency-dependent conductance measurements in selenium polycrystalline thin films show a transition with increasing frequency from low conductance to high conductance.

The dc conductivity in polycrystalline thin films is primarily controlled by interparticle barriers [12, 13]. The grain boundary trapping theory assumes the presence of trapping states at the grain boundary which capture, and therefore immobilize, free carriers. These charged states at the grain boundary create a depleted area with potential barriers which hinder the passage of carriers from a grain to the neighbouring ones [12 to 14]. Therefore, at high frequency the depletion capacitances at the grain boundaries are capacitively short-circuit (see [1], p. 243).

Thus frequency-dependent conductance measurements may indicate a transition with increasing conductance from a barrier-limited process (dc) to a bulk-limited process (ac).

The grain boundary theories show that while the dc conductivity is controlled by interparticle barriers the thermoelectric power measured is given by the prevailing scattering mechanism in the bulk [15 to 18]. Therefore ac conductivity and TEP measurements are related to the bulk, while dc conductivity is related to the grain boundaries.

As matter of fact the annealing effect under oxygen is only detected by dc conductivity measurements (Fig. 1 to 6), that is to say, the oxygen effect involves a barrier modification. Chen [19] has shown that oxygen atoms may be bonded to the end of a selenium chain. Therefore oxygen may be located at the grain boundaries [20]. Chemisorption of oxygen at the grain boundaries alters the barrier height and barrier width [21]. Many workers compare the grain boundary of polycrystalline materials to amorphous material [22], especially for selenium, this assumption is commonly accepted [23, 4]. Therefore at the grain boundary the primary source of intrinsic trapping defect states would be the dangling bonds of the broken selenium chains (D^+, D^-). The increase in conductivity probably involves that oxygen is adsorbed to the chain ends. Mott and Davis [19] suppose that oxygen is adsorbed only to the D^- centre giving negatively charged oxygen, the D^+ centre being already strongly bonded to another Se. Then the D^- centres are destroyed. Oda et al. [24] give the same conclusion: By increasing the oxygen content the density of D^- decreases and that of D^+ increases. Since selenium is a p-type semiconductor and the D^- traps are believed to act as hole traps it can be concluded that the depleted area and the potential barriers will decrease. The longer the annealing time and the higher the annealing temperature are, the higher is the decrease of the density of D^- centres and the decrease of resistivity (Fig. 1). Fig. 7 shows that a new annealing without oxygen gas implies that oxygen is desorbed. Therefore the resistivity has recovered its first value (Fig. 5).

The increase of the resistivity with temperature between 290 and 350 K cannot be easily explained by classical grain boundary theories. Two hypotheses are possible:

a) The resistivity is no longer limited by the grain boundary but by the conductivity of the grains.

b) The barrier height at the grain boundaries is modified by new phenomena.

In polycrystalline material we have (hypothesis a)

$$\varrho_\mathrm{T} = \varrho_\mathrm{c}\left(1 - \frac{\delta}{L}\right) + \varrho_\mathrm{GB}\left(\frac{\delta}{L}\right),$$

where ϱ_c is the crystallite bulk resistivity (Ω cm), ϱ_GB the grain boundary resistivity (Ω cm), δ the grain boundary thickness, and L the crystallite size.

Our hypothesis is that before oxygen annealing we have $\varrho_\mathrm{GB} \gg \varrho_\mathrm{c}$ and $\varrho_T \approx \varrho_\mathrm{GB}$ but after annealing under oxygen pressure ϱ_GB decreases and ϱ_T contains barrier and bulk components. Therefore, its temperature behaviour is affected by both components. At low temperatures ϱ_GB is predominant, while ϱ_c is at high temperature. This temperature dependence of the resistivity has been described by many workers [25, 26]. They have shown that at high temperatures the resistivity is no longer limited by the grain boundary but by the conductivity of the grains. At high enough temperature all free carriers existing in the grains have enough thermal energy to surpass the barrier. Thus the influence of the barrier on the carrier transport is eliminated and the carrier transport is only limited by the resistivity of the grains. Bonnet et al. [27] have clearly shown that an increase of the resistivity with the temperature could be explained in crystals if the charge carriers are scattered by optical modes. The same carriers being scattered in two ways, on impurity atoms and optical lattice vibrations, the electrical resistivity due to each component may be added directly to give the total resistivity. At still higher temperature carriers may be excited in the extended states of the valence or conduction band and the resistivity decreases. But the TEP S being nearly constant we can think that the hole concentration is temperature independent. Therefore the decrease of the resistivity at high temperature may be tied to another phenomenon.

Another problem is tied to the high resistivity of the film which necessitates that ϱ_c should be very high. However the electrical resistivity ϱ of trigonal selenium single crystals is rather scattered from one sample to another. Its value at room temperature lies between 10^4 to 10^8 Ω cm [1, 23, 28, 29]. To explain this scattering the main idea is that ϱ is determined by a strongly inhomogeneous distribution of hole density in the trigonal single crystals [1].

The very high value needed for ϱ_c and the difficulty to explain the decrease of ϱ at higher temperature renders this first hypothesis to be very questionable. For barrier height modification (hypothesis b) another possibility has been suggested by Abkowitz and Enck [3]. They suppose that heating in oxygen does not significantly influence the interparticle barriers above the dc temperature corresponding to the peak maximum of ϱ, but does significantly reduce the barrier height at lower temperatures. They suppose that, as mechanical straining of trigonal selenium single crystals is known to produce conductivity enhancement via a increased population of shallow acceptor states [1, 2], oxygen incorporated on or near a particle surface induces local strain and thereby creates shallow acceptors which lower interparticle barriers.

But it is very difficult to understand why the resistivity is smaller after annealing under oxygen even at high temperature. So we think as explained before that the decrease of ϱ is tied to D^- dangling bond destruction by oxygen.

It has been shown that plastic deformation of single crystals may decrease or increase the resistivity of selenium [2]. To explain the increase of ϱ with temperature one hypothesis is that oxygen incorporated at the grain boundaries induces local strains, when the temperature is high enough to create a large number of additional dislocations at the grain boundaries. Or, may be, this domain of temperature is a critical one for the oxygen adsorbed to the D^- centres. At higher temperature when local modifications are stabilized we have the classical evolution of ϱ versus T when the conductivity is controlled by the grain boundaries. Another explanation of barrier modification may be proposed.

It has been shown [1, 2, 23] that even in single crystals the conductivity cannot be explained by the usual transport mechanisms in a homogeneous material. Therefore, it is assumed that Se crystals have internal barriers which lead to an activated conductivity in spite of a constant carrier concentration (constant TEP). These barriers have a similar influence on the conductivity of a sample as grain boundaries and change their behaviour also with temperature. They alter the resistance of single crystals by orders of magnitude and are also altered by annealing. So internal barriers in the grains and their alteration by annealing should be another explanation. Experimental results — annealing under Ar or He — have shown that oxygen is required during heat treatment in order to change the resistivity, therefore internal barrier alteration with temperature should be excluded here, but not oxygen alteration. However, it has been shown [30] that the dominant diffusion mechanism in polycrystalline thin films was the grain boundary diffusion process, consequently the interparticle barriers should be more sensitive to O_2 diffusion than internal barriers in the grains.

5. Conclusion

We have shown that the decresae of the resistivity of selenium polycrystalline thin films after annealing under oxygen is tied to barrier height modulation by dangling bond (D^-) oxygen neutralization. The evolution of the resistivity versus temperature when the latter increases may be tied to a transition from grain boundary limited conductivity to bulk crystallite limited conductivity. However, at high temperature the increase of the conductivity versus the temperature, while the TEP is nearly constant, is not easy to explain, then another hypothesis is that when the temperature increases ,the oxygen induces a local modification of the grain boundaries or the internal barriers in the grains, which increases the resistivity.

References

[1] R. A. Zingaro and W. C. Cooper, Selenium, Van Nostrand/Reinhold, New York 1974.
[2] J. Stuke, phys. stat. sol. **6**, 441 (1964).
J. Stuke and K. Wendt, phys. stat. sol. **8**, 533 (1965).
[3] M. Abkowitz and R. C. Enck, J. appl. Phys. **50**, 3423 (1979).
[4] P. Burgaud, J. C. Bernède, G. Safoula, and A. Ameziane, phys. stat. sol. (a) **95**, 721 (1986).
[5] K. Napo, J. C. Bernède, G. Safoula, P. Burgaud, and A. Ameziane, Thin Solid Films **159**, 291 (1988).
[6] G. Safoula, J. C. Bernède, A. Latef, E. Rzepka, and M. Spiesser, Mater. Chem. Phys., **20**, 571 (1988).
[7] A. Bonnet, P. Said, and A. Conan, Rev. Phys. appl. **17**, 701 (1982).
[8] A. Mallouky, Thesis (3ème cycle), Nantes 1988.
[9] P. Burgaud, Thesis (3ème cycle), Nantes 1985.
[10] K. Napo, Thesis (3ème cycle), Nantes (1986).
[11] A. Ameziane, Thesis (3ème cycle), Nantes 1986.
[12] S. Y. Seto, J. appl. Phys. **46**, 5247 (1975).
[13] G. Baccarani, B. Ricco, and G. Spadini, J. appl. Phys. **49**, 5565 (1978).
[14] C. C. Lu, C. Y. Lu, and J. D. Meindl, IEEE Trans. Electron Devices **28**, 818 (1981).
C. Y. Lu, C. C. Lu, and C. S. Wang, Solid State Electronics **27**, 463 (1984).

[15] G. H. Blount, R. H. Bube, and A. L. Robinson, J. appl. Phys. **41**, 2190 (1970).
[16] C. Lemercier, phys. stat. sol. (b) **46**, K35 (1971).
[17] J. Jerhot and J. Vlcek, Thin Solid Films **92**, 259 (1982).
[18] N. F. Mott and E. A. Davis, Electronic Processes in Non-Crystalline Materials, 2nd ed., Clarendon Press, Oxford 1979.
[19] Inan Chen, J. chem. Phys. **45**, 3536 (1966).
[20] P. T. Sampath, J. chem. Phys. **45**, 3519 (1966).
[21] C. H. Wu and R. H. Bube, J. appl. Phys. **45**, 648 (1974).
[22] C. H. Seager and T. G. Castner, J. appl. Phys. **49**, 3879 (1978).
[23] K. W. Plesner, Proc. Phys. Soc. **64**, 671 (1951).
[24] O. Oda, A. Onozuka, and I. Tsuboya, J. non-crystall. Solids **83**, 49 (1986).
[25] R. Herrmann, G. Nachtwei, and W. Kraak, phys. stat. sol. (a) **83**, K207 (1984).
[26] W. Siegel, G. Kühnel, and E. Ziegler, phys. stat. sol. (a) **64**, 249 (1981).
[27] A. Bonnet, P. Said, and A. Conan, Recent Developments in Condensed Matter Physics, Vol. 3, Ed. J. T. Devreese, L. F. Lemmens, V. E. van Donen, and J. van Royen, Plenum Publ. Corp., New York 1981 (p. 37).
A. Conan, D. Delaunay, A. Bonnet, A. G. Moustafa, and M. Spiesser, phys. stat. sol. (a) **94**, 279 (1979).
A. Bonnet, A. Conan, M. Spiesser, and M. Zoaeter, J. Physique, to be published (1988).
[28] H. Lumbroso, Complèments au nouveau traité de chimie minérale — sélénium, Pascal no 7, Masson, Paris 1977.
[29] J. Helevskivi, J. Stubb, and T. Suntola, J. appl. Phys. **40**, 2923 (1969).
[30] D. Gupta and P. S. Ho, Thin Solid Films **72**, 399 (1980).

(Received May 11, 1988; in revised form September 8, 1988)

phys. stat. sol. (a) **110**, 531 (1988)

Subject classification: 61.80 and 73.40; S1.61; S4; S7.12

School of Electrical, Electronic and Systems Engineering, College of Cardiff, University of Wales[1]) *(a), Naval Research Laboratory, Washington*[2]) *(b), and RMCS Shrivenham*[3]) *(c)*

High Temperature Metallisation for GaAs Device Processing

By

D. V. MORGAN (a), H. THOMAS (a), W. T. ANDERSON (b), P. THOMPSON (b), A. CHRISTOU (b), and D. J. DISKETT (c)

A study is made of the thermal stability of a range of refractory metals and their silicides on GaAs. Particular emphasis is laid on the effects of transient annealing with temperatures up to 1050 °C. Using the technique of Rutherford backscattering the results for the metallisation systems Ta, W, $TaSi_x$, WSi_x, and VSi_x are compared and indicate the superior properties of the refractory silicides.

Die thermische Stabilität einer Reihe von schwer schmelzbaren Metallen und ihrer Silizide auf GaAs wird untersucht. Insbesondere die Einflüsse von nichtstationärer Temperung bei Temperaturen bis zu 1050 °C. Mit der Rutherfordrückstreutechnik werden die Ergebnisse für die Metallisationssysteme Ta, W, $TaSi_x$, WSi_x und VSi_x verglichen und zeigen die überlegenen Eigenschaften der schwer schmelzbaren Metallsilizide.

1. Introduction

The thermal stability of Schottky metallisation systems heated to relatively high temperatures has been studied for many years [1 to 4]. The motivation for this work has arisen partly from the need for high reliability in devices operated at elevated temperatures and partly from the need for self aligned FET gate metallisations which can withstand the high temperatures associated with device processing.

In this paper we present new results on the annealing of refractory metals and refractory silicide structures on GaAs. To complement previous studies this paper is concerned with the metallisation stability during transient optical annealing at temperatures typical and in excess of those required for ion implantation activation. The systems have been studied using Rutherford backscattering analysis (RBS) and SEM and include W, Ta, $TaSi_x$, WSi_x, and VSi_x contacted on GaAs substrates. These results are compared with related electrical studies.

In this work, two types of GaAs substrates have been used, (i) type A: undoped semi-insulating LEC, and (ii) type B: substrates doped with In in order to lower the dislocation densities. Annealing was carried out in a rapid thermal annealing optical furnace (Heatpulse 410). The time at anneal temperature was 10 s and an optical pyrometer was used in conjunction with a thermocouple to measure substrate temperature. The quartz sample chamber was flushed for 15 min with forming gas (90% N_2 and 10% H_2) and the subsequent heating of the chamber was performed using high intensity tungsten–halogen lamps. The interdiffusion was studied using RBS with a 2 MeV helium ion incident beam and a scattering angle of $\theta = 150°$.

---

[1]) POB 904, Cardiff, Wales CF1 3XE, Great Britain.
[2]) Washington, D.C, USA.
[3]) Shrivenham, Great Britain.

2. Results and Discussion

A summary of all the RBS results is presented in Table 1.

Table 1

system	substrate	850 °C	950 °C	1050 °C
Ta	A	blistered	blistered	blistered
$TaSi_x$	A	no outdiffusion	outdiffusion Ga and/or As	outdiffusion Ga and/or As
	B	no outdiffusion	no outdiffusion	no outdiffusion
W	A	no outdiffusion	no outdiffusion	no outdiffusion
WSi_x	A	no outdiffusion	no outdiffusion	no outdiffusion
	B	no outdiffusion	no outdiffusion	no outdiffusion
VSi_x	A	no outdiffusion	small outdiffusion	small outdiffusion
	B	no outdiffusion	small outdiffusion	small outdiffusion

Tantalum/GaAs annealing sequences were performed at 850, 950, and 1050 °C, and even at the lowest temperature of 850 °C the Ta metallisation became discoloured and RBS was not able to resolve interdiffused constituents. Annealing at 950 and 1050 °C, SEM studies indicated that the film became more granular and further discoloration and blistering were observed confirming the blistering and discoloration observed under the optical microscope.

Tantalum silicide/GaAs: The combination of silicon with tantalum to form a silicide $TaSi_x$ film has a dramatic and positive effect on the interface stability [5]. The RBS spectrum of the "as-deposited" sample is shown in Fig. 1. From this we see that the tantalum silicide has an enhanced Ta concentration at the outer surface. The high surface peak apart, the interface silicide has a factor x of 1.4 (in $TaSi_x$). It is normally found that annealing of the system causes the Ta and Si_x to intermix to form a uniform compound.

The first point to note is that the addition of silicon to tantalum produces a much stabler metallisation. After the 850 °C anneal the RBS spectra show no evidence of

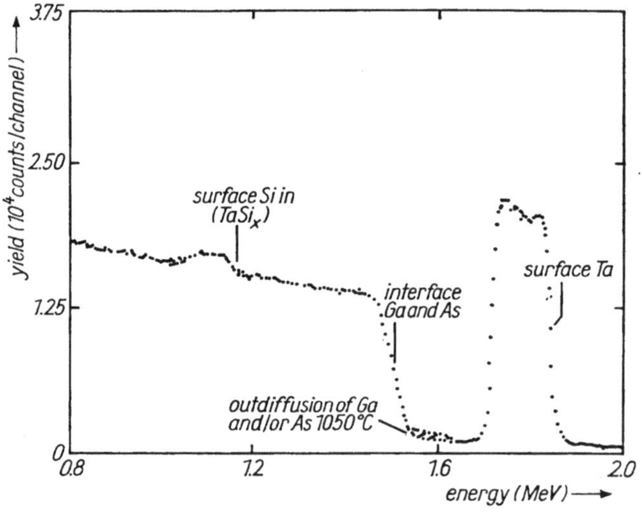

Fig. 1. The RBS spectrum for $TaSi_x$ on GaAs

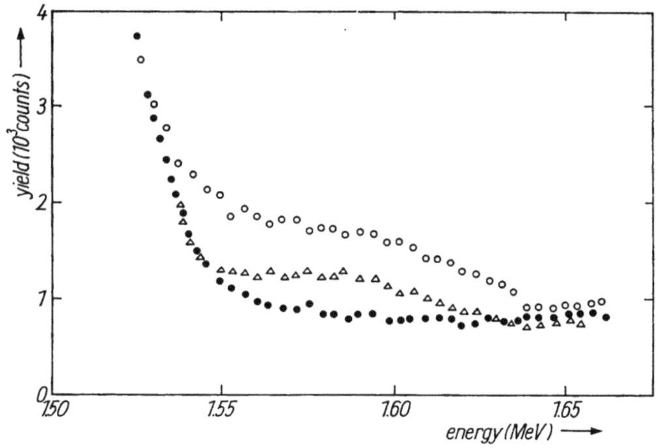

Fig. 2. An expanded plot of Fig. 1, showing the leading edge of the GaAs spectrum. ● As deposited, △ annealing at 950 °C, ○ annealing at 1050 °C

interdiffusion and SEM and optical analysis also show no evidence of peeling or visible degradation of the film. At 950 °C outdiffusion of Ga or As occurs to a measurable degree and is further enhanced at 1050 °C. Although it is not possible to distinguish between these two elements with RBS, a related study suggests that gallium is the outdiffusing species. Fig. 2 shows an enlarged plot of the RBS spectra for the as-deposited, 950 and 1050 °C annealed samples in the region corresponding to the interface GaAs to the surface GaAs. From this data a clear measure of the outdiffusion may be obtained. The concentration of Ga or As decreases from the interface and carries on through to the surface. At 950 °C the outdiffusing species have not reached the surface whilst at 1050 °C the diffusing front has fully reached the surface. At a point midway through the $TaSi_x$ film an estimate of the Ga (or As) concentration is 3 at% at 950 °C and at 1050 °C, 7 at%. These low figures for atomic percent showed only small variations in repeated runs on different samples and confirmed the general picture. The trailing edge of the Ta peak shows some qualitative evidence of inward diffusion. However, the relatively small silicon peak superimposed upon the GaAs background does not enable an estimate on the relative proportions of Si and Ta moving into GaAs to be made.

The interpretation of the RBS spectra presented above may also be attributed to the development of pin-holes in the metallisation during heating. Confidence in the diffusion interpretation is obtained from detailed SEM studies of the samples where no evidence of pin-holes has been found. It is also noted that in the evidence to be presented later on the annealing of the same metallisation on In-doped samples no "diffusion" evidence is found and the development of pin-holes should occur equally well in both cases.

As an extension to these studies the same $TaSi_x$ annealing sequence was carried out on In-doped substrates. The In doping has been found to reduce the dislocation density dramatically. The results of the annealing studies indicate that the In-doped GaAs studies do not exhibit any outdiffusion for temperatures (950 and 1050 °C) where it normally occurs. This is a significant observation and was corroborated a number of times on similarly prepared samples with reproducible results. A possible explanation of the results is that the interdiffusion between thin metallic films and GaAs is a crystalline defect assisted process and that the low defect density present in In-doped GaAs results in greater interface stability. Further work is now underway to study this effect and estimate its relevance to device production.

2.1 *Tungsten/GaAs*

In the studies of this system no evidence of the interdiffusion of W and GaAs was observed for type A substrates. The SEM study however, showed evidence of selective peeling at 1050 °C whereas film integrity was maintained at 850 and 950 °C. There was no evidence of Ga or As outdiffusion to the surface, but the high temperature tendency to peel places an upper bound to its suitability.

2.2 *Tungsten silicide/GaAs*

RBS analysis of the WSi_x system annealed on the undoped and indium-doped GaAs ($x = 1.35$) confirms the general observation obtained for $TaSi_x$, that is, the contact is stable under transient annealing up to 1050 °C with no evidence of interdiffusion. This is illustrated in Fig. 3 which shows the region of the RBS spectrum between the GaAs edge and the buried W edge in WSi_x. There is no evidence of any outdiffusion up to the 1050 °C anneal. We note, however, that for undoped GaAs which contained a heavy oxygen contamination in the WSi_x layer, which occurred during deposition, the layers did not adhere to the GaAs and were seriously degraded after the 850 °C anneal. It is clear from this observation that oxygen has a deleterious effect and that it is important to maintain clean oxygen-free conditions during film deposition and prevent contamination.

2.3 *Vanadium silicide/GaAs*

The RBS spectrum of vanadium silicide ($x = 1.9$) is superimposed on the relatively large background of the GaAs spectrum and in these circumstances it is hard to detect any inward diffusion of the vanadium. It is fortunate, however, that it does not obscure the leading edge of the GaAs spectrum. As we have seen, it is here that outdiffusion of Ga and/or As may be observed. The results for type A and type B substrates indicate a very small outdiffusion of the host species at temperatures of 950 and

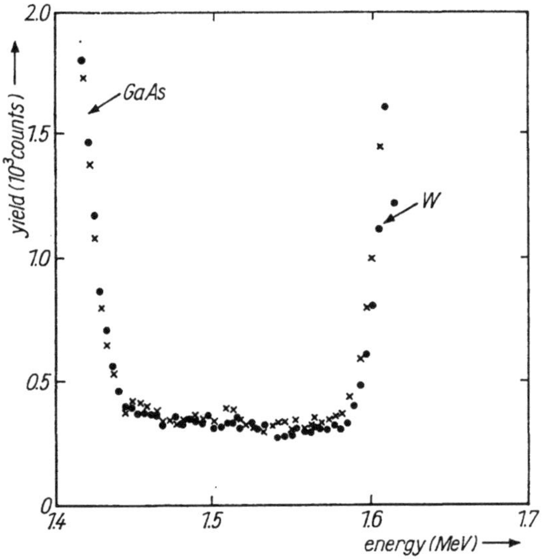

Fig. 3. The RBS spectrum showing the region between the leading edge of the GaAs spectrum and the W at the interface with GaAs. There is no evidence of interdiffusion up to annealing temperatures of 1050 °C. ● As deposited; × annealing at 1050 °C

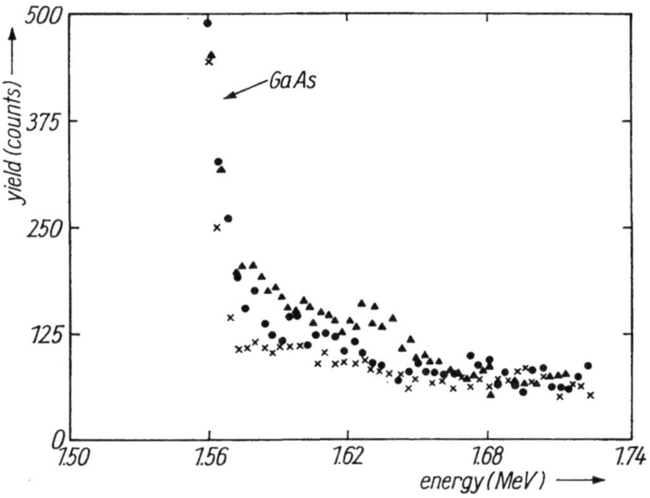

Fig. 4. The spectrum corresponding to Fig. 2 and 3, but for the GaAs/VaSi$_x$ system. A small amount of diffusion is observed at the highest temperature of 1050 °C. ● As deposited; annealed at 1050 °C; × substrate type A; ▲ type B

1050 °C as is shown in Fig. 4. This is too small to be quantified and appears to be slightly more in evidence in the indium-doped substrates. It is again observed that in both these sets of samples the "as-deposited" VSi$_x$ films were contaminated with oxygen but at a much lower level than the WSi$_x$ result discussed earlier.

In terms of device reliability the RBS data alone is not sufficient to draw firm conclusions since electrical degradation may occur at a level too low to observe by RBS. There is, however, evidence that suggests the stability observed in these experiments also reflect themselves in good electrical behavior. Haynes et al. [6] have studied the TiSi$_x$ system with x values of 0.6. Their RBS results are in close agreement to those reported here with 850 °C being the critical temperature for stability against Ga or As outdiffusion. What is significant however, is that the electrical properties of the Schottky barrier remain stable up to temperatures of 900 °C. Far from degrading however, the results indicate that the barrier height rises slightly from 0.75 to 0.78 eV — a slight improvement rather than a degradation.

3. Conclusions

The following conclusions can be drawn from this work:

1. Rapid thermal annealing up to temperatures above those needed for implant activation results in less interdiffusion than the comparable long furnace annealing times to the same temperature.

2. The replacement of the basic refractory metal with its silicide results in much improved integrity. The tendency of the films to peel away from the semiconductor is reduced and the morphology of the film is much more uniform.

3. Preliminary evidence indicates that the eventual interdiffusion of TaSi$_x$ and GaAs is initiated by the defects in the semiconductor.

4. Oxygen impurities incorporated into the silicide films during deposition inhibit good film formation and adhesion. As far as the electrical characteristics are concerned, heating up to 900 °C on TaSi$_x$ does not result in any measurable degradation of the contact.

Acknowledgements

We express our thanks to RMCS Shrivenham for the use of their Van der Graaff facilities and UWIST for support of this work. The work is partly supported by a NATO Collaborative Research Grant and partly by the EEC in ESPRIT 1270.

References

[1] C. J. Palmstrom and D. V. Morgan, GaAs Devices and Circuits, Chap. 5, Ed. M. J. Howes and D. V. Morgan, Wiley, 1985.
[2] D. A. Allan, IEEE Proc. I, Solid State and Electron Devices **133**, 18 (1986).
[3] E. Kohn, Proc. Internat. Electron Devise Meeting, IEEE, 1979 (p. 177).
[4] S. D. Mukerjee, D. V. Morgan, M. J. Howes, J. G. Smith, and P. Brook, J. Vacuum Sci. Technol. **16**, 445 (1985).
[5] D. V. Morgan, H. Thomas, W. T. Anderson, P. Thompson, A. Christou, and D. J. Diskett, Electronics Letters **23**, 1154 (1987).
[6] T. E. Haynes, C. C. Han, S. S. Lau, S. T. Picraux, and W. K. Chu, Mater. Res. Symp. Proc. **92**, 381 (1987).

(Received August 5, 1988)

phys. stat. sol. (a) **110**, 537 (1988)

Subject classification: 73.40; 73.60; S10.1

Sektion Physik und Technik elektronischer Bauelemente der Technischen Hochschule Ilmenau[1])

Charge Injection into SiO_2 Films at Fields between 1 and 3 $MV\,cm^{-1}$ after Electrical Stress

By

H. KRAUSE and H.-P. BÄR

SiO_2 films on monocrystalline silicon substrates are stressed at fields up to 10 $MV\,cm^{-1}$. After stress currents in the field range below 4 $MV\,cm^{-1}$ with current densities of 10^{-12} A cm^{-2} to larger than 10^{-10} A cm^{-2} are observed. From experimental results about current–time behaviour and simultaneously measured flatband voltage–time behaviour it is concluded that the stress generated currents are caused by tunnel injection of electrons (and possibly also holes) into trap clusters within the interface regions of the SiO_2 films. At strong action of stress microchannels become generated through whole the film where hopping conduction takes place. Currents in the lower field range flowing in the conduction band or valence band caused by locally lowered barriers or by simple emptying of flat traps can be excluded. Also moving internal charge distributed homogeneously in the SiO_2 films is not in agreement with experimental results.

SiO_2-Schichten auf einkristallinen Siliziumsubstraten werden bei Feldstärken bis zu 10 MV cm^{-1} gestreßt. Nach Streß werden Ströme im Feldbereich unterhalb 4 MV cm^{-1} mit Stromdichten von 10^{-12} A cm^{-2} bis über 10^{-10} A cm^{-2} beobachtet. Aus den experimentellen Ergebnissen zum Strom–Zeit-Verhalten und gleichzeitig gemessenem Flachbandspannung–Zeit-Verhalten ist zu schließen, daß die durch Streß generierten Ströme durch Tunnelinjektion von Elektronen (und möglicherweise auch Löchern) in Trapcluster in den Randzonen der SiO_2-Schichten verursacht sind. Bei starker Streßwirkung werden durch die Schichten hindurchgehende Mikrokanäle generiert, in denen Hoppingleitung auftritt. Ströme im unteren Feldbereich, die im Leitungsband oder Valenzband infolge lokaler Barrierenerniedrigung oder einer Entleerung flacher Traps fließen, können ausgeschlossen werden. Auch die Bewegung einer inneren, homogen in der SiO_2-Schicht verteilten Ladung ist nicht in Übereinstimmung mit den experimentellen Ergebnissen.

1. Introduction

Injection of charge and currents at high fields in SiO_2 films have been investigated for many years [1 to 10[5]]. In earlier papers [11 to 14] we have demonstrated that in the field range below 4 MV cm^{-1} in SiO_2 films a weak injection of charge also can be observed and charge transport through SiO_2 films may occur according to the hopping equation [15].

Progresses in technologies have led to SiO_2 films of high quality characterized by breakdown fields larger than 10 MV cm^{-1} also in the case of films of 50 nm and thicker and to breakdown charges in the range of Ccm^{-2}. In such high quality SiO_2 films in the lower field range currents over longer times cannot be observed. But recently we could show [16] that after strong field stress measurable currents in the lower field range were observed also at high quality oxide films. They are flowing in localized regions or microchannels that have been induced by the stress and were demonstrated

---

[1]) DDR-6300 Ilmenau, GDR.

as points of current conduction with the aid of liquid crystal technique. Stress generated currents in SiO_2 films in the field range below 4 MV cm^{-1} have also been reported by Baglee and Smayling [17].

In this work we report experimental results which will lead to better knowledge about the physical nature of currents in the lower field range that have been generated by Fowler-Nordheim injection in SiO_2 films.

2. Experimental

After careful cleaning of silicon substrates, (100) p-type, SiO_2 films have been thermally grown in dry oxygen with 3% HCl at 1000 and 1100 °C in the thickness range of 40 to 70 nm. After film growth a POA was carried out at 1000 °C in dry nitrogen, 20 min. Aluminum dots of 1 mm^2 were evaporated through a mask or dots of 0.05, 0.25, and 1 mm^2 were prepared by a usual photographical technique. Finally a PMA 15 min at 400 °C in hydrogen was performed.

On MOS capacitors prepared in this way we investigated the behaviour of currents depending on field strength and time before and after stress applied. Current measurements have been carried out as described earlier [12]. Currents decreasing with time at fixed field strengths either have been written by an x–y recorder or were measured at fixed intervals of time. Field stress up to 10 MV cm^{-1} was applied to the samples for defined times at polarity as described later, but no drastic differences have been observed at both polarities. To avoid deep depletion in the silicon substrate during stress at polarity metal plus the samples have been illuminated.

As a main point of our experiments we measured current–time characteristics in cycles of stepwise increasing and decreasing field strengths between zero and 3 MV cm^{-1}. Further we alternatingly measured current and flat-band voltage at fixed field strength (3 MV cm^{-1}) depending on time. These measurements have been performed with the aid of an automated system. Thereby the current measurement was interrupted periodically by a time controlled relay to record the flat-band voltage. This was done by measuring the high frequency capacitance with the aid of a lock-in amplifier whereby its output voltage drives a regulating circuit to find out the flat-band point. In this way the measurement of flat-band voltage was performed within about two seconds. All data were registrated by a digital voltmeter which was coupled with a thermoprinter.

The combined $j(t)$–$U_{FB}(t)$ measurements at fixed field strength have been carried out in full cycles, for example + 3 MV cm^{-1} — zero field — −3 MV cm^{-1} — zero field. All measurements have been carried out at room temperature.

3. Results

3.1 j–t measurements

The current density j at fixed field strength between 1 and 3 MV cm^{-1} decreasing with time t observed at MOS samples after electrical stress in most cases is fitted very well by

$$j(t) = a + b \exp(-t/\tau) , \tag{1}$$

where a and b are constants. This simply can be interpreted as follows. At $t \to \infty$ an approximately constant density $j_\infty = a$ flows through the oxide film. $(a + b) = j(0)$ is the current density at $t = 0$. From this follows $b = j(0) - j_\infty$ and

$$j(t) = j_\infty + [j(0) - j_\infty] \exp(-t/\tau) . \tag{2}$$

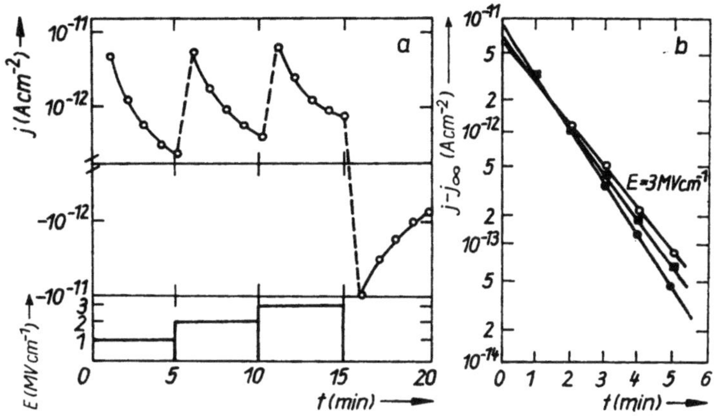

Fig. 1. a) $j(t)$ curves of a MOS sample after weak field stress (7 MV/cm, 100 s) at consecutive steps of field strength of 1, 2, 3, 0 MV/cm (metal minus) and b) corresponding $[j(t) - j_\infty]$ vs. t plots

The quantities j_∞, $j(0)$, and τ can be determined by fitting (2) to the j–t curves. The time constants mostly were found in the range from 50 to 100 s. If one applies a larger field strength in a following step one again obtains a current according to (2). If after this the sample is short circuited, a similar current but of opposite sign is observed.

We carried out j–t measurements at stressed samples as follows. After stress had been applied the sample was short circuited until no measurable current could be detected. Then at a field strength of 1 MV cm^{-1} (oxide field strength) the current was measured over 5 min. After this the field strength was raised in steps to 2 and 3 MV cm^{-1} and at every step the current–time curve was measured again. Finally the sample was short circuited and the current–time curve at zero field was observed. Typical results are demonstrated in Fig. 1 to 3. For the sample demonstrated in Fig. 2 the cycle was somewhat modified. After the step 3 MV cm^{-1} the field was decreased to the step 2 MV cm^{-1}, then increased again to 3 MV cm^{-1}, and finally decreased to zero field. Such measurements have been carried out for many samples and both polarities. The results were always very similar. A significant difference between both polarities could not be detected.

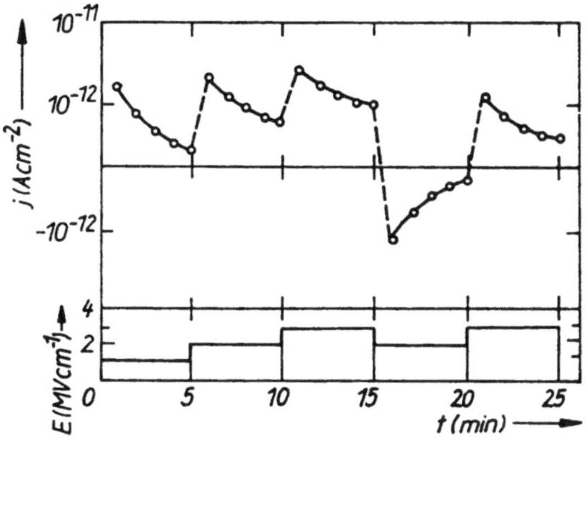

Fig. 2. $j(t)$ curves of a MOS sample after weak field stress (7 MV/cm, 60 s) at consecutive steps of field strength of 1, 2, 3, 2, 3 MV/cm (metal minus)

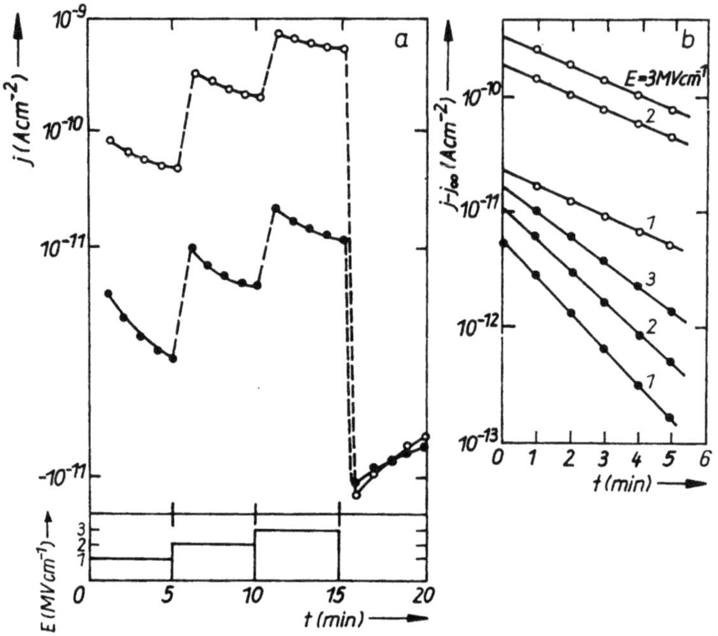

Fig. 3. a) $j(t)$ curves and b) corresponding $[j(t) - j_\infty]$ vs. t plots of a MOS sample after the first and second stress at consecutive steps of field strength of 1, 2, 3, 0 MV/cm (metal minus). ● After first stress (6.5 MV cm^{-1}, 60 s); ○ after second stress (7.5 MV cm^{-1}, 100 s)

It should be noticed from Fig. 1 and 2 that the integrals $\int j\,dt$ at any step of $\Delta E = 1$ MV cm^{-1} are nearly equal. In the case of falling step $\Delta E = -1$ MV cm^{-1} at Fig. 2 $\int j\,dt$ is approximately equal to the former step but of opposite sign, and at $\Delta E = -3$ MV cm^{-1} in Fig. 1 and 2 (zero field) $\int j\,dt$ equals the sum of the integrals $\int j\,dt$ of the former steps of field strength applied. Results calculated from Fig. 1 are given in Table 1. It has been checked that measuring cycles at samples as demonstrated in Fig. 1 and 2 could be repeated many times and were reproducible within the accuracy of our measurement.

The results shown in Fig. 3 have been obtained from a sample with lower quality of the SiO_2 film. Already the comparably weak stress of 6.5 MV cm^{-1}, 60 s has induced larger currents. A second stress of 7.5 MV cm^{-1}, 100 s has increased the current densities by about one order. Similar results from oxide films of higher quality were obtained at stress fields of 9 or 10 MV cm^{-1} (see, e.g., Fig. 5). In contrast to the results shown in Fig. 1 and 2 the integrals $\int j\,dt$ and the time constants τ are increasing with

Table 1

E (MV/cm)	$j(0)$ (10^{-12} A/cm^2)	j_∞ (10^{-13} A/cm^2)	τ (s)	$\int (j_{10} - j_\infty)\,dt$ (10^{-10} As/cm^2)
1	11	1.5	56	5.6
2	8.5	3.5	62	5.3
3	8.0	5.5	65	5.2

Table 2

	E (MV/cm)	$j(0)$ (A/cm²)	j_∞ (A/cm²)	τ (s)s)	$\int (j - j_\infty)\, dt$ (As/cm²)
after	1	7.0×10^{-12}	0.9×10^{-12}	88	5.9×10^{-10}
first	2	1.4×10^{-11}	4.0×10^{-12}	100	1.0×10^{-9}
stress	3	2.8×10^{-11}	1.2×10^{-11}	162	2.6×10^{-9}
after	1	7.6×10^{-11}	5.1×10^{-11}	186	4.7×10^{-9}
second	2	3.9×10^{-10}	1.8×10^{-10}	209	4.4×10^{-8}
stress	3	8.2×10^{-10}	4.7×10^{-10}	226	7.9×10^{-8}

field strength as demonstrated in Table 2. As a further important difference to the samples of Fig. 1 and 2 the cycles demonstrated in Fig. 3 in a repeated measurement are not reproducible, the j–t curves and also the calculated time constants are reduced to lower values. Finally from Fig. 3 the current at zero field is of opposite sign but its magnitude does not correspond to the charge injected in the foregoing part of the cycle as it has been the case for the samples of Fig. 1 and 2. Instead of this it should be noted that the discharging current after the second stress (see Fig. 3) is nearly equal to the zero field current after the first stress.

After application of still stronger stress we observed currents in the order of 10^{-10} A cm^{-2} which only decreased with time very weakly and again could be reproduced approximately in repeated measurements. Their dependence on field strength in most cases very well agreed with

$$j = A \exp\left[-\frac{B}{E^{1/4}}\right] \qquad (A, B = \text{const}) . \tag{3}$$

Equation (3) has been derived with the aid of the percolation theory [15] to govern hopping conduction in amorphous solids at field strengths corresponding to the conditions of our experiments. From experimentally determined constants $B = C^{-1}R/[eN(E_F)]^{1/4}$ we calculated trap densities N_t in the order of 10^{18} to 10^{19} cm^{-3}. (C = const of the order of unity, R Bohr radius of a site, for SiO_2 in the order of 10^{-7} cm, e electronic charge, $N(E_F)$ density of isolated states per eV close to the Fermi level).

From these results carriers moving in the conduction band or valence band can be ruled out to cause the currents observed in the lower field range. Such carriers, electrons or holes, would be swept to the electrodes by the field and cannot give a current of opposite sign at lowered field strength or at zero field. Therefore, effects of local barrier lowering or simple emptying of flat traps as discussed in the literature can be excluded.

Movable ions because of the measuring conditions and the results shown in Fig. 2 also should be very improbable. On the other hand, our results can be completely derived from the basic assumption of tunnel injection of carriers into deep traps within the insulator interface regions as proposed earlier [11].

Transitions of electrons between electrodes and isolated states within the interface regions of the SiO_2 film by tunneling or thermal-assisted tunneling (hopping) are determined by transition probabilities that are exponentially dependent on the distances of the states from the interfaces. The resulting currents either are limited by the number of empty states (of equal distance and levels within a narrow energy range) or, if this number is very large (in the order of 10^{13} cm^{-2}) by the internal field that is built up by the charged states and will equal the injecting field. A simple derivation [11] just leads to (2) which is identical with the empirical expression (1). Time constants of 50 to 70 s as calculated from Fig. 1 are related to averaged tunnel distances of 2 to 3 nm. If the injected charge Q_{inj} is limited by the growing internal field, this just leads to

$$Q_{inj} = \int_0^t j \, dt \sim \Delta E \tag{4}$$

as observed and demonstrated in Fig. 1 and 2 and Table 1. When lowering the injection field strength a current of opposite sign must result from back tunneling electrons that have been injected in the foregoing step.

On the basis of the results considered just above one may understand the action of a stronger stress as demonstrated in Fig. 3 and Table 2 as follows. Hopping transitions from interface traps to traps situated deeper in the bulk of the SiO_2 film may contribute remarkably to the injection current. In first order this may result in an increased averaged distance of interface traps and therefore larger currents and time constants are observed. At zero field part of the former injected carriers remain stored and at a following injection only a smaller current can be measured.

Generally, from the results one has to conclude that electrical stress generates deep traps in the interface regions of SiO_2 films. At stronger action of stress the generation of traps may propagate deeper into the bulk.

3.2 Simultaneously measured j–t and U_{FB}–t characteristics

If carriers are injected into stress generated deep traps in the interface regions of the SiO_2 film as discussed above, the trapped charge near to the silicon substrate (electrons at polarity metal plus or holes at metal minus) should result in a positive or negative shift ΔU_{FB} of the flat-band voltage, respectively. If the currents observed would be caused by any movable internal charge of the SiO_2 film ΔU_{FB} in each case would be of opposite sign. Therefore, from the flat-band voltage simultaneously measured with the stress generated currents additional information should be obtained.

Because of the influenced charge one has to distinguish between the current density j' measured in the outer circuit and the current density j that is really injected. If d_{ox} is the thickness of the SiO_2 film and x_0 the coordinate of the centroid of the injected charge Q_{inj}, generally [12]

$$j' = j - \frac{d}{dt}\left[\frac{d_{ox} - x_0(t)}{d_{ox}} Q_{inj}(t)\right]. \tag{5}$$

Thereby it has been supposed that the injected charge completely becomes trapped, i.e. $j \approx dQ_{inj}/dt$.

In the case of charge injected into small interface regions we may assume $x_0 \ll d_{ox}$, $x_0 \approx$ const. If the charge Q_{inj} is injected from the silicon substrate, we have

$$\frac{dQ_{inj}}{dt} = \frac{\varepsilon\varepsilon_0}{d_{ox}} \frac{dU_{FB}}{dt} \tag{6}$$

and from (5) in this approximation,

$$\frac{dU_{FB}}{dt} = j' \frac{d_{ox}^2}{\varepsilon\varepsilon_0 x_0}. \tag{7}$$

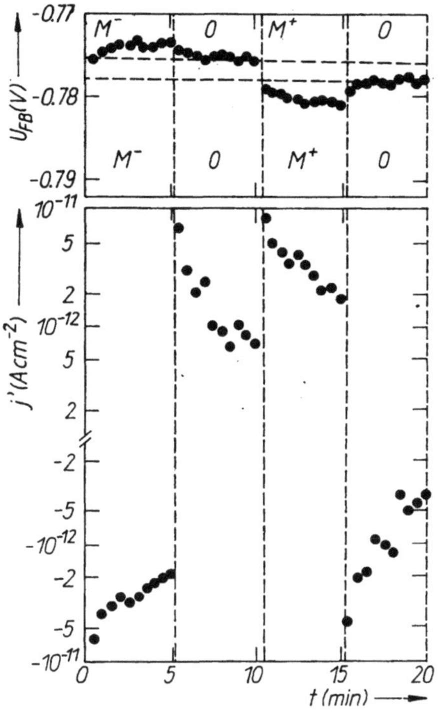

Fig. 4. Combined $j(t) - U_{FB}(t)$ results for a MOS sample measured simultaneously in the cycle -3 MV/cm $- 0 - +3$ MV/cm $- 0$

In the case of constant internal charge Q that only changes its centroid $j = 0$ and from (5) one obtains

$$\frac{dU_{FB}}{dt} = -\frac{d_{ox}}{\varepsilon\varepsilon_0} j' . \tag{8}$$

In the following both cases are considered to explain the experimental results.

In Fig. 4 typical results for a MOS sample after weak stress are demonstrated. After stress application the sample was short circuited until no measurable current could be detected. The cycle $M^-/O/M^+/O$ has been measured twice but only the results of the first measurement are drawn. When repeating the cycle the U_{FB} values were reproduced within ± 1 mV and the current values differed by less than 20%. From j–t curves at metal minus the injected current density $j'(0)$ at $t = 0$ is about 7×10^{-12} A cm^{-2}. The time constant is $\tau \approx 100$ s.

The condition of nearly complete trapping of the carriers injected should be fulfilled because the currents at zero field are nearly equal to the former injection currents of opposite sign. At dominating hole injection from the silicon substrate at metal minus, from (7) in integrated form, one should obtain a negative shift of the flat-band voltage of about 0.3 V. At dominating electron injection at metal plus a positive shift of the flat-band voltage should be larger than 0.3 V. Instead of this the measured shifts of the flat-band voltage during the measuring time of 5 min at any part of the cycle are only about 2 mV, the sign indicating a moving internal charge. But from (8) in integrated form this shift should be larger by a factor 7 or 8 in order to correspond to the measured currents. Further analysing the differentiated j–t and U_{FB}–t characteristics there is no doubt that they are far from corresponding to each other.

This discrepancy only can be solved if one assumes that the injection of charge takes place inhomogeneously into small microregions of integrated area α small compared with the area a of the sample.

Between zero field and 3 MV cm^{-1} metal plus and between zero field and 3 MV cm^{-1} metal minus there are steps of about 3 to 4 mV. Their sign indicates a fast moving internal charge. The time relation of these steps seems to indicate an additional effect that is not recorded by our current measurement.

In Fig. 5 typical results from simultaneous current–flat-band voltage measurements of a MOS sample after strong stress are shown. In this case a drastic generation of

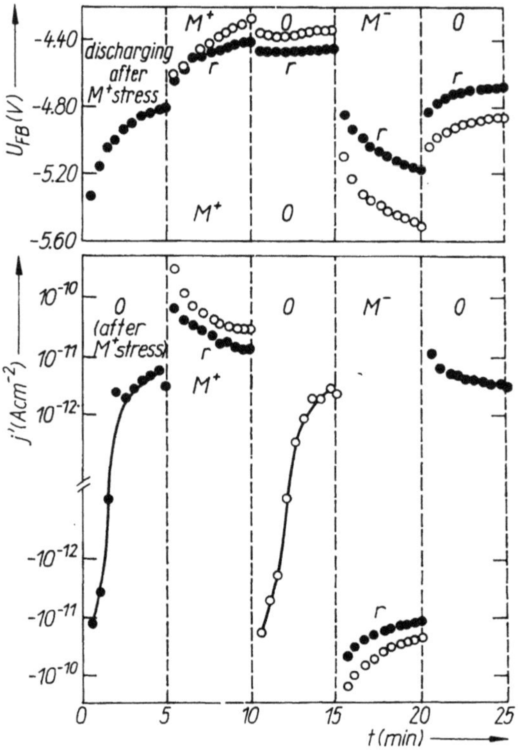

Fig. 5. Combined $j(t) - U_{FB}(t)$ results for a MOS sample after strong field stress (20 s, 9.6 MV cm^{-1}) measured simultaneously in the cycle 0 (after M^+ stress) — +3 MV/cm — 0 — −3 MV/cm — 0. Results of the repeated cycle are denoted by *r*

positive charge has occurred as is seen from the absolute values of the flat-band voltage. The signs of flat-band voltage shifts here indicate injection of charge. The whole cycle was measured twice. Curves denoted by *r* correspond to the repeated measurement. Currents at zero field did not differ in a first approximation.

We first consider the curves at zero field in Fig. 5. After stress at metal plus and after 3 MV cm^{-1} metal plus the currents go from negative to positive values while after 3 MV cm^{-1} metal minus this is not the case. This behaviour can be understood as follows. It is well known [10] that the generated positive charge can be partially discharged by back tunneling electrons. This would occur at zero field from the silicon substrate into the SiO_2 film in every case and may result in a component j_{pc} of the current density j' measured in the outer circuit. On the other hand, electrons or holes that have been injected at polarity plus or minus into interface traps at zero field will return to the adjacent electrodes. This will result in a second component j_t of the current density j'. After injection at polarity metal plus these components are of opposite sign, while after injection at polarity metal minus j_{pc} and j_t are of equal sign. Just such a behaviour of the two components is observed and shown in Fig. 5.

We are not able to understand all details of the results shown in Fig. 5 because of several difficulties. For example the currents at 3 MV cm^{-1} metal plus and metal minus of about 10^{-10} A cm^{-2} may be hopping currents in their main part and cannot simply be related to U_{FB} results. Further it seems to be questionable if the stress generated positive charge is distributed homogeneously. Another unknown fact is the fast jump of U_{FB} between zero field and positive or negative field strength applied as is also mentioned in Fig. 4. We therefore only can analyse the results in a rough estimation. In Table 3 the results from Fig. 5 at 30 s are listed. $j(\Delta U_{FB})$ means the

Table 3

part of the measuring cycle	U_{FB} (30 s) (V)	$j(\Delta U_{FB})$ (A/cm²)	j' (30 s) (A/cm²)	j_{pc} j_t
zero field after M^+ stress	+0.2	$+2 \times 10^{-11}$	-1×10^{-11}	↑ M Si ↓
+3 MV/cm	+0.05	$+5 \times 10^{-12}$	$+2 \times 10^{-10}$	↑ ↑
zero field after +3 MV/cm	+0.01	$+1 \times 10^{-12}$	-1×10^{-11}	↑ ↓
−3 MV/cm	−0.1	-1×10^{-11}	-2×10^{-10}	↓ ↓
zero field after −3 MV/cm	+0.05	$+5 \times 10^{-12}$	$+1 \times 10^{-11}$	↑ ↑

current density calculated from (7). In the last row the directions of the proposed two components j_{pc} and j_t are noticed. From the repeated cycle one has to conclude that the sample has been charged negatively what leads to positive shifts of flat-band voltages and lowered currents.

As is seen the observed current density j' in no case is in agreement with U_{FB} results. On the other hand, from Fig. 5 and Table 3 it seems to be reasonable and should not lead to any contradiction if one assumes that flat-band voltage shifts only are caused by the component j_{pc}. This again leads to the result that currents observed in the field range from 1 to 3 MV cm^{-1} in SiO_2 films after strong stress are mainly caused by tunnel transitions of carriers between stress generated interface traps in small microregions and adjacent electrodes and currents flowing in microchannels through the film.

4. Conclusions

Experimental results reported in Section 3 have shown that after Fowler-Nordheim injection in SiO_2 films currents are observed in the field range from 1 to 3 MV cm^{-1}. From the j–t behaviour of these stress generated currents they cannot flow in the conduction or valence band but agree very well with tunnel injection of carriers (electrons or holes) into deep interface traps. Therefore the electrical stress must have generated the traps.

Simultaneously measured j–t and U_{FB}–t characteristics have shown that the main part of the stress generated currents in the lower field range is not reflected by corresponding U_{FB} shifts. From this one may conclude that the currents are limited to small microregions of integrated area α small compared with the area a of the sample. Such microregions of enlarged trap density may be called interface trap clusters.

The action of stress starts with the generation of interface trap clusters. Exchange of charge between interface trap clusters and the adjacent electrodes is reproducible in first order. After stronger action of stress interface trap clusters become enlarged growing deeper into the bulk. As a consequence carriers injected in the lower field range will percolate deeper into the SiO_2 film and at reduced field strength some of them are stored. Finally at very strong action of stress microchannels of enlarged trap density become generated through the SiO_2 film where hopping conduction can occur. This is also in accordance with results from stress experiments with aid of liquid crystal technique that show localized points of current conduction [16].

It should be noted that the hopping equation in our case has to be written as

$$j = A_0 \sum_k \frac{\alpha_k}{a} \exp\left[-\frac{B_k}{E^{1/4}}\right]. \tag{9}$$

The sum has to be taken over all microchannels within the sample. They may have different cross sections α_k and different trap densities. However, it can easily be shown that (9) in first approximation can be rearranged to (3) in the most important cases. This is true if the trap densities per eV, $N_k(E_F)$, within the microchannels differ by less than one order (in the range of 10^{20} cm^{-3} eV^{-1}). If very different channels are existing, they may be divided into strong and weak channels of integrated areas α_{st} and α_w. If α_{st} is not small compared with α_w, the strong channels will dominate and therefore again (9) can be reduced to (3).

5. Discussion

For the prevailing number of MOS samples investigated we obtained results in good agreement with the results reported above, but with large spread of the thresholds of stress action. In weak samples stress action already started at field strengths of about 6 MV cm^{-1}, while also samples have been prepared where the stress actions described could only be obtained at field strengths above 9 MV cm^{-1} or just below breakdown.

It seems to be sure that this wide spread mainly results from technological influences. Until now we cannot give good instructions to minimize the effect. Careful cleaning of the silicon substrates before thermal oxidation should be a most important condition. Varying conditions of POA did not give unambiguous results. Influences of structural defects and microroughnesses of substrate slices are supposed.

From the result that the stress effect described above preferably works in localized microregions while their surroundings are not affected, one has to conclude that such microregions of lowered threshold of stress action should be regions of special structural properties. They either may be clusters of extrinsic defects as Si–H or Si–OH bands or other impurities, but also may be of intrinsic nature as stretched or disturbed Si–O bonds or other configurations that are modified by the stress.

We tried to find out if the stress generated effect reported in this paper is correlated with other known degradation effects and breakdown phenomena in SiO_2 films. First results seem to indicate that stress generated currents in the lower field range neither are directly related to breakdown field strength nor to breakdown charge. On some samples investigated we could obtain increased currents at 3 MV cm^{-1} only by raising the stress field strength but not by increasing the amount of injected charge at constant field strength. Further investigations should be necessary to get a better knowledge on these problems.

From the technological point of few the degradation effect of stress generated currents in the field range below 4 MV cm^{-1} should be of great importance because it directly will originate failures of memory cells in MOS devices.

References

[1] M. LENLINGER and E. H. SNOW, J. appl. Phys. **40**, 278 (1969).
[2] E. H. NICOLLIAN and C. N. BERGLUND, J. appl. Phys. **41**, 3052 (1970).
[3] R. J. POWELL and C. N. BERGLUND, J. appl. Phys. **42**, 4390 (1971).
[4] T. H. NING, Solid State Electronics **21**, 273 (1978).
[5] R. RADOJCIC, J. Electrochem. Soc. **130**, 1394 (1983).
[6] P. SOLOMON, J. appl. Phys. **48**, 3843 (1977).
[7] M. ITSUMI, J. appl. Phys. **52**, 3491 (1981).

[8] M. M. Heyns, R. F. De Keersmaecker, and M. W. Hillen, Appl. Phys. Letters **44**, 202 (1984).
[9] C. G. Shirley, J. Electrochem. Soc. **132**, 488 (1985).
[10] M. V. Fischetti, J. appl. Phys. **57**, 2860 (1985).
[11] H. Krause, phys. stat. sol. (a) **36**, 705 (1976).
[12] H. Krause, phys. stat. sol. (a) **72**, 607 (1982).
[13] H. Krause, M. Schell, and H.-Chr. Golinsky, phys. stat. sol. (a) **85**, 569 (1984).
[14] H. Krause, phys. stat. sol. (a) **89**, 353 (1985).
[15] N. F. Mott and E. A. Davis, Electronic Processes in Non-crystalline Materials, 2. ed., Clarendon Press, Oxford 1979.
[16] H.-P. Bär, H. Krause, and H. Schulze, 32. Internat. Koll. Ilmenau, Vol. **4**, Ilmenau 1987 (p. 151).
[17] D. A. Baglee and M. C. Smayling, Electronics Letters **21**, 224 (1985).

(Received October 5, 1987; in revised form August 19, 1988)

phys. stat. sol. (a) **110**, 549 (1988)

Subject classification: 73.60

Department for Low Temperature Physics, Roland Eötvös University[1])

Resistivity Increase in Thin Conducting Films Considering the Size Effect

By

G. Horváth and J. Bánkuti

A simple geometrical model is given for the resistivity increase of a thin conducting film considering the size effect. The results of the model agree well with the results of the Fuchs-Sondheimer non-geometrical theory.

Es wird ein einfaches geometrisches Modell für den Widerstandsanstieg einer dünnen Leiterschicht unter Berücksichtigung des Größeneffekts angegeben. Die Ergebnisse des Modells stimmen gut mit den Ergebnissen der nicht-geometrischen Fuchs-Sondheimer-Theorie überein.

1. Introduction

Because of the size effect the mean free path of conduction electrons is shortened in very thin conducting films. If the thickness of the film is commensurable with the mean free path (m.f.p.) of the electrons in a bulk material, the electrical resistivity increases because of diffusive scattering and reflection of the electrons at the film surfaces.

The theory of the size effect is elaborated by Fuchs [1] for the free-electron model and a spherical Fermi surface. This theory is developed by Price [2] for ellipsoidal Fermi surfaces. Sondheimer [3] elaborated the Fuchs theory for the explanation of the galvanomagnetic effects.

On the basis of the Fuchs-Sondheimer theory some expressions can be derived for the electrical conduction of thin conducting films, which can be used well in practice [4]. These expressions agree well with the experimental results [5 to 8].

The size effect occurs practically for all very thin films because they have an insulating (oxide) layer or an adsorbed gas layer on their surface in general, and the scattering of the electrons at the film surfaces is partly or totally diffusive [5 to 8]. The influence of the adsorbed gas layer on the resistivity of the thin metal films is examined by Finzel et al. [15].

The size effect appears in the multifilamentary superconducting composites too in their normal state if the filaments are appropriately close to each other, and the normal resistivity of the filaments is much larger than the resistivity of the matrix [9 to 14]. The resistivity increase caused by the size effect for the longitudinal resistance of the multifilamentary superconducting wires, tapes in normal state, causes a surplus resistivity [9 to 12]. A model is given by Cavalloni et al. [13, 14] to calculate the influence of the size effect on these multifilamentary systems.

In the granular superconducting composites the size effect plays a greater role in the normal state than in the continuous filamentary superconducting composites because in the granular composites there are small grains or short filaments with

---

[1]) Puskin u. 5—7, H-1088 Budapest VIII, Hungary.

random distribution [16]. In such percolative systems these filaments can be very close to each other, and so the resistivity of the very thin matrix layers between these very close filaments can hardly increase in the normal state. The normal resistivity nevertheless plays an important role in the stabilization of the superconducting state of multifilamentary wires [16].

The size effect occurs in every polycrystalline metal, too, because the resistivity of the grain boundaries is larger than the resistivity of the crystallite. The resistivity increase in this case is not significant, because the difference between the resistivity of the crystallite and the grain boundary usually is small [17 to 19].

In this work a simple geometrical model is devised to calculate the resistivity increase as a consequence of the size effect. We examine conducting films with parallel surfaces which have insulating layers or adsorbed gas layers on their surfaces. We suppose that the electrons are scattered totally diffusively on the boundary surfaces of the film. We compare the results of our model to the results of the Fuchs-Sondheimer non-geometrical theory.

2. Size Effect of a Thin Conducting Film between Insulating Layers

In the free-electron model the resistivity is

$$\varrho = m_e \bar{v}/(ne^2\lambda) \,, \tag{1}$$

where m_e, e, n, $\bar{v}$, and λ are the mass, the charge, the volume concentration, the mean velocity, and the bulk mean free path of the electrons, respectively. If the conductor has a finite extension in an insulating medium, then in the proximity of its surface the electron m.f.p. decreases because of the diffusive electron scattering on the scattering centres of the surface, and so the resistivity increases. This effect is called the size effect by us.

Our geometrical model to calculate this resistivity increase is a conducting film between two parallel insulating layers with infinite resistivity. Because of the infinite resistivity of the insulating layers the conduction electrons cannot be scattered beyond the boundaries of the film. Imagine a sphere with radius λ around the electrons, named λ-sphere in this work. In general the electrons are scattered if they have reached this sphere surface. In a homogeneous, isotropic bulk material the electron m.f.p. is isotropic because of the infinite extension. However, in a very thin conducting film between the above-mentioned insulating layers the boundary surfaces can cut a spherical calotte out of the λ-sphere. So the m.f.p. becomes direction dependent, namely in parallel to the insulating layers it does not change, but perpendicularly it decreases.

The average of this direction dependent m.f.p. can be calculated. Such a λ-sphere has a cylindrical symmetry with respect to the axis T going through the examined electron and perpendicular to the insulating layers (Fig. 1), so it is enough to make an average for the main section.

Consider an electron at a distance t_1 and t_2 from the film surfaces. In this case the average of the mean free path is

$$\bar{\lambda} = \left[\int_0^{\alpha} \lambda \, d\theta + \int_0^{\beta} \lambda \, d\theta + \int_{\alpha}^{\pi/2} (t_1/\sin\theta) \, d\theta + \int_{\beta}^{\pi/2} (t_2/\sin\theta) \, d\theta\right] \pi^{-1} \tag{2}$$

taking into account the notations of Fig. 1. Performing the integration we get from (2)

$$\bar{\lambda} = [\lambda(\alpha + \beta) - t_1 \ln \operatorname{tg}(\alpha/2) - (D - t_1) \ln \operatorname{tg}(\beta/2)] \, \pi^{-1} \,, \tag{3}$$

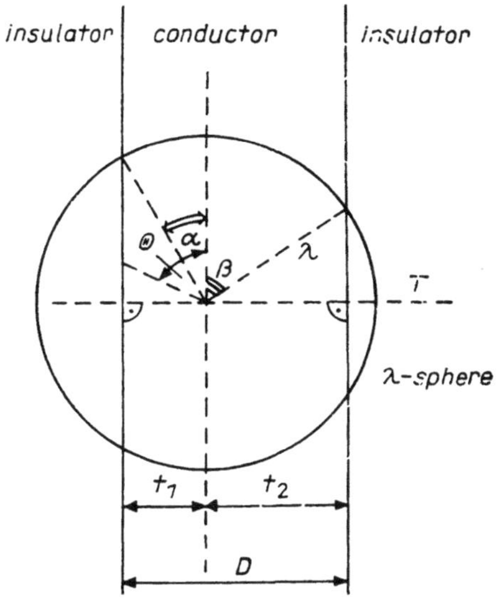

Fig. 1. The λ-sphere in a thin conducting film between two parallel insulating layers

where D is the thickness of the film. Let us introduce the following dimensionless quantities:

$$Q = D/(2\lambda)\,, \qquad x = t_1/\lambda\,, \qquad l = \bar{\lambda}/\lambda\,. \tag{4}$$

In the case of very thin films both the boundary surfaces cut the λ-sphere. In the case of thicker films such special arrangements exist when only one or none of the surfaces cuts the λ-sphere. Taking these arrangements into consideration the average of l depends on the quantities Q and x as follows:

1. if $Q > 1$ and
   - $0 \leqq x \leqq 1$, then $l = [\alpha + \pi/2 - x \ln \mathrm{tg}\,(\alpha/2)]\,\pi^{-1} \equiv l_1$
   - $1 < x < 2Q - 1$, then $l = 1$
   - $2Q - 1 \leqq x \leqq 2Q$, then $l = [\beta + \pi/2 - (2Q - x) \ln \mathrm{tg}\,(\beta/2)]\,\pi^{-1} \equiv l_2$
2. if $1/2 \leqq Q \leqq 1$ and
   - $0 \leqq x \leqq 2Q - 1$, then $l = l_1$
   - $2Q - 1 < x < 1$, then $l = [\alpha + \beta - x \ln \mathrm{tg}\,(\alpha/2) - (2Q - x) \times \ln \mathrm{tg}\,(\beta/2)]\,\pi^{-1} \equiv l_3$
   - $1 \leqq x \leqq 2Q$, then $l = l_2$
3. if $0 \leqq Q \leqq 1/2$, then $l = l_3$ (5)

If the conductor surrounded by the insulator with infinite resistivity has no parallel boundaries, we must average λ for the residual λ-sphere without the part(s) cut off by the boundary surface(s).

The functions $l(x, Q)$ and $l^{-1}(x, Q)$ are plotted in Fig. 2a and b. As we can see at larger values of Q the average of the electron m.f.p. decreases only in the proximity of the surfaces. The onset of the decrease is at the distance λ from the insulating layers. For distances larger than λ the average is equal to the bulk m.f.p. Accordingly for large Q the average resistivity increases in the proximity of the surfaces and for small Q the increase is extended over the whole film.

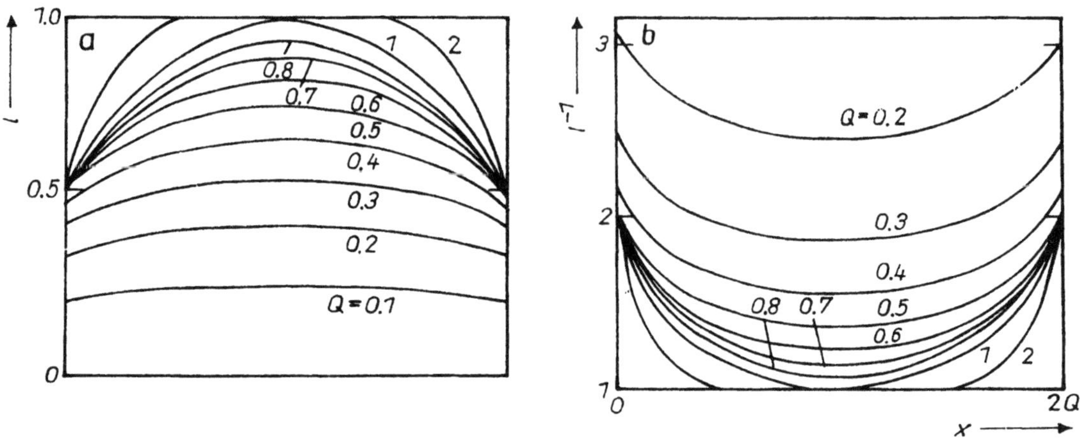

Fig. 2. a) The dependence of the dimensionless mean free path $l = \bar{\lambda}/\lambda$ on $x = t_1/\lambda$ and $Q = D/(2\lambda)$. b) The dependence of the dimensionless resistivity l^{-1} on x and Q

The average resistivity of the film is

$$\bar{\varrho}^{-1} = Q^{-1} \int_0^Q [\varrho(x)]^{-1} \, \mathrm{d}x \,. \tag{6}$$

Using (1) and (6) we get

$$\bar{\varrho}^{-1} = (\varrho Q)^{-1} \int_0^Q l(x) \, \mathrm{d}x \,, \tag{7}$$

where the functions $\varrho(x)$, $l(x)$ are symmetrical. Using (4), (5), (7) we obtain the following expressions:

1. $0 \leqq Q \leqq 1/2$

$$\bar{\varrho}^{-1} = (\varrho Q \pi)^{-1} \int_0^Q \left\{ \arcsin x + \arcsin (2Q - x) - x \ln \operatorname{tg} \left(\frac{\arcsin x}{2} \right) - \right.$$
$$\left. - (2Q - x) \ln \operatorname{tg} \left[\frac{\arcsin (2Q - x)}{2} \right] \right\} \mathrm{d}x \equiv$$
$$\equiv (\varrho Q \pi)^{-1} \int_0^Q F(x) \, \mathrm{d}x.$$

2. $1/2 < Q < 1$ (8)

$$\bar{\varrho}^{-1} = (\varrho Q \pi)^{-1} \left\{ \int_0^{2Q-1} \left[\pi/2 + \arcsin x - x \ln \operatorname{tg} \left(\frac{\arcsin x}{2} \right) \right] \mathrm{d}x + \right.$$
$$\left. + \int_{2Q-1}^Q F(x) \, \mathrm{d}x \right\} \equiv (\varrho Q \pi)^{-1} \left\{ \int_0^{2Q-1} G(x) \, \mathrm{d}x + \int_{2Q-1}^Q F(x) \, \mathrm{d}x \right\}.$$

3. $Q \geqq 1$

$$\bar{\varrho}^{-1} = (\varrho Q)^{-1} [\pi^{-1} \int_0^1 G(x) \, \mathrm{d}x + \int_1^Q \mathrm{d}x] \,.$$

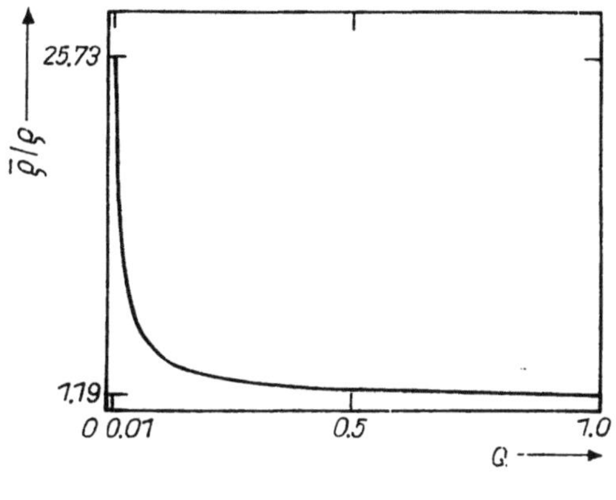

Fig. 3. The dependence of $\bar{\varrho}/\varrho$ on Q

Integrating (8), the average resistivities are

1. $0 < Q < 1/2$

$$\bar{\varrho}^{-1} = (\varrho Q \pi)^{-1} \Big\{ 3(1 - 4Q^2)^{1/2}/2 - (1 - Q^2)^{1/2} - 1/2 + 2Q \arcsin(2Q) - - 2Q^2 \ln \left[\frac{1 - (1 - 4Q^2)^{1/2}}{2Q} \right] \Big\} .$$

2 to 3. $1/2 \leqq Q$

$$\bar{\varrho}/\varrho = 1 + (2\pi Q - 1)^{-1} . \tag{9}$$

The calculated $\bar{\varrho}/\varrho$ is plotted against Q in Fig. 3.

On the basis of the Fuchs-Sondheimer non-geometrical theory, the resistivity ratio

$$\bar{\varrho}/\varrho \approx 1 + 3/(16Q) = 1 + 0.1875/Q ; \qquad Q \gg 1 , \tag{10}$$

$$\bar{\varrho}/\varrho \approx -4/(3Q \ln Q) = -1{,}3330/(Q \ln Q) ; \qquad 0 < Q \ll 1$$

can be obtained [1 to 4] if the scattering of electrons is totally diffusive at the film surfaces. From (9) we get

$$\bar{\varrho}/\varrho \approx 1 + 1/(2\pi Q) = 1 + 0.1590/Q ; \qquad Q \gg 1 , \tag{11}$$

$$\bar{\varrho}/\varrho \approx -\pi/(2Q \ln Q) = -1.5710/(Q \ln Q) ; \qquad 0 < Q \ll 1 .$$

It can be seen from (11) that the results of our simple geometrical model agree well with the results of the Fuchs-Sondheimer non-geometrical theory.

3. Conclusions

The size effect occurs in a lot of important systems: in very thin conducting films, in continuous filamentary and granular superconducting composites, and in polycrystalline metals. This effect increases the resistivity more or less significantly in these systems. The resistivity increase is described well by the Fuchs-Sondheimer non-geometrical theory. The results of our simpler, geometrical model for the resistivity increase agrees well with the results of the Fuchs-Sondheimer theory.

References

[1] K. Fuchs, Proc. Cambridge Phil. Soc. **34**, 100 (1938).
[2] P. J. Price, IBM J. Res. Developm. 4, 152 (1960).
[3] E. H. Sondheimer, Phys. Rev. **80**, 401 (1950).
[4] K. L. Chopra, Thin Films Phenomena, McGraw-Hill Book Co., New York 1969.
[5] R. Nossek, Z. Naturf. **16A**, 1162 (1961).
[6] K. L. Chopra, L. C. Bobb, and M. H. Francombe, J. appl. Phys. **34**, 1699 (1963).
[7] M. S. P. Lucas, Appl. Phys. Letters **4**, 73 (1964).
[8] K. L. Chopra and M. R. Randlett, J. appl. Phys. **38**, 3144 (1967).
[9] V. M. Drobin, E. I. Dyachkov, I. S. Khukhareva, V. G. Luppov, and A. Nichitiu, Cryogenics **22**, 115 (1982).
[10] D. P. Lazar, N. M. Vladimirova, V. M. Drobin, E. I. Dyachkov, I. S. Khukhareva, Cryogenics **26**, 152 (1986).
[11] A. E. White, M. Tinkham, and W. I. Skocpol, Phys. Rev. Letters **48**, 1752 (1982).
[12] F. Nakane, T. Uchiyama, S. Nakamura, S. Miyake, and D. Ito, Proc. Internat. Cryogenic Mater. Conf., Kobe/Japan 1982 (p. 203).
[13] C. Cavalloni and R. Monnier, Helv. phys. Acta **55**, 669 (1982).
[14] C. Cavalloni, K. Kwasnitza, and R. Monnier, Appl. Phys. Letters **42**, 734 (1983).
[15] H. U. Finzel, E. Schmiedel, and P. Wissmann, Appl. Phys. A **42**, 87 (1987).
[16] Ch. Lobb, Techn. Rep. **16** (1980).
[17] R. A. Brown, J. Phys. F 7, 1477 (1977).
[18] F. Lormand, Phil. Mag. **B44**, 389 (1981).
[19] F. Lormand, J. Physique **43**, 283 (1982).

(Received July 27, 1988)

phys. stat. sol. (a) **110**, 555 (1988)

Subject classification: 72.20; 71.55; S5.12

Institute of Semiconductor Physics,
Academy of Sciences of the Lithuanian SSR, Vilnius[1]) *(a)*
and Institute of Physics, Academy of Sciences of the Latvian SSR, Salaspils[2]) *(b)*

Self-Generated Chaos in n-Ge after Electron Irradiation

By

J. K. Požela (a), Z. N. Tamaševičienė (a), A. V. Tamaševičius (a),
J. K. Ulbikas (a), and G. V. Bandurkina (b)

Evidence is presented of self-generated chaos in n-Ge irradiated with high-energy electrons. Instability occurs above some critical field $E_{cr} \approx 10$ V/cm under positive differential conductivity. The phase space dimension k_0 is investigated as a function of the dc voltage applied to the sample. Two distinct types of oscillations: i) self-generated chaos of low dimensionality ($k_0 \leqq 4$), and ii) oscillations with indeterminately large phase space dimension ($k_0 > 10$) are observed. This phenomenon is discussed in terms of recombination instability in the spatially nonuniform system.

В n-Ge, облученном быстрыми электронами, обнаружены хаотические автоколебания тока. Неустойчивость возникает на сублинейном участке вольтамперной характеристики выше критической напряженности поля ≈ 10 V/cm при положительной дифференциальной проводимости. Изучена зависимость размерности фазового пространства k_0 от приложенного напряжения. Наряду с колебаниями, характеризуемыми малой размерностью ($k_0 \leqq 4$), обнаружены сложные временные реализации с $k_0 > 10$. Результаты обсуждаются в рамках модели рекомбинационной неустойчивости в пространственно неоднородной системе.

1. Introduction

During the last few years accumulation of the observations of nonlinear instabilities leading to a dynamical chaos has been evident. It is well known that chaotic oscillations similarly to random noise are characterized by chaotic time sequences, broadband continuous power spectra, but have extremely high noise temperatures (10^{12} to 10^{13} K). The main difference between these two types of noise is that random noise is governed by statistical laws, while chaotic oscillations are described by completely deterministic dynamics with only a few degrees of freedom. Such chaotic behaviour of a dynamical system corresponds to the motion in the phase space along trajectories attributed to a strange attractor. The dimension of the strange attractor provides the number of degrees of freedom involved in the oscillations [1].

One can determine the dimension of a strange attractor by different methods [1]. However, there are some difficulties in determining this quantity from the experimental data blurred with the random noise always existing in real systems [2]. In many experimental cases, nevertheless, it is sufficient to estimate the minimum embedding dimension, i.e. the minimum dimension of the phase space required to place the attractor [3 to 5]. This quantitative measure is an integer and is easier to obtain.

In semiconductors chaos has been observed in such physical phenomena as filamentation [6], photoconductivity [7, 8], Hall effect [9], electrical breakdown [10, 11],

[1]) K. Poželos 52, SU-232600 Vilnius, USSR.
[2]) SU-229021 Salaspils-1, USSR.

and others. In most studies on chaos in semiconductors the main attention was turned to such characteristics as snapshots of time sequences, phase portraits, and power spectra. Classification of these results allowed to identify the features typical for different systems, e.g. universal scenarios of routes to chaos, but it had not led to an unambiguous discrimination between dynamical chaos and random noise. The phase space dimension measurements make such a discrimination possible. Moreover, this quantitative characteristic is very useful for an analytic description of chaotic instabilities.

Chaotic oscillations in semiconductors in most cases are associated with impurities or other defects, such as Si in GaAs [6], B and Ga in Ge [7, 10], Ni in Ge [8], Te in InSb [9], Cr or EL2 in GaAs [11, 12].[3])

In this paper we report the results of a study on chaotic oscillations in n-Ge irradiated with high-energy electrons. The dose of the irradiation can be easily selected to obtain the desired concentration of the introduced defect centres and to optimize the conditions for the rise of the effect.

2. Experimental

Experiments were carried out on single crystals of Sb-doped Ge with dislocation density of $10\ \text{cm}^{-2}$ and $5 \times 10^{3}\ \text{cm}^{-2}$. The donor concentration was about $2 \times 10^{14}\ \text{cm}^{-3}$ corresponding to the resistivity of $10\ \Omega\text{cm}$ at room temperature. The samples were rectangular shaped ($5 \times 1 \times 1\ \text{mm}^3$) with Sn contacts deposited on the $1 \times 1\ \text{mm}^2$ surfaces. As shown in Fig. 1 the defects were introduced only in the central part of the sample with a length of about 2 mm. Such a configuration of irradiation was picked over to save regions with the initial ϱ_0 to eliminate the contact effects in high electric fields. The irradiation was performed with a flux of 3.5 MeV electrons of $1.56 \times 10^{12}\ \text{cm}^{-2}\ \text{s}^{-1}$. Cumulative dose Φ_e was varied from 10^{13} to $5 \times 10^{17}\ \text{cm}^{-2}$.

Probe contacts of 50 μm diameter were used to investigate the electric field distribution along the sample and the spatial coherence of the oscillations. To study the electrical instability the samples were immersed in liquid nitrogen and illuminated with an incandescent lamp. The dc voltage U_0 across the sample served as a control parameter. Besides the snapshots of time sequences and power spectra, the ac signal $U(t)$ appearing across the load resistor R was analysed by means of a special electronic instrument [5] for determining the phase space dimension k_0. The main idea of the method realized in this instrument is the determination of the minimum number of conditions sufficient to specify the state of the system [15]. One can obtain the number of independent quantities for such a specification by constructing the k-th order conditional probability distributions of the variable $U(t)$ [5],

$$P^{(k)} = P(U| \tau_1, \tau_2, \dots \tau_k; U^*)\ ,$$

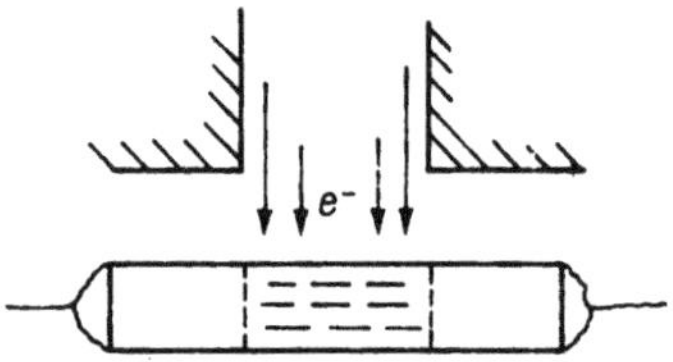

Fig. 1. Arrangement for sample irradiation

[3]) It should have been mentioned that the phenomenon of high-level noise due to impurity states in semiconductors is very old (see, e.g. [13, 14]), though only in recent years the methods of dynamical chaos have been applied.

which is defined as the probability of observing U provided that U^* has been observed at a time τ_1 before, also at a time τ_2 still before, and so on. The dimension of the phase space sufficient to place the attractor is just the minimum number of conditions k yielding a simple sharp distribution of U [15].

3. Results

The electrical instability was observed in the samples irradiated over some minimum cumulative dose $\Phi_{e\,min}$. The value of $\Phi_{e\,min}$ depended only on the initial ϱ_0 and amounted to 10^{15} cm^{-2} for $\varrho_0 = 10\ \Omega$cm. The dislocation density had no influence on $\Phi_{e\,min}$. It is well known that radiation induced defects lead to the compensation of shallow donors. The instability occurred just in highly compensated material (Fig. 2). Some optimal cumulative dose of radiation for a given ϱ_0 is obvious. The total dose of above 10^{16} cm^{-2} (for $\varrho_0 = 10\ \Omega$cm) led, apparently, to the p-type conversion of the material and to an increase of conductivity (Fig. 2).

The compensated samples exhibited an essential nonlinearity in high electric fields (Fig. 3). The critical voltage for the onset of the oscillations U_{cr} was about 2 V, independently on the load resistor R and the dislocation density. The critical electric field in the irradiated part of the sample was about 10 V/cm. Such electric fields are sufficient for hot electron effects in n-Ge at 78 K. It must be emphasized that the static differential conductivity remained positive ($dI/dU_0 > 0$) throughout the investigated field range.

We found that the electric field distribution along the irradiated part of the sample was inhomogeneous (Fig. 4). As the investigated material exhibits highly nonlinear properties one can expect differences in the evolution of the instability in the spatial regions of low and high fields. To confirm this assumption we determined whether or not the chaotic oscillations were spatially coherent. The phase portraits shown in Fig. 4c were obtained from pairs of probe contacts placed in different regions of the sample. The phase portraits (U_2, U_4), (U_2, U_5), (U_4, U_5), obtained from points characterized by different electric fields show an essential phase shift between the signals, thus indicating the oscillations to be spatially incoherent along the irradiated part of the sample. There was no phase shift in the points with the same field value (U_2, U_3) as well as between the signals in the irradiated and unirradiated parts of the sample (U_2, U_1), (U_5, U_6). Evidently the instability arises in the irradiated part, while the other parts of the sample play the role of the ohmic "contacts". Moreover, such results can be obtained only in a spatially nonuniform system.

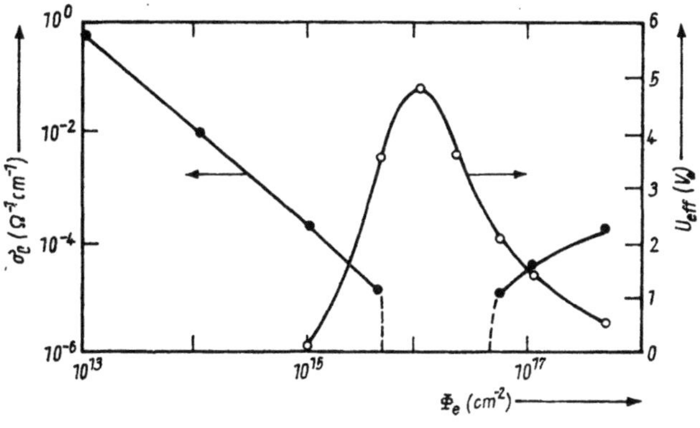

Fig. 2. Conductivity σ_0 (●) and effective noise level U_{eff} (○) vs. cumulative dose Φ_e at 78 K, $\varrho_0 = 10\ \Omega$ cm

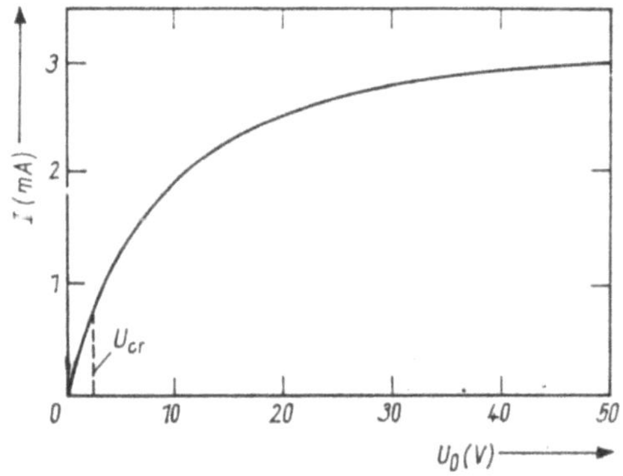

Fig. 3. Current–voltage characteristic of an illuminated sample. The onset of the oscillations is marked; $\Phi_e = 10^{16}$ cm^{-2}

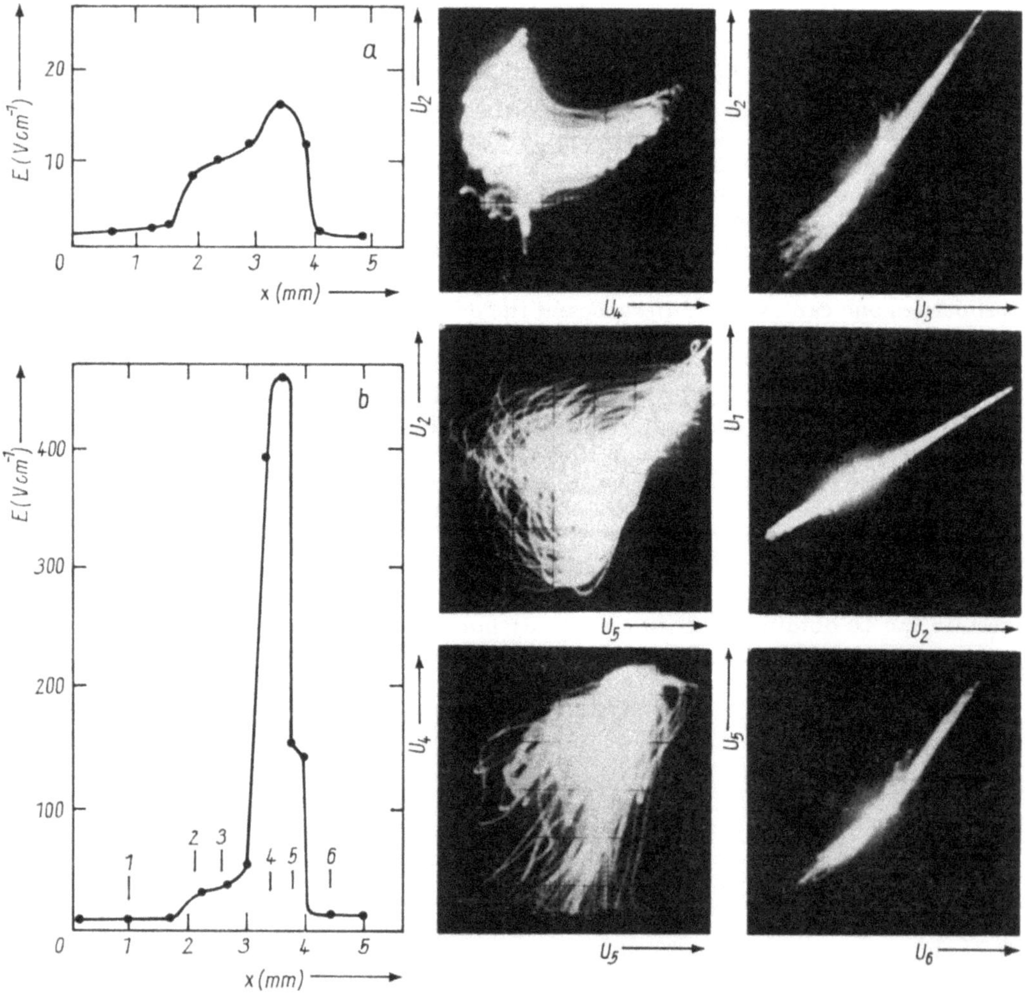

Fig. 4. Electric field distribution along the sample for a) $U \geqq U_{cr}$, b) $U \gg U_{cr}$, c) phase portraits U_i vs. U_j with i and j corresponding to the probe positions

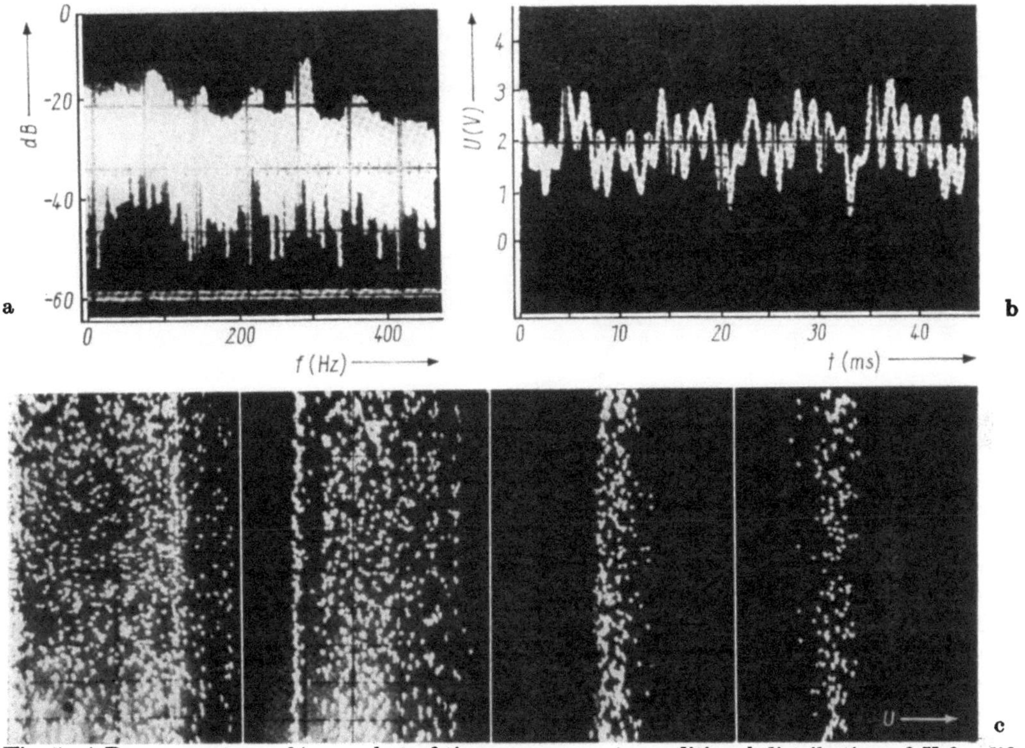

Fig. 5. a) Power spectrum, b) snapshot of time sequence, c) conditional distribution of U for different number of conditions k increasing from 1 to 4 (the case of $k_0 = 3$), $R = 20\ \text{k}\Omega$

Close to the critical voltage U_{cr} the oscillations were characterized by a low amplitude (a few mV) of the signal and low level (−70 dB) of the noise spectra. A slight increase of the applied voltage led to the jump of the amplitude up to a few V (Fig. 5b) and a noise level in the power spectra up to −20 dB (Fig. 5a). Such a development of chaos was typical of all samples independently of their initial ϱ_0 and the dislocation density.

No clear route to chaos has been observed in our experiments. Nevertheless, some similarity to the Ruelle-Takens-Newhouse [16] scenario, i.e. the appearance of two

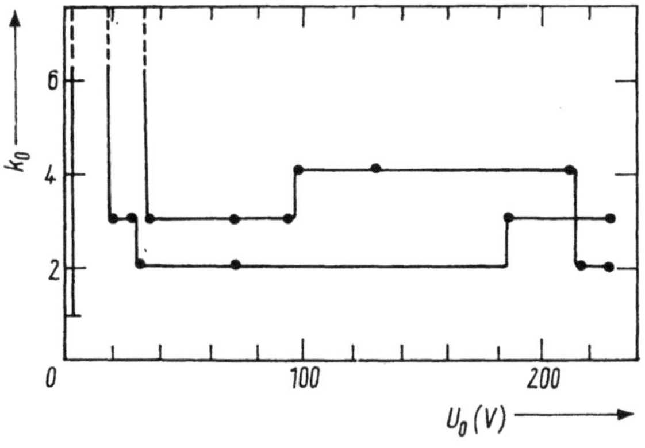

Fig. 6. Phase space dimension k_0 as a function of applied voltage U_0 for two samples

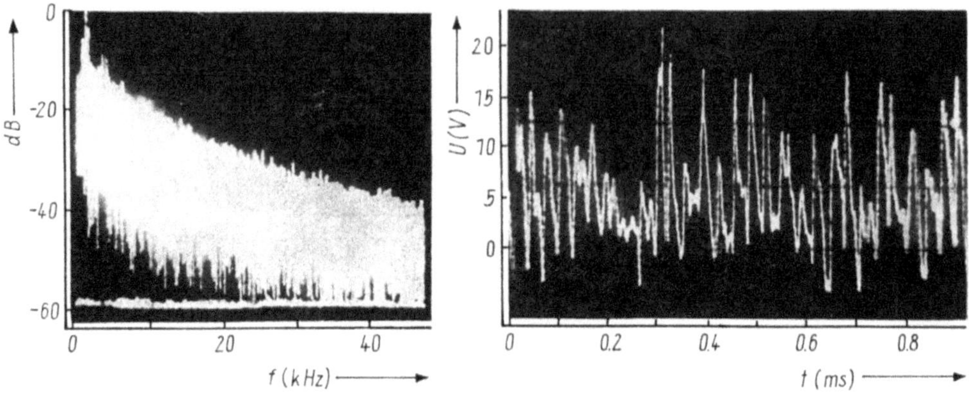

Fig. 7. a) Power spectrum, b) snapshot of the time sequence in the case $k_0 > 10$, $R = 20\ k\Omega$

incommensurate frequencies in the threshold region of generation was observed. Such behaviour can be associated with the component of high-dimensional noise inherent in the system. Although, the Ruelle-Takens-Newhouse bifurcations are known to remain stable at small noise [16], apparently large noise can lead to the destruction of the scenario.

To characterize the dynamical state of the investigated system we measured the dimension of the phase space as described above. The experiments showed that only a few degrees of freedom were enough to specify the variable $U(t)$. Fig. 5 shows the power spectrum, the typical waveform, and the conditional distributions of U for various k. A relatively narrow distribution of U appeared for $k = 3$ and kept so with increasing number of conditions k. Evidently, three conditions are needed to specify the state of the system, i.e. $k_0 = 3$ for a given case. The distributions of U for $k = 3$ and $k = 4$ had a certain width due to the random noise existing in the investigated system [5, 15].

The typical dependence of k_0 on the control parameter U_0 obtained for a number of samples is shown in Fig. 6. The phase space dimension k_0 remained low in a wide range of the applied voltage U_0.

The light intensity influenced only the amplitude of the oscillations $U(t)$ but not the phase space dimension k_0. In a narrow interval of light intensity, however, some samples exhibited a highly complicated behaviour (Fig. 7). In such cases the phase space dimension was indeterminately large ($k_0 > 10$) over all investigated voltages.

4. Discussion

The defects in n-Ge produced by γ-quanta [17] or high-energy (0.5 to 4 MeV) electron irradiation are responsible for the appearance of at least two deep levels in the energy gap. These defects are negatively charged centres [17, 18], characterized by a large capture cross-section for holes and relatively small sections for electrons. Moreover, the strong energy dependence of the recombination rate of the electrons leads to nonlinear current voltage characteristics [18] and in some cases evokes the recombination instability.

As follows from the experimental data the irradiated Ge is a spatially nonuniform system in the case of high electric fields. For a qualitative explanation of the observed chaotic oscillations we consider the following simplified model. The active part of the sample can be represented as a series chain, consisting of the N elements, each of them characterized by a uniform local electric field E_i and other parameters independent of the coordinates. Under these conditions a set of ordinary differential equations describes the dynamics of the local free electron concentration n_i, the electrons trapped on the defects m_{1i}, m_{2i}, and the hole concentration p_i,

$$\frac{\mathrm{d}n_i}{\mathrm{d}t} = g - C_{n1}(E_i)\,(M_{1i} - m_{1i})\,n_i - C_{n2}(E_i)\,(M_{2i} - m_{2i})\,n_i\,,$$

$$\frac{\mathrm{d}m_{2i}}{\mathrm{d}t} = C_{n2}(E_i)\,(M_{2i} - m_{2i})\,n_i - C_{p2}m_{2i}p_i\,, \tag{1}$$

$$\frac{\mathrm{d}p_i}{\mathrm{d}t} = g - C_{p1}m_{1i}p_i - C_{p2}m_{2i}p_i\,,$$

$$n_i + m_{1i} + m_{2i} = p_i + N_d^*\,,$$

$$E_i = \frac{V}{d}\,\frac{R_i}{R + R_s}\,, \qquad R_s = \sum_{i=1}^{N} R_i\,, \qquad R_i = \frac{d}{en_i\mu_n S}\,; \qquad i = 1, \dots, N\,.$$

In (1) g determines the electron–hole pair generation rate due to the optical excitation, $C_{n1,2}$, $C_{p1,2}$ are the recombination coefficients for electrons and holes, respectively, M_{1i} and M_{2i} denote the defect concentrations, N_d^* is the effective concentration of shallow donors including the compensation by multicharged centres, V the voltage of the dc source, d the length of the i-th element, μ_n the mobility of electrons, S the cross section area of the sample.

In equilibrium all the electrons excited from donors occupy the lowest defect level, say M_1: $m_1^0 = N_d^*$. For relatively low excitation by light ($n, m_2, p \ll N_d^*$) this level is not recharged, i.e. $m_1 \approx N_d^*$. Due to the rapid trapping of the holes ($C_{p1,2} \gg C_{n1,2}$) the conductivity appears to be quasimonopolar ($n \gg p$). The hole concentration can be estimated from the steadystate solution of $\mathrm{d}p/\mathrm{d}t = \dots = 0$: for $C_{p2} \ll C_{p1}$ [17] $p \approx g/C_{p1}/N_d^*$.

Similarly to the model presented in [8] we approximate the field dependent recombination coefficients in the following form:

$$C_{n1}(E_i) = C_{n1}(0)\,(1 + aE_i)\,, \qquad C_{n2}(E_i) = C_{n2}(0)\,(1 + bE_i^2)\,,$$

where $C_{n1,2}(0) \approx 10^{-12}\ \mathrm{cm^3\,s^{-1}}$ [17]. The parameter a can be evaluated from the dc current–voltage characteristic, and b must be chosen to fit the onset of the instability of (1) at appropriate fields.

We begin the analysis of (1) with the inspection of the frequency dependence of the differential conductivity $\sigma_d = \mathrm{d}j/\mathrm{d}E_i$ for the i-th element. The real part of the conductivity which is of interest has a very cumbersome form. However, in the case when $C_{p2} \ll C_{p1}$ it can be simplified to

$$\mathrm{Re}\,\frac{\sigma_d(\omega)}{\sigma_0} \approx 1 - \frac{\omega_1'\omega_3(\omega_1\omega_3 - \omega^2) + \omega^2(\omega_1' + \omega_2')\,(\omega_1 + \omega_2)}{(\omega_1\omega_3 - \omega^2)^2 + \omega^2(\omega_1 + \omega_2)^2}\,E_0\,, \tag{2}$$

where $\sigma_0 = en_0\mu_n$, $\omega_1 = C_{n1}(M_1 - N_d^*)$, $\omega_2 = C_{n2}(M_2 - m_{20})$, $\omega_3 = C_{n2}n_0$, $\omega_{1,2}' = \mathrm{d}\omega_{1,2}/\mathrm{d}E|_{E=E_0}$; n_0, m_{20}, E_0 are the steady-state values of the relevant variables.

For $\omega = 0$ $\mathrm{Re}\,\sigma_d > 0$, while in a certain frequency range (Fig. 8) it becomes negative. The obtained result well explains the possibility of spontaneous oscillations under positive dc characteristic.

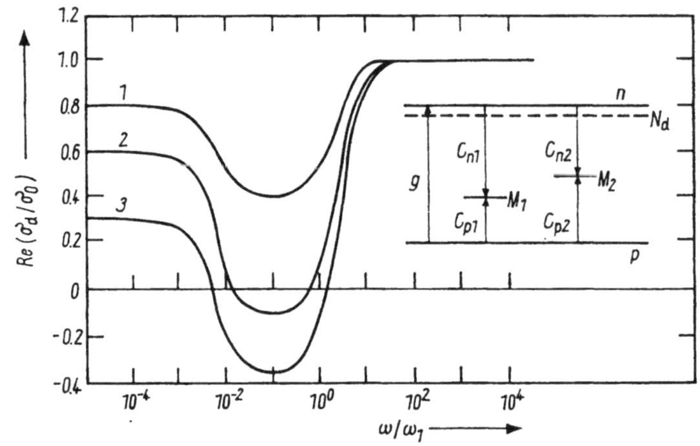

Fig. 8. Real part of the differential conductivity Re σ_d/σ_0 as a function of frequency ω for different values of electric field. (1) $E_0 = 10$, (2) 30, (3) 100 V/cm. $a = 0.02\ \text{V}^{-1}\ \text{cm}$, $b = 0.01\ \text{V}^{-2}\ \text{cm}^2$, $\omega_2/\omega_1 = 1$, $\omega_3/\omega_1 = 0.01$. The insert represents schematically the electron and hole transitions

It must be noted that the recombination instability can arise under certain conditions in a uniform sample ($N = 1$) as well [19]. However, such a system can display only simple periodic oscillations.

The linear analysis of (1) ($N = 2$) indicates the steady state of the system to be unstable above some critical voltage $U_0 > U_{cr}$ in the range where Re $\sigma_d < 0$.

This unstable system in the nonlinear stage is analogous in a sense to that of two coupled sine-wave oscillators, which are known to exhibit chaotic behaviour [20]. The maximum phase space dimension of the system, described by (1) with $dp/dt = 0$ and $N = 2$, is evidently $2N$, i.e. $= 4$. However, due to the synchronization effects the actual numbers of degrees of freedom can occur to be less than $2N$, e.g. 3 or even 2 (the limit cycle). The dimensionality of the system increases with N, while the dimensionality of the strange attractor appearing in the phase space depends on the parameters of the system. In a wide range of the parameter values the dimension of the strange attractor d in the coupled oscillatory system remains relatively low ($d \ll 2N$) [21, 22], but in some cases it can approach $2N$. The same happens evidently with the minimum space phase dimension k_0.

5. Conclusion

n-Ge, compensated with radiation defects, displays an electrical instability leading to chaotic current behaviour. The observed phenomenon can be explained in terms of the recombination instability under spatially nonuniform conditions. The chaotic oscillations are associated with a low-dimensional strange attractor in the phase space.

Acknowledgements

We would like to thank U. A. Ulmanis for discussion and A. M. Namajūnas for help in measuring the phase space dimensions.

References

[1] G. MAYER-KREISS (Ed.), Dimensions and Entropies in Chaotic Systems, Springer-Verlag, Berlin 1986.

[2] N. B. ABRAHAM, A. M. ALBANO, B. DAS, G. DE GUZMAN, S. YONG, R. S. GIOGGIA, G. P. PUCCIONI, and J. R. TREDICCE, Phys. Letters A114, 217 (1986).

[3] Yu. I. Neimark and P. S. Landa, Stochastic and Chaotic Oscillations, Izd. Nauka, Moscow 1987 (p. 236) (in Russian).
[4] A. Čenys and K. Pyragas, Phys. Letters **A129**, 227 (1988).
[5] A. Namajūnas, J. Požela, and A. Tamaševičius, Phys. Letters **A131**, 85 (1988).
[6] K. Aoki, T. Kobayashi, and K. Yamamoto, J. Phys. Soc. Japan **51**, 2373 (1982).
[7] W. Teitsworth, R. M. Westervelt, and E. E. Haller, Phys. Rev. Letters **51**, 825 (1983).
[8] S. B. Bumelienė, Yu. K. Pozhela, K. A. Pyragas, and A. V. Tamaševičius, Physica (Utrecht) **B134**, 293 (1985).
[9] D. G. Seiler, C. L. Littler, R. J. Justice, and P. W. Milonni, Phys. Letters **A108**, 462 (1985).
[10] J. Peinke, A. Mühlbach, R. P. Huebener, and J. Parisi, Phys. Letters **A108**, 407 (1985).
[11] Z. N. Bartashevich, Yu. K. Pozhela, K. A. Pyragas, A. V. Tamaševičius, and Yu. K. Ulbikas, in: Gallium Arsenide Research 1987, Ed. E. Lendvay, Trans. Tech. Publ., Aedermannsdorf 1987.
[12] G. N. Maracas, W. Porod, D. A. Johnson, and D. K. Ferry, Physica (Utrecht) **B134**, 276 (1985).
[13] M. Pilkuhn, Z. Naturf. **16a**, 182 (1961).
[14] I. V. Karpova, S. G. Kalashnikov, O. V. Konstantinov, V. I. Perel, and G. V. Tsarenkov, in: Proc. 9th Internat. Conf. Physics of Semiconductors, Part 2, Ed. S. M. Ryvkin, Izd. Nauka, Leningrad 1968 (p. 959).
[15] N. H. Packard, J. P. Crutchfield, J. D. Farmer, and R. S. Shaw, Phys. Rev. Letters **45**, 712 (1980).
[16] J.-P. Eckmann, Rev. mod. Phys. **53**, 643 (1981).
[17] B. G. Streetman, J. appl. Phys. **37**, 3145 (1966).
[18] A. V. Latyshev, V. F. Stelmakh, and V. D. Tkachev, Fiz. Tekh. Poluprov. **5**, 1650 (1971).
[19] V. L. Bonch-Bruevich, Fiz. Tekh. Poluprov. **3**, 357 (1969).
[20] T. Yamada and H. Fujisaka, Progr. theor. Phys. (Kyoto) Suppl., No. 64, 269 (1979).
[21] A. V. Gaponov-Grekhov, M. I. Rabinovich, and I. M. Starobinets, Zh. eksper. teor. Fiz. Pisma **39**, 561 (1984).
[22] I. S. Aranson, A. V. Gaponov-Grekhov, and M. I. Rabinovich, Zh. eksper. teor. Fiz. **89**, 92 (1985).

(Received March 14, 1988; in revised form August 2, 1988)

phys. stat. sol. (a) 110, 565 (1988)

Subject classification: 72.20; 61.70; S8.15

Department of Metallurgy and Institute of Materials Science, University of Connecticut, Storrs[1])

The Photoplastic Effect in II–VI Compounds

Mercury Cadmium Telluride

By

J. PELLEGRINO[2]) and J. M. GALLIGAN

Experiments demonstrating the photoplastic response in mercury cadmium telluride crystals are presented. These experiments show the presence of a series of time delays prior to the change in stress when $(Hg_{1-x}Cd_x)Te$ crystals are irradiated with light. The change in stress is observed to be positive or negative. In addition the application of a voltage — constant in magnitude or varying temporally — affects the plastic response of the deforming crystal. All of these results are explained by a model in which charged point defects interact coulombically with charged dislocations. The electrons and or holes introduced by the light diffuse to either point defects or dislocations, leading to changes in the interaction between point defects and dislocations.

Experimente werden angegeben, die die photoplastische Response in Quecksilber–Kadmiumtelluridkristallen demonstrieren. Diese Experimente zeigen eine Reihe von Zeitverzögerungen vor der Spannungsänderung, wenn $(Hg_{1-x}Cd_x)Te$-Kristalle mit Licht bestrahlt werden. Es werden positive oder negative Spannungsänderungen beobachtet. Zusätzlich beeinflußt das Anlegen einer elektrischen Spannung — konstant oder zeitlich veränderlich — die plastische Response des sich deformierenden Kristalls. Alle diese Ergebnisse werden mit einem Modell erklärt, in dem geladene Punktdefekte über Coulombkräfte mit geladenen Versetzungen wechselwirken. Die durch das Licht eingeführten Elektronen und Löcher diffundieren entweder zu Punktdefekten oder Versetzungen, und führen zu Änderungen in der Wechselwirkung zwischen Punktdefekten und Versetzungen.

1. Introduction

Light impinging on a semiconductor crystal, while being plastically deformed, results in substantial changes in stress, $\Delta\sigma$. Studies of this process, termed photoplasticity [1] have primarily been carried out in translucent electronic and opto-electronic materials [2]. As a result the electron–hole pairs generated by the impinging light have been formed throughout the bulk of the crystal and, more probably at or near dislocations. In the present experiments which deal with the photoplastic process in crystals of mercury cadmium telluride the electron–hole pairs are mainly generated at or near the surface of the crystals because of HgCdTe's opacity, so that the electrons or holes must diffuse to the dislocations (or other defects) in order to interact. The choice of mercury cadmium telluride is fortunate in this regard since the present experiments provide substantial new and clarifying evidence on the nature of the photoplastic process.

In particular the present experiments show that the photoplastic process is affected, in part, by the surface state of the material, the temperature so as to be consistent with electrons and/or holes diffusing to dislocations or other defects, and by the

[1]) 97 North Eagleville Road, Storrs, CT 06268, USA.

[2]) Currently at the Naval Research Laboratory, Washington, D.C., USA.

presence of a voltage. In addition the influence of plastic deformation on the electrical properties of mercury cadmium telluride are presented.

These experiments have suggested a search for transient effects in other photoplastic materials, such as CdS where, indeed, related yield point phenomena have been observed [3].

2. Experimental

Experiments on the photoplasticity of mercury cadmium telluride $(Hg_{1-x}Cd_x)Te$ with $x \approx 0.3$ were carried out as follows: The crystals were deformed in a laboratory scale compression machine. The stress sensing members of the machine included two independent sets of strain gages, arranged in separate bridges, so that an overall stress–strain curve and differential stress could be measured simultaneously. The crystal while being deformed was illuminated with a halogen bulb; this light passed through a Jarrel-Ash spectrometer, allowing the selection of monochromatic light of a given wavelength. In additional experiments, light of a fixed wavelength, in the form of a 5 mW helium–neon laser, was used to irradiate the crystals. The quartz lamp, in contrast to the laser, produced a measured intensity of the order of $\mu W/mm^2$ at the sample surface, considerably less than the laser's intensity, which was three orders of magnitude larger.

In addition to studying the influence of light on the plasticity of these crystals we also examined the effect of an applied voltage on the plastic response of a crystal. Electrical leads were attached to the sample by soldering copper leads to opposite sides of a crystal with pure indium as the joining material. The copper–indium–mercury cadmium telluride joint, which formed an ohmic contact, was applied by heating the constituent materials in vacuum. Finally, experiments at different temperatures were carried out by surrounding a deforming sample with a small heater. The temperature of the sample was measured by placing a thermocouple in intimate contact with the deforming sample.

3. Results and Discussion

The primary measurement in the present set of experiments consisted of the observation of a time delay, t, before the onset of a change in stress, $\Delta\sigma$. This is illustrated in Fig. 1. This time delay as well as the change in stress has the following characteristics.

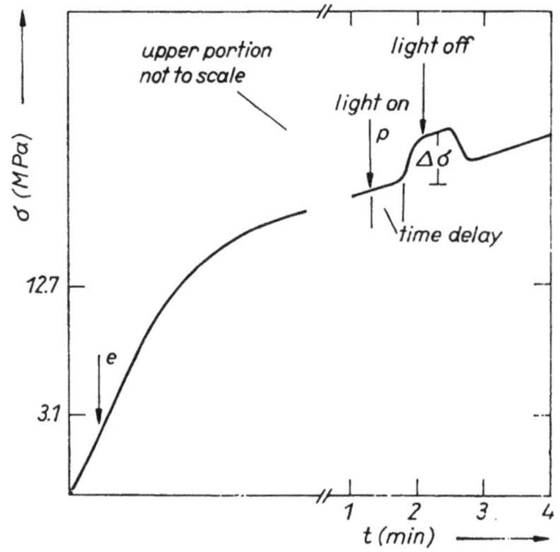

Fig. 1. The observed photoplastic effect in mercury cadmium telluride. Note that a change in stress is only observed when the crystal is deforming in a plastic manner. Note, also, that the change in stress occurs after an easily measurable time delay t

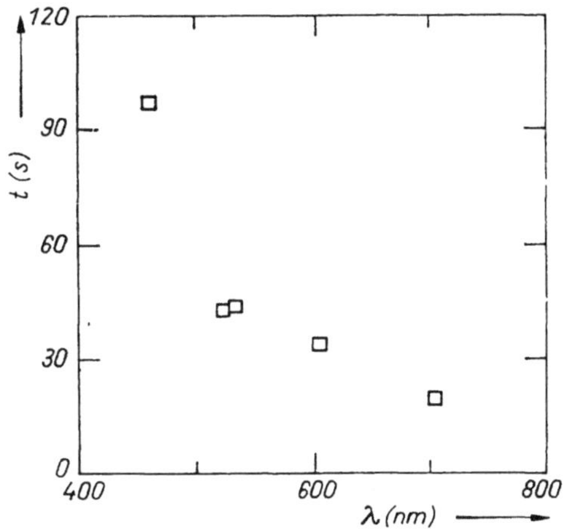

Fig. 2. Time delay, as defined in Fig. 1, as a function of the wavelength of the light

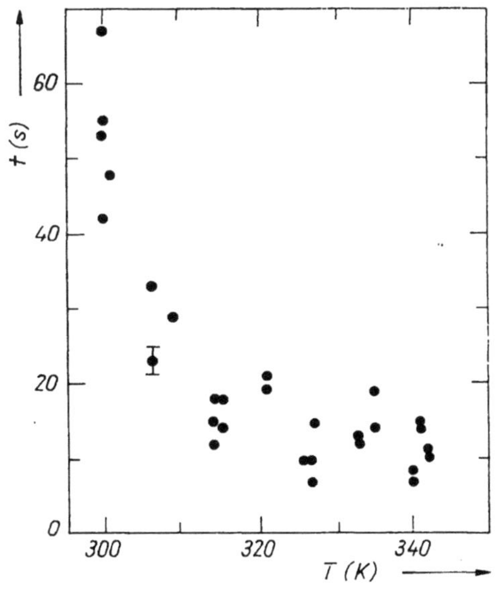

Fig. 3. The time delay as a function of deformation temperature

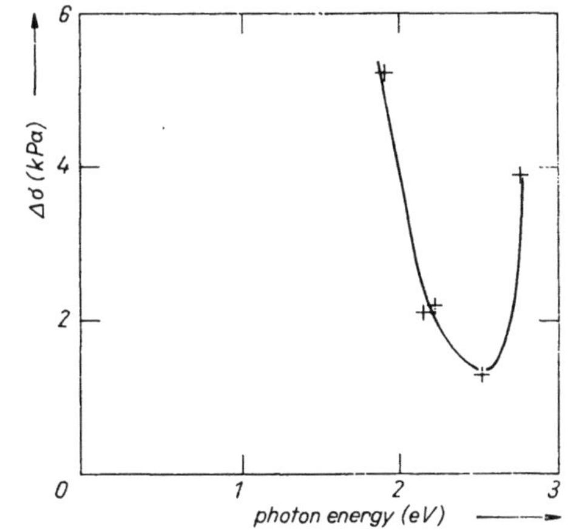

Fig. 4. The change in stress as a function of photon energy

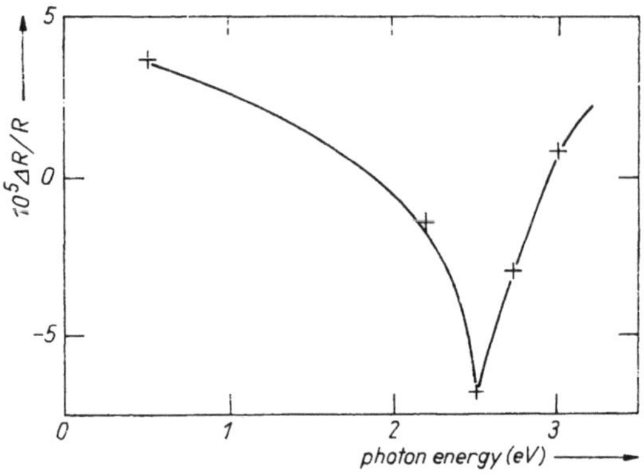

Fig. 5. The electroreflectance for a mercury cadmium telluride crystal (after Meritani et al. [4])

1. The time delay, as well as the attendant change in stress, are only observed in the region where the crystal deforms plastically, Fig. 1 arrow marked p. No such change occurs in the elastic range; Fig. 1 arrow marked e.

2. The time delay depends upon the wavelength of light used i.e. the time delay increases with increasing light energy, Fig. 2.

3. The observed time delay is shorter the higher the temperature, Fig. 3, showing that the photoplastic response of a crystal involves a thermally activated process.

4. The photoplastic related change in stress, $\Delta\sigma$, shows a minimum as a function of energy, Fig. 4. For comparison the electroreflectance of mercury cadmium telluride ($Hg_{1-x}Cd_xTe$ with $x = 0.3$) is shown in Fig. 5. Note that both the change in stress and the electroreflectance show a minimum *at the same energy*. This illustrates the influence of the surface on the photoplastic process.

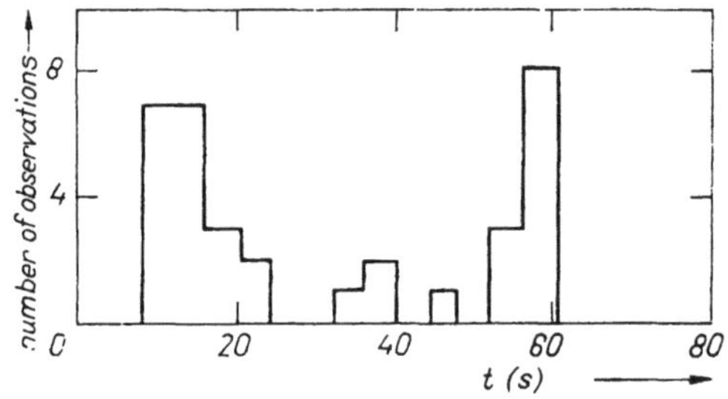

Fig. 6

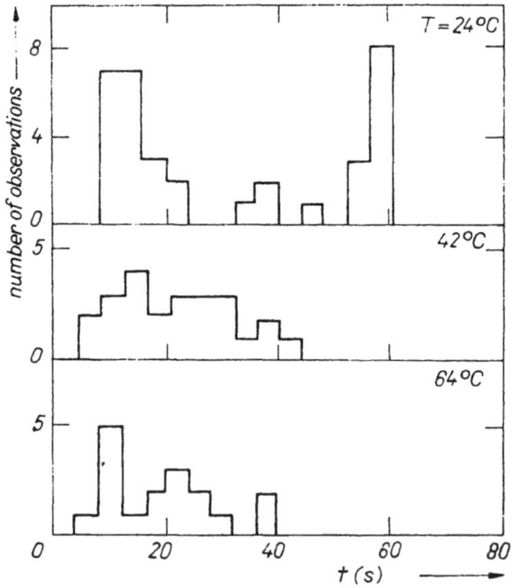

Fig. 7

Fig. 6. Time delays show distinct clustering in time; in this figure the measure numbers of time delays are compared to the time after the light is turned on

Fig. 7. The temperature dependence of the time delay measurements, measured at a number of different temperatures

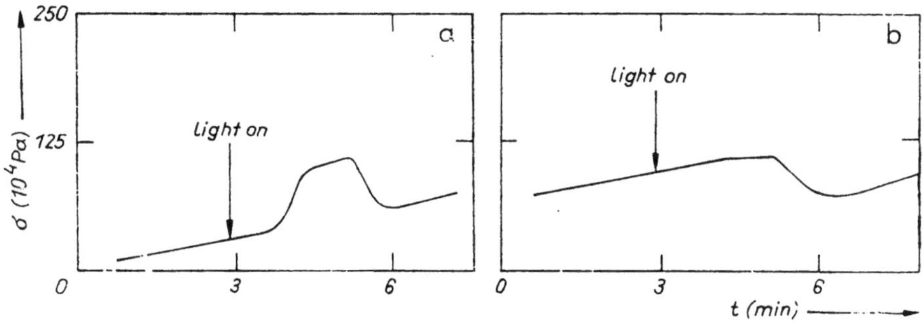

Fig. 8. The change in stress is observed to be either a) positive or b) negative

5. The time delays show four distinct clusters in time, Fig. 6. In this experiment (or set of experiments) a crystal is deformed, light impinges on the crystal and any change in stress is measured as well as the associated time delay, t. The measurement is repeated a number of times, typically about thirty measurements per run. The various observed times, t_i, are compared with the number of times $N(t_i)$ a time delay is observed. Clearly the time delays are not random, nor do they fit a single distribution function, again Fig. 6.

6. As the temperature increases the centroid of the time delay clusters shifts to shorter times, Fig. 7, consistent with point 3 above.

7. The change in stress in any given experiment, for a given crystal can be positive or negative, Fig. 8. This observation of a positive or a negative change in stress in conjunction with a time delay in the photoplastic process has not been observed before.

8. The application of a positive voltage, relative to ground, compared to that of a negative voltage shifts the time delay clusters such that the time delays show a smaller spread in the range of time delays, Fig. 9a. The application of a negative voltage, in contrast, spreads out the time delay clusters by about 100% compared to positive charge case, Fig. 9b.

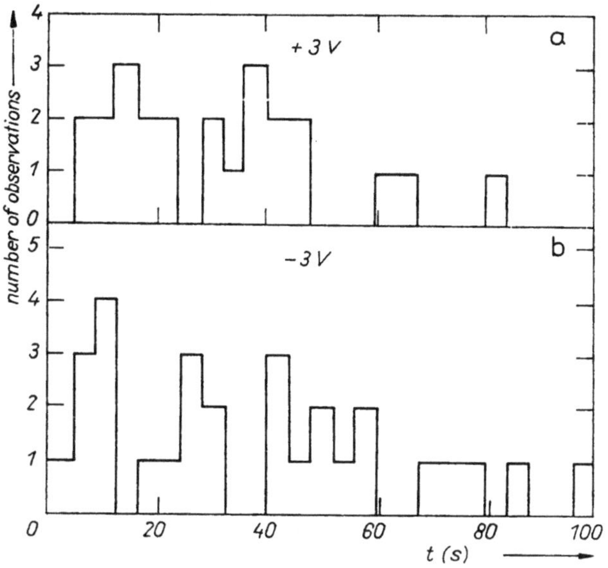

Fig. 9. The influence of an applied a) positive and b) negative voltage on the observed time delay clusters

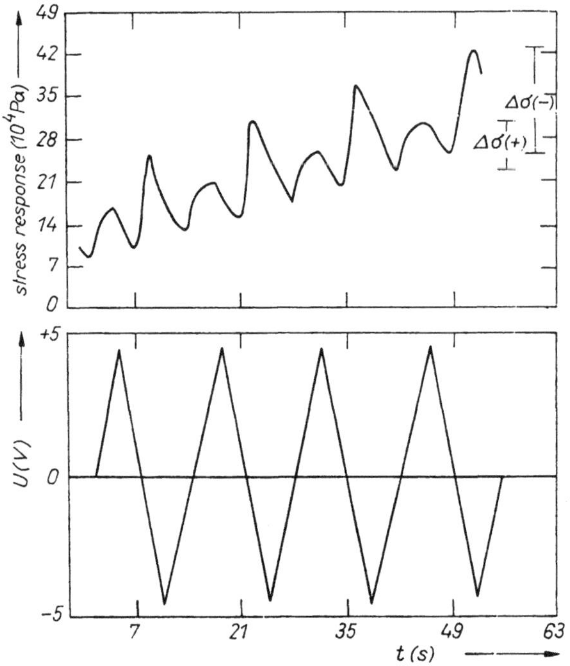

Fig. 10. The influence of a time dependent voltage on the plastic behavior of a mercury cadmium telluride crystal

9. In addition to measuring the time delays and changes in stress we have also measured the influence of an applied voltage on the overall plasticity of mercury cadmium telluride. In this experiment we apply a slowly varying voltage, with a maximum of +5 V and a minimum of −5 V, and measure any change in stress. As shown in Fig. 10 the application of a voltage, plus or minus, leads to a suppression of plasticity, i.e. the crystal deforms elastically; after this increase in stress then the crystal again deforms plastically. Note further that a negative voltage leads to a

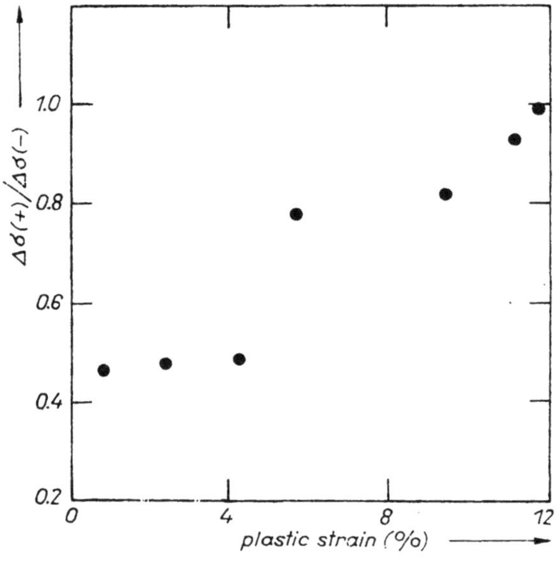

Fig. 11. The ratio of $\Delta\sigma(+)/\Delta\sigma(-)$ as a function of plastic strain

larger range of stresses, $\Delta\sigma(-)$ over which the plasticity is suppressed, than the range of stresses for a positive voltage, $\Delta\sigma(+)$, again Fig. 10. The ratio $\Delta\sigma(+)/\Delta\sigma(-)$ is however strain dependent and it approaches unity as the strain increases, Fig. 11.

10. It is interesting to note that resistance measurements, carried out on a number of crystals, while a crystal is deformed plastically, show a similar pattern, Fig. 12. That is for small strains, about $\approx 4\%$, both $\Delta\sigma(+)/\Delta\sigma(-)$, Fig. 11, and the crystal resistance, Fig. 12, are roughly independent of strain. At higher strains $\Delta\sigma(+)/\Delta\sigma(-)$ goes towards unity, while the crystals resistance increases by a factor of about two.

The results described above show that photoplasticity is observed in mercury cadmium telluride $(Hg_{1-x}Cd_x)Te$ with $x = 0.3$ and that there is a time delay associated with this change in stress. The time delay, t, is temperature dependent and this process occurs with an activation energy of 0.4 to ± 0.2 eV/atom, Fig. 13. A simple model which is consistent with these observations goes as follows: While a crystal of mercury cadmium telluride is irradiated with light there is a constant difference in excess carriers between the surface of the crystal, and its interior, ΔC. Such a concentration difference provides the driving force $D\,\Delta C/x$, where D is the diffusion coefficient of the pertinent species and x is the distance over which ΔC is maintained. The driving force is proportional to the rate of transport of these carriers, dx/dt, so that $-(dx/dt) = D\,\Delta C/x$ which gives, on integration, $-x^2/2 = \Delta t\,\Delta C$ or $t \sim 1/D$. The diffusion coefficient, D, is expected to obey an Arrhenius type of relationship, $D = D_0 \exp(-\Delta U/kT)$ where D_0 is a constant and ΔU is the activation energy barrier, T the temperature, and k Boltzmann's constant. The measured time delays are plotted according to this relationship in Fig. 13,; the activation energy observed is in the range of 0.4 ± 0.2 eV/atom.

In this regard note that a comparison of the wavelength dependence of the time delay, Fig. 2, with the change in stress, Fig. 4, shows that the time delay decreases monotonically with increasing wavelength, but $\Delta\sigma$ shows a minimum with wavelength. This comparison shows that while the change in stress is affected by the surface

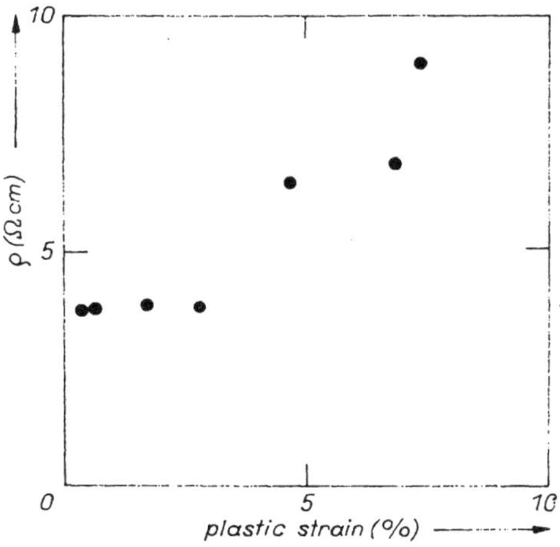

Fig. 12

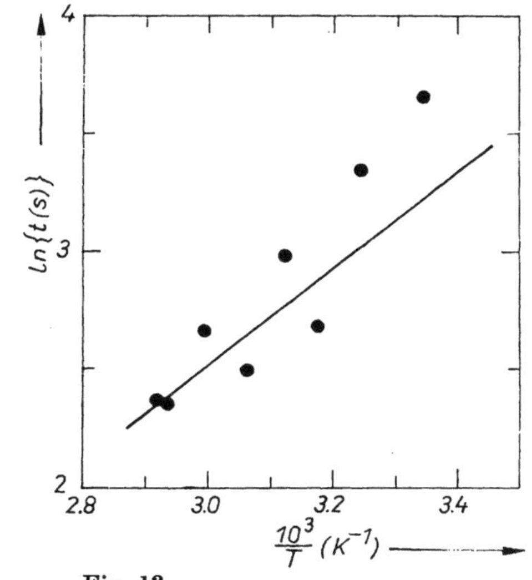

Fig. 13

Fig. 12. The resistivity as a function of plastic strain for a mercury cadmium telluride crystal

Fig. 13. The log of the time delays vs. temperature. The line drawn is consistent with an activation energy of 0.4 ± 0.2 eV/atom state (i.e. fewer of the electrons and holes reach defect centers at a particular wavelength) the diffusion to defect centers is not altered, viz. the time delay is independent of the surface state of the material. This comparison demonstrates clearly that the photoplastic process involves the diffusion of the electrons and/or holes to defect centers in the interior of the material. Note also that the higher the concentration of carriers, for a given distance over which diffusion occurs, the larger the driving force, leading to less time for diffusion. This result of a simple diffusion calculation is also consistent with the observed time delay, Fig. 2.

Consider now the observation of at least four distinct time delays: This observation is consistent with point defects interacting with dislocations in a coulombic manner, as shown below. Further, such an interaction is consistent with the observation that the change in stress can be, and is, positive or negative. If point defects and dislocations are charged positively or negatively and have their charge changed, then the coulombic interaction between point defects and dislocations would also be altered. Since electrons and/or holes can raise or lower this charge interaction then it is to be expected that the change in stress can be positive or negative. In addition if charged dislocations have a compensating charge sheath about them [5], then the rate of arrival of electrons or holes at a dislocation core would be expected to have a characteristic response time. This is so because different species would have to spend differing amounts of time surmounting particular barriers, consistent with the thermally activated process described above. Also, in agreement with observation roughly one half of the changes in stress would be hardening and one half softening.

The presence of a barrier which prevents electrons or holes from reaching annihilation centers in the crystal is consistent with a thermally activated process. This view is also consistent with the observation of a minimum in $\Delta\sigma$, due to a surface state, Fig. 4: Electrons and/or holes introduced by the light must leave the surface area, diffuse within the crystal and then reach an annihilation center. Part of the time delay will be related to diffusion in the crystal lattice, while part will be related to overcoming a barrier.

These arguments show that the photoplastic process involves the diffusion of electrons and holes to defects in the interior of the crystal. In prior studies of photoplasticity the electrons and holes may have been formed at or near defect centers. The pulsing experiments in CdS show, however, that time delays are observable in other materials [3].

The influence of a voltage on the observed time delays, Fig. 9 and 10 is also of importance and relevant to this model since a negative voltage clearly reveals the presence of four time delay clusters. The fact that a positive voltage affects the time delay clusters in a different manner than does a negative voltage shows that differently charged species are involved in the time delays. Consider the case of a positive voltage, Fig. 9: There are two main time delay peaks – one centered at $\approx$18 s another centered at about 40 s, with two smaller peaks at 60 s and at 80 s, but no trace of a fourth peak. The time delay clusters in the case of minus three volts shows that the missing peak is buried in the peak centered at $\approx$40 s. Note, Fig. 9 that the time average for the peak observed with a positive voltage is about 40 s. The average of the two time delay peaks – clearly revealed when a negative voltage is applied, Fig. 10 – is about 40 s! This agreement is well within the scatter of the results, and demonstrates that there are four separate time delays in the cases where a positive and a negative voltage is applied to the crystal. The presence of four time delay peaks in consistent, again, with a coulombic interaction, one involving electrons and/or holes interacting with positively or negatively charged entities.

All of the above results are in agreement with a model in which charged point defects are interacting with charged dislocations. The time delays, as well as the time delay clustering, all point to a diffusion of charged entities, leading to a change in charge on defects in the interior of the crystal. The observation of two separate yield points, superimposed on a general increase in stress in a photoplastic experiment in CdS [3] suggests that the present model is applicable to other semiconducting compounds.

In conclusion we have presented results on the photoplasticity of mercury cadmium telluride. These results show that photoplasticity in this material occurs with a time delay, which corresponds to a thermally activated process; the activation barrier for this process is 0.4 eV/atom. We present a model which is consistent with all the aspects of photoplasticity in mercury cadmium telluride. The model is also consistent with recently observed transient aspects of photoplasticity in cadmium sulfide [3].

Acknowledgements

It is a pleasure to acknowledge very useful conversations with Professor P. Haasen and Dr. J. Tregilgas. The superb machine shop work of The Institute of Materials Science was instrumental in carrying out this research. Special thanks to Lane Witherell and Madeline Nasansky for the manuscript preparation.

References

[1] See, for example, a recent review: Y. Osipyan, V. F. Petrenko, A. Zaretskii, and R. Whitworth, Adv. Phys. **35**, 115 (1986).
[2] Y. Osipyan and I. B. Savchenko, Soviet Phys. – Solid State **14**, 1723 (1973).
[3] M. Grondolsky and J. M. Galligan, to be published.
[4] A. Meritani, K. Taniguchi, C. Hamaguchi, and J. Nakai, J. Phys. Soc. Japan **34**, 79 (1973).
[5] T. Figielski, phys. stat. sol. **9**, 555 (1965); **14**, 381 (1966).

(Received February 16, 1988; in revised form August 2, 1988)

phys. stat. sol. (a) **110**, 575 (1988)

Subject classification: 72.20 and 78.20; 71.55; S8.16

Centro de Estudios de Semiconductores, Facultad de Ciencias, Universidad de Los Andes, Mérida[1])

Electrical and Optical Properties of n- and p-Type $CuInTe_2$

By

S. M. WASIM and J. G. ALBORNÓZ

The electrical properties between 80 and 300 K and the optical absorption at room temperature of both n- and p-type $CuInTe_2$ are studied. The n-type sample is obtained by annealing p-type $CuInTe_2$ in the presence of indium for a prolonged period of time. The p-type samples are grown by programmed directional freezing technique with slight excess of indium in the stoichiometry. Two defect levels with $E_D = 58$ meV and E_A around 15 meV are identified from the analysis of the temperature dependence of the carrier concentration. The origin of these levels, consistent with the covalent bonding model, is attributed to In_{Cu} and V_{Te}, respectively. From a theoretical fit to the experimental data, the density of states effective mass of the electrons is found to be $m_n^* = 0.16m_e$. The estimated values of the conduction and valence-band deformation potentials are in agreement with the general trend observed in other I–III–VI_2 compounds. Using the model proposed for the band-gap shrinkage, the energy gap of n-type $CuInTe_2$ in the dilute limit is calculated to be 1.026 eV.

On a étudié les proprietés électriques entre 80 et 300 K, et l'absorption optique à température ambiante de $CuInTe_2$ type n et type p. L'échantillon type n a été obtenu par recuit et celui type p par mise en présence de l'Indium durant une période de temps prolongé. On a fait croître les échantillons type p par la technique de refroidissement directionnel programmé avec un léger excès d'Indium dans la stoechiométrie. Par l'analyse de la dépendance de la température de concentration des porteurs, on a identifié deux niveaux de défauts avec $E_D = 58$ meV et E_A environ 15 meV. L'origine de ces niveaux, compatible avec le modèle de liaison covalente, est attribué à In_{Cu} et a V_{Te}, respectivement. Après un ajustement théorique aux données expérimentales, la densité des états de la masse effective des électrons est de $m^* = 0{,}16m_e$. Les valeurs estimées des potentiels de déformation des bandes de conduction et de valence sont en accord avec la tendance générale observée dans les composés I–III–VI_2. En utilisant le modèle proposé pour la contraction de la bande du gap, le résultat de gap d'énergie de $CuInTe_2$ type n dans la limite diluée est de 1,026 eV.

1. Introduction

The ternary compound $CuInTe_2$ belongs to the I–III–VI_2 family of semiconductors [1]. This group of materials has received considerable attention in recent years due to the potential for opto-electronic applications, leading to efficient thin film solar cells [2]. Single crystals of $CuInTe_2$ grown under stoichiometric conditions and with slight excess of cations in the stoichiometry by vertical Bridgman zone growth and programmed directional freezing (PDF) techniques have mainly been reported to be p-type conducting [3]. No change in the conductivity type is observed by annealing in Cu, In, Zn, and Cd [4]. However, p-type samples of this material grown by the closed tube chemical vapour transport technique which were later heat treated in Cd atmosphere [5] and thin films annealed in vacuum [6] give n-type behaviour. An n-type sample, embedded in a p-type ingot grown by PDF technique, is also reported [7].

---

[1]) Mérida 5101, Venezuela.

From the study of the effect of annealing in In on the electrical properties of $CuInTe_2$ samples, it is suggested [3] that the p-type behaviour of this material, which evaporates incongruently, can be explained consistently by using the covalent bonding model [8]. According to this model V_{Te}, being a donor in the ionic model, should behave as an acceptor. It was also proposed [3] that if the covalent bonding model is applicable to $CuInTe_2$, samples grown with slight excess of In in the stoichiometry or very thin samples annealed in In for a prolonged period of time, when the concentration of In atoms on Cu sites is greater than the Te vacancies ($N_{In_{Cu}} > N_{V_{Te}}$), should give n-type conductivity. Hence, in the present paper, in continuation of our earlier work, we report on the electrical and optical properties of both p-type $CuInTe_2$ samples grown with 0.03 at% excess of In and an n-type sample which was later obtained after annealing at 660 °C in In for 45 days. Electrical data published by Neumann et al. [7] on an n-tpye sample isolated from an overall p-type ingot are also analysed.

2. Experimental Methods

Single crystals of $CuInTe_2$ with excess of In were grown by the directional freezing technique. To investigate the possible existence of any secondary phase, DTA spectra of samples taken mainly from the central part of the ingot were obtained. Three peaks, the first at 788 °C corresponding to the melting point of the compound, the second at 669 °C of origin not yet established, and the third at 657 °C related to the transition from zincblende to chalcopyrite structure, are observed. This is in complete agreement with the phase diagram of In-rich $CuInTe_2$ reported by Palatnik and Rogacheva [9].

As-grown samples of approximate dimensions of $0.4 \times 0.4 \times 0.04$ cm^3, cut from the ingot and having p-type conductivity, were polished and etched in a (7:3) solution of $K_2Cr_2O_7$ and H_2SO_4. Their electrical conductivity and Hall mobility were measured between 77 and 300 K by the van der Pauw method. The thickness of these samples was then reduced to about 0.1 mm or less for the optical absorption study at room temperature by a Cary 17I spectrometer. Later, pure In was evaporated on both sides of the same samples in a vacuum of about 10^{-6} Torr which were then annealed at 660 °C for 45 days in an evacuated quartz ampoule. The n-type conductivity of the annealed samples, as observed by a thermal probe, did not change even after several etchings.

3. Experimental Results and Discussion

3.1 Carrier concentration

The hole concentrations p of samples MB1 and MB2, taken from different parts of the ingot, are plotted in Fig. 1a as a function of temperature. The same value of p of MB1 and MB2 at any given temperature between 300 and 150 K indicates the relatively homogeneous nature of the ingot. An acceptor ionization energy around 16 and 9 meV obtained from the slope of the curves in Fig. 1a and given in Table 1 can be compared with similar values reported [10] by other authors for stoichiometric $CuInTe_2$. The origin of this shallow acceptor is attributed by them to either V_{In} or Cu_{In}. However, in the present case of In-rich $CuInTe_2$ and consistent with the covalent bonding model [3], this level could also possibly be due to V_{Te}.

The variation with temperature of the concentration of the n-type sample MB1(In) which was obtained after annealing MB1 in In for a prolonged period of time is also shown in Fig. 1a. It is observed that n first decreases with temperature up to about 150 K. A similar behaviour, also indicated in the same figure, is reported [7] by Neumann et al. in the limited temperature range between 300 and 200 K. However, in

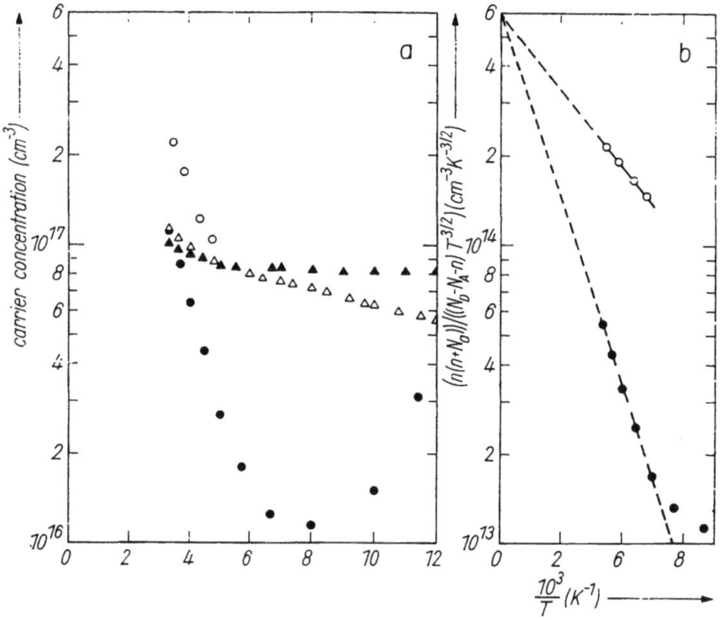

Fig. 1. a) The variation of the carrier concentration with temperature of as-grown p-type (△ MB1, ▲ MB2), n-type (○ [7], ● MB1(In)), samples of $CuInTe_2$. b) A plot of $n(n + N_A)/[(N_D - N_A - n)\, T^{3/2}]$ vs. $10^3/T$ of n-type (○ [7], ● MB1(In)) samples of $CuInTe_2$. The corresponding values of N_A and N_D are given in Table 1

the case of MB1(In) at temperatures below 150 K, the carrier concentration is found to increase with decreasing temperature. The appearance of a minimum in the temperature dependence of the charge carrier concentration, in general, is attributed to the impurity band conduction. However, it is shown by Klotynsh [11] that this type of anomaly can also be explained for an n-type material by considering the simultaneous existence of two impurity levels, one in the conduction band and the other donor

Table 1

Carrier concentration, mobility, and energy gap at 300 K, and donor and acceptor concentration and ionization energies of n-type and p-type $CuInTe_2$ samples. The ionized impurity concentration is also given

samples	n or p (300 K) ($10^{17}\,cm^{-3}$)	μ (300 K) (cm^2/Vs)	N_A (cm^{-3})	N_D (cm^{-3})	N_i (cm^{-3})	E_A or E_D (meV)	E_g (eV)
MB1 p-type	1.10	34.60	5.86×10^{18}*)	5.74×10^{18}*)	1.60×10^{19}*)	16	0.94
MB2 p-type	1.03	35.70	5.50×10^{17}*)	5.40×10^{18}*)	1.09×10^{19}*)	9	0.94
MB1(In)	1.08	189.20	7.95×10^{17}*)	9.05×10^{17}*)	1.70×10^{18}*)	58	0.97
n-type			6.30×10^{17}**)	1.03×10^{18}**)	1.66×10^{18}*)		
[7]	2.10	355.0	2.50×10^{18}**)	3.20×10^{18}**)	5.50×10^{18}**)	25	—
n-type							

*) Calculated from mobility data.
**) Calculated from concentration data.

level in the band gap. Starting from lower temperatures, first a decrease in the charge concentration originating from the impurity level in the conduction band is expected with increasing temperature. When the temperature is raised to a certain value T_m, the donor level in the band gap becoming active, starts to dominate and the charge carrier concentration increases with temperature.

To estimate the activation energy of the donor level in samples MB1(In) and that of [7], the variation of the electron concentration in the high temperature region between 300 and 150 K is analyzed by the following expression for nondegenerate statistics of a single level [12]:

$$\frac{n(n + N_A)}{(N_D - N_A - n)} = \frac{N_d}{\beta} \exp(-E_D/kT), \tag{1}$$

where N_D and N_A are donor and compensating acceptor densities, E_D and β are the donor activation energy and impurity degeneracy, respectively. N_d is the density of states of the conduction band given by $2(2\pi m_n^* kT/h^2)^{3/2}$, where m_n^* is the electron density of state effective mass. By choosing the proper values of N_D and N_A, as given in Table 1, for the two samples, a linear plot of $n(n + N_A)/[(N_D - N_A - n) T^{3/2}]$ vs. $10^3/T$, as shown in Fig. 1b, is obtained. From the slope, the ionization energy is found to be 58 and 25 meV for MB1(In) and the sample of [7], respectively. On the other hand, the ionization energy of the donor level calculated by the hydrogenic approximation $E_D = 13.6 m_n^*/m_e \varepsilon_0^2$, with $m_n^* = 0.155$ m_e and $\varepsilon_0 = 9.66$ (to be discussed later) gives a value of 22.6 meV. This agrees well with 25 meV obtained from the analysis in this work.

It is reasonable to assume that these values of E_D have the same origin and the difference can be attributed to the screening effects. However, because of the limited number of data points, no conclusive information about E_D in the dilute limit can be obtained from the E_D versus $N_D^{1/3}$ plot.

From the knowledge of the initial stoichiometry of the sample and the annealing conditions, two possible origins for this defect level, namely In_i and In_{Cu}, can be expected. Although the formation energy of the electrically active intrinsic defects in $CuInTe_2$ is not known, comparing with the value proposed by Neumann [13] for In_i and In_{Cu} in $CuInSe_2$, it can be suggested that In_{Cu} is the most likely origin of the defect level observed in the present samples.

It is interesting to observe that the linear plot in Fig. 1b for the data of the two samples, having different origins, gives very nearly the same intercepts. Using the relation $2(2\pi m_n^* k)^{3/2}/\beta h^3$, with $\beta = 1/2$, the effective mass is calculated to be $m_n^* = (0.155 \pm 0.005)$ m_e. This value can be compared with $0.09 m_e$ and estimated $0.16 m_e$ reported for n-types $CuInSe_2$ [14] and $CuInS_2$, [15], respectively.

3.2 Carrier mobility

The mobility data of the charge carriers of both the p-type and n-type samples are plotted as a function of temperatures in Fig. 2. To get an estimate of the strength of different scattering mechanisms, the total mobility is calculated using Mathiessen's approximation,

$$\mu^{-1} = \mu_i^{-1} + \mu_{ac}^{-1} + \mu_{npo}^{-1} + \mu_{po}^{-1}, \tag{2}$$

where the terms μ_i, μ_{ac}, μ_{npo}, and μ_{po} represent the mobilities of the charge carriers due to the scattering by ionized impurities, acoustic-lattice modes, nonpolar optical modes, and polar optical modes, respectively.

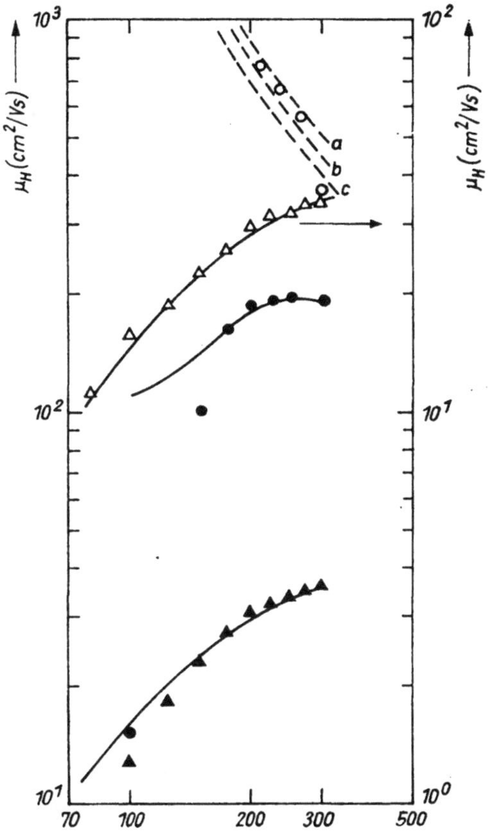

Fig. 2. The variation of the Hall mobility with temperature of as-grown p-type (△ MB1, ▲ MB2) and n-type (○ [7], ● MB1(In)) samples of $CuInTe_2$. The curves show a theoretical fit to the data with the parameters given in the text and in Table 1. Dashed curves (a), (b), and (c) are obtained with $m_n^* = 0.15m_e$, $0.16m_e$, and $0.17m_e$, respectively

Based on the work of Wiley and DiDomenico applied earlier [19] in the case of $CuInSe_2$, in the present analysis the contribution due to the scattering by polar optical modes in p-type and that of nonpolar optical modes in n-type $CuInTe_2$ is not considered. A distinction is also made between the nondegenerate s-like conduction band and the p-like valence bands.

For the p-type samples, the Hall mobility due to ionized impurity scattering was calculated according to the Brooks-Herring formula [16],

$$\mu_i = \left(\frac{2}{300}\right) \frac{2^{7/2}\varepsilon_0^2(kT)^{3/2}}{\pi^{3/2}e^3(m_p^*)^{1/2}\, N_i f(x)}\ \mathrm{cm^2/Vs}\,, \tag{3}$$

where N_i is the ionized impurity density and $m_p^* = 0.78m_e$ represents the hole density of states effective mass. The function $f(x)$ is given by the relation $\ln(1+x) - /x(1+x)$, where

$$x = \frac{6\varepsilon_0 m_p^*(kT)^2}{\pi e^2 \hbar^2 p}\,. \tag{4}$$

The combined effect of the acoustic and nonpolar optical mode scattering contributions to the mobility is calculated using the expression [16].

$$\mu_{\mathrm{acnpo}} = \mu_{\mathrm{ac}} S(\vartheta, \eta, T)\,. \tag{5}$$

Here ϑ is the optical phonon characteristic temperature and $\eta \approx (E_{\mathrm{acnpo}}/E_{\mathrm{ac}})^2$, where E_{npo} and E_{ac} are the nonpolar optical and acoustic mode deformation potentials. An analytical approximation to $S(\vartheta, \eta, T)$ is found to be

$$S(\vartheta, \eta, T) \approx (1 + A\eta)^{-1}\,, \tag{6}$$

where $A = HZ/e^Z - D$, H and D being constants for each value of η and $Z = \vartheta/T$.

The mobility due to acoustic mode scattering was calculated according to the relation

$$\mu_{ac} = \left(\frac{2}{300}\right) \frac{(8\pi)^{1/2}\, eh^4 \varrho u^2}{3E_{ac}^2 (m_p^*)^{5/2} (kT)^{3/2}}\ \text{cm}^2/\text{Vs}\,, \tag{7}$$

where $\varrho = 6.09$ m/cm³ is the density and u the longitudinal velocity of sound.

From the recently measured [17] value of the Debye temperature $\vartheta_D = 195$ K, u is calculated to be 2.04×10^5 cm/s. With this value of u together with $\varepsilon_0 = 9.66$, $\varepsilon_\alpha = 7.25$, $\vartheta = 271.5$ K as in [18] and H, D, and η as 1.34, 0.914, and 4, respectively [19], in (2) to (7), the mobility data of samples MB1 and MB2 is fitted with N_i and E_{ac} as adjustable parameters. A good fit between 150 and 300 K is obtained with $E_{ac} = 3.8$ eV and N_i given in Table 1. This value of E_{ac} is in good agreement with the valence band deformation potential of 3.5 eV of the binary analogue of $CuInTe_2$. The higher value of 5.35 eV, reported earlier [4], could be related to the fact that the contribution of the nonpolar optical phonon scattering was not considered in the calculations.

To analyze the data of n-type samples, the expression used in [16] for the mobility due to the scattering of electrons by polar-optical phonons is

$$\mu_{po} = 25.4 \frac{T^{1/2}}{m_n^{*3/2} \vartheta} \left(\frac{1}{\varepsilon_\alpha} - \frac{1}{\varepsilon_0}\right)^{-1} (e^Z - 1)\, G(Z)\ \text{cm}^2/\text{Vs}\,, \tag{8}$$

where $Z = \vartheta/T$ and $G(Z)$ is a tabulated function which may be approximated, in the range 120 to 300 K, by $0.48 \exp(0.18\vartheta/T)$.

The increase in mobility with decreasing temperature of n-$CuInTe_2$ [7], as shown in Fig. 2, indicates that the dominant scattering mechanisms in this sample are mainly due to acoustic phonons and polar optical modes. Thus, to make a theoretical fit to this set of data only these two mechanisms are considered. From the valence band deformation potential of 3.8 eV obtained earlier for the p-type samples and using a rough rule-of-thumb estimate [15], the corresponding conduction band deformation potential is assumed to be 9.8 eV. With this value of E_{ac}, the total mobility of n-type $CuInTe_2$ is calculated for different values of m_n^*. This is shown in Fig. 2 by curves a, b, and c. A good fit to the data is observed with $m_n^* = 0.16 m_e$. This is in excellent agreement with the value of $m_n^* = 0.155 m_e$ obtained from the analysis of the electron concentration data.

Because of the temperature dependent characteristic of the mobility data of MBI(In), in addition to μ_{po} and μ_{ac}, the scattering by ionized impurities was also considered in the analysis. In this case, using $m_n^* = 0.16 m_e$, an acceptable fit between 150 and 300 K could be obtained with $N_i = 1.7 \times 10^{18}$ cm^{-3} and $E_{ac} = 16.6$ eV. This is also shown in Fig. 2 by the dotted curve. The departure of this curve from the experimental data occurs in the temperature region where the predominant contribution from the impurity level of the conduction band to the free carrier concentration is proposed. An additional scattering mechanism for the charge carriers originating from this level, which should have a much stronger temperature dependence than the ionized impurity term, is required to explain the sharp decrease in the mobility data below 150 K.

The deformation potential of 16.6 eV for MB1(In) is relatively large as compared to 9.8 eV for the sample of [7]. A wide range of E_{ac} from 13 to 55 eV is reported [16] for n-type $CuInSe_2$. It has been proposed [20] that while the hole–hole scattering does not influence the mobility of p-type $CuInSe_2$, the e–e scattering plays an important role in determining the mobility of n-type samples. It is also suggested [21] that a higher value of conduction band deformation potential obtained from the theoretical fit to the mobility data is due to the compensating effect when the e–e scattering is not considered in the calculation. Under this condition, it is argued that the adjusted E_{ac} value should increase with the electron concentration if the e–e scattering contribution is ignored in the curve fitting. However, in the present case of n-type $CuInTe_2$ with a limited number of samples, we find that E_{ac} decreases with increasing electron concentration. A close examination of the data of n-type $CuInSe_2$ samples in Fig. 1 and 2 and Table V of [16] also gives a similar indication. Thus, it is proposed that to explain the mobility data of n-type $CuInTe_2$ in the high temperature region, in addition to the acoustic and optical polar mode scattering, the effect of e–e scattering where the relaxation time should depend on both the concentration and the degree of compensation, is perhaps necessary.

3.3 Optical absorption

Measurements of the absorption spectrum at room temperature were made on the samples MB1 and MB1(In) with a Cary 17I spectrometer. The expression $\alpha = [\ln (I/I_0)]/d$ where d is the sample thickness, is used to calculate the absorption coefficient α. A plot of $(\alpha h\nu)^2$ versus $h\nu$ is shown in Fig. 3. This is to check from the relation $\alpha = A[(h\nu - E_g)^{1/2}/(h\nu)]$ the validity of the well-established direct band gap of $CuInTe_2$. Also from the extrapolation of these curves to $(\alpha h\nu)^2 = 0$, the energy gap is calculated to be 0.94 and 0.97 eV for MB1 and MB1(In), respectively. This is in agreement with the values between 0.92 and 1.06 eV reported in [22].

Recently attempts have been made to explain the measured value of E_g in $CuInSe_2$ in terms of the band gap narrowing effects caused either by the ionized impurity induced screening [23] or due to the quantum mechanical penetration of the band states in the band gap, originating from the potential gradient of the charged defects [24]. For a doubly ionized defect, Rincon [25] has given the following expression to explain the band gap shrinkage in $CuInSe_2$ in terms of the ionized impurity concentration and the compensation ratio c,

$$E_g(N_i, c) = E_g(0, 0) - \gamma[8e^2\hbar b^{2/3}/\varepsilon_0 m_n^{*1/2}]^{2/3} N_i^{4/9} \text{ eV}, \tag{9}$$

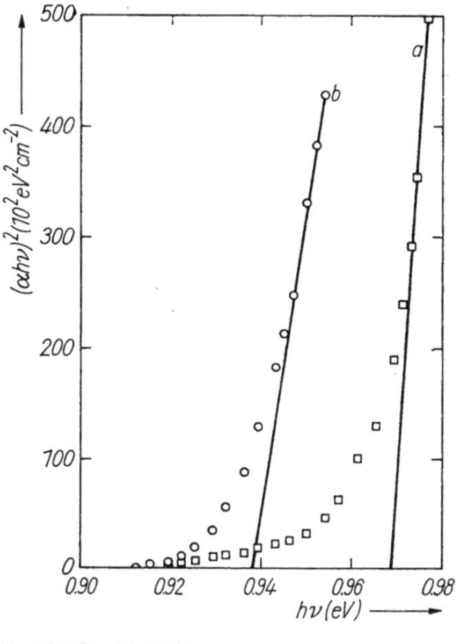

Fig. 3. The dependence of $(\alpha h\nu)^2$ on photon energy at 300 K in n-type and p-type $CuInTe_2$. (a) and (b) correspond to samples MB1(In) and MB1, respectively

where $E_g(0, 0)$ is the energy gap of a highly nondegenerate sample of the same material with vanishing free carrier and impurity concentrations, $b = (1 - c)/(1 + c)$ and $\gamma =$ const, approximately equal to 1.5.

Using (9) with $m_n^* = 0.16m_e$, $c = 0.88$, obtained from n and N_i of the mobility data and $\gamma = 2.06$ from [25], we find $E_g(0, 0) = 1.001$ eV. On the other hand, with $c = 0.61$ and $N_i = N_A + N_D = 1.66 \times 10^{18}$ cm^{-3} calculated from the analysis of the electron concentration data, we get $E_g(0, 0) = 1.026$ eV. This is in excellent agreement with $E_g = 1.04$ eV reported earlier from optical and electrical conductivity measurements [5, 26, 27]. This gives additional evidence that In_{Cu} is the predominant defect level in n-type $CuInTe_2$ since in (9) the possible, doubly ionized donor defect level is In_{Cu}.

When the same method is applied to the p-type MB1 sample with $c = 0.98$ and $N_i = 1.6 \times 10^{19}$ as in Table 1, a rather large value of $\gamma = 9$ is needed to obtain $E_g(0, 0) = 1.026$ eV. No appreciable decrease in γ could be obtained by considering a triply ionized V_{In} defect level in the formulation. This discrepancy could perhaps be attributed to the fact that either the compensation c used in the calculation is high and or several dominant acceptor levels in different charged states are present in this sample.

In conclusion, single crystals of $CuInTe_2$ grown with slight excess of In in the stoichiometry by the directional freezing technique and initially having p-type conductivity can be converted to n-type by prolonged annealing in indium. Fro melectrical measurements the shallow acceptor level around 15 meV in the p-type sample is attributed to V_{Te}, whereas the donor level around 58 meV is suggested to be due to In_{Cu}. The density of states effective mass of the electron is estimated to be $0.16m_e$ from both the free carrier concentration and the mobility data. Although the valence band deformation potential is found to be 3.8 eV, the conduction band deformation potential varies from one sample to another. This is not unusual in n-type semiconductors if the e–e scattering is not considered in the analysis. From the band gap of 0.97 eV in n-type $CuInTe_2$ and the model proposed by Rincon [25], $E_g = 1.026$ eV is calculated in the dilute limit. However, the p-type data give a relatively smaller value of $E_g(0, 0)$, most probably due to the existence of defect levels in different charge states not included in the formulation.

Acknowledgements

This work was supported by the Consejo de Desarrollo Cientifico y Humanístico (C.D.C.H.) of the Universidad de Los Andes and also by the Consejo Nacional de Investigaciones Científicas y Tecnólogicas (CONICIT). The authors wish to thank Mr. F. Sánchez Pérez and Ms. C. Kauman for technical assistance.

References

[1] L. I. Berger and V. D. Prochukhan, Ternary Diamond-Like Semiconductors, Consultant Bureau Enterprises, Inc., New York 1969 (p. 47).
[2] J. D. Meakin, SPIE **543**, 108 (1985).
[3] S. M. Wasim, G. Sánchez Porras, and H. Neumann, Solid State Commun. **54**, 239 (1985).
[4] S. M. Wasim, G. S. Porras, and R. D. Tomlinson, phys. stat. sol. (a) **71**, 523 (1982).
[5] E. Gombia, F. Leccabue, C. Pelosi, D. Seuret, and O. Vigil, Progr. Crystal Growth Charact. **10**, 225 (1984).
[6] L. L. Kazmerski and Y. J. Juang, J. Vacuum Sci. Technol. **14**, 769 (1977).
[7] H. Neumann, R. D. Tomlinson, E. Nowak, E. Elliott, and L. Haworth, Crystal Res. Technol. **16**, K112 (1981).
[8] H. Neumann, Crystal Res. Technol. **18**, 483 (1983).

[9] L. S. PALATNIK and E. I. ROGACHEVA, Soviet Phys. – Doklady **12**, 503 (1967).
[10] L. I. HAWORTH, I. S. AL-SAFFAR, and R. D. TOMLINSON, phys. stat. sol. (a) **99**, 603 (1987).
[11] E. KLOTYNSH, phys. stat. sol. (a) **21**, 377 (1974).
[12] J. S. BLAKEMORE, Semiconductor Statistics, Pergamon Press, 1962.
[13] H. NEUMANN, Seminar Centro de Estudios de Semiconductores, Universidad de Los Andes, Mérida, Venezuela, November 1984; see also ref. 16.
[14] H. WEINERT, H. NEUMANN, H.-J. HÖBLER, G. KÜHN, and NGUYEN VAN NAM, phys. stat. sol. (b) **81**, K59 (1977).
[15] D. C. LOOK and J. C. MANTHURUTHIL, J. Phys. Chem. Solids **37**, 173 (1976).
[16] S. M. WASIM, Solar Cells **16**, 289 (1986).
[17] K. BOHMHAMMEL, P. DEUS, G. KÜHN, and W. MÖLLER, phys. stat. sol. (a) **71**, 505 (1982).
[18] S. M. WASIM, phys. stat. sol. (a) **75**, K69 (1983).
[19] J. D. WILEY and M. DIDOMENICO JR., Phys. Rev. B **2**, 427 (1970).
[20] R. T. BATE, R. D. BAXTER, F. J. REID, and A. C. BEER, J. Phys. Chem. Solids **26**, 1205 (1965).
[21] C. RINCÓN, J. Phys. Chem. Solids **49**, 391 (1988).
[22] S. M. WASIM, A. LACRUZ VIELMA, and C. RINCÓN, Solid State Commun. **51**, 935 (1984).
[23] H. NEUMANN and R. D. TOMLINSON, Solid State Commun. **57**, 591 (1986).
[24] D. REDFIELD, Phys. Rev. **130**, 916 (1963).
[25] C. RINCÓN, Solid State Commun. **64**, 15 (1987).
[26] S. M. ZALAR, J. Electrochem. Soc. **113**, 230 (1966).
[27] V. P. CHERNYAVSKII, YU. I. POLYGALOV, and A. S. POPLAVONI, Ukr. fiz. Zh. **17**, 1535 (1972).

(Received July 25, 1988)

phys. stat. sol. (a) **110**, 585 (1988)

Subject classification: 72.20

Departamento de Fisica Aplicada, Facultat de Física, Burjassot[1])

A Method for Measuring the Resistivity of a Layered Semiconductor Perpendicular to the Layers

By

F. POMER and J. NAVASQUILLO

The current distribution within a layered semiconductor sample when a potential difference is applied to two contacts deposited on opposite faces is obtained from the numerical solution for the potential. From this solution a method for determining the resistivity of the sample perpendicular to the layers is derived, when the resistivity along the layers is known.

Aus der numerischen Lösung für das Potential wird die Stromverteilung in einer geschichteten Halbleiterprobe bestimmt, an der an zwei auf entgegengesetzten Flächen angebrachten Kontakten eine Potentialdifferenz gelegt wird. Aus dieser Lösung wird eine Methode zur Bestimmung der Leitfähigkeit der Probe senkrecht zu den Schichten abgeleitet, wenn der Widerstand in Schichtrichtung bekannt ist.

1. Introduction

Layered semiconductors have been widely studied in the last few years and a theoretical background accounting for their general properties has been established [1]. The transport properties of these semiconductors for conduction along the layers have been analyzed in many instances [2 to 9] but their transport properties for conduction across the layers have received less attention [10, 11]. This is not casual because, as the samples have to be very thin, reliable measurements of resistivity $\varrho_{||}$ and mobility $\mu_{||}$ when transport takes place across the layers are difficult to carry through. In [11] a four-contact method for measuring $\varrho_{||}$ has been described, which is based in two assumptions: first, the confidence that the potential difference measured between the corresponding contacts did not differ very much from the potential difference between the current contacts, and second, that the current density is uniform.

In this work we shall solve numerically the potential problem corresponding to conduction across the layers of a layered semiconductor for different ratios of its two conductivities and for several values of the sample side length and of the contacts. All this information would be properly plotted in order to give a method for determining the resistivity $\varrho_{||}$ of the material when the resistivity parallel to the layers $\varrho_{\perp}$ is known by an independent measurement.

2. The Current Distribution

Let us consider a parallelepiped of a layered semiconductor with dimensions a, b, c. The side of length b is perpendicular to the layers and two metallic contacts of dimensions d, c ($d < a$) are deposited on the two faces normal to b. Taking cartesian co-

---

[1]) E-46100 Burjassot (València), Spain.

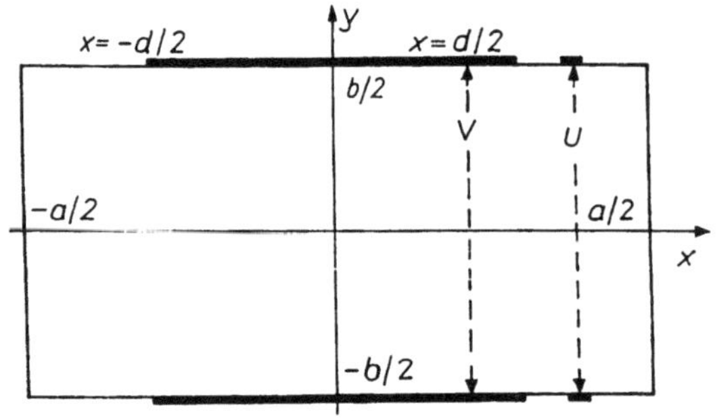

Fig. 1. Geometry of the sample and the contacts. V is the applied potential difference. U the measured potential difference

ordinates and axes x, y, z parallel to the sides a, b, c, respectively, the current distribution depends only on x, y. The electric potential Φ satisfies the equation

$$r \frac{\partial^2 \Phi}{\partial x^2} + \frac{\partial^2 \Phi}{\partial y^2} = 0 , \tag{1}$$

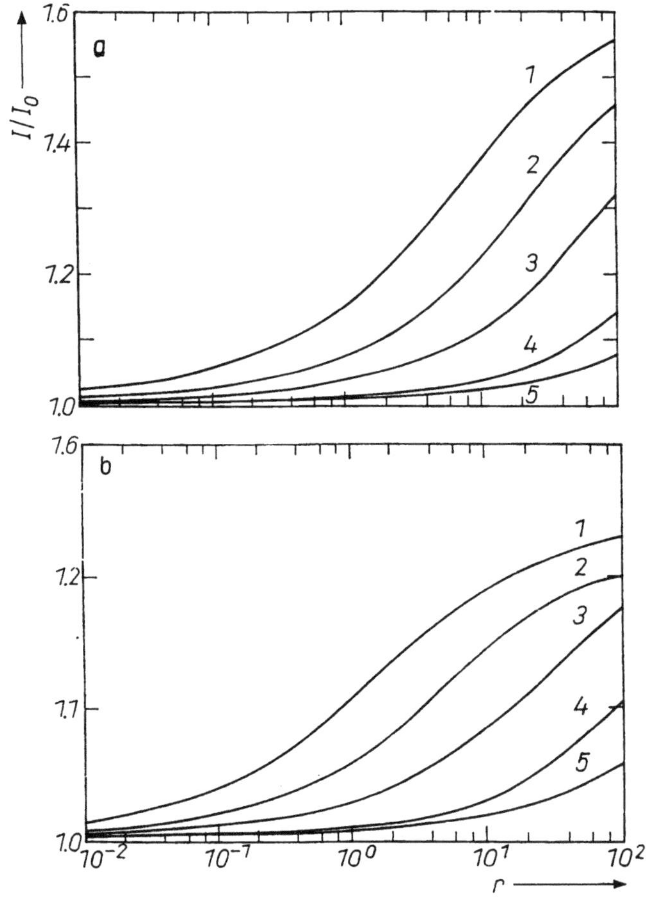

Fig. 2 Normalized current vs. resistivity ratio for a) $d/a = 0.6$ and b) 0.8, and for (1) $a/b = 5$, (2) 10, (3) 20,(4). 50, (5) 100

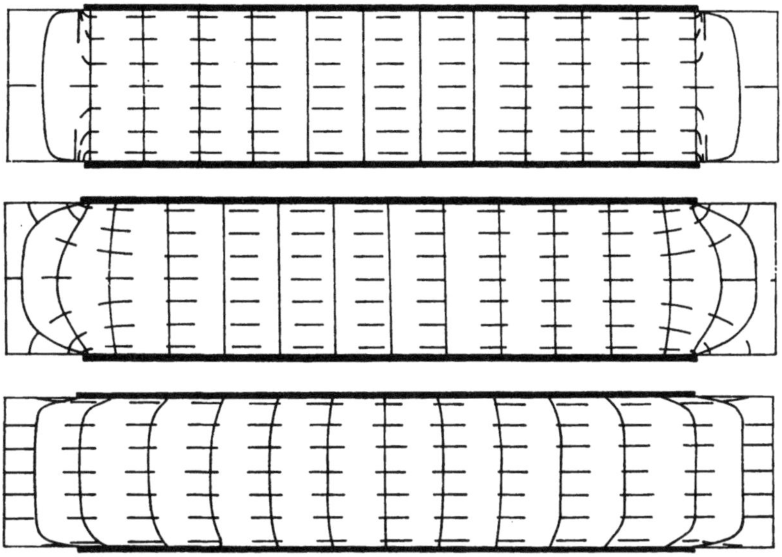

Fig. 3. Equipotential lines (dashed lines) and stream lines (solid lines) for $d/a = 0.8$ and a) $r = 0.01$ b) 1, and c) 100. The equipotentials correspond to the values 0.05, 0.2, 0.35, 0.5, 0.45, 0.8, 0.95 V. The current lines have been chosen arbitrarily

where $r = \varrho_{\|}/\varrho_{\perp}$ is the ratio of the resistivities parallel and perpendicular to the y-axis, and Φ is subjected to the boundary conditions (see Fig. 1)

$$\left.\begin{aligned}\Phi(x, -b/2) &= -V/2\\ \Phi(x, \ \ b/2) &= \ \ \ V/2\end{aligned}\right\} \quad \text{if} \quad |x| < d/2,$$

$$\frac{\partial \Phi}{\partial y} = 0 \quad \text{for} \quad y = \pm b/2 \quad \text{and} \quad d/2 < |x| < a/2\,, \tag{2}$$

$$\frac{\partial \Phi}{\partial x} = 0 \quad \text{for} \quad x = \pm a/2\,.$$

In order to find the potential, a rectangular net has been chosen on the rectangle where the solution is searched and the Gauss-Seidel method with a relaxation procedure

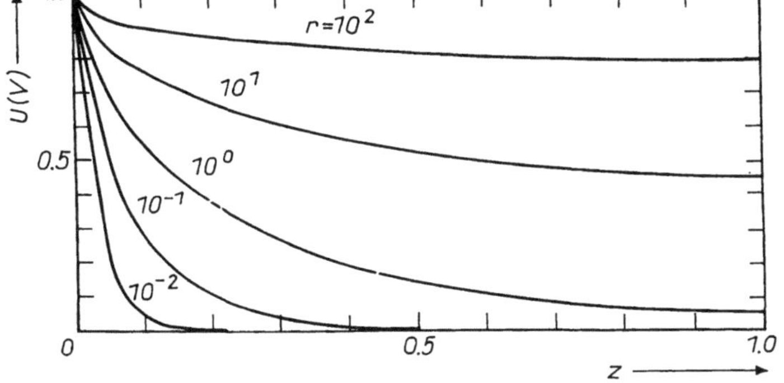

Fig. 4 Normalized potential difference as a function of the normalized distance to the contacts, z. (At the ordinate read $u(z)/\mathrm{V}$)

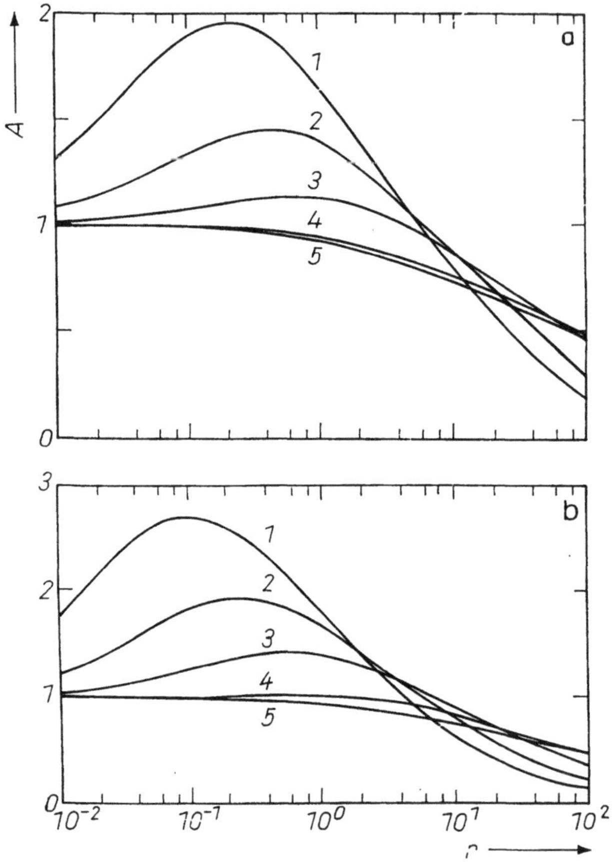

Fig. 5. Coeffizient A vs. resistivity ratio for a) $d/a = 0.6$ and b) 0.8, and for (1) $a/b = 5$, (2) 10, (3) 20, (4) 50, (5) 100

has been applied [12]. The symmetry properties of the potential have been taken into account in order to reduce the number of points where the potential is to be found. Specifically, these properties are

$$\Phi(x, y) = \Phi(-x, y) = -\Phi(x, -y) = -\Phi(-x, -y) \tag{3}$$

and allow the problem to be solved in a quadrant of the cross section of the sample. The number of iterations necessary to reach convergence depends on the geometry of the sample. We have found that after 1500 iterations a satisfactory approach to the true potential is achieved.

The potential problem has been solved for a great variety of geometrical and transport parameters, namely $a/b = 5$, 10, 100; $d/a = 0.8$, 0.6, and $r = 0.01$, 0.02, 0.05, 0.1, 0.2, 0.5, 1, 2, 5, 10, 20, 50, 100. Only two values of the length of the contacts have been taken because it is expected that the greater the ratio d/a is, the better is the accuracy of the measurements. Numerical results confirm this expectation. In practice, the d/a value is limited by the need of leaving space enough for the potential contacts.

In Fig. 2 the ratio I/I_0, where I is the measured current and $I_0 = \sigma_{\parallel} dcV/b$, has been plotted versus r, for several a/b values and for $d/a = 0.6$ and 0.8.

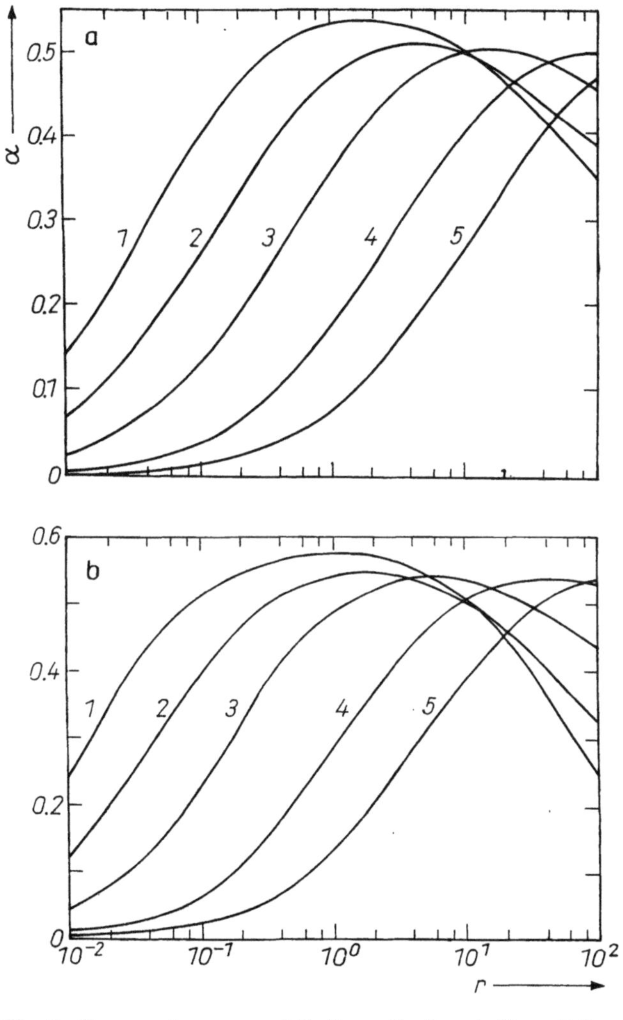

Fig. 6. Exponent α vs. resistivity ratio for a) $d/a = 0.6$ and b) 0.8, and for (1) $a/b = 5$, (2) 10, (3) 20, (3) 50, (5) 100

A first insight into the influence of the anisotropy on the current distribution can be obtained through the equipotential and stream lines which are shown in Fig 3 for three r values. In order to get $\varrho_{||}$ we also need to know the behaviour of the potential difference U between two opposite points on the surfaces $y = \pm b/2$ as a function of their distance to the current contacts. This is shown in Fig 4 for the case $a/b = 5$ and $d/a = 0.8$. For fixed a/b U drops more quickly when $d/a = 0.6$ than it does for fixed d/a when a/b increases.

Moreover, for fixed a/b and d/a, U decreases very slowly for low r values, and very quickly for large r values. The function $U(x)$ must behave near the contacts like $V - Az^{\alpha}$ [13] with $bz = x - d/2$ and A and α real constants depending on a/b, d/a, and r. This behaviour is in fact observed in our results which also show that for points not close to the contacts $U(x)$ has a different form. For this reason we have tried to

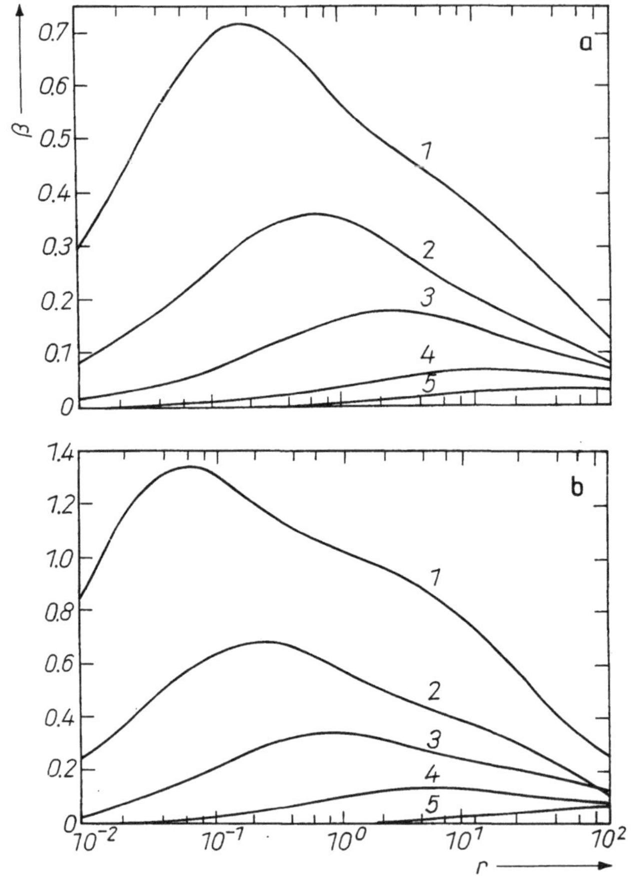

Fig. 7. Exponential factor β vs. resistivity ratio for a) $d/a = 0.6$ and b) 0.8, and for (1) $a/b = 5$, (2) 10, (3) 20, (4) 50, (5) 100

approximate the relative potential difference by a function of the type

$$\frac{U(x)}{V} = 1 - A z^{\alpha} e^{-\beta z} \tag{4}$$

fitting the parameters A, α, and β by the least-squares method. The results are very satisfactory and the values of A, α, and β have been plotted versus r for several values of a/b and for $d/a = 0.8$ and 0.6 (Fig. 5, 6, 7).

3. Discussion

The geometrical features of the field lines and of the stream lines are comprised between those corresponding to the two extreme cases $\sigma_{\parallel} = 0$, $\sigma_{\perp} \neq 0$ and $\sigma_{\parallel} \neq 0$, $\sigma_{\perp} = 0$. In the first case no current flows between the electrodes and, inside the sample, the electric field has only a y-component. This condition is accomplished because of the surface charge density that appears on the free regions of the surfaces of the sample where the contacts are made. The potential difference for any two points on each of these surfaces is $U = V$. On the other hand, when $\sigma_{\|} \neq 0$ and $\sigma_{\perp} = 0$ the current density has only a y-component and the electric field has a constant y-component which gives rise to the potential difference V, and an x-component that takes opposite values on the upper and lower halves of the medium. This component only produces a fall of the potential difference from V to $U(x)$ that depends on the geometry of the sample and the contacts. When measurements are made on a material with $\sigma_{\|} \ll \sigma_{\perp}$ $(r \gg 1)$ U will be very close to V, but I will be greater than I_0 because there are stream lines out of the volume comprised between the contacts and the zone where the current density is uniform has an effective width greater than d. Thus the simple relation $I = \sigma_{\|} dcV/b$ cannot be applied for determining $\sigma_{\|}$. For a material with $\sigma_{\|} \gg \sigma_{\perp}$ $(r \ll 1)$ I/I_0 is near unity but U departs from V even for points close to the contacts. This effect is especially marked for very thin samples.

Our results can be used in order to get a good accuracy in $\varrho_{\|}$ measurements when $\varrho_{\perp}$ is known. The procedure could be the following: the measured values I and U are to be taken as the zero-order approximation for I_0 and V, respectively. With them and the known $\varrho_{\perp}$, the zero-order approximation r_0 for r is obtained: $r_0 = dcU/bI\varrho_{\perp}$. This r_0 value gives, by means of the curves of Fig 2 that correspond to the geometry of the sample, a more accurate value I_{01} for I_0 and, by means of the curves of Fig. 5, 6, 7, the parameters A, α, and β that allow the determination of a first-order approximation V_1 for V. Now the procedure can be repeated until sufficient accuracy is reached. When the lengths a, b, and d are not in the ratios for which the curves have been plotted, a linear interpolation is needed.

4. Conclusions

The equation for the electric potential inside a slab of a layered semiconductor with two current contacts covering partially two opposite faces, has been numerically solved for a large set of the parameters entering in the problem, namely the ratio of the sides, the ratio of the length of the contacts and one of the sides and the ratio of the two resistivities of the material. With a proper normalization the electric current and the potential difference between two opposite points on the surfaces in contact with the electrodes have been computed and plotted as a function of the resistivity ratio for all the cases. These curves show that when $r \gg 1$ the potential difference for points near the contacts is close to the potential difference between these ones but the effective width of the contacts is greater than their actual width. Conversely, when $r \ll 1$ the current density is almost uniform between the electrodes but the potential difference falls off very rapidly when the potential points separate the contacts. In any case, the transverse resistivity $\varrho_{\perp}$ has to be known in order to find the parallel resistivity $\varrho_{\|}$. A procedure for obtaining $\varrho_{\|}$ from $\varrho_{\perp}$ and the curves that summarize our results has also been specified.

When these conclusions are applied to the measurements reported in [11], it is found that the resistivity ratios given there are lower than the actual ones. For instance, for a typical sample of $5 \times 5 \times 0.5$ mm^3 with 4×4 mm^2 contacts and a distance of 0.1 mm, between the potential points and the contacts, ratios $r = 10$, 100 should be changed to 17 and 140, respectively. The general features of the behaviour of r versus T^{-1} remain valid as also does the interpretation of the experimental results.

References

[1] R. FIVAZ and PH. SCHMID, Optical and Electrical Properties of Materials with Layered Structures, Reidel, Dordrecht (The Netherlands) 1976.
[2] R. FIVAZ and E. MOOSER, Phys. Rev. **163**, 743 (1967).
[3] A. H. KIPPERMAN and C. J. VERMIG, Nuovo Cimento **63B**, 29 (1969).
[4] PH. SCHMID and J. P. VOITCHOVSKY, phys. stat. sol. (b) **65**, 249 (1974).
[5] V. ANGELLI, C. MANFREDOTTI, R. MURRI, and L. VASENELLI, Phys. Rev. B **17**, 3221 (1978).
[6] S. SHIGETOMI, T. IKARI, Y. KOGA, and S. SHIGETOMI, Japan. J. appl. Phys. **20**, L343 (1981).
[7] C. DE BLASI, G. MICOCCI, A. RIZZO, and A. TEPORE, Phys. Rev. B **27**, 2429 (1983).
[8] A. SEGURA, F. POMER, A. CANTARERO, W. KRAUSE, and A. CHEVY, Phys. Rev. B **29**, 5708 (1984).
[9] R. CINGOLANI, L. VASANELLI, and A. RIZZO, Nuovo Cimento **6D**, 383 (1985).
[10] S. M. ATAKISHIEV and G. A. AKHUNDOV, phys. stat. sol. **32**, K33 (1969).
[11] F. POMER, X. BONET, and A. SEGURA, to be published.
[12] J. P. NOUGIER, Méthodes de calcul numérique, Masson, Paris 1985.
[13] J. D. JACKSON, Classical Electrodynamics, Wiley, New York 1975 (p. 77).

(Received March 22, 1988; in revised form September 19, 1988)

phys. stat. sol. (a) **110**, 593 (1988)

Subject classification: 72.20; S12.1

Institute of Molecular Physics, Polish Academy of Sciences, Poznań[1]) (a), Central Research Institute for Physics, Hungarian Academy of Sciences, Budapest[2]) (b), and Centre de Recherche Paul Pascal, Domaine Universitaire de Bordeaux I, Talence[3]) (c)

Mircowave Conductivity and Thermoelectric Power of Some Ternary Salts of TCNQ and Iodine

By

R. Świetlik (a), G. Mihàly (b), and P. Delhaès (c)

For a series of ternary compounds made with TCNQ, iodine, and quaternary ammonium the complex dielectric constant at microwave frequency (9.1 GHz) and the thermoelectric power are measured in the temperature range 80 to 300 K. Above a "Peierls-like" transition the thermoelectric power results are discussed in terms of an intrinsic semiconductor with Hubbard gap. For interpretation of the microwave results in the high-temperature phase an interrupted strand model can be applied, whereas in the low-temperature phase the localized electron concept is important.

Pour une série des sels ternaires composés du TCNQ, d'iode et d'ammonium quaternaire, on a étudié, en fonction de la temperature (80 à 300 K), le pouvoir thermoélectrique et la constante diélectrique à une fréquence microonde (9,1 GHz). On a associé le comportement du pouvoir thermoélectrique au-dessus de la transition de "type de Peierls" à celui d'un semiconducteur presentant un gap de Hubbard. Pour interpréter les résultats microondes dans la phase haute-temperature on peut appliquer le modele, de segments coupés alors que dans la phase basse-temperature l'influence de la localization des électrons peut être importante.

1. Introduction

The various substituted ammonium cations form an isomorphous series of ternary compounds with tetracyanoquinodimethane (TCNQ) and iodine. About ten compounds with general formula $M^+(TCNQ)^{2/3-}\ (I_3^-)_{1/3}$ where M^+ is a monocation and $M^{2+}(TCNQ)_2^{2/3-}\ (I_3^-)_{2/3}$ where M^{2+} is dication were synthetized [1 to 4]. The stoichiometry of these compounds is such that each TCNQ carries a mean charge equal to 2/3 of an electron, so the band is 1/3 filled in the electron independent approximation. The interest in this group of materials results mainly from the fact that for different crystals the conducting TCNQ chains appear to be regular and have nearly the same stacking, but the interchain distance is different. The interchain distance increases with the cation size without any essential change of the TCNQ chain geometry. Therefore, this series gives unique opportunity to study the effect of interstack interactions on the physical properties of quasi-one-dimensional conductors. In particular, the influence of Coulomb interactions between stacks on their transport properties and on the "metal–insulator" transition temperature was established [3, 4].

The most intensive investigations have been performed on the basic salt of the series TMA–TCNQ–I (TMA-trimethylammonium): crystal structure, electric transport, magnetic susceptibility, EPR, infrared and Raman spectra, specific heat, elastic

[1]) Smoluchowskiegeo 17/19, PL-60-179 Poznań, Poland.

[2]) P.O.B. 49, H-1525 Budapest, Hungary.

[3]) F-33405 Talence Cédex, France.

properties were studied [3 to 22]. In TMA–TCNQ–I crystals the electrical conductivity $\sigma(T)$ measured along the stacking axis (b-axis) is around 20 Ω^{-1} cm^{-1} at room temperature and exhibits "metal-like" behaviour with an anomaly at $T \approx 150$ K [3, 4, 13 to 15]. The nature of the "metal-like" state above 150 K and especially the phase transition at 150 K were a matter of controversy. The transition at 150 K was interpreted as a metal–insulator transition associated with Peierls-like lattice distortion [2 to 4] or a semiconductor–semiconductor transition [12 to 14, 16, 17, 22]. The intensive structural studies [7 to 9, 11] have shown that in TMA–TCNQ–I crystals below 150 K a modulated structure exists with the periodicity 3b (due to the 1/3 filled band) and above 150 K precursor effects in agreement with the Peierls mechanism. However, in an interpretation one should take into account the particular nature of this compound; i.e. the coupling between fluctuations of the order parameter of the organic chain distortion and iodine lattice fluctuations. Therefore, this transition can be classified as a "generalized" Peierls transition [4, 14].

Apart from TMA–TCNQ–I also the crystals IPrDMA–TCNQ–I, IPentDMA–TCNQ–I and TMHDA–$TCNQ_2$–I_2 were intensively studied [3, 4, 15] (IPrDMA — isopropyldimethylammonium, IPentDMA — isopentyldimethylammonium, TMHDA — tetramethylhexamethylenediammonium). The physical properties of these salts are very similar to the basic salt of the series TMA–TCNQ–I. As the size of the ammonium cation increases, the temperature of the "metal–insulator" transition decreases to 139, 138, and 118 K for the series IPrDMA–TCNQ–I, IPentDMA–TCNQ–I, and TMHDA–$TCNQ_2$–I_2, respectively. For these crystals the measurements of electrical conductivity, static magnetic susceptibility, EPR spectra, and specific heat have been made and compared with analogous results for TMA–TCNQ–I, but no investigations of thermoelectric power and microwave conductivity were performed [3, 4, 15].

In TMA–TCNQ–I crystals a long-range order was found in the iodine arrays organized in I_3^- chains but it is also possible to synthetize crystals with disordered I_3^- chains [4, 15]. The "ordered" TMA–TCNQ–I crystals exhibit "metal-like" properties above 150 K and "disordered" crystals do not exhibit such a behaviour, but the phase transition is observed in both kinds of crystals. For IPrDMA–TCNQ–I, IPentDMA–TCNQ–I, and TMHDA–$TCNQ_2$–I_2 only "ordered" crystals were synthetized.

In this paper, we report on microwave measurements of the complex dielectric constant of IPrDMA–TCNQ–I and IPentDMA–TCNQ–I crystals in the temperature range 80 to 300 K. Moreover, we report measurements of the Seebeck coefficient of TMA–TCNQ–I, IPrDMA–TCNQ–I, IPentDMA–TCNQ–I, and TMHDA–$TCNQ_2$–I_2 in the same temperature range.

2. Experimental and Results

Single crystals of the studied salts were prepared by TCNQ oxidation of quaternary ammonium iodides as described earlier [1, 3, 4].

The measurements of the dielectric constant and microwave conductivity were carried out for IPrDMA–TCNQ–I and IPentDMA–TCNQ–I crystals at a frequency of 9.1 GHz in the temperature range 80 to 300 K. Typical dimensions of the crystals were ($3 \times 0.2 \times 0.1$ mm^3) with the long axis being the stacking axis. The cavity perturbation technique was used [23]. The needle-shaped crystal was placed at the maximum of electric field of a strongly undercoupled high quality resonant cavity. Relative changes of the resonance frequency δ and the absorption width Δ when the sample was inside and outside the cavity was measured. When the skin depth was larger than the thickness of the sample (low conductivity regime) the real and imaginary parts of the microwave conductivity were determined using cavity perturbation

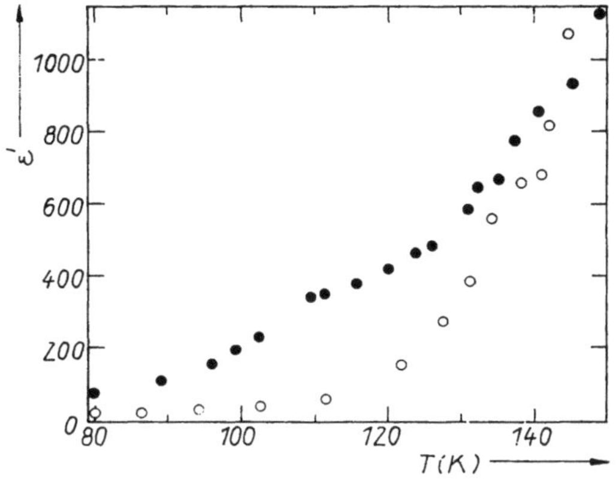

Fig. 1. Real part of the dielectric constant as a function of temperature for IPrDMA-TCNQ-I (○) and IPentDMA-TCNQ-I (●) salts

formulas [23]. When the skin depth was smaller than the maximum sample dimension perpendicular to the electric field (high conductivity regime) only the real part of the conductivity could be determined [24]. At low temperature both IPrDMA-TCNQ-I and IPent-DMA-TCNQ-I crystals were in the low conductivity regime, therefore, it was possible to determine the real part of the dielectric constant ε' and the microwave conductivity $\sigma(\omega)$ (Fig. 1 and 2). Above about 150 K the crystals were in the high conductivity regime and $\sigma(\omega)$ was measured only (Fig. 2).

The thermoelectric power was measured on single crystals along the stacking direction using the technique of alternating temperature gradient. The temperature difference between the electrodes was maintained below 0.5 K. Small, needle-shaped crystals were selected for the measurement. The reproducibility of the results between different samples of the same compound at room temperature was usually better than about 10% and even better at lower temperatures. The Seebeck coefficient $S(T)$ as a function of the temperature for different salts is shown in Fig. 3. For all compounds we observed similar temperature dependence of the Seebeck coefficient in the high-temperature range and similar influence of the phase transition. $S(T)$ above the phase transition temperature is well fitted by the classical equation for semiconductors,

$$S = -(k_B/e)\left[-(A/T) + E\right], \tag{1}$$

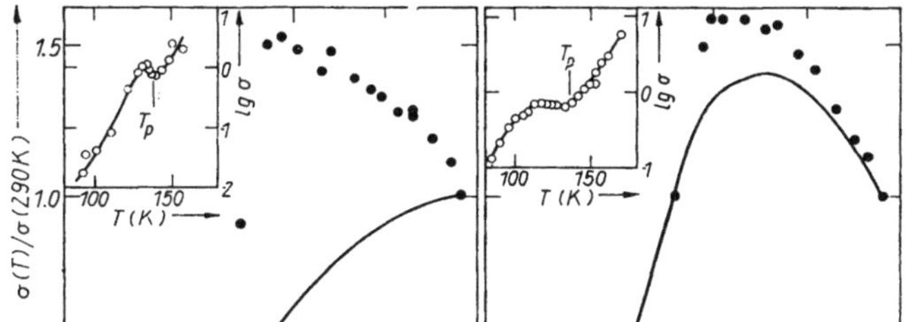

Fig. 2. Microwave conductivity (●) in comparison with dc conductivity (solid line) from [15] for a) IPrDMA-TCNQ-I and b) IPentDMA-TCNQ-I. The inset shows the microwave conductivity in the vicinity of the phase transition

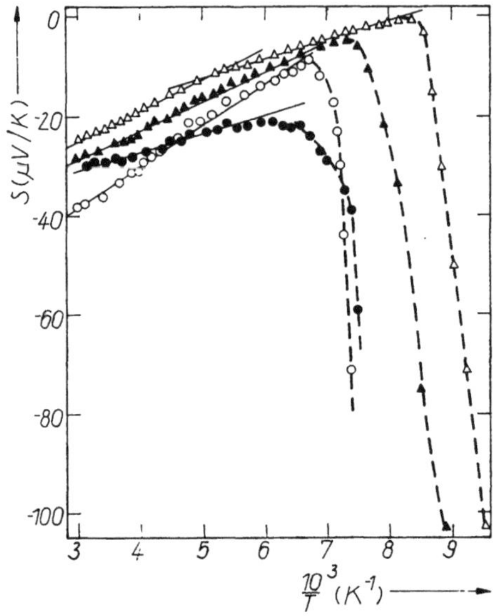

Fig. 3. Thermoelectric power vs. T^{-1} for different salts. The solid lines are fits of (1) to the data points using the parameters of Table 1. Dashed lines are drawn as guides for the eye only (△ TMHDA-TCNQ$_2$-I$_2$, ▲ IPrDMA-TCNQ-I, ● IPentDMA-TCNQ-I, ○ TMA-TCNQ-I)

for TMA-TCNQ-I, IPrDMA-TCNQ-I, and IPentDMA-TCNQ-I. The results for TMHDA-TCNQ$_2$-I$_2$ were also fitted by (1) but two temperature ranges were distinguished because of the change of the slope at about 190 K. The fitted curves are shown by the solid lines in Fig. 3 and the parameters of the fit of (1) to the data are summarized in Table 1.

3. Discussion

At 80 K the microwave dielectric constant ε' for IPrDMA-TCNQ-I and IPentDMA-TCNQ-I is about ≈ 30 and ≈ 80, respectively, and increases with temperature (Fig. 1). Our preliminary results for TMHDA-TCNQ$_2$-I$_2$ showed that $\varepsilon' \approx 110$ at 80 K and exhibits an analogous temperature dependence. These results can be compared with ε' for TMA-TCNQ-I salt; in [13] $\varepsilon' \approx 40$ at $T < 60$ K but other studies showed that $\varepsilon' \approx 10$ at 100 K [5, 6].

Although at the phase transition temperature T_p ε' does not exhibit an abrupt change (Fig. 1), one cannot claim that the phase transition has no influence on the dielectric constant. ε' was measured in a small temperature range above T_p, where

Table 1

Thermoelectric power parameters of (1) and temperatures of the phase transition

salt	A (K)	E	T_p (K) from TEP	T_p (K) from σ_{dc} ([15])
TMA-TCNQ-I	96	0.73	149	150
IPrDMA-TCNQ-I	69	0.54	138	139
IPentDMA-TCNQ-I	45	0.49	140–150	136
TMHDA-TCNQ$_2$-I$_2$	65*), 46**)	0.47*), 0.37**)	117	118

*) Temperature range 190 to 300 K.
**) Temperature range 125 to 190 K.

the pretransitional effects can be important. On the other hand, the accuracy of estimation of ε' and microwave conductivity decreases just above T_p because the area is approached where the skin depth becomes comparable with the thickness of the sample and the experiment becomes insensitive to the real part of the dielectric constant (change of the calculation regime). A similar remark can be made on the discrepancy between dc and microwave conductivity seen in Fig. 2 below about 160 K.

The microwave conductivity at room temperature was similar to the dc conductivity and equaled to about $20\,\Omega^{-1}\text{cm}^{-1}$ for both IPrDMA–TCNQ–I and IPentDMA–TCNQ–I. In contrast to IPentDMA the dc conductivity for IPrDMA does not exhibit a "metal-like" temperature dependence [15], but the microwave conductivities are qualitatively similar in both salts (Fig. 2); this proves the similarity of both salts from the microscopic point of view. The different behaviour of IPrDMA–TCNQ–I crystals could be related to the crystal quality which is of great importance for the dc experiment. A small increase of microwave conductivity below the phase transition temperature can be related to pretransitional effects (insets of Fig. 2 a and b).

In quasi-one-dimensional organic systems with large Coulomb interactions the observed dielectric properties can be explained in terms of the localized electron model [25]. In this model the ground state at absolute zero corresponds to a regular distribution of electrons along the sites of the linear chain. The electric field mixes the ground state and an unoccupied state localized on a neighbouring site in the chain thus creating a dipole moment. As is also discussed below, in the case of ternary salts the Coulomb interactions are important and one can assume the localized electron picture, therefore, this approach could be used in the interpretation of ε' in our salts.

At high temperatures (above about 180 K) the discrepancy between dc and microwave conductivity is in agreement with the interrupted strand model. In this model ε' depends on the effect of the perpendicular conductivity $\sigma_\perp$ between the chains; i.e. when $\sigma_\perp$ decreases, ε' increases [26]. The experimental evidence for ternary salts shows that ε' at low temperatures increases when the interaction between strands decreases in agreement with the interrupted strand model.

The thermoelectric power $S(T)$ is negative in the whole temperature range and exhibits similar temperature dependence for every compound (Fig. 3). Above the phase transition temperature a semiconducting behaviour is observed. The fundamental point is about the origin of the gap, which in our opinion is a Hubbard gap. Below the phase transition temperature the absolute value of $S(T)$ increases abruptly, also in agreement with the semiconducting behaviour. From the temperature dependence of the Seebeck coefficient it is possible to determine the phase transition temperature T_p. The T_p determined in this way is in good agreement with the previous determinations except for IPentDMA–TCNQ–I. For the IPentDMA salt the phase transition is smeared out, therefore, it is difficult to estimate T_p more accurately; the smearing can be due to defects or impurities. The intrachain interactions are almost constant in the whole series and the interchain interactions decrease with the size of the cation. From Table 1 we can conclude that the parameters A and E in (1) decrease when the interchain interactions decrease.

In [14] the Seebeck coefficient data for the TMA–TCNQ–I crystal above 150 K were interpreted in terms of an intrinsic semiconductor with a one-dimensional tight-binding band and strong Coulomb interactions. The electronic correlations were not only introduced for the interpretation of $S(T)$. The static magnetic susceptibility of these salts tends to show that there is a sizeable electron–electron repulsion [4]. As the electronic correlations are important and $S(T)$ for every compound is proportional to T^{-1}, we assume that the semiconductor concept is applicable to the studied salts above the phase transition temperature.

For an intrinsic semiconductor with strong Coulomb interactions the thermoelectric power is described by (1), where $A = [(b - 1)/(b + 1)]\ (\Delta/k_B)$, Δ is the energy gap, and $b = \mu_e/\mu_h$ the ratio of electron to hole mobilities [14]. The constant E is related to the electron and hole masses, the heats of carrier transport and (when the strong one-site Coulomb repulsion should be taken into account) a constant contribution related to Coulomb correlations [$S \approx -60\ \mu V/K = -(k_B/e) \ln 2$]. In [14] a mobility ratio of electrons to holes in TMA–TCNQ–I of $b \approx 0.8$ was calculated from the temperature dependent term of (1), taking into account an energy gap of $\Delta = 0.14$ eV estimated in [16] from optical reflectivity measurements. This calculation cannot be performed for other salts because the energy gap is unknown.

In [14] $S(T)$ was only studied for "disordered" crystals as results from the electric conductivity $\sigma(T)$. On the other hand, in [13] the $\sigma(T)$ was "metal-like" in the high-temperature range ("ordered" crystals) and $S(T)$ was also proportional to T^{-1} as expected for semiconductor. In [14] the $S(T)$ above $T = 150$ K was fitted by (1) with parameters: $A = 110$ K and $E = 0.67$, whereas our parameters are $A = 96$ K and $E = 0.73$ (Table 1). This difference can be related to the fact that we investigated "ordered" crystals.

Taking into account the strong Coulomb correlations we can assume that the gap observed above the phase transition is a Hubbard gap, observed in TMA–TCNQ–I in an optical experiment [16]. Below the phase transition there is a periodic lattice distortion due to a Peierls-type transition which opens a gap at the Fermi level. The resulting effect is a Peierls-Hubbard gap. It has been shown that in the limit of infinitely strong intra-atomic Coulomb repulsion ($U \to \infty$) the Hubbard chain can distort to become a Peierls band insulator [27]. The wavelength of the Peierls distortion for strong one-site repulsion ($U \to \infty$) is different from that obtained for free electrons ($U = 0$). However, in the case of one-third band filling (as in our case) the periodicities of distortion are the same in both cases, therefore, it is difficult to distinguish between $U = 0$ or $U \to \infty$ limits.

In the ternary salts of TCNQ and iodine the I_3^- anions interact strongly with TCNQ chains thus inducing an external periodic potential. As was shown in [28], the external periodic potential smears out the Peierls transition around the transition temperature T_p and both the energy gap and the periodic lattice distortion are not zero above T_p (they vary slowly towards limiting values at high temperature). Therefore, the Hubbard gap above T_p in the studied salts can be perturbed by the electrostatic interactions of I_3^- anions.

Acknowledgements

One of us (R.S.) is very grateful to Prof. A. Graja for the critical reading of the manuscript and useful discussion. The work in IMP PAS was supported by the Project CPBP-01.14.

References

[1] P. Dupuis, S. Flandrois, P. Delhaès, and C. Coulon, J. Chem. Soc. Chem. Commun. 8, 328 (1978).

[2] S. Flandrois, C. Coulon, P. Delhaès, and P. Dupuis, in: Molecular Metals, Ed. W. E. Hatfield, Plenum-Press, New York 1979 (p. 67).

[3] C. Coulon, S. Flandrois, P. Delhaès, C. Hauw, and P. Dupuis, Phys. Rev. B 23, 2850 (1981).

[4] C. Coulon, Ph. D. Thesis, Bordeaux University 1982.

[5] A. Cougrand, S. Flandrois, P. Delhaès, P. Dupuis, D. Chasseau, J. Gaultier, and J. L. Miane, Mol. Cryst. liquid Cryst. 32, 165 (1976).

[6] A. FILHOL, M. ROVIRA, C. HAUW, J. GAULTIER, D. CHASSEAU, and P. DUPUIS, Acta Cryst. **B35**, 1652 (1979).
[7] A. FILHOL and J. GAULTIER, Acta Cryst. **B36**, 592 (1980).
[8] A. FILHOL, B. GALLOIS, J. LAUGIER, P. DUPUIS, and C. COULON, Mol. Cryst. liquid Cryst. 84, 17 (1982).
[9] T. GRANIER and R. AYROLES, C.R. Acad. Sci. (France) **294**, 303 (1982).
[10] P. COPPENS, P. LEUNG, K. E. MURPHY, P. R. van TILBORG, A. J. EPSTEIN, and J. S. MILLER, Mol. Cryst. liquid Cryst. **61**, 1 (1980).
[11] B. GALLOIS, J. GAULTIER, T. GRANIER, R. AYROLES, and A. FILHOL, Acta Cryst. **B41**, 56 (1985).
[12] M. A. ABKOWITZ, A. J. EPSTEIN, C. H. GRIFFITHS, J. S. MILLER, and M. L. SLADE, J. Amer. Chem. Soc. **99**, 5304 (1977).
[13] M. A. ABKOWITZ, J. W. BRILL, P. M. CHAIKIN, A. J. EPSTEIN, M. F. FROIX, C. H. GRIFFITHS, W. GUNNING, A. J. HEEGER, W. A. LITTLE, J. S. MILLER, M. NOVATNY, D. B. TANNER, and M. L. SLADE, Ann. New York Acad. Sci. **313**, 459 (1978).
[14] T. IKARI, S. JANDL, M. AUBIN, and K. D. TRUONG, Phys. Rev. B **28**, 3859 (1983).
[15] C. COULON, P. DELHAÈS, S. FLANDROIS, J. AMIELL, E. BONJOUR, and P. DUPUIS, J. Physique **46**, 783 (1985).
[16] D. B. TANNER, J. E. DEIS, A. J. EPSTEIN, and J. S. MILLER, Solid State Commun. **31**, 671 (1979).
[17] A. GRAJA, E. SOPA, and L. DOBJÁŠOVÁ, phys. stat. sol. (a) **73**, K237 (1982).
[18] A. GRAJA and S. FLANDROIS, Acta phys. Polon. **A61**, 347 (1982).
[19] J. E. ELDRIDGE, N. FORTIER, F. E. BATES, and J. S. MILLER, Canad. J. Phys. **61**, 1271 (1983).
[20] A. D. BANDRAUK, K. D. TRUONG, C. CARLONE, and S. JANDL, Chem. Phys. Letters **95**, 78 (1983).
[21] A. D. BANDRAUK, K. D. TRUONG, C. CARLONE, and S. JANDL, J. Physique **44**, C3–1473 (1983).
[22] J. W. BRILL, A. J. EPSTEIN, and J. S. MILLER, Phys. Rev. B **20**, 681 (1979).
[23] L. I. BURAVOV and I. F. SHCHEGOLEV, Prib. i Tekh. Eksper. **4**, 171 (1971).
[24] M. COHEN, S. K. KHANNA, W. J. GUNNING, A. F. GARITO, and A. J. HEEGER, Solid State Commun. **17**, 367 (1977).
[25] I. F. SHCHEGOLEV, phys. stat. sol. (a) **12**, 9 (1972).
[26] M. WEGER and M. KAVEH, J. Phys. C **12**, 2567 (1979).
[27] J. BERNASCONI, M. J. RICE, W. R. SCHNEIDER, and S. STRÄSSLER, Phys. Rev. B **12**, 1090 (1975).
[28] L. K. HANSEN and K. CARNEIRO, Solid State Commun. **49**, 531 (1983).

(Received July 19, 1988)

phys. stat. sol. (a) **110**, 601 (1988)

Subject classification: 75.40; S1.4

Institute for Low Temperature and Structure Research, Polish Academy of Sciences[1]) *(a),*
International Laboratory of High Magnetic Fields and Low Temperatures[2]) *(b), Wrocław, and II. Physikalisches Institut der Universität zu Köln*[3]) *(c)*

Thermal Properties of $NdIn_3$

By

A. CZOPNIK (a), H. MÄDGE (b), R. POTT[4]) (c), and B. STALIŃSKI (a)

The specific heat of $NdIn_3$ in the temperature range 2.3 to 27 K and in the magnetic field up to 5 T as well as the thermal expansion between 1.5 K and room temperature are measured. The low-temperature behaviour of $NdIn_3$ is governed by the energy gap in the spin wave spectrum. The specific heat at the antiferromagnetic transition reveals anomalously large critical exponents. It is shown that the specific heat anomaly at T_N contains contributions from two closely lying transitions. The influence of the magnetic field on the energy gap, on the first-order transitions found in the antiferromagnetic state, and on the Néel peak is studied. Thermal properties in the paramagnetic state are analyzed basing on the available CF data.

An $NdIn_3$ werden im Temperaturbereich 2,3 bis 27 K und im Magnetfeldern bis 5 T die spezifische Wärme und im Bereich zwischen 1,5 K und Raumtemperatur die thermische Ausdehnung gemessen. Das Tieftemperaturverhalten wird von einer Energielücke im Spinwellenspektrum bestimmt. Die spezifische Wärme wird in der Nähe der antiferromagnetischen Ordnung durch ungewöhnlich hohe kritische Exponenten beschrieben. Es wird gezeigt, daß die Anomalie in der spezifischen Wärme bei T_N aus zwei nah beieinander liegenden Übergängen besteht. Untersucht wird der Einfluß des Magnetfeldes auf die Energielücke, die beiden Übergänge 1. Art im antiferromagnetischen Ordnungszustand und auf den Néel-Peak. Die thermischen Eigenschaften im paramagnetischen Zustand werden auf der Grundlage der zur Verfügung stehenden Kristallfeldparametern analysiert.

1. Introduction

The physical properties of $NdIn_3$ (a cubic $AuCu_3$-type structure) have been studied in a series of works [1 to 8]. The compound orders antiferromagnetically at $T_N = 5.90$ K through a continuous transition [6]. Due to magnetic structure changes two extra first-order transitions at $T_{c1} = 4.65$ K and $T_{c2} = 5.13$ K appear [6]. Free energy calculations in terms of bilinear exchange interaction and cubic crystal field show [9] that these two successive transitions may occur only in the case when the $\Gamma_8^{(1)}$ quadruplet is the CF ground state and the first excited level, the Γ_6 doublet, lies in a precisely limited energy region. With increasing temperature the magnetic structure transforms from the single-$\bar{k}$(collinear), through a double-$\bar{k}$(biaxial with moments along the twofold axes in the basal plane) to a triple-$\bar{k}$(quadriaxial with moments along the threefold axes) alignment. However, this prediction may not be true; near T_N such interactions as the magnetoelastic and/or quadrupolar coupling are able in principle to stabilize other magnetic structures than the calculated one.

---

[1]) Pl. Katedralny 1, PL-50-950 Wrocław, P.O. Box 937, Poland.
[2]) Próchnik Str. 95, PL-53-529 Wrocław, Poland.
[3]) D-5000 Köln 41, FRG.
[4]) Present address: Bayer AG, D-5090 Leverkusen, FRG.

By means of inelastic neutron scattering two sets of CF parameters of $NdIn_3$ have been determined [3]: $x = 0.43$ and $W = 1.06$ K or $x = 0.31$ and $W = 0.91$ K. Most likely the first set is the proper one [7 to 9], however, both of them give a nearly sixfold degenerated ground state (exact but accidental degeneracy takes place at $x_c = 0.38$ where crossing of $\Gamma_8^{(1)}$ and Γ_6 levels occurs) and an excited $\Gamma_8^{(2)}$ quadruplet well above. Such degeneracy of the CF ground state allows the cubic model of Kim et al. [10, 11] to operate. According to this model we may expect even the antiferromagnetic transition of $NdIn_3$ to be a tricritical-like one.

In the present work the results of the specific heat and thermal expansion measurements of $NdIn_3$, mainly on monocrystalline samples, are reported.

2. Experimental Details

Single crystals of $NdIn_3$ were grown from the melt with concentration of 90 at% In and 10 at% Nd. The purity of the components was 5N and 3N for indium and neodymium, respectively. The cooling rate was increased from 0.5 K/h at the beginning to 4 K/h at the end of the growing process.

Polycrystalline samples were prepared by arc melting the components on a water-cooled copper hearth in a very pure argon atmosphere.

The specific heat has been measured by using an adiabatic calorimeter [12] in magnetic fields up to 5 T.

The relative length change has been measured by a capacitive method [13]. The linear thermal expansion coefficient has been calculated by a spline fit to the relative length change [14].

We have used rectangular-shaped single crystals ($(7 \times 2.1 \times 1.5)$ mm^3, 179 mg and $(4.1 \times 2.1 \times 1.5)$ mm^3 for the specific heat and thermal expansion measurements, respectively) whose edges were oriented along quaternary axes with an accuracy better than $\pm 2°$.

The polycrystalline sample, used for specific heat measurements, had a mass of about 9.6 g.

3. Results and Discussion

3.1 Specific heat

The specific heat of the polycrystalline sample of $NdIn_3$ plotted as C/T versus T^2 versus H is shown in Fig. 1. In this narrow temperature range three transitions appear. All of them have a magnetic origin. We have already reported [6] that at $(5.90 \pm \pm 0.01)$ K the compound undergoes a continuous transition to the antiferromagnetic state whereas the transitions at $T_{c1} = 4.65$ K and $T_{c2} = 5.13$ K are first-order ones due to changes of the magnetic structure. In the latter transitions the rising slopes are much sharper than the decreasing ones; they are nearly vertical whereas the decreasing ones form soft tails. We characterize the sharpness of the transitions by the half-widths: 70 and 50 mK at T_{c1} and T_{c2}, respectively.

The discontinuity of the order-order transitions was well established by supercooling effects (Fig. 2). The entropy jumps give latent heats of about 5.4 and 3.1 J/mol for the transitions at T_{c1} and T_{c2}, respectively (the value of the latent heat of the transition at T_{c1} reported previously [6] is not correct).

A magnetic field diminishes the height of the anomalies and shifts them to lower temperatures; in a field of 5 T, the transition temperatures T_{c1}, T_{c2}, and T_N are equal to 4.00, 4.47, and 5.40 K, respectively. The first-order anomalies become higher

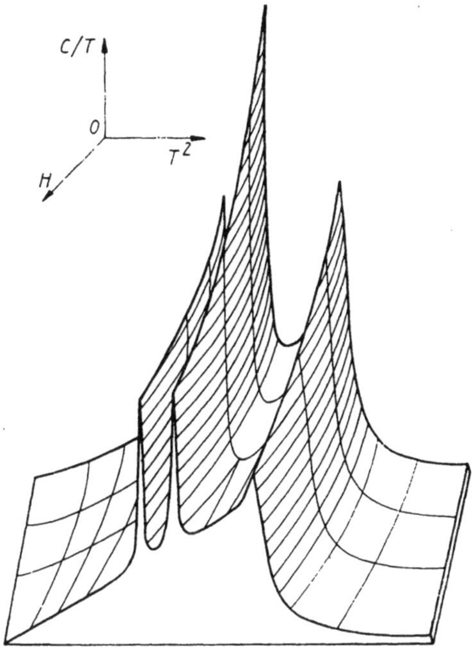

Fig. 1. Temperature and magnetic field dependence of the specific heat of polycrystalline NdIn

and sharper in the specific heat of the single crystal. The specific heat reaches 300 and 180 J/mol at T_{c1} and T_{c2}, respectively. The half-widths of these transitions are one order of magnitude smaller than those of polycrystalline samples.

The specific heat anomaly at the antiferromagnetic transition is shown in Fig. 3 for the single crystal and the polycrystalline sample. From a log–log plot (Fig. 4) of the magnetic specific heat (correction for the lattice and electronic contributions has been made assuming their sum to be equal to the specific heat of $LaIn_3$ [15]) versus the reduced temperature $\varepsilon(\varepsilon = |T - T_N|/T_N)$ a power law divergence $C = A\varepsilon^{-\alpha}$ is observed with $\alpha = 0.96 \pm 0.05$ for $0.006 < \varepsilon < 0.2$ and $\alpha' = 0.26 \pm$

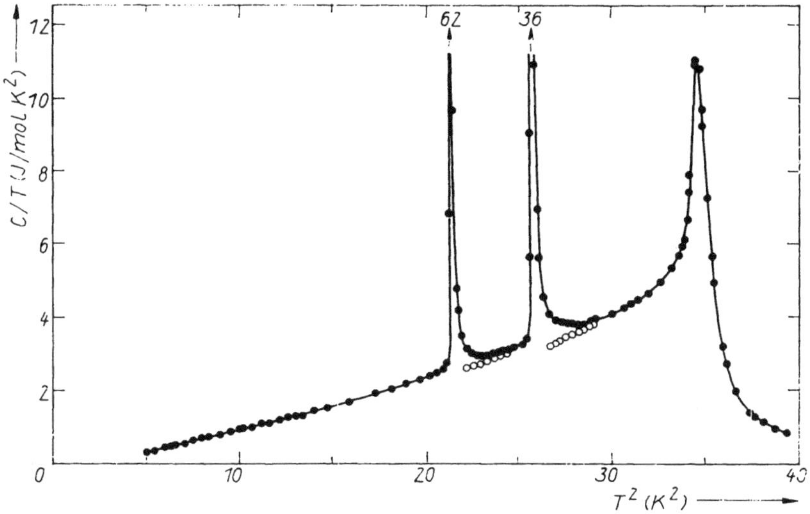

Fig. 2. Specific heat of the $NdIn_3$ single crystal in zero magnetic field. Open circles show supercooling effects at the transitions

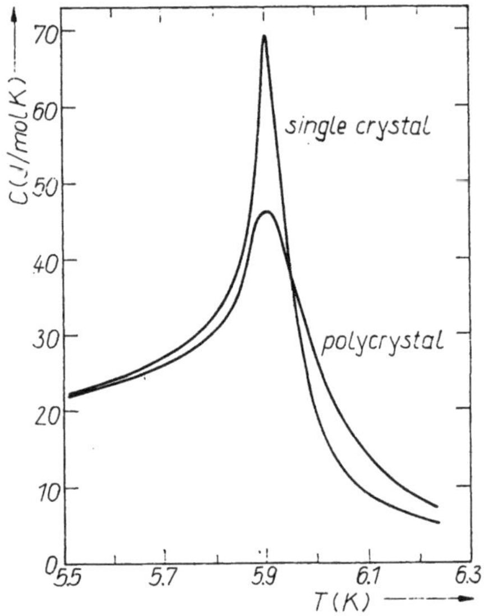

Fig. 3. Specific heat of the $NdIn_3$ single crystal at antiferromagnetic transition. For comparison the specific heat of polycrystalline $NdIn_3$ is also shown

± 0.05 for $0.001 < \varepsilon < 0.1$ above and below the maximum, respectively. It turns out that the maximum contains contributions from two close-lying transitions; the specific heat diverges at $T_N = 5.900$ K and $T'_N = 5.890$ K above and below the maximum, respectively. The best fit to the data allows to determine T_N and T'_N with the accuracy ± 1 mK. The magnetic field applied along the ⟨001⟩ axis separates these transitions distinctly when its strength exceeds 4 T (Fig. 5). The magnetic field has also a strong influence on the effective critical exponent α; in the field in which both transitions are well separated α reaches its usual value. All the behaviour of $NdIn_3$ at the antiferromagnetic transition is very similar to that found recently for isostructural $ErIn_3$ [16] and long before for HoSb [17 to 22]. All these compounds have a nearly sixfold degenerated CF ground state. According to Kim et al. [10, 11] this is the necessary condition for the appearance of large critical exponents in the specific

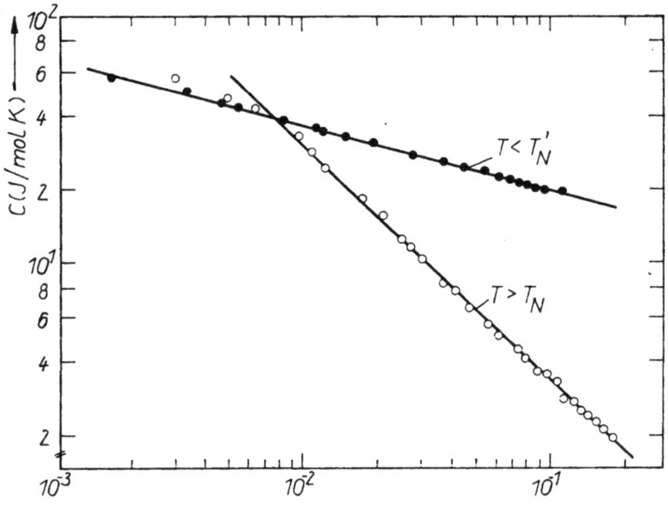

Fig. 4. log-log plots of the magnetic specific heat C_m vs. the reduced temperature $\varepsilon = |T - T_N|/T_N$ for the single crystal of $NdIn_3$. A fit to the data may be obtained only in the case when one assumes the existence of two close-lying transitions at $T_N = 5.900$ K and $T'_N = 5.890$ K

heat. A further similarity follows from the occurrence of two close-lying transitions: with decreasing temperature at T_N a continuous transition to an intermediate phase and then at T'_N a tricritical-like transition to a collinear magnetic structure with magnetic moments aligned along a fourfold axis. Jensen et al. [22] have shown that in the case of HoSb with decreasing temperature at T_N a continuous transition to the ordered state with magnetic moments pointing along threefold axes and then at T'_N a tricritical-like transition to the magnetic phase with moments aligned along a fourfold axis occur. An analogous explanation has been assumed for $ErIn_3$: at T_N there is a continuous transition from the paramagnetic state to the $3\bar{k}$ structure and then at T'_N a tricritical-like transition to the collinear magnetic structure with magnetic moments aligned along a fourfold axis. Also in the case of $NdIn_3$ it seems that the magnetic moments in an intermediate phase are aligned along the threefold axes. Really, when the magnetic field is applied along the ⟨111⟩ axis the separation is not observed (Fig. 6). Therefore, we propose the following sequence of the transitions:

$$1\bar{k} - (T_{c1}) \rightarrow 2\bar{k} - (T_{c2}) \rightarrow 1\bar{k} - (T'_N) \rightarrow 3\bar{k} - (T_N) \rightarrow P\,.$$

The specific heat anomalies around the Néel point in a magnetic field of 5 T applied along the ⟨001⟩ and ⟨111⟩ axes are shown in Fig. 6. The entropy of the transition in the field parallel to the ⟨001⟩ axis is higher than in the field applied in the ⟨111⟩ direction. This allows the sample to be cooled by the adiabatic rotation from the

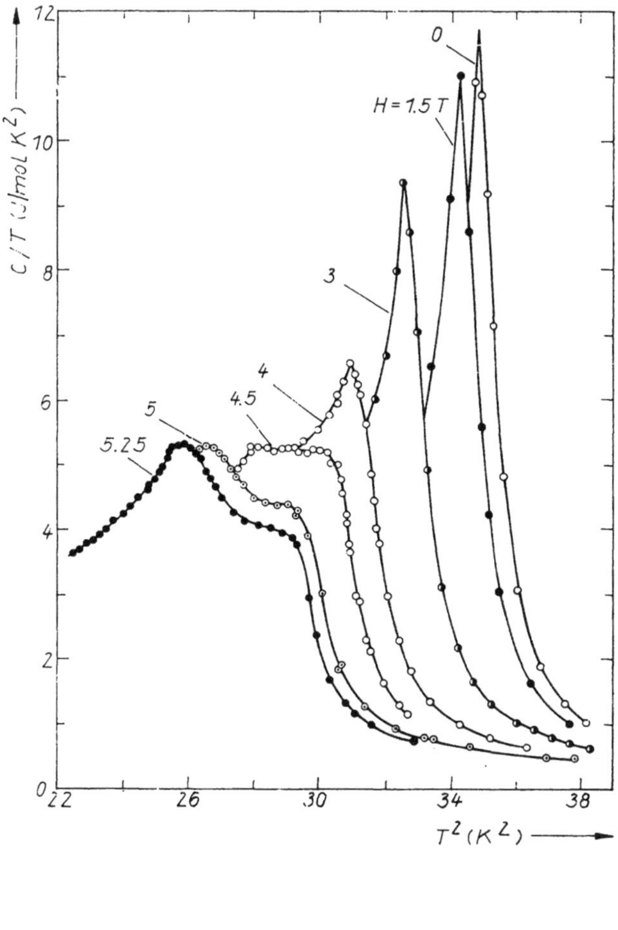

Fig. 5. Influence of the magnetic field applied in the ⟨001⟩ direction on the antiferromagnetic transition

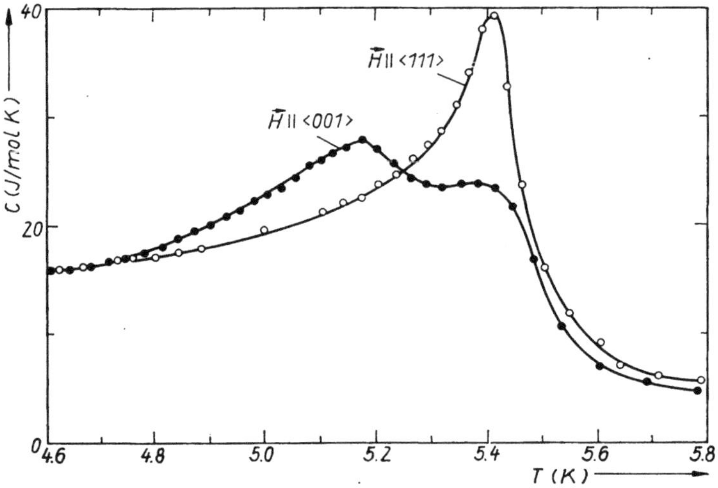

Fig. 6. Specific heat around the antiferromagnetic transition of $NdIn_3$ in a field of 5 T applied along the ⟨001⟩ and ⟨111⟩ directions

⟨001⟩ axis to the ⟨111⟩ one. From the entropy difference (Fig. 7) we estimate the energy of the transition to the paramagnetic state in the field of 5 T parallel to the ⟨001⟩ axis to be 19.3 J/mol higher than in the field parallel to the ⟨111⟩ axis.

Well below the Néel temperature the magnetic contribution to the specific heat of $NdIn_3$ (Fig. 2) does not follow a T^3-dependence. In fact, the spin wave theory (see e.g. [23, 24]) predicts such a simple temperature dependence only for an antiferromagnet without an energy gap in the spin wave spectrum. In general, thermal excitations of antiferromagnons undergo through the energy gap, leading to the exponential term in the specific heat

$$C_m = BT^3 \exp(-\Delta/T) \quad (1)$$

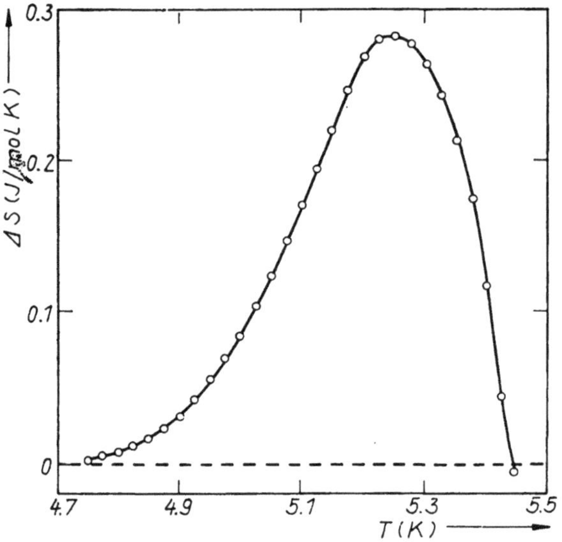

Fig. 7. Entropy difference between antiferromagnetic transitions with a 5 T magnetic field in the ⟨001⟩ and ⟨111⟩ directions

with the energy gap

$$\Delta = (2H_eH_a + H_a^2)^{1/2}\,, \tag{2}$$

where H_e and H_a are the exchange and magnetocrystalline anisotropy fields, respectively. For metamagnets $H_a \geqq H_e$ and Δ may be large.

After the usual correction for the lattice C_l and electronic C_e contributions (assuming $C_l + C_e = C(LaIn_3)$ [15]), we have established that at temperatures fairly below the Néel point the magnetic specific heat of $NdIn_3$ in a zero magnetic field (Fig. 8) is well described by (1) with Δ equal to 3.5 and 2.5 K for single crystal and polycrystalline sample, respectively. The lower value of Δ may only be attributed to higher internal stresses in the polycrystalline sample. In the following we will discuss C_m of the single crystal only.

The spin-wave energy state of an antiferromagnet is twofold degenerated. The magnetic field applied along the easy axis removes the degeneracy (see e.g. [23, 24]

$$\Delta(H) = \Delta_0 \pm g\mu_B JH\,, \tag{3}$$

where Δ_0 is the gap in the absence of the magnetic field, and other symbols have their usual meaning. Denoting H_{c0} as the critical magnetic field, which destroys completely the gap at 0 K, we have then

$$\Delta(H) = \Delta_0 \pm \Delta_0 H/H_{c0}\,. \tag{4}$$

Taking into account the thermal population of both branches we obtain the temperature and magnetic field dependence of the magnetic specific heat in the form

$$C_m = BT^3 \exp(-\Delta_0/T) \cosh(\Delta_0 H/H_{c0}T) \tag{5}$$

with $\Delta_0 = 3.5$ K, $H_{c0} = 5.72$ T, and $B = 0.28$ the above formula describes quite well the magnetic specific heat of $NdIn_3$ in the field applied in the ⟨001⟩ direction

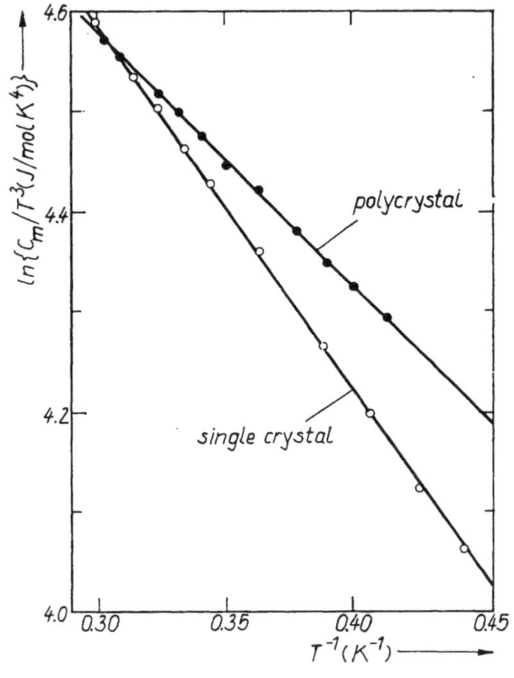

Fig. 8. ln (C_m/T^3) vs. T^{-1} plot for single crystal and polycrystalline $NdIn_3$

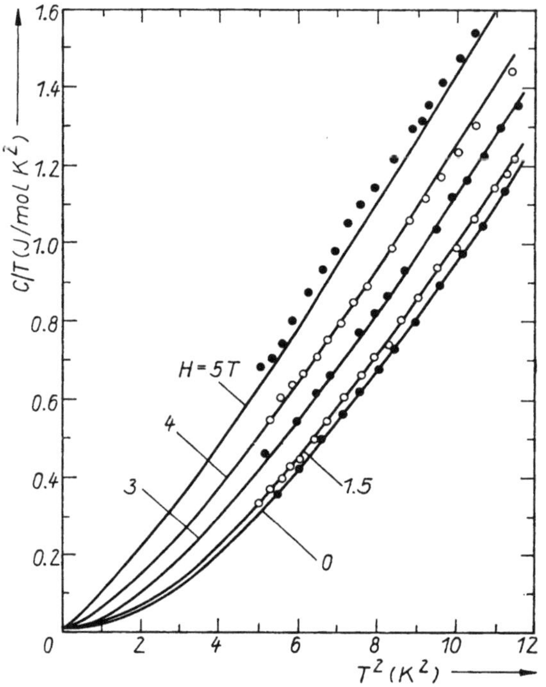

Fig. 9. Low-temperature part of the specific heat of $NdIn_3$ in the magnetic field applied in the ⟨001⟩ direction. The solid lines are calculated as explained in the text

(Fig. 9). The fit becomes a little worse only for high magnetic fields. It is probably because we have neglected an influence of the field on the Allen-Bradley thermometer [12].

The effective energy gap Δ_{eff}, determined from the $\ln(C_m/T^3)$ versus T^{-1} plot at temperature near 2.5 K, is shown in Fig. 10. Its nonlinear field dependence has its origin, of course, in the thermal population also of the upper branch of the spin-wave spectrum. In fact, we may approximate the exact formula (5) by (1) in a limited temperature range. The solid line (calculated) in Fig. 10 gives the effective gap at 2.5 K and as is seen it fits quite well to the experimental data. For comparison we also show in this figure the expected field dependence of Δ_{eff} at higher temperatures. The dependences have only an illustrative character because of the proximity to the Néel point. From the intersections of the field dependences of Δ_{eff} with the H-axis we may obtain the critical fields H_{cT} destroying the gap at a given temperature (inset in Fig. 10).

In the paramagnetic range the specific heat of $NdIn_3$ differs distinctly from the specific heat of nonmagnetic $LaIn_3$ [15] (Fig. 11). It is, for the most part, due to the Schottky contribution to the specific heat of $NdIn_3$. From inelastic neutron scattering and complementary experiments two sets of the CF parameters were proposed [3]: $x = 0.43$ and $W = 1.06$ K or $x = 0.31$ and $W = 0.91$ K. Both give a nearly (within $\approx$7 K) sixfold degenerated ground state formed from $\Gamma_8^{(1)}$ and Γ_6 levels and an excited $\Gamma_8^{(2)}$ level about 66 K above. Most likely the first set is the proper one [7 to 9]. The Schottky contribution connected with this splitting has been subtracted from C_{tot}. The lattice specific heat C_l of $NdIn_3$ (correction for the electronic contribution has been made assuming it equal to that of $LaIn_3$ [15]) is smaller than that of $LaIn_3$. In the temperature range 10 to 27 K, C_l cannot be described by one Debye specific

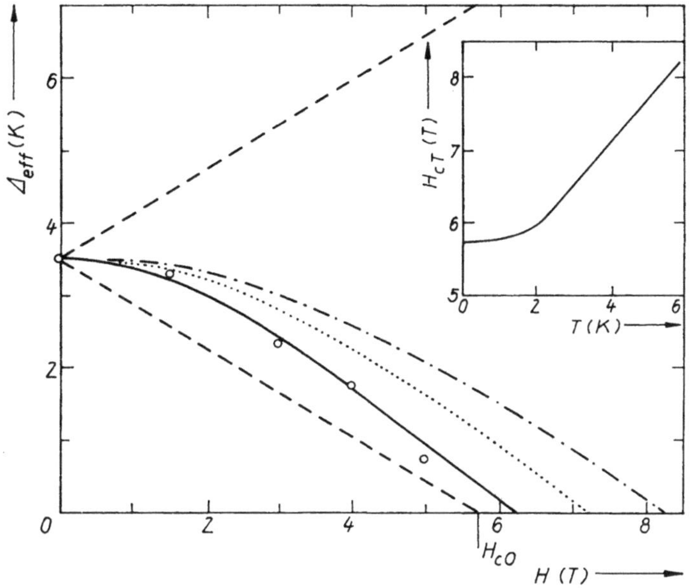

Fig. 10. Field dependence of the effective energy gap Δ. The solid line gives the fit at 2.5 K. For comparison the expected dependences at 0 (– – – –), 4.2 (· · · · ·), and 5.9 K (—·—) are shown also. The inset shows the calculated temperature dependence of the critical field which destroys the energy gap at a given temperature

heat function; θ_D varies from 170 to 180 K and does not reach the high-temperature limit. For comparison, C_l of $LaIn_3$ above 20 K may be described by the specific heat function with $\theta_D = 168.5$ K [15].

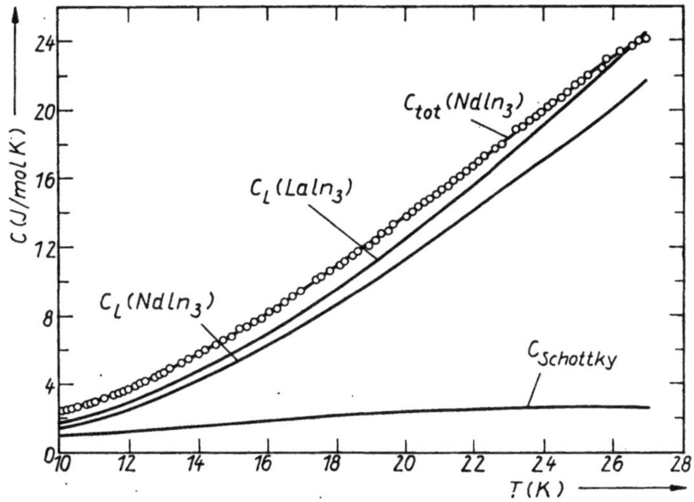

Fig. 11. Specific heat of $NdIn_3$ in the paramagnetic range: as measured and corrected for the Schottky contribution. For comparison the specific heat of $LaIn_3$ [15] is also shown

3.2 Thermal expansion

In the ordered state the thermal expansion has been measured along two equivalent fourfold axes denoted as c- and a-axes, being the longer and shorter edges of the single crystal, respectively (Fig. 12a and b). The thermal expansion anomalies are observed at temperatures where peaks of the specific heat occur; they are much more pronounced along the c-axis. From the data for c- and a-axes we estimate the relative magnitude of a tetragonal distortion $((c - a)/a)$ in the antiferromagnetic state to be about 33×10^{-6} at 1.5 K.

At temperatures well below the Néel point the linear thermal expansion coefficient α_c is negative and depends strongly on temperature, whereas α_a is nearly zero in a wide temperature range (Fig. 13). This behaviour seems to be consistent with a one-antiferromagnetic-domain structure of the single crystal where the magnetic moments are aligned along the c-axis.

We assume the background thermal expansion due to the lattice and the conduction electrons to be negligible in this temperature range. Because of the complicated behaviour, the thermal expansion of $LaIn_3$ at low temperatures [15] cannot serve as a background for other $REIn_3$ compounds. In Fig. 13 we compare the thermal expan-

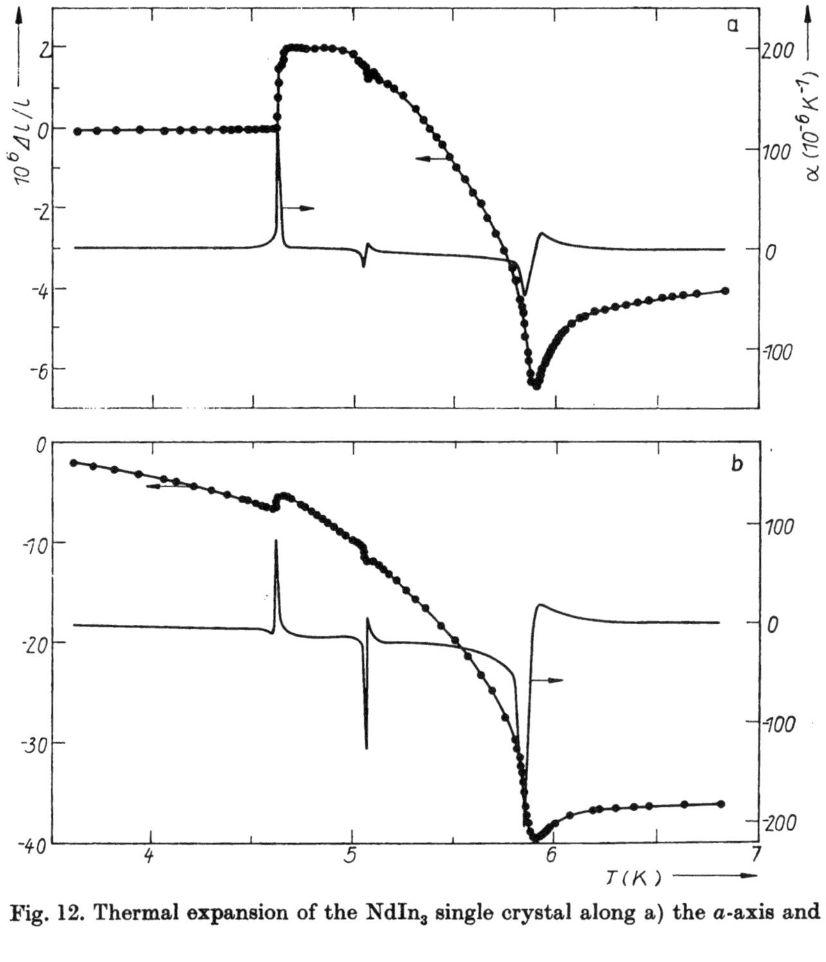

Fig. 12. Thermal expansion of the $NdIn_3$ single crystal along a) the a-axis and b) the c-axis

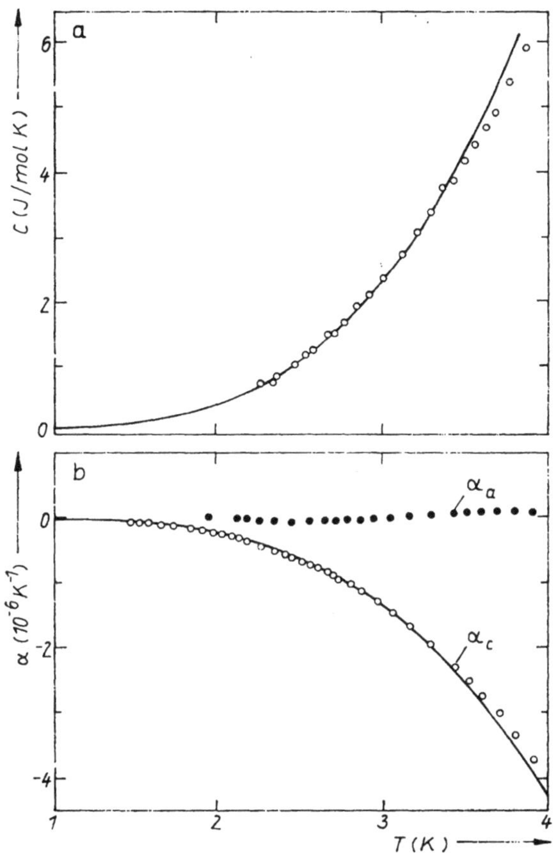

Fig. 13. Low-temperature part of a) the magnetic specific heat C_m and b) the linear thermal expansion coefficients α_c and α_a of $NdIn_3$. The solid lines are calculated as explained in the text

sion coefficient α_c with the magnetic specific heat C_m. The solid line in the upper part of the figure represents the fit of $C_m = 0.28T^3 \exp(-3.5/T)$ J/mol K. The α_c coefficient is proportional to C_m and well described by the relation $\alpha_c = \gamma_m C_m / c_B V$ (the solid line in the lower part of the figure) with the Grüneisen parameter $\gamma_m = -3.05$. The bulk modulus $c_B = 84.7$ GPa was determined by Hillebrands [25].[5])

Because of the harmonic approximation the conventional spin-wave theory does not predict a thermal expansion. Therefore, the spin-wave spectrum of $NdIn_3$ is most likely renormalized due to magnetoelastic interactions. The ionic 4f charge distribution is nonspherical and its quadrupole momentum is tied to the angular momentum vector. In the magnetically ordered state the quadrupoles are also ordered. Thus the excitations come from a spin-quadrupole coupled system.

At the antiferromagnetic transition point the thermal expansion as well as the specific heat show a strong variation. Jumps of the volume thermal expansion coefficient $\Delta\beta$ and specific heat ΔC are coupled with a pressure effect on the ordering temperature through the Ehrenfest formula,

$$dT_N/dp = VT_N \, \Delta\beta/\Delta C \, .$$

In the case under consideration, $\Delta\beta = \Delta\alpha_c + 2\,\Delta\alpha_a = -2.9 \times 10^{-4}\ K^{-1}$ and $\Delta C = 64$ J/mol · K and hence $dT_N/dp = -1.7$ K GPa^{-1}. On the other hand, $dT_N/dp = \gamma T_N / c_B$ and from here the Grüneisen parameter $\gamma = -24.6$. Such a large value of

[5]) Bulk elastic moduli $c_r = 29.8$ GPa, $c_t = 34.4$ GPa, $c_l = 130.6$ GPa and hence $c_B = 84.7$ GPa of $NdIn_3$ have been determined by Brillouin scattering in the way as in the paper of R. Mock and G. Güntherodt [30].

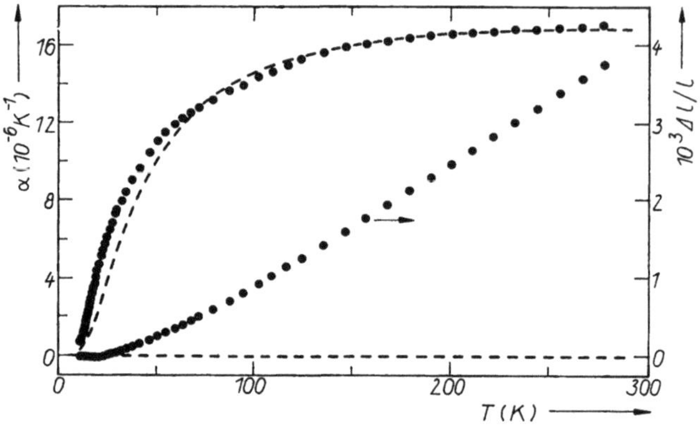

Fig. 14. Linear thermal expansion coefficient and relative length change in the paramagnetic phase of $NdIn_3$. The dashed line gives the linear thermal expansion coefficient of $LaIn_3$ after [15]

the Grüneisen parameter is most likely connected with singularities of the antiferromagnetic transition.

In the paramagnetic phase (Fig. 14), the value of the linear thermal expansion coefficient α of $NdIn_3$ exceeds distinctly that of $LaIn_3$ in a wide temperature range. The difference between the curves in Fig. 14 is attributed to the CF effect. In fact, the maximum of $\Delta\alpha = \alpha(NdIn_3) - \alpha(LaIn_3)$ lies at a temperature close to the maxima of the calculated Schottky specific heats (Fig. 15).

The CF contribution to the volume thermal expansion coefficient, β_{CF}, is (see e.g. [26, 27])

$$\beta_{CF} = (R/c_B T^2) \left(\langle E_i^2 \gamma_i \rangle - \langle E_i \rangle \langle E_i \gamma_i \rangle\right),$$

where $\gamma_i = -(\mathrm{d} \ln E_i/\mathrm{d} \ln V)$ and $\langle\rangle$ is a statistical average over the energy levels. The coefficient $\beta_{CF}(T)$ is proportional to $C_{Schottky}(T)$,

$$C_{Schottky} = (R/T)^2 \left(\langle E_i^2 \rangle - \langle E_i \rangle^2\right),$$

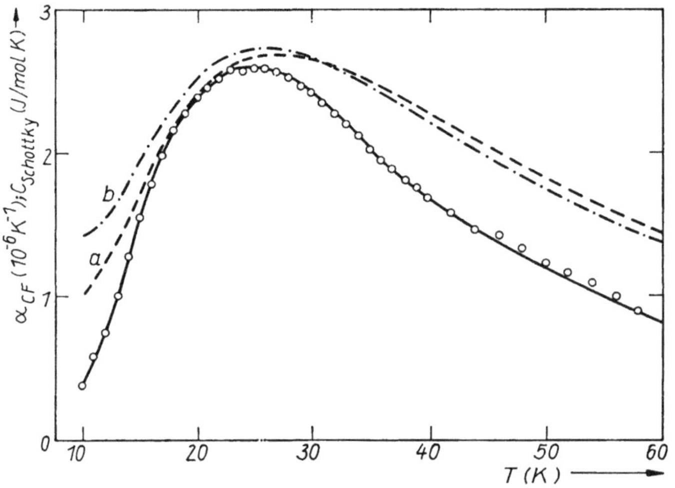

Fig. 15. Comparison of $\Delta\alpha = \alpha(NdIn_3) - \alpha(LaIn_3)$ with the calculated Schottky specific heat. (a) $x = 0.43$, $W = 1.06$ K; (b) $x = 0.31$, $W = 0.91$ K

if the Grüneisen CF γ_i parameters of all thermally populated energy levels are the same or in a two-level system of splitting (where $\gamma_{CF} = \beta_{CF} c_B / C_{Schottky}$).

The CF splitting observed for $NdIn_3$ may be approximated by a two-level system: 1. below 10 K where the thermal population of the $\Gamma_8^{(2)}$ level lying 66 K above the ground state may be neglected and 2. above $\approx$20 K where the couple of $\Gamma_8^{(1)}$ and Γ_6 levels may be treated as a nearly sixfold degenerated ground state. The proportionality of β_{CF} to $C_{Schottky}$ is observed only in a limited temperature range near the maximum (Fig. 15) and we estimate γ_{CF} to be about 15.2, which means the Grüneisen parameter of the $\Gamma_8^{(2)}$ level with respect to the couple of the $\Gamma_8^{(1)}$ and Γ_6 levels.

The lack of the proportionality of β_{CF} to $C_{Schottky}$ below 10 K is due to an influence of the antiferromagnetic transition.

4. Conclusions

$NdIn_3$ turns out to be an unusual antiferromagnet undergoing four successive phase transitions in zero magnetic field between $1\bar{k}$, $2\bar{k}$, and $3\bar{k}$ structures. The range of the stability of these structures seems to be the result of a subtle competition between the exchange, crystal field and magnetoelastic, and or/ pair quadrupolar interactions. If we restrict our consideration to the first two terms only [9] we are not able to describe the observed behaviour as a whole. Also, to assure the continuity of the transition from the paramagnetic to the antiferromagnetic state of a system with sixfold degenerated CF ground state, with large critical exponents in the specific heat, an antiferroquadrupolar coupling has to be assumed [11]. Indeed, Morin et al. [28, 29] have pointed out that the antiferroquadrupolar pair interaction seems to be characteristic of the cubic $AuCu_3$-type structure compounds.

Although, the occurrence of an extra transition at T'_N lying very close to T_N, not expected by the cubic model [10, 11], seems to be an intrinsic feature of systems with an accidental sixfold degeneracy of the CF ground state.

Acknowledgements

We thank Dr. Z. Żołnierek for a critical reading of the manuscript and Dr. J. Ulner for useful discussions in connection with this work.

References

[1] K. H. J. Buschow, H. W. De Wijn, and A. M. van Diepen, J. chem. Phys. **50**, 137 (1969).
[2] P. Lethuillier, J. Pierre, G. Fillion, and B. Barbara, phys. stat. sol. (a) **15**, 613 (1973).
[3] P. Lethuillier and J. Chaussy, J. Physique **37**, 123 (1976).
[4] P. Lethuillier, J. Pierre, K. Knorr, and W. Drexel, J. Physique **36**, 329 (1975).
[5] B. Staliński, A. Czopnik, N. Iliev, and T. Mydlarz, J. Physique **40**, 5C–149 (1979).
[6] A. Czopnik, N. Iliev, B. Staliński, C. Bazan, H. Mädge, and R. Pott, Physica (Utrecht) **B130**, 259 (1985).
[7] Z. Kletowski, P. Markowski, and B. Staliński, Solid State Commun. **57**, 293 (1986).
[8] T. Hiraoka and T. Kamihira, J. Phys. Soc. Japan **56**, 1837 (1987).
[9] A. Czopnik, J. Kowalewski, M. Hackemer, B. Staliński, and P. Weidner, submitted to Solid State Commun.
[10] D. Kim, P. M. Levy, and L. F. Uffer, Phys. Rev. **B12**, 989 (1975).
[11] D. Kim, P. M. Levy, and J. J. Sudano, Phys. Rev. B **13**, 2054 (1976).
[12] K. Bohmhammel, H. Mädge, and G. Wolf, Cryogenics **47**, 553 (1977).
[13] R. Pott and R. Schefzyk, J. Phys. E **16**, 444 (1983).
[14] R. Schefzyk, Diploma Thesis, Universität zu Köln, 1980.

[15] A. Czopnik, R. Pott, H. Mädge, N. Iliev, and B. Staliński, phys. stat. sol. (a) **93**, K137 (1986).
[16] A. Czopnik, H. Mädge, and B. Staliński, phys. stat. sol. (a) **107**, K151 (1988).
[17] H. Taub and S. J. Williamson, Solid State Commun. **13**, 1021 (1973).
[18] H. Taub, S. J. Williamson, W. A. Reed, and F. S. L. Hsu, Solid State Commun. **15**, 185 (1974).
[19] H. Taub and C. B. R. Parente, Solid State Commun. **16**, 857 (1975).
[20] L. F. Uffer, P. M. Levy, and M. J. Sablik, Solid State Commun. **15**, 191 (1974).
[21] N. Hessel Andersen, J. Jensen, and O. Vogt, J. Magnetism magnetic Mater. **15 to 18**, 13 (1980).
[22] J. Jensen, N. Hessel Andersen, and O. Vogt, J. Phys. C **13**, 213 (1980).
[23] F. Keefer, in: Hdb. Phys., Vol. 18/2, Ed. S. Flügge and H. P. J. Wijn, Springer-Verlag, 1966.
[24] A. I. Akhiezer, V. G. Bariakhtar, and S. V. Peletminski, Spinovye volny, Izd. Nauka, Moskva 1967.
[25] B. Hillebrands, private information.
[26] T. H. B. Barron, J. G. Collins, and G. K. White, Adv. Phys. **29**, 609 (1980).
[27] H. R. Ott and B. Lüthi, Z. Phys. B **28**, 141 (1977).
[28] P. Morin, D. Schmitt, and E. Du Tremolet De Lacheisserie, J. Magnetism magnetic. Mater. **30**, 257 (1982).
[29] P. Morin, J. Rouchy, M. Giraud, and A. Czopnik, J. Magnetism magnetic Mater. **67**, 95 (1987).
[30] R. Mock and G. Güntherrodt, J. Phys. C **17**, 5635 (1984).

(Received April 29, 1988)

phys. stat. sol. (a) **110**, 615 (1988)

Subject classification: 76.30; 61.70; S10.1

Departement Natuurkunde, Katholieke Universiteit Leuven[1])

MgO Powder Containing Low Concentrations of Isotopically Pure $^{52}Cr^{3+}$ Ions

Its Application as ESR Marker

By

A. STESMANS[2]) and G. DE VOS

The small width (≈ 0.02 mT for $T < 150$ K) and low g value (≈ 1.9800) of the ESR signal observed from low concentrations ($\lessapprox 1$ ppm) of $^{52}Cr^{3+}$ ions embedded in MgO suggests that this material may be very useful as a general ESR intensity and g marker. The 9.55% naturally abundant ^{53}Cr ($I = 3/2$) isotope, however, complicates the former single-line spectrum with a hyperfine (HF) quartet. The low-field HF components often hinder (frequent) measurements of weak spectra in the near-$g = 2.0023$ region. Based on ^{52}Cr ion implantation in 6 N pure Mg foil a method is presented which eliminates the HF spectrum. The resulting MgO: $^{52}Cr^{3+}$ powder only exhibits one narrow line of which ESR analysis shows it to be a very convenient marker. Regarding the physical properties of the $^{52}Cr^{3+}$ ions, weakly asymmetric line shapes emerge, being more pronounced the higher the observational frequency and/or the lower T. From this, the existence of a non-negligible $u\beta S^3 B$ term in the effective spin Hamiltonian of cubic-site Cr^{3+} impurities in MgO is inferred. Independent measurements of $u(T)$ and $g(T)$ (g being the isotropic Zeeman-term spectroscopic splitting factor) reveal both quantities to exhibit an identical T dependence arising from the implicit thermal lattice expansion (i.e., u and $g \propto d^5$; d is the Cr^{3+} site-to-nearest-neighbor-oxygen distance). This finding, which disagrees with former results, is in accordance with the simple crystal field model. Also in accordance with this model, no g anisotropies are observed.

Die geringe Breite ($\approx 0{,}02$ mT für $T < 150$ K) und der niedrige g-Wert ($\approx 1{,}9800$) des bei niedrigen Konzentrationen (<1 ppm) von $^{52}Cr^{3+}$-Ionen in MgO beobachteten ESR-Signals weist darauf hin, daß dieses Material als allgemeiner ESR-Intensitäts-Standard und g-Marker sehr brauchbar sein sollte. Das mit 9,55% natürlicher Häufigkeit vorliegende $^{53}Cr(I = 3/2)$-Isotop kompliziert jedoch das ursprüngliche Ein-Linienspektrum mit einem Hyperfein-(HF)-Quartett. Die Niederfeld-HF-Komponenten behindern häufig (immer) Messungen von schwachen Spektren im Bereich um $g = 2{,}0023$. Auf der Grundlage einer ^{52}Cr-Ionenimplantation in 6 N reiner Mg-Folie wird eine Methode angegeben, die das HF-Spektrum eliminiert. Das resultierende MgO: $^{52}Cr^{3+}$-Pulver zeigt nur eine schmale Linie, dessen ESR-Analyse zeigt, daß sie ein sehr geeigneter Marker ist. Bezüglich der physikalischen Eigenschaften des $^{52}Cr^{3+}$-Ions tritt eine schwach asymmetrische Linienform auf, die bei höheren Beobachtungsfrequenzen und/oder niedrigeren T ausgeprägter ist. Daraus wird die Existenz eines nicht vernachlässigbaren $u\beta S^3 B$-Terms im effektiven Spin-Hamiltonian von Cr^{3+}-Störstellen auf Plätzen mit kubischer Symmetrie in MgO abgeleitet. Unabhängige Messungen von $u(T)$ und $g(T)$ (g ist der isotrope spektroskopische Zeeman-Aufspaltungsfaktor) zeigen, daß beide Größen eine identische T-Abhängigkeit aufweisen, die von der impliziten thermischen Gitterausdehnung (d. h., u und $g \propto d^5$, d ist der Abstand des Cr^{3+}-Platzes zum nächsten Sauerstoffnachbarn). Dieses Ergebnis, das mit früheren Ergebnissen nicht übereinstimmt, befindet sich in Übereinstimmung mit einem simplen Kristallfeldmodell. Ebenfalls in Übereinstimmung mit diesem Modell werden keine g-Anisotropien beobachtet.

---

[1]) Celestijnenlaan 200D, B-3030 Leuven, Belgium.

[2]) Supported by the National Fund for Scientific Research (Belgium).

1. Introduction

Generally, the behavior of Cr^{3+} ions in MgO:Cr^{3+} single crystals is well understood [1 to 3]. The trivalent chromium ions are known to replace the divalent magnesium ions substitutionally. Concomitantly, Mg^{2+} ion vacancies are formed to compensate for the extra positive charge on the Cr^{3+} ions and the particular position of these vacancies with respect to the Cr^{3+} ions determine the details of the electron spin resonance (ESR) spectrum. If the Cr^{3+} concentration is low ($\leqq 10$ ppm)[3]), there are no vacancies in the near neighborhood of the Cr^{3+} ions [1, 2] and the crystalline field is purely cubic. The dominant very pure cubic field splits the sevenfold orbitally-degenerate ground state level $^4F_{3/2}$ of the three d-electrons into a low singlet and two higher triplets. The main part of the single crystal spectrum then is a strong narrow signal with spectroscopic splitting factor $g \approx 1.9800$ (for $T < 300$ K) [2, 3] due to $^{50}Cr^{3+}$, $^{52}Cr^{3+}$, and $^{54}Cr^{3+}$ (all with nuclear spin $I = 0$ and natural abundance of 4.3, 83.8, and 2.4%, respectively). This signal, even though it originates from Cr^{3+} impurities at "perfectly" cubic sites, shows weak anisotropies [4, 5]. This is ascribed to a weak splitting term (see, e.g., [6 to 10] $u\beta S^3B$ (β and $\boldsymbol{B}$ represent the Bohr magneton and the externally-applied magnetic induction, respectively) occurring in the spin-Hamiltonian for states of spin $S > 3/2$. It originates from magnetization-density distortions on the ion of hexadecapolar nature, due to indirect-crystalline field influence. The central signal is flanked with a hyperfine (HF) quartet also centered at $g = 1.9800$ and with HF-splitting constant $a = 1.785$ mT. Each signal of the quartet has an intensity of about one-fourtieth of the main line. This quartet originates from $^{53}Cr^{3+}$ ions ($I = 3/2$, 9.5% natural abundance) at cubic sites.

At higher Cr^{3+} concentrations the local symmetry around the Cr^{3+} ion can be either orthorhombic or tetragonal. This depends on whether the Mg^{2+} ion vacancies are formed in a nearest or next-nearest neighbor position, respectively. Such centers give rise to additional anisotropic signals and, if all three types of local symmetry would prevail in equal numbers, a complicated ESR spectrum would result. When the crystal field is tetragonal the g-tensor has principle g-values [11] of $g_{\parallel} = 1.9788 \pm$ $\pm$ 0.0001 and $g_{\perp} = 1.9820 \pm 0.0001$ at RT [12]. For low Cr^{3+} concentrations, i.e., $[Cr^{3+}] < 10$ ppm, the linewidths are small; viz., peak-to-peak width as measured on the first-derivative absorption spectrum ($dP_{\mu a}/dB$) [13] $\Delta B_{pp} \approx 0.02$ mT for $T < 150$ K. For $[Cr^{3+}] > 100$ ppm [1] the line broadens to > 0.5 mT at RT. The line broadening at higher Cr^{3+} concentrations is ascribed to residual strains and subgrain misorientation in the disturbed crystalline field [4, 14, 15].

It thus follows that MgO:Cr^{3+} of sufficiently low Cr^{3+} content exhibits a neat weakly-anisotropic ESR spectrum, mainly comprising a strong, narrow central line at $g \approx 1.9800$. The g-anisotropy can be "eliminated" by resorting to powdered samples, as will be outlined. Hence, in principle, this material could well serve as a marker (both for intensity and g). However, a somewhat unfortunate feature is the MgO:$^{53}Cr^{3+}$ HF quartet. While this HF structure could be very useful for field-sweep calibration it is often felt — particularly the high-g components — to be a nuisance in trying to analyse weak signals in the near-$g = 2.0023$ region, which region is mostly investigated by ESR. This is particularly true if both the unknown and the MgO:Cr^{3+} standard are placed together in the microwave cavity and if no extra precautions are taken to enhance signal discrimination, e.g., create a constant shift between the B-values felt by both samples, dual-sample cavity technique, etc. [16]. For accurate intensity calibration the latter techniques appear to be inappropriate anyway.

[3]) All concentration specifications in this work refer to weight ratios.

The properties characterizing an ideal ESR calibration standard are well known [17]; a) it should exhibit a single, narrow symmetric signal, the signal should be b) sufficiently intense, i.e., the amount of material needed to generate a sufficiently strong resonance signal should be as small as possible, c) isotropic, d) difficult to saturate, e) stable in time, and f) of low magnetization such that eventual corrections due to demagnetizing fields are negligible; g) for intensity calibration purposes its magnetic susceptibility χ should either be of the Curie (Brillouin) or the Pauli paramagnetic type such that the number of spins involved can be correctly determined. As will be shown MgO:Cr^{3+} meets most of the enumerated requirements very well, except for its multi-line spectrum (HF signals) and saturation properties.

The present work discusses a simple method *to eliminate the $^{53}Cr^{3+}$-induced HF quartet* in the MgO:Cr^{3+} spectrum. Post-implantation annealing of high-purity Mg foil implanted with ^{52}Cr ($I = 0$) allowed to produce MgO:Cr^{3+} powder freed from the HF structure. (As the resonance line is nearly isotropic, the MgO:Cr^{3+} powder is expected to have the same ESR properties as the single crystal. Therefore it should be a very suitable marker.) The ESR properties of this powder have been carefully analysed, both at X and K band, in the temperature range 4.2 K $\leqq T \lessapprox$ 300 K.

2. Effective Spin-Hamiltonian of Cubic-Site $^{52}Cr^{3+}$ Impurities in MgO

The ESR spectrum of $^{52}Cr^{3+}$ impurities in MgO residing in a crystal field of cubic (octahedral) symmetry can be described by the generalized electronic spin-Hamiltonian [4, 6 to 10]

$$H = g\beta \boldsymbol{S}\cdot\boldsymbol{B} + u\beta\{S_x^3B_x + S_y^3B_y + S_z^3B_z - \tfrac{1}{5}(\boldsymbol{S}\cdot\boldsymbol{B})[3S(S+1)-1]\}\,, \quad (1)$$

in which higher-order Zeeman terms have been omitted. The $^{52}Cr^{3+}$ ion has an orbital singlet ground state of effective spin $S = 3/2$; x, y, z, refer to the cubic $\langle 100\rangle$ crystal axes. The second term on the right-hand side of (1) will be referred to as the $u\beta S^3B$ or Bleaney-Koster-Statz [6, 7] (BKS) term. It is an intrinsic "crystal-field" effect and produces an angular dependence of the eigenvalues of (1) and, accordingly, of the allowed electron spin transitions. In the absence of this term only the isotropic Zeeman term $g\beta\boldsymbol{S}\cdot\boldsymbol{B}$ would remain and the allowed ESR transitions, i.e., $M = +3/2 \rightarrow +1/2$; $-3/2 \rightarrow -1/2$ and $+1/2 \rightarrow -1/2$ (M being the electronic magnetic quantum number), all coincide resulting in a single ESR line of isotropic g-value ≈ 1.9800. With the BKS term operative the two transitions $\pm 3/2 \rightarrow \pm 1/2$ still do coincide but, generally, do not coincide with the $+1/2 \rightarrow -1/2$ transition. Hence, in principle, a two-line spectrum would emerge. Neglecting terms of order $(u/g)^2$, (1) gives the resonance fields B_{01} and B_{02} for both lines as

$$B_{01} = \frac{hf}{g\beta}\left(1 + \frac{9u}{5g}\,p\right) \equiv \frac{hf}{g_e\beta} \quad \text{for} \quad M\colon \frac{1}{2} \rightarrow -\frac{1}{2}, \quad (2)$$

$$B_{02} = \frac{hf}{g\beta}\left(1 - \frac{6u}{5g}\,p\right) \quad \text{for} \quad M\colon \pm\frac{3}{2} \rightarrow \pm\frac{1}{2}, \quad (3)$$

in which h is Planck's constant, f the observational microwave frequency, and $p = 1 - 5(l^2m^2 + n^2l^2 + n^2m^2)$; l, m, and n represent the direction cosines of B with respect to the cubic crystal axes. The quantity p determines the angular dependence of the electron spin transition. g_e is the "effective" g-value as read from the resonant field B_{01}. However, as clearly stated elsewhere, the $\pm 3/2 \rightarrow \pm 1/2$ transitions are very sensitive to inhomogeneous line broadening (see, e.g., [8]), which is accepted to be the reason for not observing this signal in a general ESR experiment. On the contrary, the central $+1/2 \rightarrow -1/2$ line is not broadened in first order [8]. The physical broadening mechanisms are believed to be the strain fields within the crystal (modulating the fine structure parameters or g-value) or either electric fields and field gradients. The presence of a small, though not negligible, $u\beta S^3B$ term in the spin-Hamiltonian was suggested from a few experimental reports [4, 5]. The g-value (and also u; cf. later) shows an implicit temperature (volumetric) dependence as a result of thermal expansion. As stated before, the sevenfold orbitally-degenerate ground state level $^4F_{3/2}$ of the three 3d electrons at the Cr^{3+} cation is split by the surrounding octahedral crystal field into a lowest singlet (A_2) and two higher triplets (T_1 and T_2). In the absence of spin–orbit interaction, i.e., λ S.L. (λ being the spin–orbit coupling parameter), this isolated singlet would exhibit "spin-only" magnetism, the spin system being described by the isotropic Zeeman term $g\beta \boldsymbol{B} \cdot \boldsymbol{S}$ in which now $g = 2.002319$, the free-electron value. However, due to the presence of the λ S.L. interaction, some of the orbital moment of the first excited triplet (T_2) is mixed into the ground state which primarily causes an isotropic shift of g away from g_0, given as (see, e.g., [8] and references therein)

$$g = g_0 \left(1 - \frac{4\lambda}{\Delta}\right), \tag{4}$$

in the crystalline-field approximation. Herein, Δ represents the singlet–triplet energy separation which is sensitive to local variations in the cation surroundings; it is a measure of the cubic moment (hexadecapole) of the crystalline potential at the Cr^{3+} site and varies as [18, 19]

$$\Delta \propto \langle r^4 \rangle \, d^{-5}; \tag{5}$$

d is the distance between the Cr^{3+} site and the nearest-neighbor oxygen atoms, r is the spherical coordinate vector, and $\langle r^4 \rangle$ represents the average value of r^4 over the radial part of the 3d orbitals. Hence, if neglecting any volumetric influences on $\lambda/\langle r^4 \rangle$, we expect $g - g_0 \sim d^5$. Relying upon the thermal-expansion data of MgO, this relation was shown to closely describe the measured *implicit* $g(T)$ variation in the range $77\ \text{K} \leqq T \leqq 860\ \text{K}$. In this interpretation, the local thermal dilation in the vicinity of the substitutional Cr^{3+} ions was supposed to be essentially unchanged from that of the MgO crystal, which assumption apparently appeared well justified. No evidence has been found for the existence of an *explicit*, phonon-induced T dependence in g, which has been explained in the effective point charge model by the cancellation of quadrupolar lattice vibration terms imposed by the cubic surroundings [15]. Hence, the expected T dependence of g may be represented by

$$g - g_0 = -k \left(\frac{d(T)}{d_0}\right)^5, \tag{6a}$$

$$g - g_0 = -k\left(1 + \int_{273}^{T} \alpha(T) \, \mathrm{d}T\right)^5, \tag{6b}$$

where d_0 is the interionic spacing at $T = 273$ K, k a constant, and $\alpha(T)$ the linear thermal expansion of MgO.

Literature contains much less explicit considerations on u than on g. However, from the treatment in [8, 6], u is generally expected to exhibit a similar T-dependence as g. Hence, we find

$$u(T) = h + i \left(\frac{d(T)}{d_0}\right)^5, \tag{7}$$

h and i being constants.

Unlike former work, $g(T)$ and $u(T)$ have now been (accurately) determined from the $M = 1/2 \to -1/2$ line essentially *independently*, without having to rely on the fitting of some theoretical (g, u, T, p) relationship to the experimental data. A broad T-range has been scanned and the presence of the $u\beta S^3B$ term is well affirmed.

3. Experimental Technique

3.1 Sample preparation

About 4 cm^2 of 6 N pure[4]) Mg foil of thickness $d = 61 \pm 3\,\mu m$ has been single-side implanted with 8.6×10^{15} ^{52}Cr ions cm^{-2} at 60 keV. To prevent as much as possible the influence of other impurities — some of which, when placed in MgO, are known to generate strong ESR signals also (see, e.g. [1]) — the Mg foil surface has been carefully cleaned in a HNO_3–H_2O mixture at RT followed by abundant rinsing in H_2O–Di and drying in acetone (MOS quality) before ion implantation. Post-implantation oxidation was applied resulting in MgO powder doped with $^{52}Cr^{3+}$ ions. The oxidation set-up comprised a conventional furnace equipped with a high-purity double-walled quartz insert, which could be evacuated to below 13 mPa and which could be easily moved in or out the oven as necessary.

If the implanted ^{52}Cr would homogeneously distribute throughout the Mg foil and if *all* ^{52}Cr would show up as substitutional $^{52}Cr^{3+}$ impurities in the MgO powder, then after oxidation, the applied dose would result in a nominally homogeneous concentration $[^{52}Cr^{3+}] = (42 \pm 4)$ ppm. Since both the ESR linewidth and the amount of prevailing noncubic crystalline-field surroundings for the substitutional Cr^{3+} sites — the latter generating additional lines — depend strongly on the local Cr^{3+} concentration, a well-defined homogeneous, low Cr^{3+} concentration is prerequisite to obtain only one central line of as narrow a ΔB_{pp} as possible. However, since ion implantation piles up all ^{52}Cr atoms in a surface layer of thickness < 100 nm the preparation method must aim at the *homogenization* of the Cr^{3+} density to be successful. "Various" oxidation procedures — essentially three — were applied, mainly differing only in the Mg-foil supporting material used during heat treatment. For this the original Mg foil was cut into four pieces.

In a first procedure, a cylindrically-shaped Mo crucible (of inner dimensions; diameter: 5 mm and height: 8 mm) was taken as Mg-foil support. This metal is generally applied for Mg heat treatments. It is an obvious choice since Mg is known not to alloy with Mo in any detectable amount [20]. This Mo holder was firmly etched in $HCl:H_2O_2$ (1:1) at RT, followed by the usual rinsing procedure as mentioned above, and subsequently fired (outgassed) at 860 °C for 2 h in a vacuum (pressure $p_v < 13$ mPa). A piece of Mg foil — firmly folded together in order to promote ^{52}Cr density homogenization upon melting — was placed in the Mo crucible in the furnace insert. Then a heat treatment at $T \approx 600$ °C for about 10 min in 5.5 N pure, dry Ar ambient at $p_{Ar} \approx 0.1$ MPa was applied. This was followed by the oxidation treatment. The quartz insert was filled with 5 N pure, dry O_2 ($p_{O2} \approx 0.1$ MPa) at RT and subsequently inserted in the oven being at $T_0 \approx 600$ °C. The oxidation rate of Mg at 600 °C is rather slow. Hence, T_0 was slowly increased (allowing again for ^{52}Cr impurity diffusion) to $T_0 = (655 \pm 10)$ °C, typically ≈ 8 min after having slided in the insert Suddenly a short lighting emerged from the furnace betraying that the Mg had ignited, the Mg foil thus oxidizing at once. Apparently, Mg ignites upon the onset of melting (melting point $T_m = 651$ °C).

[4]) The Mg foil was prepared by many rolling and careful etching sequences until the final thickness was reached. The metallic impurity content of the starting material was ≤ 1 ppm as determined by emission spectroscopy.

The experimental results presented furtheron (namely, the observation of only one narrow ESR line) show that the Cr distribution has been well homogenized. This may have several reasons.

Firstly, there are the homogenizing thermal treatments in both Ar and O_2 at $T_0 \approx 600$ °C for about 8 min each. However, since the diffusion constant D_{Cr} of Cr in Mg is not known yet [21], the efficiency of these treatments cannot be evaluated. Secondly Mg apparently ignites upon (or immediately after) melting and the formation of a liquid droplet just before oxidation will thoroughly stir the Cr impurities.

There are a few practical experiences which might be worth while to mention:

1. Trying to apply a homogenizing anneal at about 400 °C with $p_v < 13$ mPa appeared impossible since Mg then evaporates at a high rate, which is clearly to be avoided.

2. The Mo crucible, after oxidation, comes out heavily attacked (oxidized). This should not surprise in view of the applied temperatures and the 0.1 MPa O_2 ambient. It is covered with black–blue pellets which may peel off and some caution in assembling the produced MgO powder is advised.

Although the ^{52}Cr density homogenization over the Mg foil appears to be very satisfactory, it comes out (cf. spin density data) that only *a small fraction of the implanted dose*, i.e., $(4.1 \pm 0.9)\%$, is retrieved as ESR active sites. Among others, this might indicate that the particular oxidation treatment followed allows much of the ^{52}Cr atoms to leave the Mg foil. Referring to the heavily-attacked Mo crucible and the known solubility [20] of Cr in Mo, the Mg–Mo contact was suspected. To check this, some items in the oxidation procedure were altered. In a second oxidation, the treatment times were considerably shortened. In a third procedure, a pure quartz crucible was used instead of Mo, although it is known that Mg reacts with Si at higher temperatures to form Mg–Si compounds [20]. Fourthly, a Mg foil was suspended on a 6 N pure Pt wire[5]) (diameter 0.5 mm) whose ends were supported by a cylindrically-shaped quartz holder, so that the only contact of Mg with another solid was a tiny contact with the Pt wire. For all these oxidation procedures, ESR revealed the same efficiency, namely $(4.1 \pm 0.9)\%$, for turning the implanted ^{52}Cr into ESR-active sites. This clearly shows that the idea of Cr escaping from the Mg foil is to be abandoned. Instead it is concluded that the oxidation also forms some other Cr compounds, e.g., Cr_2O_3, not resulting in substitutional Cr^{3+} sites in the MgO. Also, surface segregation of Cr could have occurred. Finally, some Cr impurities might be involved in other ESR-active MgO:Cr structures, not contributing to the central line and possessing a much broader ΔB_{pp} such that these structures remain undetectable. A main conclusion from this is that if one wants to fabricate MgO powder of a certain Cr^{3+} content in this way, the efficiency factor for creating ESR-active sites has to be calculated in advance, before implanting ions.

After the oxidation treatment, the MgO powder was carefully assembled; a few mg was encapsulated in a sheet of PTFE to form a sample.

3.2 ESR observations

The ESR experiments were done at 20.9 GHz in the range between $4.2\text{ K} \leqq T < 300$ K using a conventional homodyne reflection-type spectrometer equipped with signal averaging facilities. Due to the large spin–lattice relaxation time T_1 ($\approx 10^{-4}$ s at 77 K) of the MgO:$^{52}Cr^{3+}$ signal, the ESR measurements had to be performed at low incident hf power P_μ ($\approx 5\,\mu$W) to avoid saturation. Supplementary X-band (9 GHz) saturation and lineshape measurements were made at $T \approx 150$ and 300 K.

---

[5]) Obtained from the National Bureau of Standards, Washington, USA.

Special attention has been paid to obtain accurate $g(T)$ data. For this, an auto-tracking NMR Gaussmeter (of 0.5 μT accuracy) has been incorporated for B calibration. The observational K-band frequency f was determined using an EIP 371-type f counter via zero-beating against the first harmonic of an X-band signal; the obtained accuracy is ± 0.2 MHz.

Generally, exact measurement of B at the sample site remains a problem even when experimenting with a magnet of perfectly homogeneous field, if the ESR sample and the NMR probe do not "coincide" in space. Practically, the NMR probe is placed outside the cavity (dewar) during experiments. Hence, eventual demagnetizing influences from impurities in the cavity walls [17], dewar walls, etc., are not accounted for. Generally, these do not receive sufficient attention. Hence, we used a high-accuracy g-marker, i.e., neutron-irradiated LiF (LiF:Li) of $g_{Li} = 2.00226$ [22] which was mounted tightly to the sample. By measuring f and supposing that g_{Li} is exactly known, the LiF:Li resonant field B_0(Li) (at the zero crossing of $dP_{\mu a}/dB$) could be exactly determined. Subsequently, the field distance between B_0(Li) and B_0(SS) of the signal studied was measured by the NMR fieldmeter, resulting in accurate values of g(SS) relative to g_{Li}. As stated elsewhere, the T-independent g_{Li} is accurately known [23] and is a very reliable g-marker.

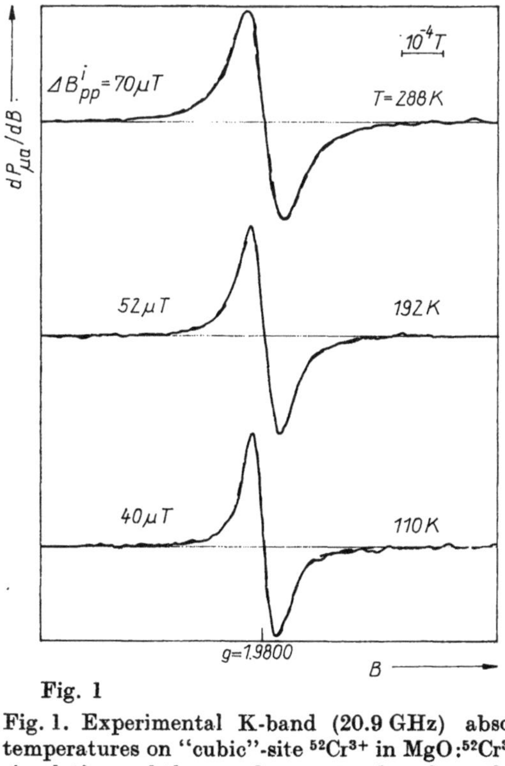

Fig. 1

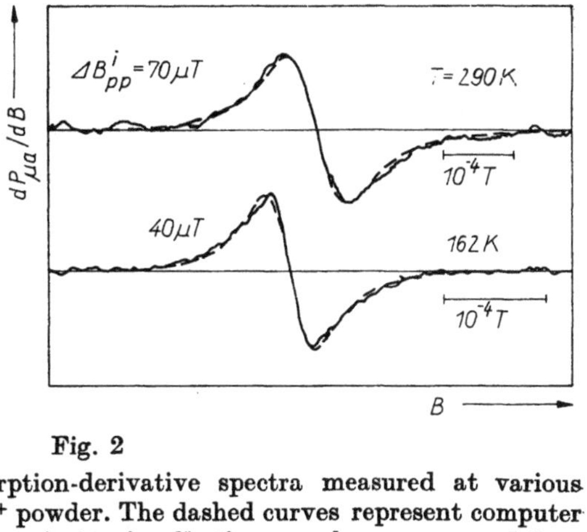

Fig. 2

Fig. 1. Experimental K-band (20.9 GHz) absorption-derivative spectra measured at various temperatures on "cubic"-site $^{52}Cr^{3+}$ in MgO:$^{52}Cr^{3+}$ powder. The dashed curves represent computer simulations of the powder spectra based on the anisotropic effective g_e-value; $g_e = g - 1.8pu$, in which $p = 1 - 5(l^2m^2 + n^2l^2 + m^2n^2)$, while l, m, and n represent the direction cosines of $\boldsymbol{B}$ with respect to the crystal cubic axes

Fig. 2. X-band (9.0 GHz) spectra measured at 290 and 162 K for "cubic-site" $^{52}Cr^{3+}$ substitutional centers in MgO:$^{52}Cr^{3+}$ powder. The dashed curves represent computer-calculated simulations based on the powder-spectrum interpretation invoking the $u\beta S^3B$ term anisotropy. The linewidth ΔB^i_{pp} used for the intrinsic broadening line is also quoted

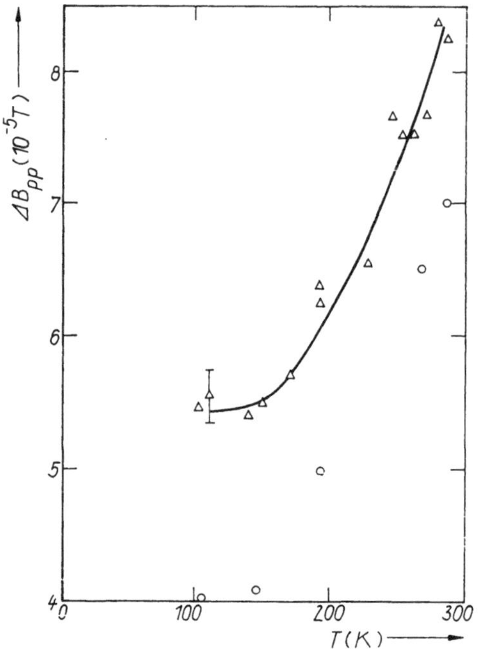

Fig. 3. K-band ESR peak-to-peak linewidth (represented by triangles) of $^{52}Cr^{3+}$ impurities with cubic crystalline-field surroundings in MgO:$^{52}Cr^{3+}$ powder plotted vs. T. The circles represent the intrinsic linewidth ΔB^{i}_{pp} as obtained from powder-spectra computer simulations of the measured lineshapes, from where the growing relative importance of the inhomogeneous broadening is noticed. The latter arises from the existence of an anisotropic, non-negligible $u\beta S^3 B$ interaction in combination with the powdered nature of the sample. The solid line is only meant to guide the eye

4. Experimental Results and Interpretation

4.1 Observed ESR spectrum and lineshape analysis

ESR measurements could only detect one signal of which some typical spectra observed at K- and X-band are shown in Fig. 1 and 2, respectively; the characteristic features of this line, that is, narrow ΔB_{pp}, g-value, lineshape, isotropic behavior, etc., are those of the well-known central line ascribed to the Cr^{3+} impurities in MgO of local cubic symmetry. No other signals — a fortiori no HF components — could be traced, although a broad B-range, including the $g = 2.0023$ region has been carefully investigated. Based on the known spectrometer sensitivity, we deduced that the material studied does not contain other ESR-active centers of concentration $> 3 \times \times 10^{13}$ g^{-1} of MgO and per mT linewidth. Consequently, this MgO appears to be very useful as an ESR calibration standard.

The K-band linewidth is shown in Fig. 3 as a function of T. At RT, this value agrees with the smallest ones [1, 13] reported in literature (see also Table 1). The latter ones are obtained for the lowest-doped MgO:Cr^{3+} single-crystalline samples studied; however, no impurity levels were specified but, likely, $[Cr^{3+}] < 10$ ppm. The concentration of cubic site Cr^{3+} impurities in our samples has been determined from the "central" line ESR signal relative to a known Al_2O_3:Cr^{3+} ruby standard[6]). For all four oxidation methods, the spin concentration came out equal within experimental error giving $[^{52}Cr^{3+}] = (1.7 \pm 0.5)$ ppm (corresponding to $(2.0 \pm 0.6) \times \times 10^{16}$ $^{52}Cr^{3+}$ g^{-1} of MgO), which value copes with literature. However, below RT, ΔB_{pp} starts to deviate; it is systematically broader than the best literature single-crystalline data. The deviation increases with decreasing T, as exemplified in Table 1. Also, while the lineshape at RT is nearly perfectly Lorentzian, the line grows more and more asymmetric at lower temperature, as may be inferred from Fig. 1. The K-

[6]) Obtained from the National Bureau of Standards, Washington, USA.

Table 1

Linewidth and T_1-data obtained for MgO:Cr^{3+} powder at two frequencies, i.e. 9.0 and 20.9 GHz. ΔB_{pp} represents the peak-to-peak width of the absorption-derivative ESR line; ΔB^i_{pp} is the intrinsic Lorentz linewidth used in the powder-spectra simulation. T_1-values are averages over the three methods used to analyse the saturation (i.e., signal height versus P_μ) data and are derived based on ΔB^i_{pp} and the known value of B_1

T (K)	linewidth (10^{-4} T)				T_1 (10^{-6} s)		
	ΔB_{pp} ($\pm 0.02 \times 10^{-4}$ T)			ΔB^i_{pp}	($\pm 0.2 \times 10^{-6}$ s)		
	9.0 GHz	20.9 GHz	X band*)		9.0 GHz	20.9 GHz	X band*)
100		0.55	0.21	0.40		10.1	8.7
145	0.37	0.56	0.24	0.40	1.6	3.5	1.4
193		0.63	0.31	0.50		1.4	0.8
287	0.72	0.83	0.70	0.70	0.24	0.24	0.24

*) Data from [13], measured on cubic-site Cr^{3+} in monocrystalline MgO of low concentration, i.e., $[Cr^{3+}] < 15$ ppm.

band asymmetry, as measured by the difference in height between the high and low-field peaks of the $dP_{\mu a}/dB$ curve, increases from 5% at RT towards 12% at 110 K. X-band data, however, behave differently. The line becomes symmetric while it narrows from 0.056 mT (K-band) to 0.037 mT (X-band) at 150 K. This reveals the presence of an inhomogeneous linewidth contribution ΔB^G_{pp}. The line narrowing relative to K-band data increases with decreasing temperature. The observed inhomogeneous line broadening with attendant asymmetric lineshapes in these powder samples do strongly refer to the existence of a slightly anisotropic g (or at least g_e)-tensor, resulting in *powder-pattern lineshapes* [24]. The fact that the signal is symmetric and of equal width for both observational frequencies at RT then means (see, e.g., [25, 26]) $\Delta B^G_{pp} \ll \Delta B_{pp}$, the latter being the intrinsic linewidth; the intrinsic width with attendant Lorentzian shape is observed. At low T, e.g., $T \approx 100$ K, while ΔB^i_{pp} is narrowing [11], the situation is almost unaltered for X-band, still having $\Delta B^G_{pp} \ll \Delta B^i_{pp}$. At K-band, however, with 20.9/9 times larger ΔB^G_{pp} value, the asymmetric line reveals $\Delta B^G_{pp} \approx \Delta B^i_{pp}$. As a result, an inhomogeneously-broadened signal with non-Lorentzian shape (powder pattern) is observed. The details of the overall shape depend on the specific "g_e-tensor" involved and the powder-effect influence on the lineshape increases with decreasing ΔB^i_{pp}.

4.2 Lineshape fittings; presence of the $u\beta S^3B$ term

Evidently, the anisotropy of the $u\beta S^3B$ term in the Hamiltonian (1) (see also (2)) became very suspected as originating the effective g_e-tensor anisotropies, i.e., the observed powder pattern. Alternatively, the observed effect might also be interpreted as due to the existence of a weakly anisotropic, oblate axially symmetric g-tensor [24]. Both onsets have been tested, relying upon the powder spectra interpretation outlined above.

Firstly, computer calculated simulations to the observed powder spectra were carried out ignoring the existence of the u term, from where a g-tensor with $g_{\|} = 1.97973$ and $g_{\perp} = 1.97988$ has been derived. Although reasonable fittings could be produced in this way over the whole T-range covered, the results were felt to be of little satisfaction. Indeed, the fitting quality appeared much too sensitive to the particular ΔB^{i}_{pp} values used. If not chosen large enough, multi-line structure easily developed, especially in the low-T range, clearly not in agreement with observations (cf. Fig. 1 and 2). Trying to reduce this by reducing the amount of g-anisotropy then made the fitting at RT almost impossible. On the other hand, the second ansatz based upon the anisotropic u-term easily resulted in perfect fittings over the whole T-range. The u-values are derived from the observed (asymmetric) *lineshape* by fitting computer calculated powder spectra. For each microcrystallite in the powder, having a certain crystal orientation (i.e., p) relative to $\boldsymbol{B}$, the observed B_{01} (i.e., apparent g) is given by (2). Hence, over the total set of microcrystallites, *the inhomogeneous contribution* ΔB^{G}_{pp} *to* ΔB_{pp} *is determined by the particular value of* u. At each T, u and ΔB^{i}_{pp} are adjusted to obtain the best fitting to the experimental lineshape. The u data thus obtained are shown in Fig. 4. The value found at 77 K, i.e., $u(77\text{ K}) = (5.7 \pm 0.2) \times 10^{-5}$ compares favorably with the value $(4.6 \pm 0.2) \times 10^{-5}$ measured by the strain-modulated ESR technique [4] on MgO:Cr^{3+} crystals. The ΔB^{i}_{pp} results are shown in Fig. 3.

It can easily be shown that the average u-term induced shift of the central line away from g, is exactly averaged out to zero for the powder spectrum, i.e., *the zero crossing of the* $dP_{\mu a}/dB$ *powder spectrum exactly corresponds to* g. Hence, from the studied MgO powder of low [$^{52}Cr^{3+}$] and narrow ΔB_{pp}, accurate $g(T)$ data can "directly" be read from the spectrum, independently of u-"readings". One has not to go through a fitting scheme relying upon a certain (g, u, T, p) relationship as was the case, e.g., in work on c-MgO:Cr^{3+}. The resulting $g(T)$ data are shown in Fig. 5 for $4.2\text{ K} \leqq T \leqq 273\text{ K}$; they supplement the previous [15] high-T data. At 77 K, we find $g(77\text{ K}) = 1.980008 \pm 0.00001$ in perfect agreement with the data from Henning and den Boef [4], (i.e., 1.9799 ± 0.0001) and Walsh [15] et al. (i.e., 1.9800). At 273 K we find $g(273\text{ K}) = 1.97988 \pm 0.00001$ in agreement again with [15] (i.e., 1.9799).

As mentioned, $g(T)$ and $u(T)$ are expected to exhibit an implicit T-dependence. Hence, assuming that there is no significant explicit T-dependence, the $g(T)$ and $u(T)$ data were fitted by (6b) and (7), respectively, relying upon the thermal expansion data for MgO [27]. The results are shown by the solid curves in Fig. 5 and 4, respec-

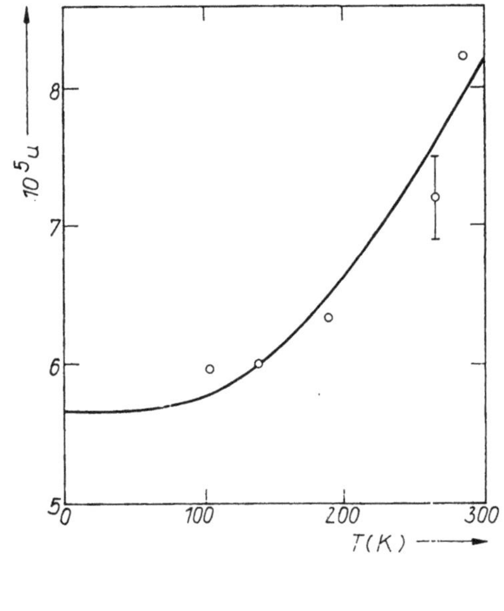

Fig. 4. The u data coping with the $u\beta S^3B$ term in the effective spin-Hamiltonian for MgO: $^{52}Cr^{3+}$ as obtained from computer simulation of the observed powder spectra. The solid curve is a fit through the data points calculated from the relation $u = h + i(d/d_0)^5 = h + i(1 + \int_{273}^{T} \alpha\, dT)^5$, where $h = -0.003347$ and $i = 0.003424$

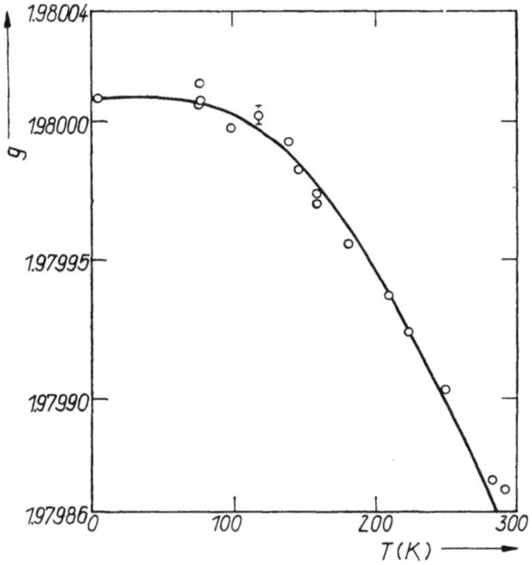

Fig. 5. The isotropic Zeeman term g-factor measured as a function of temperature of cubic-site $^{52}Cr^{3+}$ substitutional impurities in MgO powder. The solid curve is calculated from the relation $g = g_0 - k(1 + \int_{273}^{T} \alpha \, dT)^5$, with $k = 0.02244$, where the thermal expansion coefficient α has been taken from [27]

tively, from where satisfactory fittings are recognized. These are given as

$$g = g_0 - k(d/d_0)^5 = 2.002319 - 0.02244(d/d_0)^5 \,, \tag{8a}$$

$$u = h + i(d/d_0)^5 = -0.003347 + 0.003424(d/d_0)^5 \,. \tag{8b}$$

This result is considered as an a posteriori justification of the applied fitting procedure, which is based on powder spectra generated by u-term anisotropies. Relation (8a) is in agreement with former data [15].[7])

From these findings, several conclusions may be drawn. Firstly, the T-dependences of both g and u are well accounted for by the implicit effect due to thermal lattice expansion of MgO. Regarding g, this points out that the local lattice dilatation around the Cr^{3+} ions is essentially identical to the one of the host MgO crystal [15]. No explicit phonon-induced effect needs to be invoked to explain the T-dependence. Secondly, the parameters g and u show the *same* T-dependence as expected. These T-dependences disagree with previous $u(T)$ and $g(T)$ measurements on single crystals over a smaller T-range reporting both g and $u \sim T$, for which no explanation was offered [5]. This unfortunate conclusion is felt to have arisen partly from the limited T-range covered. Moreover, the "significantly" differing absolute g-values reported by these authors might indicate that they did not properly take into account demagnetization effects.

---

[7]) This agreement probably refers to the fact that Walsh et al. [15] have also measured on powdered $MgO:Cr^{3+}$ samples, although it has not been clearly stated. If not, for a general orientation of $\boldsymbol{B}$ with respect to the crystal cubic axes, i.e., $p \neq 0$, their $g(T)$ measurement would have suffered from the $u\beta S^3 B$ term, thus effectively having measured $g_e(T)$. However, in the range $4.2 \text{ K} \leq T \leq 300 \text{ K}$, a maximum interference would occur at RT for $p = 1$, giving a maximum error (cf. Fig. 5) on g of $\approx -1.4 \times 10^{-4}$. This might just have been on the edge of their experimental accuracy, or alternatively, their "large" spread in the RT g-data (cf. Fig. 2 of [15]) might just point to the influence of various crystal orientations investigated. In case they measured on single crystals, one should be aware of possible larger u-induced disturbances in the higher-T range ($T > 300$ K). However, their experimental accuracy slightly decreases with increasing T, such that after all, the u effect might just not have been resolved.

Thus, the cubic-site Cr^{3+}-in-MgO case appears fully described by (1); g stays isotropic within experimental accuracy, the whole anisotropy observed being accounted for by the $u\beta S^3 B$ term. With the above assumptions all observed spectra, both at X- and K-band, over all temperatures studied could be perfectly fitted both what concerns the precise linewidth and overall shape, of which some examples (dashed lines) are shown in Fig. 1 and 2 for various T-values at K- and X-band. The significance of this has been commented on extensively before [28]. The accurate measurement of g (and u) might be extremely difficult if not invoking improved NMR field-marking methods or an accurate, perfectly isotropic g-marker of g-value near 1.98. It may thus be stated that the measured $g(T)$ and $u(T)$ dependences in the T-range covered of the "cubic" site Cr^{3+} could be revealed thanks to the incorporated g-marker and the powdered nature of the sample; *the correct g is supplied at once by the zero crossing of the line, while u is "read" from the powder-effect lineshape.*

$\Delta B^{\mathrm{i}}_{\mathrm{pp}}$ at low T is found to be slightly larger than the smallest ones reported in literature on single crystals (cf. [13]) as shown in Table 1. This larger intrinsic linewidth is ascribed to residual strains and/or slightly higher $^{52}Cr^{3+}$ concentrations.

4.3 Saturation measurements ; T_1 determination

The high sensitivity to microwave-absorption induced saturation of the Cr^{3+} signal was striking, in agreement with previous T_1 data [13] of single-crystalline MgO:Cr^{3+}. For reasons of comparison between the powdered samples produced in this work and the mono-crystalline MgO:Cr^{3+} properties, and also because the knowledge of the saturation properties of a calibration standard is felt essential, extensive saturation measurements have been carried out.

T_1 can easily be derived from signal height versus P_μ data if assuming to apply Bloch's phenomenological saturation theory [29], which describes homogeneously broadened Lorentzian lines. Clearly then, for our X-band measurements, it should be no problem to find the correct T_1 at RT. However, for K-band at lower T-values, the growing relative importance of the inhomogeneous part in ΔB_{pp} needs some caution. The saturation behavior of an inhomogeneously broadened line, i.e., as arising from a single Gaussian [25] or Lorentzian [26] distribution convoluted by a Lorentzian broadening function, is well analysed; we may refer to this as the symmetric cases since all lineshapes involved are symmetric. However, the saturation characteristics of the asymmetric case, namely for a powder-effect induced, inhomogeneously-broadened line, has not yet been considered such that relying on the Castner plots [25] is not permitted. Nonetheless, the powder-effect case is likely to be well described by the Bloch case provided the inhomogeneous interference remains small. The induced error on the thus derived T_1 will depend on $\Delta B^{\mathrm{i}}_{\mathrm{pp}}/\Delta B^{\mathrm{G}}_{\mathrm{pp}}$. Relying upon this, K-band T_1-values were also derived at lower T.

T_1 has been determined using three different analysing methods of the measured signal height versus P_μ data [16]. The first one is based on the relation

$$y'_m = \frac{KB_1}{(1 + \gamma^2 B_1^2 T_1 T_2)^{3/2}}, \tag{9}$$

where y'_m is the peak-to-peak height of the $\mathrm{d}P_{\mu\mathrm{a}}/\mathrm{d}B$ curve assuming the spectrometer to detect in the linear regime, $2\,B_1$ is the peak amplitude of the linearly polarized rf field felt by the sample, γ the gyromagnetic ratio, K a constant, and $T_2 = 2/(\sqrt{3}\,\gamma\,\Delta B^{\mathrm{i}}_{\mathrm{pp}})$ the spin–spin relaxation time. T_1 is obtained from the slope of the curves $(y'_m/KB_1)^{2/3}$ versus P_μ $(\sim B_1^2)$ relying upon the known values of B_1 and $\Delta B^{\mathrm{i}}_{\mathrm{pp}}$. In a second method T_1 is derived from the averaged values of tabulated parameters. Finally, a third procedure relies upon the maximum in the y'_m versus B_1 plot, at which maximum one has $(1 + \gamma^2 B_1^2 T_1 T_2)^{-1} = 2/3$. As expected, the three methods (being based on the same theory) lead to equal results within experimental error; the average values are shown in Table 1. The T_2-values used are obtained from the ΔB_{pp}^{i} data (also quoted in Table 1) as derived from the powder-spectra simulations. For reasons of comparison, the T_1-data obtained by pulsed-saturation methods on single-crystalline MgO:Cr^{3+} of lowest Cr^{3+} concentration so far studied are included in Table 1 also. Clearly, at RT, our K-band results are in agreement with literature. At lower T-values, however, slightly deviating (higher) T_1-values are found, as expected. This is caused by the growing interference of the powder-effect type inhomogeneous broadening.

Regarding X-band T_1 measurements, e.g., for $T = 150$ K, good agreement with literature is obtained. This, again, is as expected and clearly shows that the inhomogeneous broadening part is negligible at X-band in the T-range considered.

We also mention that at RT (cf. Table 1) T_1 is found to be B-independent in agreement with the known main spin-lattice relaxation mechanism [13] operative in this T-range.

4.4 Susceptibility

The relative K-band signal intensity $I_r \equiv I_{Li}/I_{Cr}$ ($\propto \chi_{Cr}$, the MgO:Cr^{3+} susceptibility) is shown in Fig. 6 for the range 4.2 K $\leq T <$ 300 K. I_{Cr} and I_{Li} represent the areas under the absorption curves of MgO:$^{52}Cr^{3+}$ and LiF:Li, respectively; the LiF:Li behaves Pauli-paramagnetic. The triangles represent values as calculated from

$$I_r^{-1} = \frac{y'_m \, \Delta B_{pp}^2}{y'_{m,Li} \, \Delta B_{pp,Li}^2}, \tag{10}$$

while the circles give results obtained by numerical integration. The latter have been included to check the former. Indeed, (10) relies upon identical lineshapes pertaining to both samples studied. Clearly, regarding K-band data, this is only correct at RT. At lower T, the cubic-site Cr^{3+} signal starts to develop powder-spectrum properties (LiF:Li exhibits the Lorentzian lineshape over the whole T-range covered). As can be seen in Fig. 6 the relation $I_r \propto T$ holds within experimental scatter, which demonstrates the paramagnetic character of the Cr^{3+} ions in the T-range studied. The error is mainly caused by inaccuracies in the determination of the linewidth. Apparently, for

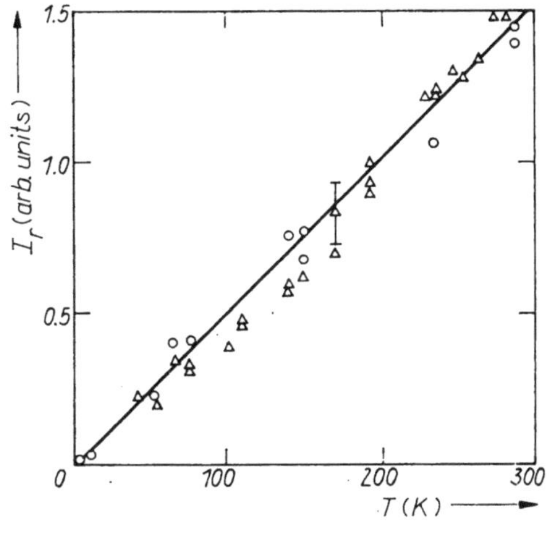

Fig. 6. T-dependence of the inverse of the intensity of the cubic-site MgO:$^{52}Cr^{3+}$ ESR signal at 20.9 GHz relative to the T-independent intensity of a LiF:Li Pauli-paramagnetic standard. The triangles are calculated from (10) while circles are obtained by numerical double integration of the respective $dP_{\mu a}/dB$ spectra. The solid line is a least-squares fit through the data points showing $\chi_{Cr}(\sim I_r^{-1}) \propto T^{-1}$ from where the paramagnetic character (i.e., non-interacting sites) of the studied centers is concluded

f's up to K-band the limited powder-effect induced lineshape distortions at lower temperatures do not introduce much error in determining I_r using (10). Hence, (10) can safely be applied to determine χ_{Cr}.

5. Conclusion

MgO:Cr^{3+} powder doped with low concentrations of isotopically-pure $^{52}Cr^{3+}$ ions has been produced which exhibits ESR characteristics comparable to the best single-crystalline MgO:Cr^{3+} data. As a result the $^{53}Cr^{3+}$-induced HF quartet, inevitably present in MgO:Cr^{3+} doped with naturally-abundant Cr, has been eliminated. The conversion efficiency of $^{52}Cr^{3+}$ ions implanted in Mg foil into ESR-active sites $^{52}Cr^{3+}$ in the oxidation methods followed is found to be $(4.1 \pm 0.9)\%$, resulting in powder samples of $[^{52}Cr^{3+}] = (1.7 \pm 0.5)$ ppm. This material shows very satisfactory properties as regards its use as ESR standard. Only one narrow line of $g \approx 1.9800$ is observed; the material does not contain other ESR-active centers beyond concentrations of 3×10^{13} g^{-1} mT^{-1}. The signal is stable in time and sufficiently strong. Its susceptibility is Curie-type ($\chi_{Cr} \sim T^{-1}$) in the range 4.2 K $\leqq T \leqq 300$ K. A drawback, however, could be its strong saturability, viz., in cases where for sensitivity reasons a maximum P_μ is required. The small lineshape distortion developing for $T < 300$ K and $f > 9$ GHz should not be troublesome.

The physical characteristics of the cubic-site $^{52}Cr^{3+}$ line have been further analysed. The existence of anisotropies in the observed line-shapes has been demonstrated, which, from powder-spectra simulations, was shown to be caused by the anisotropic $u\beta S^3B$ term. Convincing evidence has been provided for the presence of this term in the effective spin-Hamiltonian describing the cubic-site Cr^{3+}-in-MgO case. No evidence for anisotropies in g has been found within experimental accuracy. The T-dependences of g and u have been determined independently revealing an identical thermal lattice-expansion influence (T-dependence), i.e., $\propto d^5$ for both observables. This is in agreement with the predictions of simple crystal field theory. Finally, g-values of improved accuracy have been obtained for the range 4.2 K $\leqq$ $\leqq T \leqq 300$ K. The observed relaxation times T_1 and T_2 are in agreement with the mono-crystalline data.

References

[1] J. E. WERTZ and P. V. AUZINS, Phys. Rev. **106**, 484 (1957).
[2] W. LOW, Phys. Rev. **105**, 801 (1956).
[3] B. HENDERSON and J. E. WERTZ, Adv. Phys. **17**, 749 (1968).
[4] J. C. M. HENNING and J. H. DEN BOEF, Phys. Rev. **B18**, 61 (1978).
[5] R. S. DE BIASI, J. magnetic Resonance **44**, 479 (1981).
[6] B. BLEANY, Proc. Phys. Soc. **73**, 939 (1959).
[7] G. F. KOSTER and H. STATZ, Phys. Rev. **113**, 445 (1959).
[8] A. ABRAGAM and B. BLEANY, Electron Paramagnetic Resonance of Transition Ions, Clarendon Press, Oxford 1970.
[9] W. GEHLHOFF, phys. stat. sol. (b) **80**, 549 (1977).
[10] HUANG WU-HAN, LIN FU-CHENG, and ZHU JI-KANG, Proc. Phys. Soc. **84**, 661 (1964).
[11] D. H. DICKEY and J. E. DRUMHELLER, Phys. Rev. B **1**, 3582 (1970).
[12] S. A. MARSCHALL, J. A. HODGES, and R. A. SERWAY, Phys. Rev. **136**, A1024 (1964).
[13] R. L. HARTMAN, J. S. BENNETT and J. G. CASTLE, Phys. Rev. B **1**, 1946 (1970).
[14] P. V. AUZINS and J. E. WERTZ, J. chem. Phys. **43**, 1229 (1965).
[15] W. M. WALSH, JR., J. JEENER, and N. BLOEMBERGEN, Phys. Rev. **139**, A 1338 (1965).
[16] See, e.g., C. P. POOLE, Electron Spin Resonance, Interscience Publishers, New York 1967.
[17] See, e.g., G. FEHER, Phys. Rev. **114**, 1219 (1959) and Ref. [8].

[18] J. Becquerel and W. Opechowski, Physica (Utrecht) **6**, 1039 (1939).
R. Finkelstein and J. H. Van Vleck, J. chem. Phys. **8**, 790 (1940).
[19] W. M. Walsh, Jr., Phys. Rev. **122**, 762 (1961).
[20] M. Hansen, Constitution of Binary Alloys, McGraw-Hill Publ. Co., New York 1958.
[21] See, e.g., R. C. Weast (Ed.), Handbook of Chemistry and Physics, CRC Press, Inc., Boca Raton 1982 (p. F53).
[22] Y. W. Kim, R. Kaplan, and P. J. Bray, Phys. Rev. **117**, 740 (1960).
[23] R. T. Schumacher and W. E. Vehse, J. Phys. Chem. Solids **24**, 297 (1963).
N. S. Van der Ven, Phys. Rev. **168**, 787 (1968).
[24] See, e.g., J. E. Wertz and J. R. Bolton, Electron Spin Resonance, McGraw-Hill Publ. Co., New York 1972, and references therein.
[25] T. G. Castner, Phys. Rev. **115**, 1506 (1959).
[26] S. Hasegawa and S. Yazaki, Thin Solid Films **55**, 15 (1978).
[27] G. H. White and O. L. Anderson, Phys. Rev. **37**, 430 (1966).
[28] D. L. Griscom, J. non-crystall. Solids **13**, 251 (1973).
[29] See, e.g., A. Abragam, The Principles of Nuclear Magnetism, Clarendon Press, Oxford 1962, and references therein.

(Received May 30, 1988)

phys. stat. sol. (a) **110**, 631 (1988)

Subject classification: 78.20; 72.40; S8.16

A. F. Ioffe Physico-Technical Institute, Academy of Sciences of the USSR, Leningrad[1])

Fundamental Optical Absorption Edge in $MnGa_2Te_4$ Single Crystals

By

G. A. MEDVEDKIN, YU. V. RUD, and M. A. TAIROV

A study is made of the optical properties of oriented $MnGa_2Te_4$ crystals in the region of the fundamental absorption edge. The energy gap width for the temperatures 77, 300, and 370 K is determined to be $E_G = 1.635$, 1.52, and 1.50 eV. The spectral response $\alpha(\hbar\omega)$ is found to follow Urbach's rule throughout the temperature range studied, the slope of the absorption edge remaining constant ($\alpha > 10^2\ cm^{-1}$). Crystal annealing with subsequent rapid cooling results in a shift of the absorption edge longward by 25 meV with the exponential form of $\alpha(\hbar\omega)$ prevailing over the range $T = 77$ to 370 K. An analysis shows the optical absorption in the region of the fundamental edge to be a sum of the effects coming from the density-of-states tails, local scattering centers associated with a high vacancy concentration, and electron–phonon interaction. Optical linear dichroism of the absorption edge of $MnGa_2Te_4$ single crystals with pseudotetragonal structure is revealed and studied. The single crystals are established to be optically uniaxial, their optical transmission dichroism being negative. It is shown that the minimal direct optical transitions in $MnGa_2Te_4$ are allowed in the $\boldsymbol{E} \parallel \boldsymbol{c}$ polarization in the temperature range 77 to 370 K, the crystal-field splitting of the valence band increasing with temperature.

Исследованы оптические свойства в области фундаментального края поглощения ориентированных кристаллов $MnGa_2Te_4$ Определена ширина запрещенной зоны $E_G = 1{,}635$; 1,52 и 1,50 eV при температурах 77, 300 и 370 K. Установлено, что спектры $\alpha(\hbar\omega)$ во всем температурном диапазоне следуют правилу Урбаха, причем крутизна края поглощения сохраняется постоянной ($\alpha > 10^2\ cm^{-1}$). Отжиг кристаллов с последующим быстрым охлаждением приводит к смещению края поглощения в длинноволновую область на 25 meV с сохранением экспоненциального характера $\alpha(\hbar\omega)$ во всем диапазоне $T = 77$ до 370 K. Анализ результатов показывает, что оптическое поглощение в области фундаментального края определяется суммарным эффектом — хвостами плотности состояний, локальными рассеивающими центрами, связанными с высокой концентрацией вакансий и электрон–фононным взаимодействием. Обнаружен и изучен оптический линейный дихроизм края поглощения монокристаллов $MnGa_2Te_4$ с псевдотетрагональной структурой. Установлены оптическая одноосность и отрицательный знак дихроизма оптического пропускания монокристаллов. Показано, что минимальные прямые оптические переходы в $MnGa_2Te_4$ разрешены в поляризации $\boldsymbol{E} \parallel \boldsymbol{c}$ в диапазоне температур 77 до 370 K, а кристаллическое расщепление валентной зоны увеличивается с ростом температуры.

1. Introduction

The ternary compounds $Mn–III_2–VI_4$ belong to a class of magnetic semiconductors which have been enjoying in recent years considerable attention in connection with the possibility of revealing unusual combinations of optical, magnetic, and semiconducting properties [1 to 4]. Compounds of this class form along the quasibinary joins of the Mn–III–VI ternary systems with cations in different valence states.

[1]) Politekhnicheskaya 26, SU-194021 Leningrad, USSR.

In $MnGa_2Te_4$, the Mn^{2+} cation is in a not fully valent state, i.e. the degree of its oxidation is less than the group number in the Periodic Table. At the same time the second cation, Ga^{3+}, is in a fully valent state, its degree of oxidation corresponding to the group number. The formation of the ternary compound, in accordance with the reaction $Ga_2Te_3 + MnTe = MnGa_2Te_4$, does not follow the simple valence principles expounded in [5 to 7] since it is determined by the heterodesmic character of the bonding [8].

Just as most of the compounds in the $II–III_2–VI_4$ group, $MnGa_2Te_4$ belongs to ternary adamantine semiconductors, has a tetrahedral coordination and an ordered defect structure. It is characteristic of substances of this class to manifest crystal lattices of different types belonging to several syngonies, to wit, tetragonal (defect chalcopyrite or thiogallate $I\bar{4}$, defect stannite $I\bar{4}2$ m, pseudocubic $P\bar{4}2m$), trigonal (layer structure R3m), cubic (spinel Fd3m), rhombic (Fddd) [3]. Hence the crystals from the broad class $II–III_2–VI_4$ may reveal, besides optical isotropy, also uniaxial or biaxial optical anisotropy. For the tetragonal structures there is only one optical axis which coincides with a fourth order crystal axis. As shown for the chalcopyrite ($I\bar{4}2d$) crystals [9, 10], the optical anisotropy depends predominantly on the tetragonal compression τ, and is not observed in some crystals with $c/a = 2.0$. For the $II–III_2–VI_4$ crystals, the quasi-isotropic structure ($c/a = 2.0$) is found in the $II–In_2Te_4$ compounds (II = Zn, Cd, Hg), $ZnGa_2Se_4$, $ZnIn_2Se_4$, $HgGa_2Te_4$ [3, 4].

We have already reported preliminary information on the optical and optoelectronic properties of $MnGa_2Te_4$ [11 to 13]. The present paper deals with a detailed analysis of the fundamental absorption edge of $MnGa_2Te_4$ single crystals as a function of the heat treatment of crystals, temperature, and polarization of the incident radiation.

2. Experimental

Single crystals of $MnGa_2Te_4$ were grown by directed crystallization of a stoichiometric melt, the purity of the component elements Mn, Ga, and Te being not worse than 99.999 wt%. No deliberate doping with chemical elements was attempted. X-ray diffraction analysis and microhardness measurements showed the material to be single-phase. Measurements on a JXA-5A electron microprobe analyzer confirmed the formula of the compound to be $MnGa_2Te_4$. The single crystals obtained had a clearly pronounced cleavage plane and could be cleaved well in the direction perpendicular to [001].

Electric conductivity and Hall constant measurements showed the $MnGa_2Te_4$ crystals to be p-type and to have a resistivity of 10^3 to 10^5 Ωcm, free hole concentration $p = 10^{13}$ cm^{-3}, and mobility $\mu_h \approx 1$ cm^2/Vs at $T = 300$ K.

The single crystals of $MnGa_2Te_4$ were oriented by means of the X-ray diffraction and optical polarization techniques. The **c** axis was oriented to within $\pm 1°$. Plates ranging in thickness from 125 to 10 μm and with the orientations (100) and (010) were lapped and rinsed in ethanol, while (001) plates were obtained by cleavage and were not subjected to any additional treatment. The $Cu–MnGa_2Te_4$ Schottky barriers were produced by depositing chemically a copper layer from an aqueous solution of blue vitriol: $CuSO_4 + In = In_2(SO_4)_3 + Cu\downarrow$. The ITO-$MnGa_2Te_4$ heterojunctions were obtained by cathode sputtering of indium and tin oxides in the rate 9:1, the layers being ≈100 nm thick. The current carrying leads to the surface barrier structures were soldered with pure indium.

The spectral response of the optical transmittance T and reflectance R of the crystals, as well as the photoresponse of the Schottky barriers and heterostructures were measured on a КСВУ-3М computerized spectrometer with a subsequent mathematical processing using a СПЕКТР-КСВУ code. Optical and photoelectric responses could also be measured on a spectrometer based on an SPM-2 prism monochromator. The optical absorption coefficient was calculated by the expression

$$\alpha = \frac{1}{d} \ln \left(A + \sqrt{A^2 + R^2}\right) , \tag{1}$$

where $A = (1 - R)^2/2T$, d is the sample thickness (125, 40, 35, 30, 15, and 10 μm).

The temperature behavior was studied in a Dewar flask with a mixture of liquid nitrogen and ethanol serving as cryogenic liquid, the temperature being varied by adjusting properly the liquid composition. As a polarizer we used an IR polaroid ($\Delta\lambda = 700$ to 2500 nm) whose transmission axis, just as the optical axis of the crystal was aligned to within $\pm 1°$. The spectral resolution of the instrumentation was 1 to 2 meV.

3. Crystal Structures

The X-ray diffraction measurements suggest that the substructure of $MnGa_2Te_4$ can be represented by a defect sphalerite-type face-centered cubic lattice with a slight deformation along one of the axes. $CdIn_2Se_4$ is known to have a pseudocubic structure P$\bar{4}$2m close to it [7, 14] (Fig. 1a). For $CdIn_2Se_4$, the lattice parameter ratio $c^*/a = 1.00$ (c^* is the cubic lattice parameter), the cation sublattice having 25% of atomic vacancies. The lattice has a pronounced layered structure with ordered vacancies (one each for a Cd layer, Fig. 1a). For $MnGa_2Te_4$, reflections with even h, k, l multiples of four, to which all strong reflections belong yield tetragonal symmetry, although one observes also "superstructural" reflections. Within the pseudotetragonal structure the parameters of the substructural cell in $MnGa_2Te_4$ are $a = b = 0.6001 \pm 0.0002$ nm, and $c = 1.2090 \pm 0.0005$ nm. The weaker "superstructural" reflections evidence a lowering of the symmetry to monoclinic, although no differences in a and b is observed, and the angle $\beta = 90°$. The true dimensions of the unit cell can in this case be obtained by doubling a and b and magnifying c^* fourfold (Fig. 1b). Since structures with tetrahedral coordination and different kinds of atoms are typically not centrosymmetric, the observed "superstructure" may be assigned to space group $C_c \equiv C_s^4$. Similar to analogous crystals with tetragonal structures I$\bar{4}$2d, I$\bar{4}$, I$\bar{4}$2m, P$\bar{4}$2m, the uniaxial deformation of the lattice of $MnGa_2Te_4$ can be evaluated by means of the parameter of tetragonal compression along [001]: $\tau = (1 - c/2a) \times 100\% =$

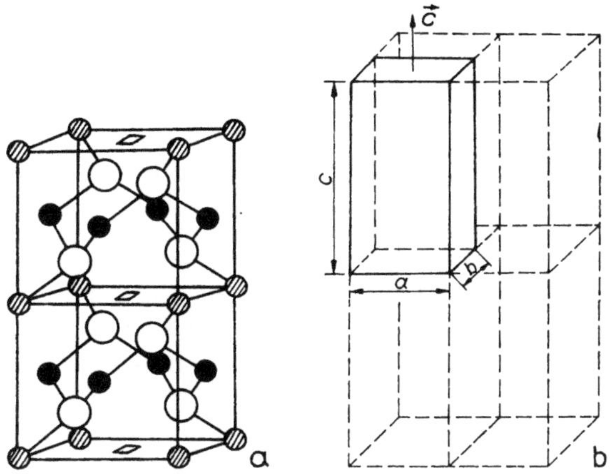

Fig. 1. a) Pseudocubic lattice of $CdIn_2Se_4$ (P$\bar{4}$2m). Open circles Se, full circles In, hatched circles Cd. b) Substructural unit cell of $MnGa_2Te_4$ (solid lines). The optical c axis of $MnGa_2Te_4$ is shown in the crystallographic orientation for the tetragonal syngony. The "superstructural" unit cell of $MnGa_2Te_4$ corresponds to the monoclinic syngony C_c (dashed lines)

Table 1

Tetragonal compression $\tau = (1 - c/2a) \times 100\%$ and the sign of the linear absorption dichroism in quasi-isotropic crystals

compound	space group	$-\tau$ (%)	P_α
$CuInS_2$	$I\bar{4}2d$	0.79	+
$CuInSe_2$	$I\bar{4}2d$	0.48	−
$CuInTe_2$	$I\bar{4}2d$	0.065	−
$MnIn_2Te_4$	$I\bar{4}2m$	0.08	−
$MnGa_2Te_4$	$C_c(P\bar{4}\bar{2}m)$	0.73	+

$= -0.73\%$. This value agrees in amplitude and sign with τ for the compounds $MnIn_2Te_4$ ($I\bar{4}2m$) and $CuInVI_2$ ($I\bar{4}2d$) (see Table 1).

Because of the ionic radius of Ga^{3+} being smaller than that of In^{3+} (0.062 and 0.092 nm, respectively), the probability for an atom to escape from the Ga sublattice in $MnGa_2Te_4$ is higher than that for In in $MnIn_2Te_4$. The ratio $r_{Mn^{2+}}/r_{Te^{2-}} = 0.43$ in these compounds being fairly large, an increase of the coordination number of Te from 4 to 6 and a formation of octahedral in place of tetrahedral voids becomes energetically preferrable [15]. Note also that the cleavage plane in $MnGa_2Te_4$ is perpendicular to the [001] direction, just as is the case with II–VI hexagonal crystals. Hence hexagonal (rhombohedral) patterns leading to polytypes may appear in the layered structure of $MnGa_2Te_4$ (Fig. 1b) [15]. Note that no third order axis has been revealed in the X-ray diffraction studies of $MnGa_2Te_4$. As is shown below, the optical properties of $MnGa_2Te_4$ single crystals lend themselves to a description in terms of uniaxial anisotropy and, thus, the crystal structure of this compound may in the first approximation be considered as pseudotetragonal.

4. Absorption and Reflection in Natural Light

Fig. 2 presents spectral responses of the optical absorbance of the crystals and photo-e.m.f. of the structures at 300 K. The long wavelength tail of fundamental absorption (curve 1) is described satisfactorily by Urbach's rule [16, 17]

$$\alpha = \alpha_0 \exp\left[-\frac{\sigma}{kT}(E_0 - \hbar\omega)\right], \tag{2}$$

where α_0 and E_0 are constants of the semiconductor material. Urbach's parameter $\sigma = 0.22$ to 0.25 at $T = 300$ K corresponds to indirect (or pseudodirect) optical transitions in the crystal. Expression (2) describes well the absorption edge over about two decades: $\alpha = 2 \times 10^2$ to 8×10^3 cm^{-1}. Since the free carrier concentrations in $MnGa_2Te_4$ do not typically exceed $p = 10^{13}$ cm^{-3} at 300 K [11], the light absorption in this compound can be described in the way as is done for doped semiconductors with a nondegenerate carrier gas having Gaussian fluctuations of the r.m.s. potential γ [18]. A remarkable feature of the structure of $MnGa_2Te_4$ (Fig. 1b) is a high concentration of inherent lattice defects, vacancies (5×10^{21} cm^{-3}), and of probable antistructural interchanges of the cations, Mn $\rightleftarrows$ Ga, or of the cations with vacancies, Mn, Ga, and V, as well as an alternation in the cation layers (polytypes) [14, 15]. This complex ensemble of inherent defects in the $MnGa_2Te_4$ lattice produces a random Coulomb field of charged local centers $E = ze/\varepsilon r^2$. The contribution to the absorbance of the concentration fluctuations z of the atoms Mn, Ga and structural vacancies in a p-type semiconductor can be expressed as was done in [18]:

$$\alpha \sim \exp\left(-\frac{z^2}{N_t r^3} - \frac{r\sqrt{m_h(E_G - \hbar\omega)}}{\hbar}\right), \tag{3}$$

where $N_t = N_A + N_D$ is the total concentration of the charged acceptors and donors, and r is the linear dimension of the fluctuation.

Experimental data on the spectral response of the absorbance in the range $\alpha > 8 \times 10^3\ cm^{-1}$ in the ternary compound were obtained by means of a surface barrier structure in the form of a Schottky barrier and a heterojunction. When a Cu–$MnGa_2Te_4$ structure is illuminated from the side of the metal, or an ITO-$MnGa_2Te_4$ heterojunction, from that of the oxide, the response of the photo-e.m.f. corresponds to photoactive absorption in the crystal (Fig. 2, curves 3 and 2), since the photo-generated charge carriers become separated in the space charge region of the semiconductor. The long-wavelength growth of $U_{ph}(\hbar\omega)$ has, similar to the $\alpha(\hbar\omega)$, an exponential form with a smooth transition to saturation for $\hbar\omega > 2.0$ eV (curve 3). The fall-off of the photo-e.m.f. amplitude in the ITO–$MnGa_2Te_4$ heterojunction at photon energies >2.3 eV is associated with interference in the transparent anti-reflection layer of the ITO.

We carried out a measurement of the optical constants of $MnGa_2Te_4$ on an ЛЭФ-3M laser ellipsometer at a wavelength $\lambda = 632.8$ nm to derive the absolute value of α in the region $\hbar\omega > E_G$. In analogy with $CuInSe_2$ [19], the model used was an isotropically absorbing crystal with a transparent oxide film ($n = 2.12$). For a $MnGa_2Te_4$ crystal with a polished specular surface the complex refractive index is $N = n - ik = 3.34 - i0.83$ for an oxide thickness of 4.7 nm. The optical absorbance calculated by the expression $\alpha = 4\pi k/\lambda$ was used to construct $\alpha(\hbar\omega)$ curves in the region $\hbar\omega > E_g$.

In a heavily doped semiconductor with indirect optical transitions the momentum conservation can be ensured either by the involvement of phonons or, for instance, due to electron–electron or impurity scattering [20]. The scattering probability is in this case proportional to the number of scatterers N_S, no participation of phonons being required. In the semiconductor $MnGa_2Te_4$ the number of ordered and antistructural defects acting as scatterers is very large. Therefore we used the following expression for the absorbance of a nondegenerate semiconductor [20]:

$$\alpha(\hbar\omega) = A_1 N_S(\hbar\omega - E_G)^2\,, \tag{4}$$

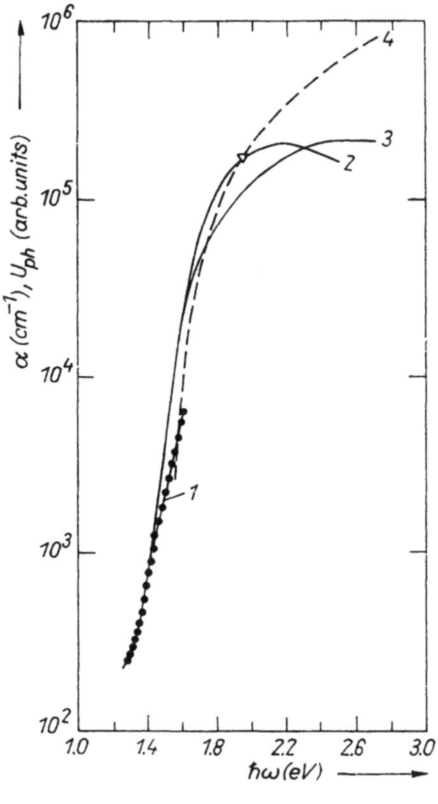

Fig. 2. Edge absorption of $MnGa_2Te_4$ for 300 K; ●, ▽, (1) to (3) experiment, (4) calculation by equation (4) in parabolic approximation, ● optical absorption, (2) photo-e.m.f. of ITO-$MnGa_2Te_4$ heterojunction, (3) photo-e.m.f. of Cu–$MnGa_2Te_4$ Schottky barrier; ▽ ellipsometric data

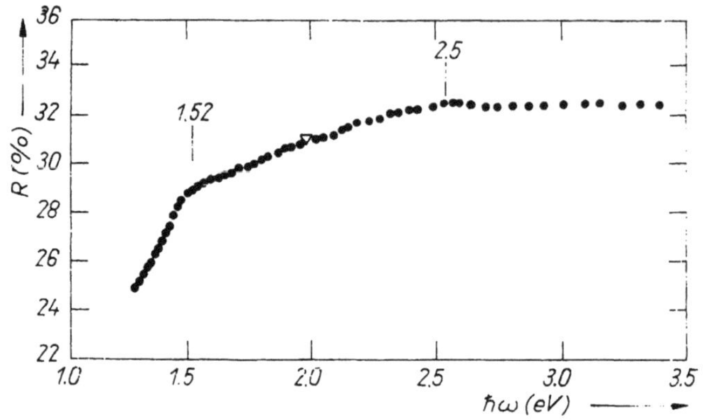

Fig. 3. Optical reflection spectra of $MnGa_2Te_4$ single crystals for 300 K (●); ▽ ellipsometric data

where A_1 is the material constant which cannot presently be calculated accurately since a number of fundamental parameters of $MnGa_2Te_4$, such as the momentum matrix element, and the electron and hole effective masses are not known. Therefore to compare the calculations with experiment, we used the values of α derived from the ellipsometric data (point ∇ in Fig. 2).

The best agreement of (4) with the experimental data is obtained with $E_g^{opt} = 1.50$ eV at 300 K (curve 4). As seen from Fig. 2, the simple parabolic approximation describes well the experiment in the intermediate region of optical absorption, $\alpha = 4 \times 10^3$ to $2 \times 10^5\ cm^{-1}$. Deep in the fundamental band, for $\hbar\omega > 2.0$ eV, one observes considerable disagreement with the experimental spectra of photoactive absorption (curves 2, 3) which implies a more complex than parabolic law for the energy bands of $MnGa_2Te_4$. The long wavelength tail in $\alpha(\hbar\omega)$ likewise deviates from the theoretical curve 4 and is described by an exponential form of type (3) or (2). The value of the optical width of the gap derived from the parabolic approximation is, as a rule, somewhat smaller than the E_g obtained from Van Hove's singular points, for instance, from reflectance spectra.

Fig. 3 presents a spectral response of the optical reflectance of $MnGa_2Te_4$ obtained in unpolarized light at $T = 300$ K. In the long wavelength region one observes a fast growth of $R(\hbar\omega)$ followed by a sharp rollover at the energy $\hbar\omega = 1.52$ eV. This singular point corresponds to the energy of strong interband transitions in $MnGa_2Te_4$ and can be compared with the energy gap width of this semiconductor. As the photon energy increases still further, the reflectance amplitude grows gradually up to 32 to 33% reaching a diffuse maximum in the region of ≈ 2.5 eV. Throughout the range 1.3 to 3.4 eV the reflectance of $MnGa_2Te_4$ has a positive dispersion as is the case, for example, with the known III–V compounds [21]. The value of R at $\hbar\omega = 1.96$ eV derived from an independent ellipsometric measurement (point ∇) agrees well with the spectral response $R(\hbar\omega)$ (Fig. 3). The reflectance spectrum of $MnGa_2Te_4$ is smoother than that of $MnIn_2Te_4$, a compound representing a chemical analog of $MnGa_2Te_4$ and differing from it in crystal structure. This may imply a greater structural disorder as is the case, for instance, with polycrystalline $MnIn_2Te_4$ films [22].

5. Temperature Behavior of Edge Absorption

As the temperature is varied within $T = 370$ to 77 K, the fundamental absorption edge of $MnGa_2Te_4$ shifts to shorther wavelengths following Urbach's rule (2). Fig. 4

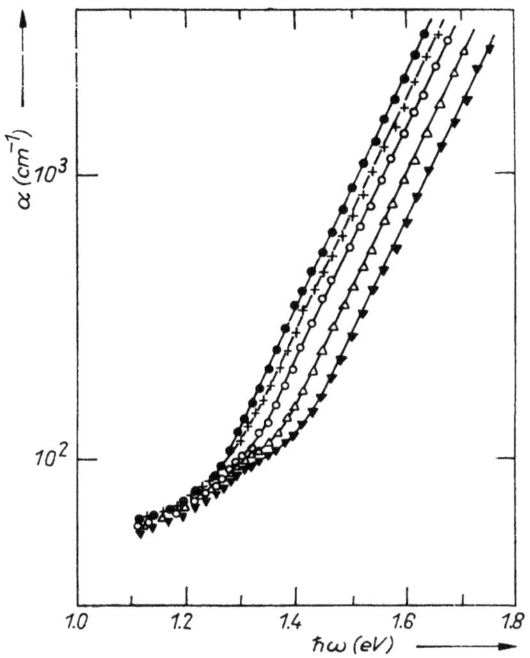

Fig. 4. Shift of the exponential absorption edge of $MnGa_2Te_4$ in the range $T = 370$ to 77 K, crystal thickness 125 and 30 μm

displays the results of temperature dependence measurements of $\alpha(\hbar\omega)$ made on samples 125 and 30 μm thick. The experimental points obtained for $\alpha > 10^2$ cm^{-1} lie on a straight line in log scale for temperatures from 370 to 77 K. The curves displace with varying temperature parallel to one another so that the focal point lies at infinity. A similar temperature-induced shift of the absorption edge was obtained also on samples produced in different ways and with thicknesses down to 15 μm. Depending on the actual growth conditions in the various modifications of the directed crystallization technique, the absorption edge may vary in slope, $s = \mathrm{d}(\ln\alpha)/\mathrm{d}(\hbar\omega)$, within 7 to 10 eV^{-1}, for undoped $MnGa_2Te_4$ single crystals. Note that the major features of the spectral response $\alpha(\hbar\omega)$ persist, as can be seen from Fig. 4. For $\hbar\omega <$ < 1.3 eV, the long wavelength tail flattens still more, the spectral response $\alpha(\hbar\omega)$ becoming practically temperature-independent. This region corresponds to absorption involving levels of the inherent defects in $MnGa_2Te_4$ whose high concentration in the energy gap of the semiconductor may result in the formation of an impurity band. This is supported also by the low mobility of holes in $MnGa_2Te_4$ which may be accounted for by impurity band conduction. The energy of optical carrier activation from this band may be evaluated as $E_{act} = 0.2$ to 0.4 eV (Fig. 4). These features in the spectra $\alpha(\hbar\omega)$ suggest that the major contribution to the formation of the energy spectrum in the ternary semiconductor comes from the inherent lattice defects which have loosened the crystal structure and produce density-of-states tails in the energy gap of $MnGa_2Te_4$.

The temperature behavior of the energy gap of $MnGa_2Te_4$ single crystals derived from the spectral responses $\alpha(\hbar\omega)$ is presented in Fig. 5 (curve 1). The values of E_g were obtained from measurements of $\alpha(\hbar\omega)$ at a level of $\approx 7 \times 10^3$ cm^{-1} made on samples ≈ 10 μm thick. An analysis of these spectra by (4) yields values of E_g smaller than the ones given above. However, the course of the $E_g(T)$ relationship persists (curve 1) showing that the temperature coefficient of the energy gap for crystals from different production lots lies in the range $\mathrm{d}E_g/\mathrm{d}T = -(4.5$ to $4.8) \times 10^{-4}$ eV/K. This value is close to the corresponding coefficient for $MnIn_2Te_4$ [22] and is about three times that for the analogous crystals $CuInVI_2$ [19, 23] having a same small tetragonal compression $\tau \leq 0$ (see Table 1).

In order to study the effect of inherent defects on the energy spectrum of $MnGa_2Te_4$, the samples were thermally annealed in vacuum at $T = 700$ °C for 30 min with a subsequent rapid cooling ($v \approx 10^3$ K/min). When compared with as-grown material,

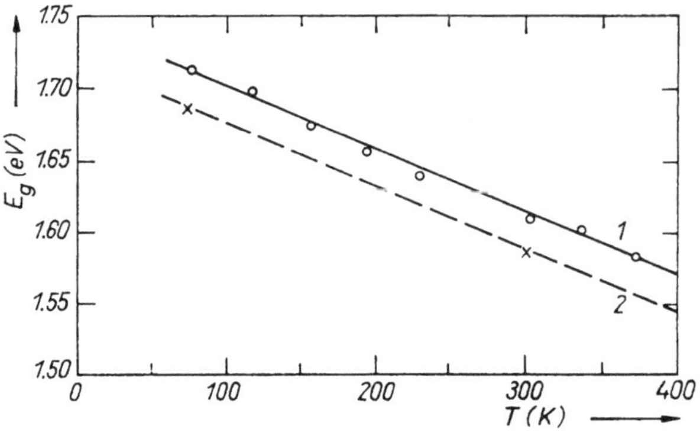

Fig. 5. Temperature dependence of the energy gap of $MnGa_2Te_4$ in unpolarized light, orientation (001), crystal thickness 10 μm. The value of E_g determined is at a level $\alpha = 7 \times 10^3$ cm^{-1}. Points: experiment, ○ before, × after thermal treatment

these samples revealed a parallel displacement of the absorption edge to longer wavelengths by ≈25 meV without any change in the major features of $\alpha(\hbar\omega)$.

The temperature dependence of E_g of $MnGa_2Te_4$ single crystals with frozen-in thermal defects is similar to that presented in Fig. 5 (curve 2). Thus, one can conclude that a predominant contribution to the formation of the edge absorption in $MnGa_2Te_4$ comes from the inherent lattice defects of the ternary semiconductor.

Urbach's parameter describing the exponential long wavelength tail of $\alpha(\hbar\omega)$ seen in unpolarized light in $MnGa_2Te_4$ obeys the relation [16]

$$\sigma = \sigma_0 \left(\frac{2kT}{\hbar\omega}\right) \text{th} \left(\frac{\hbar\omega_p}{2kT}\right), \tag{5}$$

where σ_0 is the maximum value of Urbach's parameter, $\hbar\omega_p$ is the energy of phonons participating in the electron–phonon interaction. Throughout the temperature range studied, the parameter σ depends linearly on T (Fig. 6, curve 1). Introducing thermal defects into $MnGa_2Te_4$ crystals does not affect this relationship. The theoretical curve approximates well the experiment (points +) for $\sigma_0 = 0.6$ and $\hbar\omega_p = 120$ meV. Since thermal treatment does not result in qualitative changes in the edge absorption mechanism, one may conclude that the inherent lattice vacancies in $MnGa_2Te_4$ dominate over the electron–phonon mechanism.

6. Polarization Studies of Optical Absorption

By (1), the dominant contribution to the polarization dependence of the optical absorption comes from transmittance T. The anisotropy in R affects the amplitude of α in a much lesser degree. Therefore in our polarization measurements the possible anisotropy in the reflection coefficient was disregarded.

The spectra of $\alpha_\parallel$ and $\alpha_\perp$ for the corresponding polarizations $\boldsymbol{E} \parallel \boldsymbol{c}$ and $\boldsymbol{E} \perp \boldsymbol{c}$ were calculated by (1) from the experimental data on $T_\parallel$, $T_\perp$, and R. The splitting of the optical edge absorption spectra of $MnGa_2Te_4$ in linearly polarized light turns out to be not large, $\Delta\alpha = (\alpha_\parallel - \alpha_\perp) \approx 200$ cm^{-1}, which is much smaller than that for its direct-gap analog $MnIn_2Te_4$ [22]. The exponential form of the $\alpha(\hbar\omega)$ curves persists in both natural (Fig. 2) and linear polarized light. The absorption edge of single crystals $MnGa_2Te_4$ seen in $\boldsymbol{E} \parallel \boldsymbol{c}$ polarization turns out to be shifted towards longer wavelengths compared to $\boldsymbol{E} \perp \boldsymbol{c}$ polarization. The steepness of $\alpha(\hbar\omega)$ is larger in the parallel polarization implying that the minimum transitions in $MnGa_2Te_4$, unlike manganese and indium telluride, occur primarily under $\boldsymbol{E} \parallel \boldsymbol{c}$. Therefore the linear dichroisms of $MnGa_2Te_4$ and $MnIn_2Te_4$ have opposite signs (see Table 1) which is another indication of a strong difference in the crystal structure between the two compounds. However, the existence of an optical anisotropy in both compounds characterized by quasi-isotropic principal translations in their lattices evidences a substantial role of ordered lattice vacancies and local electrostatic field effects.

Fig. 6 presents Urbach's parameters derived from the $\alpha(\hbar\omega)$ curves obtained in linearly polarized light as a function of T. The parameters σ are essentially different for $\boldsymbol{E} \parallel \boldsymbol{c}$ and $\boldsymbol{E} \perp \boldsymbol{c}$ polarizations: indeed, $\sigma_\perp$ is linearly dependent on temperature (curve 3) and approximates a superlinear form for $T < 260$ K, while $\sigma_\parallel$ follows a sublinear relationship (curve 2). Calculation by (5) yields the following constants: $\sigma_{0,\parallel} = 0.39$, $\hbar\omega_{p,\parallel} = 74$ meV and $\sigma_{0,\perp} = 0.50$, $\hbar\omega_{p,\perp} = 120$ meV, which describe satisfactorily the experiment (points in Fig. 6). This difference can be attributed to an anisotropy both of the phonon spectrum and of local scatterers in the crystal. Thus, the results displayed in Fig. 4 and 6 suggest that the fundamental absorption edge in $MnGa_2Te_4$ originates from several mechanisms, namely, electronic transitions to the density-of-states tails, electron–phonon interaction, and scattering from local centers, such as structural vacancies and antistructural defects.

We carried out a complete set of polarization measurements in a crystallographic orientation corresponding to the tetragonal syngony [15]. Fig. 7 presents spectral responses of the linear dichroism in optical transmittance for three directions: [100], [010], and [001] obtained at $T = 300$ K. $MnGa_2Te_4$ single crystals do not exhibit any dichroism in the direction [001] throughout the range studied, $\hbar\omega = 1.25$ to 1.75 eV (Fig. 7, points ○). At the same time for the directions [100] and [010] the spectra $P_T(\hbar\omega)$ are totally similar (points × and ●). Hence the direction [001] defines the optical axis $\boldsymbol{c}$ in the crystal, suggesting that by the character of its linear dichroism the semiconducting compound $MnGa_2Te_4$ may be attributed to optically uniaxial crystals.

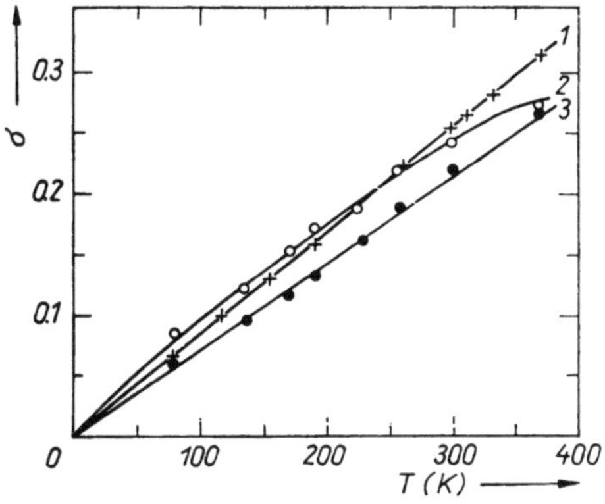

Fig. 6

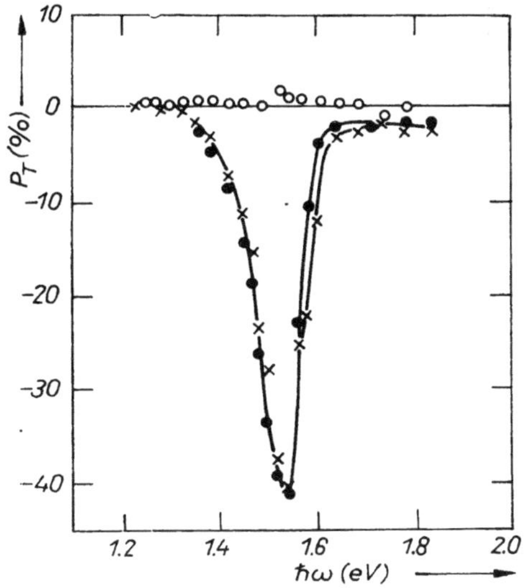

Fig. 7

Fig. 6. Temperature dependence of Urbach's parameters for (1) unpolarized light (+) and two polarizations, (2) $\boldsymbol{E} \parallel \boldsymbol{c}$ (○) and (3) $\boldsymbol{E} \perp \boldsymbol{c}$ (●). Points: experiment, (1) to (3) calculation by (5)

Fig. 7. Linear dichroism spectra of the transmittance of $MnGa_2Te_4$ in three orientations. Light propagation along ○ [001]; × [100]; ● [010]; $d = 40\mu m$; $T = 300$ K

Fig. 8 presents spectra of the linear dichroism of the optical absorbance and transmittance of $MnGa_2Te_4$ plates of different thickness. P_α reaches a maximum $\approx +8\%$ at an energy $\hbar\omega = 1.55$ eV (curve 1). The quantities P_α and P_T are close in magnitude and sign to the corresponding parameters of the chalcopyrite crystal $CuInS_2$. In the region of strong optical absorption, $\alpha_{\parallel,\perp} d > 1$, the dichroism of optical transmittance depends nonlinearly on $\Delta\alpha = (\alpha_\parallel - \alpha_\perp)$ (or P_α) [19]

$$P_T = \frac{|T_\parallel - T_\perp|}{T_\parallel + T_\perp} = \frac{1 - \exp(\Delta\alpha d)}{1 + \exp(\Delta\alpha d)}. \tag{6}$$

As the crystal thickness increases from 15 to 35 μm, the maximum amplitude of P_T grows, in accordance with (6), from 30 to 48% (curves 2, 3), the lineshape of the linear dichroism acquiring a more symmetrical bell-shaped form. Therefore the curves $\alpha_\parallel(\hbar\omega)$ and $\alpha_\perp(\hbar\omega)$ have an equal slope in the corresponding spectral region, no features in the form of local levels being revealed. By going over to small thicknesses, one can reach the energies of the minimum interband transitions in $MnGa_2Te_4$. As follows from Fig. 8, curve 3 is made up of two $P_T(\hbar\omega)$ bands, namely, the dominant maximum lies at 1.52 eV while the short-wavelength shoulder is centered near 1.57 eV. For a crystal of a larger thickness, $d = 35$ μm (curve 2), these features are spectrally not resolved, since for $\hbar\omega > 1.55$ eV the amplitude of P_T falls off sharply because of the growing absorbance. A similar shape of the $P_T(\hbar\omega)$ profile was obtained by us for small thicknesses of $MnIn_2Te_4$, 15 and 10 μm [22], where the short wavelength maximum was found to correspond to a higher-lying interband transition. The minimum transitions in the region 1.4 to 2.0 eV for $MnGa_2Te_4$ are sufficiently well approximated by expression (4) for indirect optical transitions (Fig. 2) as well as the pseudodirect transitions in the chalcopyrite II–IV–V_2 crystals [10]. Therefore the notations of Fig. 8 are similar to those of the pseudodirect transitions A′, B′, C′ in the analogous crystals. The data obtained on $MnGa_2Te_4$ single crystals permitted identification of the following pseudodirect (indirect) transitions: A′ = 1.52 eV (‖), B′ = 1.57 eV (‖, ⊥), and C′ = 1.66 eV (⊥) at $T = 300$ K. The selection rules for these transitions turn out to be opposite to those for the direct transitions A, B, C in $MnIn_2Te_4$ and $HgIn_2Te_4$ having a defective stannite structure [22, 24]. Such a change in selection rules is known to exist for the chalcopyrite II–IV–V_2 crystals [25], however, for $MnGa_2Te_4$ no calculation of the band structure has yet been made.

As the temperature varies from 370 to 77 K, the spectral profile of the transmittance dichroism shifts toward higher photon energies with a coefficient dE_g/dT. As seen

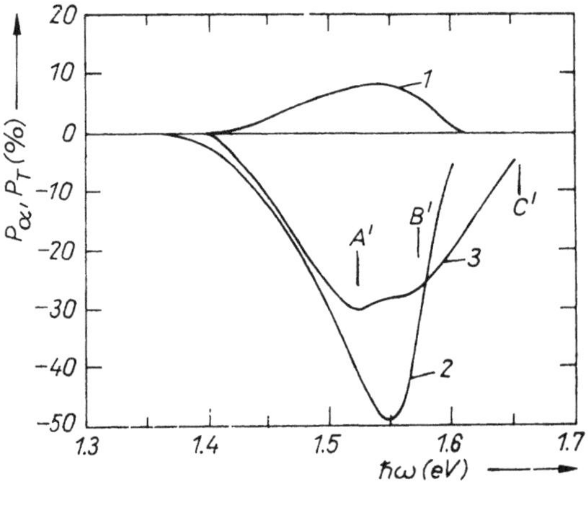

Fig. 8. Linear dichroism spectra of $MnGa_2Te_4$ for different crystal thicknesses: (1), (2) $d = 35$; (3) 15 μm; $T = 300$ K

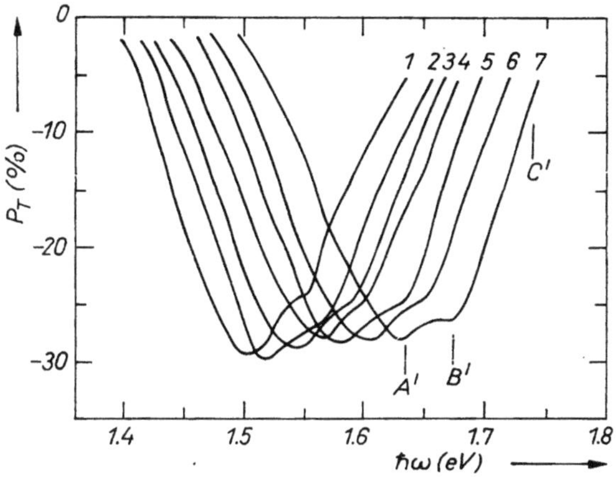

Fig. 9. Temperature-induced shift of $P_T(\hbar\omega)$ spectra of $MnGa_2Te_4$ in the range 370 to 77 K; $d = 15$ μm. (1) $T = 370$, (2) 300, (3) 268, (4) 224, (5) 167, (6) 138, (7) 77

from Fig. 9, the $P_T(\hbar\omega)$ profile has a complex shape for all temperatures and consists of two bands corresponding to pseudodirect (indirect) optical transitions in $MnGa_2Te_4$. Hence the valence band splitting in $MnGa_2Te_4$ is observed at all temperatures throughout the range 77 to 370 K. Using the quasi-cubic model of Hopfield [26] developed for the II–VI wurtzite crystals and subsequently extended to cover the chalcopyrite [10], defective stannite [14], and other tetragonal structures [24], one may calculate the band parameters Δ_{cr} and Δ_{so} for $MnGa_2Te_4$ within the framework of this model. By [10], the valence band splitting Δ_{cr} and Δ_{so} is given by the expression

$$E_{1,2} = -\tfrac{1}{2}(\Delta_{so} + \Delta_{cr}) \pm \tfrac{1}{2}[(\Delta_{so} + \Delta_{cr})^2 - \tfrac{8}{3}\Delta_{so}\Delta_{cr}]^{1/2} . \quad (7)$$

Here the quantities E_1 and E_2 determine the energies of the levels with irreducible representation Γ_7 relative to the Γ_6 level. The valence band splitting and minimum pseudodirect (indirect) transition energies for $MnGa_2Te_4$ are presented in Table 2 for three temperatures.

Fig. 10 displays the temperature behavior of the A′, B′, C′ transition energies and of the amplitude of $P_T(A')$ and $P_T(B')$ derived from the linear dichroism spectra (Fig. 9). The energies A′, B′, C′ vary linearly in the temperature range 77 to 370 K, so that the spin–orbit splitting practically does not change while the crystalline splitting grows with increasing T. Since Δ_{cr} is related with the tetragonal compression through [10]:

$$\Delta_{cr} = -3b\left(1 - \frac{c}{2a}\right), \quad (8)$$

where b is the deformation potential of the valence band in a cubic analog, one may assume that the tetragonal distortion of the $MnGa_2Te_4$ lattice should grow with increasing temperature. No X-ray diffraction data on the temperature-induced

Table 2
Band parameters of $MnGa_2Te_4$ (in eV)

(K)	A′	B′	C′	$-\Delta_{cr}$	Δ_{so}
77	1.635	1.68	1.74	0.07	0.06
300	1.52	1.57	1.66	0.10	0.06
370	1.50	1.55	1.64	0.10	0.06

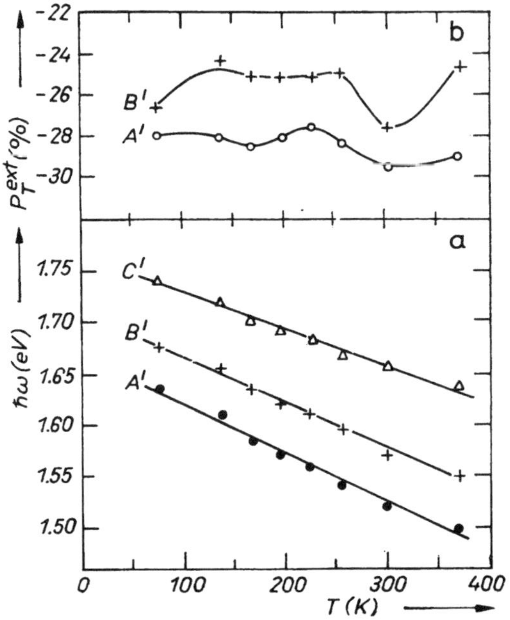

Fig. 10. Temperature behavior of the features in the $P_T(\hbar\omega)$ spectra corresponding to optical transitions A', B', and C' (a) and of the extrema $P_T^{ext}(\hbar\omega)$ for the transitions A' and B' (b)

variation of the lattice parameters of this compound being available, the above hypothesis awaits experimental verification.

Thus the fundamental absorption edge in $MnGa_2Te_4$ originates from the density-of-states tails and anisotropic scattering centers due to the high concentration of inherent lattice defects, as well as from the electron–phonon interaction. Single crystals of this ternary compound, despite their quasi-isotropic crystal structure, represent clearly pronounced anisotropic materials with one optical axis and a negative linear dichroism of optical transmission.

Acknowledgement

The authors are grateful to Dr. A. A. Vaipolin for X-ray diffraction data on the crystals studied.

References

[1] E. L. NAGAEV, Physics of Magnetic Semiconductors, Izd. Nauka, Moscow 1979 (in Russian).
[2] Proc. FIAN, Magnetic Semiconductors, Vol. 139, Izd. Nauka, Moscow 1982 (in Russian).
[3] A. V. NOVOSELOVA and V. B. LAZAREV (Ed.), Handbook Physicochemical Properties of Semiconducting Materials, Izd. Nauka, Moscow 1979 (in Russian).
[4] LANDOLT-BÖRNSTEIN, Physics of Ternary Compounds, Ed. O. MADELUNG, Vol. 17h, Springer-Verlag, 1985.
[5] H. GRIMM and A. SOMMERFELD, Z. Phys. 36, 36 (1926).
[6] C. H. L. GOODMAN, J. Phys. Chem. Solids 6, 305 (1958).
[7] N. A. GORYUNOVA, Complex Diamond-Like Semiconductors, Soviet Radio, Moscow 1968 (in Russian).
[8] V. D. PROCHUKHAN, Classification of Ternary Semiconductors according to the Character of Interaction in the Ternary Systems, Proc. Vth All-Union Conf. Ternary Semiconductors and Their Applications, Shtiintsa, Kishinev 1987 (p. 9).
[9] G. A. MEDVEDKIN, YU. V. RUD, and M. A. TAIROV, Polarization Photosensitivity Effects in the Ternary Semiconductors II–IV–V_2, Leningrad, PTI Preprint No. 1185, 1987.

[10] J. L. Shay and J. H. Wernick, Ternary Chalcopyrite Semiconductors: Growth, Electronic Properties and Applications, Pergamon Press, 1975.
[11] R. N. Bekimbetov, A. A. Vaipolin, N. N. Konstantinova, L. V. Kradinova, G. A. Medvedkin, V. D. Prochukhan, Yu. V. Rud, and M. A. Tairov, Crystal Res. Technol. **22**, K238 (1987).
[12] R. N. Bekimbetov, G. A. Medvedkin, V. D. Prochukhan, Yu. V. Rud, and M. A. Tairov, Zh. tekh. Fiz., Pisma **13**, 1040 (1987).
[13] R. N. Bekimbetov, Yu. V. Rud, and M. A. Tairov, Fiz. Tekh. Poluprov. **21**, 1916 (1987).
[14] A. Miller, A. MacKinnon, and D. Weaire, Solid State Phys. **36**, 119 (1981).
[15] M. P. Shaskolskaya, Crystallography, Vysshaya Shkola, Moscow 1976 (in Russian).
[16] M. V. Kurik, phys. stat. sol. (a) **8**, 9 (1971).
[17] B. L. Gelmont, V. I. Perel, and I. N. Yassievich, Fiz. tverd. Tela **25**, 727 (1983).
[18] B. I. Shklovskii, and A. L. Efros, Electronic Properties of Doped Semiconductors, Springer Series in Solid State Sciences, Vol. **45**, Springer-Verlag, 1984.
[19] G. A. Medvedkin, Yu. V. Rud, and M. A. Tairov, phys. stat. sol. (b) **144**, 809 (1987).
[20] J. Pankove, Optical Processes in Semiconductors, Prentice-Hall, Inc., Englewood Cliffs, New Jersey 1971.
[21] Yu. I. Ukhanov, Optical Properties of Semiconductors, Izd. Nauka, Moscow 1977 (p. 368) (in Russian).
[22] G. A. Medvedkin, V. D. Prochukhan, Yu. V. Rud, and M. A. Tairov, to be published.
[23] G. A. Medvedkin, Yu. V. Rud, and M. A. Tairov, Optika i Spektroskopiya, to be published (1988).
[24] A. MacKinnon, A. Miller, and G. Ross, Proc. IIIrd Internat. Conf. Ternary Compounds, Edinburgh, 1977, Inst. Phys. Conf. Ser. **35**, 171 (1977).
[25] A. Shileika, Optical Studies of the Band Structure of $A^2B^4C_2^5$ Compounds, in: Electrons in Semiconductors, Vol. 1, Mokslas, Vilnius 1978 (p. 143).
[26] J. J. Hopfield, J. Phys. Chem. Solids **15**, 97 (1960).

(Received June 7, 1988)

phys. stat. sol. (a) **110**, 645 (1988)

Subject classification: 78.65; S12

Faculty of Engineering, Yokohama National University[1])

Uniaxial Stress-Effect on Visible and Near-Infrared Absorption Spectra of Si(phthalocyaninato) $(OH)_2$ Epitaxial Films

By

M. YUDASAKA, M. TANAKA, Y. KUWAE[2]), K. NAKANISHI, and S. KURITA

Piezotransmission measurements are applied to determine the changes of absorption spectra of Si(phthalocyaninato) $(OH)_2$ deposited films due to uniaxial stress. The changes of the spectra under the compressive forces along the fiber-axis of the deposited films indicate that there are four molecules in a unit cell.

Piezotransmissionsmessungen werden zur Bestimmung der Änderung von Absorptionsspektren von aufgedampften Si(phthalocyaninato) $(OH)_2$-Schichten infolge uniaxialer Spannung angewendet. Die Änderungen der Spektren unter Kompressionskräften in Richtung der Fiberachsen der abgeschiedenen Schichten zeigen, daß sich vier Moleküle in einer Einheitszelle befinden.

1. Introduction

Thin films of Si(phthalocyaninato) $(OH)_2$ (abbreviated as Si(Pc) $(OH)_2$ hereafter) deposited on mica substrates have fiber-like structures [1, 2]. Their absorption spectra are reported to show strong dependence on the polarization directions of incident light. When the polarization direction is parallel to the fiber axis, absorption maxima appear around 1.41 and 1.65 eV. When the polarization direction is perpendicular to the fiber-axis (c-axis), maxima appear around 2.0, 1.75, and 1.38 eV. Although we gave them some consideration in earlier work, the details were not clarified [3].

Piezomodulation spectroscopy, in which one measures changes of reflectance or transmission due to periodic stress, is a well-known experimental technique for inorganic semiconductors. It was established in the middle of the sixties and has yielded a lot of successful results on the electronic structures of semiconductors [4]. Absorption coefficients of semiconductors are too large to detect transmission in the VIS region, so that piezoreflectance is usually applied. However, piezotransmission can be applied in some cases, for example, for color centers in alkali halides [5]. Piezotransmission gives the same information on the pressure dependence of electronic structures as piezoreflectance does. In addition information from piezotransmission is more direct and clearer than that from piezoreflectance, because the latter always has some ambiguity due to the Kramers-Kronig transformation. Eckhardt and co-workers firstly applied piezoreflectance to molecular crystals and showed the piezomodulation

[1]) Tokiwadai 156, Hodogaya-ku, Yokohama 240, Japan.

[2]) Present address: Canon Incorporation, Morinosato, Atsugi-shi, Kanagawa-ken, Japan.

spectroscopy to be a powerful method even for organic compounds [6]. $Si(Pc)(OH)_2$ film is a suitable sample to measure piezotransmission, because it can be prepared with a definite fiber axis and with high transmissivity.

The aim of this paper is to develop the interpretation for the absorption spectra of $Si(Pc)(OH)_2$ deposited films by measuring piezotransmission spectra with uniaxial stress parallel to the fiber axis.

2. Experimental

The films were formed on mica substrates by the vacuum evaporation method described before [2]. The thickness of the films was about several hundred nm. The absorption spectra were measured with a Shimazu UV-160 spectrometer. In the piezotransmission measurements, a halogen lamp was used as light source in combination with grating monochromator and polarizer. Uniaxial stress was applied to the sample along the fiber axis by using a transducer in the same way as shown in Fig. 1 of [7]. The transducer was driven by an alternating voltage (frequency $\approx$ 80 Hz) with an amplitude of 2000 V. Transmitted light was detected by a photomultiplier. The signal was fed into a dc voltmeter and a lock-in amplifier which yielded the average transmission, I, and the change of transmission induced by the uniaxial stress, ΔI, respectively. ΔI corresponds to $I(p) - I(-p)$, where $I(p)$ and $I(-p)$ are transmission under compression and expansion, respectively. The ratio, $\Delta I/I$, was sent to a digital voltmeter interfaced to a personal computer. The Lambert-Beer law, $I/I_0 = \exp(-\alpha d) = \exp(-D_{opt})$, relates $\Delta I/I$ to the change in optical density, $\Delta D_{opt} = -\Delta(\ln(I/I_0))$. In the case of an isolated absorption band $\alpha(E)$, one can expect that ΔD_{opt} is proportional to $\alpha(E)$ when the transition probability changes, and to $d\alpha(E)/dE$ when the absorption band shifts rigidly [8].

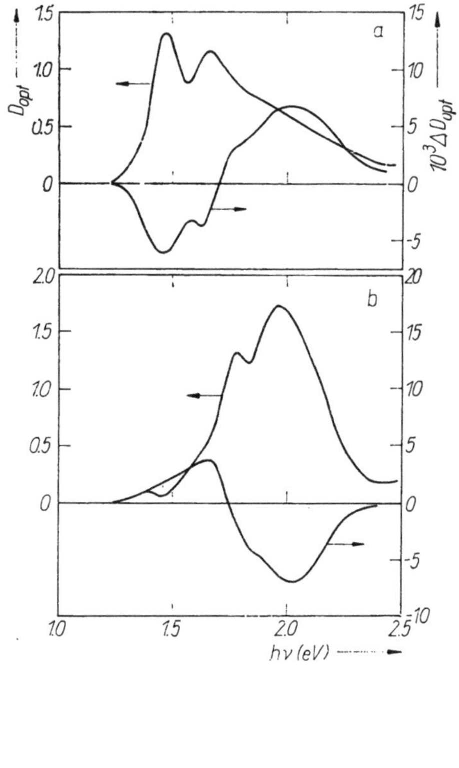

Fig. 1. Spectra of D_{opt} and ΔD_{opt} of the $Si(Pc)(OH)_2$ epitaxial film. Polarization directions of incident light are a) parallel and b) perpendicular to the c-axis of the film

Table 1

Energies of the maxima, minima, and shoulders of D_{opt} and ΔD_{opt} spectra in Fig. 1. The symbols ↑ and ↓ indicate that D_{opt} increases and decreases by compressive stresses, respectively

parallel			perpendicular		
D_{opt} (eV)	ΔD_{opt} (eV)		D_{opt} (eV)	ΔD_{opt} (eV)	
	↑ 2.02	e↑	1.96	↓ 2.02	f↓
	↑ 1.76	h↑	1.78	↓ 1.82	g↓
1.65	↓ 1.63	d↓		↑ 1.66	b↑
1.47	↓ 1.47	a↓	1.39	↑ ?	c↑

For explanation of the symbols (a to h), see Fig. 2.

3. Results and Discussion

The D_{opt} and ΔD_{opt} spectra of the Si(Pc) $(OH)_2$ films are shown in Fig. 1. The positive value of ΔD_{opt} indicates that D_{opt} increases under compressive stress along the c-axis. The polarization direction of the incident light is parallel to the c-axis in Fig. 1a and perpendicular in Fig. 1b.

The energies of the maxima, minima, and the shoulders of the D_{opt} and ΔD_{opt} spectra are listed in Table 1. The symbols, ↑ and ↓ in the ΔD_{opt} columns mean that the sign of ΔD_{opt} is positive and negative, respectively. It is apparent from Fig. 1 and Table 1 that by compressive stress along the c-axis, the absorption spectra in parallel polarization increase around 1.76 and 2.02 eV, but decrease around 1.47 and 1.63 eV. The large maxima of ΔD_{opt} around 2.02 eV and the shoulder around 1.76 eV indicate that two absorption bands appear around 2.02 and 1.76 eV under compression.

It is recognized in perpendicular polarization in Fig. 1b and Table 1 that the bands near 1.96 and 1.78 eV have a tendency to decrease in intensity. The value of ΔD_{opt} around 1.39 eV induced by the stress is difficult to describe convincingly but it seems to be slightly positive. The maximum of ΔD_{opt} around 1.66 eV in Fig. 1b shows that an absorption maximum appears around 1.66 eV under compression along the c-axis.

The absorption spectra and their changes induced by uniaxial stress indicate that there are eight absorption bands as seen in Table 1. They can be considered to result from the Davydov splitting of the $\pi \rightarrow \pi^*$ exciton bands. Here the π^* states are doubly degenerated in a single Si(Pc) $(OH)_2$ molecule [9], but are split into two states in the crystal state [3] giving rise to two absorption bands. Each state splits into four states as a result of the Davydov splitting if the unit cell has four Si(Pc) $(OH)_2$ molecules. In discussing the Davydov splitting due to the interactions among the four molecules in a unit cell and its effects on the absorption spectra, it is convenient to use the transition moments [10], because the length and the direction of the sum of the four transition moments reflect the absorption intensity and the active polarization of the incident light, respectively. As the Si(Pc) $(OH)_2$ molecular plane makes a small angle with the c-axis [3], one of the two $\pi \rightarrow \pi^*$ transition moments of a molecule is almost parallel to the c-axis (|| transition), while the other one is almost perpendicular (⊥ transition) [3]. The combinations of the four || transition moments and the four ⊥ transition moments are shown in Fig. 2a and b, respectively. In Fig. 2, the four || transition moments of Si(Pc) $(OH)_2$ molecules are along the side of the stretched four-cornered pyramid and the four ⊥ transition moments are along

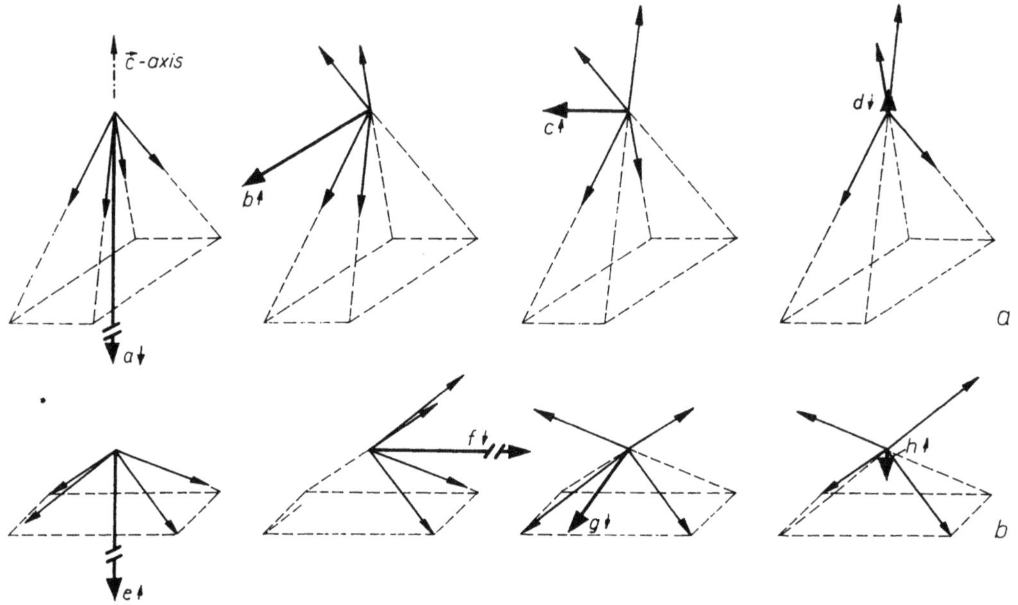

Fig. 2. Net dipole transition moments (thick arrows, a to h) for various combinations of the four monomer transition dipoles (thin arrows) of a) || transitions and b) ⊥ transitions. Small upward and downward arrows beside symbols, ↑ and ↓, indicate that the net transitions dipoles increase and decrease by the compressive stresses, respectively

the side of the compressed four-cornered pyramid. The bottom of the pyramid is a parallelogram which leads to two different angles between the transition moments and the c-axis. Here the length of the four || transition moments are considered to be approximately equal. The sum of the four transition moments (thick arrows) determine the allowed dipole transitions labeled a to h in Fig. 2 a and b. The allowed dipole transitions are in good correspondence to the absorption bands seen in Fig. 1, as demonstrated in Table 1 by assigning the observed eight absorption bands to the allowed dipole transitions a to h. The agreement is qualitatively good also for the absorption intensities. Further it is easy to propose a crystal structure of Si(Pc) $(OH)_2$ which is in accordance with the results obtained so far [2, 3]. The unit cell is predicted to have two columns of Si(Pc) $(OH)_2$ units, which is a common feature of phthalocyanines [10], one column having a period of two molecules. We think that the two molecules in a period are staggered, but are making the same angle with the column axis (c-axis). Such a staggered packing along the column axis is seen in other phthalocyanines as well [12].

Besides the tetramer model, we have also considered the dimer model and trimer model. It is difficult to explain the experimental results in terms of the dimer model because there are at least five absorption bands in the spectra (Fig. 1). The trimer model is difficult to pertain to the Si(Pc) $(OH)_2$ crystals because it is impossible to construct a reasonable crystal structure having a unit cell containing three molecules.

An explanation for the spectral changes under stress is provided by the notion that the angles made by the Si(Pc) $(OH)_2$ molecular planes and the c-axes change or that the molecular plane rotates around the normal of the molecular plane by the compressive stress along the c-axis, resulting in an increase of the parallel component of the perpendicular transition moment and a decrease of the parallel components of the parallel transition moment. The effects of these changes of the transition moments are shown in Fig. 2. The upward arrows ↑ beside the symbols in Fig. 2 indicate that the transition moments increase by the compressive stress and vice versa. It is apparent in Table 1 that the arrows of column ΔD_{opt} and the arrows beside the letters from a to h coincide.

In analyzing the ΔD_{opt} spectra we neglect the overall shift of the absorption bands under compression. In general, the energy shift of an absorption band appears as differential-type profile in the ΔD_{opt} spectra. In the ΔD_{opt} spectra of Fig. 1 a and b, the differential type of profiles seem to exist in the energy range from 1.6 to 1.9 eV. In this interpretation, the number of absorption bands are judged to be three in each polarization case in Fig. 1. Apparently this leads to the trimer model which gives three absorption bands in each polarization case. However, as is described above, it is very hard to make models of the Si(Pc) $(OH)_2$ crystal structure with three molecules in a unit cell.

References

[1] T. Kobayashi, H. Kurata, and N. Uyeda, Bunshi Kouzou Touronkai, 806 (1986) (in Japanese, available from The Chemical Society of Japan).
[2] M. Yudasaka, M. Kawai, S. Kurita, K. Nakanashi, and Y. Kuwae, Thin Solid Films **151**, L114 (1987).
[3] M. Yudasaka, Y. Kuwae, M. Kawai, K. Nakanishi, and S. Kurita, Thin Solid Films **157**, 189 (1988).
[4] M. Cardona, Modulation Spectroscopy, Chap. 3, Academic Press 1969.
[5] G. Ascarelli and A. Barone, Nuovo Cimento **37**, 818 (1965).
[6] J. Merski and C. J. Eckhardt, J. chem. Phys. **75**, 3691, 3705, 3719, 3731 (1981).
[7] M. Tanaka, W. Kaneko, S. Kurita, A. Yamada, and H. Fukutani, J. Phys. Soc. Japan **56**, 1197 (1987).
[8] C. H. Henry, S. E. Schnatterly, and C. P. Schlichter, Phys. Rev. **137**, A583 (1965).
[9] E. Ciliberto, K. A. Doris, W. J. Pietro, G. M. Reisner, D. E. Ellis, I. Fragala, F. J. Herbstein, M. A. Ratner, and T. J. Marks, J. Amer. Chem. Soc. **106**, 7748 (1984).
[10] M. Pope and C. E. Swenberg, Electronic Processes in Organic Crystals, Chap. I to D, Oxford University Press, New York 1982.
[11] T. Kobayashi, Y. Fujiyoshi, F. Iwatsu, and N. Uyeda, Acta cryst. **A37**, 692 (1981).
[12] T. Kobayashi and N. Uyeda, J. Crystal Growth **84**, 589 (1987).

(Received April 7, 1988; in revised form August 2, 1988)

phys. stat. sol. (a) **110**, 651 (1988)

Subject classification: 78.20; S10.15

Department of Solid State Physics,
Indian Association for the Cultivation of Science, Calcutta[1])

Fundamental Absorption Edge in Bismuth–Vanadate Glasses

By

A. GHOSH

Optical measurements of fundamental absorption edge of bismuth–vanadate glass films containing 70 to 95 mol% V_2O_5 are made in the wavelength range 200 to 700 nm. The absorption edge in these glasses obeys the relation $\alpha \sim (\hbar\omega - E_{opt})^2/\hbar\omega$. The optical band gap, (E_{opt}) is determined to be 2.30 to 2.15 eV, depending on compositions of the glass and is found to decrease with increase of V_2O_5 content in the glass. The absorption coefficient is found to depend exponentially on the photon energy, which is explained to be due to the exponential distribution of the localized states in the optical gap.

Im Wellenlängenbereich von 200 bis 700 nm werden optische Messungen der Grundgitterabsorptionskante von Wismutvanadatglasschichten mit 70 bis 95 Mol% V_2O_5 durchgeführt. Die Absorptionskante dieser Gläser befolgt die Beziehung $\alpha \sim (\hbar\omega - E_{opt})^2/\hbar\omega$. Die optische Bandlücke (E_{opt}) wird zu 2,30 bis 2,15 eV bestimmt, abhängig von der Zusammensetzung der Gläser, und es wird gefunden, daß sie mit steigendem V_2O_5-Gehalt der Gläser abnimmt. Der Absorptionskoeffizient hängt exponentiell von der Photonenenergie ab, was durch die exponentielle Verteilung der lokalisierten Zustände in der optischen Bandlücke erklärt wird.

1. Introduction

Transition metal oxides such as V_2O_5, when heated with glass forming oxides like P_2O_5, TeO_2, Bi_2O_3, etc. form semiconducting glasses on quenching the melt [1 to 5]. Low valency transition metal ions are produced due to loss of oxygen from the melts. Electron–phonon interaction in these glasses is strong enough to form small polarons and electrical conduction occurs by hopping of small polarons between transition metal ions of different valency states [6]. Electrical properties of the vanadate glasses have been studied extensively [7 to 10]. However, optical properties of the transition metal oxide glasses have received little attention in most reports on the V_2O_5–P_2O_5 glasses [11 to 15]. A study on the blown films of V_2O_5–P_2O_5 glass was reported covering the range from far-infrared to near-ultraviolet, presenting in detail the vibrational absorption bands and the fundamental absorption edge [12]. The fundamental absorption edge of these glasses and of crystalline V_2O_5 near 2.35 eV [12, 16, 17] fits the condition for direct forbidden transitions,

$$\alpha = \frac{A(\hbar\omega - E_{opt})^{3/2}}{\hbar\omega}, \tag{1}$$

much better than that for direct allowed transitions,

$$\alpha = \frac{B(\hbar\omega - E_{opt})^{1/2}}{\hbar\omega}, \tag{2}$$

[1]) Jadavpur, Calcutta 700032, India.

where α is the absorption coefficient and E_{opt} the optical band gap. A broad absorption tail extends from the absorption edge to the lattice vibrational absorption bands at 7 µm in the case of the glasses in contrast to the crystalline V_2O_5 [12]. The fundamental absorption edge shifts to higher energies with decreasing V_2O_5 content and occurs at slightly higher energies than in crystalline V_2O_5. The broad absorption tail and the optical band gap were related to the high concentration of V^{4+} ions [11, 13, 18].

In an earlier paper [5], the electrical properties and microstructure of the bismuth–vanadate glasses were reported. Electrical conduction in these glasses was shown to be due to polaronic hopping in the adiabatic regime. The purpose of the present paper is to study the fundamental absorption edge of these bismuth–vanadate glasses containing 70 to 95 mol% V_2O_5.

2. Experimental Details

The samples were prepared by mixing appropriate amounts of V_2O_5 (Reanal, Hungary) and Bi_2O_3 (Merck, Germany). The starting compositions were 95 to 70 mol% of V_2O_5 (Table 1). The mixture was melted in a quartz crucible, kept in an electrical

Table 1

Values of some physical parameters for the bismuth–vanadate glasses from optical absorption spectra at room temperature

glass composition (mol%)	E_{opt} (eV)	C (10^5 cm^{-1} eV^{-1})	ΔE (eV)	E_e (eV)
70% V_2O_5–30% Bi_2O_3	2.30	2.70	0.51	0.68
75% V_2O_5–25% Bi_2O_3	2.27	2.57	0.57	0.65
80% V_2O_5–20% Bi_2O_3	2.25	2.40	0.63	0.63
85% V_2O_5–15% Bi_2O_3	2.22	2.26	0.70	0.62
90% V_2O_5–10% Bi_2O_3	2.18	2.15	0.81	0.60
95% V_2O_5–5% Bi_2O_3	2.15	1.80	0.92	0.57

furnace for 2 h, in presence of air and at constant temperature of 900 °C, and stirred occasionally to achieve homogeneity. A silica tube of diameter 0.5 cm was then dipped into the melt. A small amount of the melt was gathered on the end of the tube and was finally blown into air to get glass films. The amorphous nature of the films was confirmed by X-ray diffraction (Philips, PW 1050/51) and scanning electron microscopy (Hitachi, S-145A). The thickness of the films was measured with a sigma comparator. The optical absorption spectra of the glass films were recorded in a Cary model 17D spectrometer in the wavelength range 200 to 700 nm at room temperature (300 K).

3. Results and Discussions

The optical absorption spectra of three compositions of the bismuth-vanadate glass films containing 70, 80, and 90 mol% of V_2O_5 are shown in Fig. 1 as a function of wavelength in the visible range. This figure shows that there are broad absorption edges with broad absorption tails in the spectra of these glasses in contrast to those of crystalline V_2O_5 [12], and also that the position of the absorption edge shifts to lower energy with the increase of V_2O_5 content in the glasses. Mott and Davis [19] have shown that the absorption edge in amorphous semiconductors should be of the

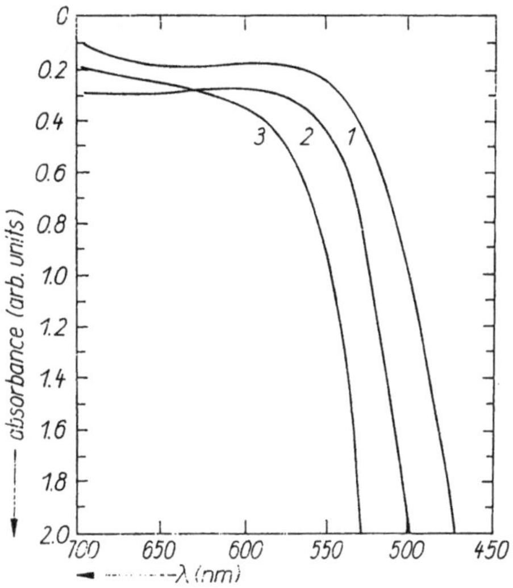

Fig. 1. Optical absorption spectra of bismuth–vanadate glass at room temperature in the visible range. V_2O_5 content: (1) 90, (2) 80, (3) 70 mol%

form

$$\alpha = C\,\frac{(\hbar\omega - E_{opt})^2}{\hbar\omega}\,. \tag{3}$$

Equation (3) was found to hold good in many amorphous semiconductors [19].

Now, Fig. 2 shows the plot of $(\alpha\hbar\omega)^{1/2}$ against photon energy for the bismuth–vanadate glasses. The straight lines drawn through each set of data are indications of the good fit to the criterion given by (3). It is observed from Fig. 2 that the slopes of the lines vary only slightly with the composition of the glass.

The intersections of the dashed lines in Fig. 2 with the horizontal axis determine the optical band gap, E_{opt}. The values of E_{opt} (Table 1) vary in the range 2.30 to 2.15 eV depending on glass composition and decrease with the increase of V_2O_5 content in the glass. The variations of E_{opt} and the slopes with the compositions of the glass are shown in Fig. 3.

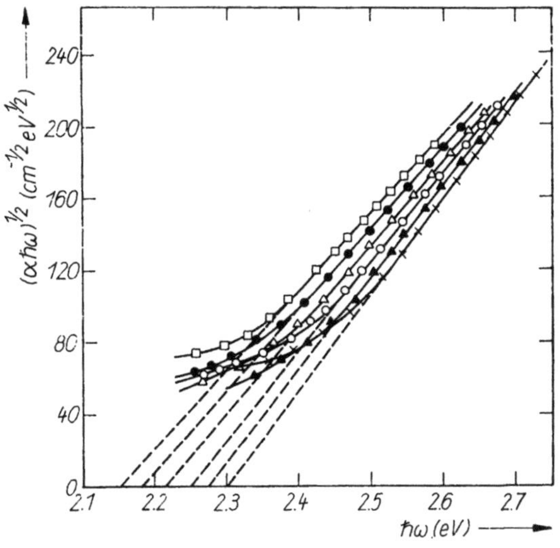

Fig. 2. Plot of $(\alpha\hbar\omega)^{1/2}$ vs. photon energy ($\hbar\omega$) of bismuth–vanadate glass. V_2O_5 content: □ 95; ● 90; △ 85; ○ 80; ▲ 75; × 70 mol%

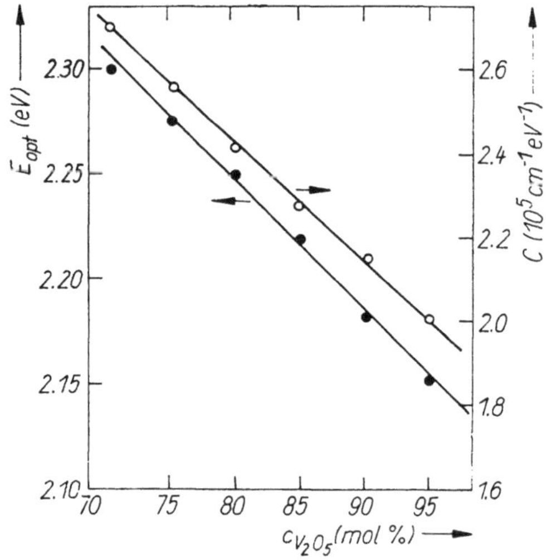

Fig. 3. Variation of the optical gap (●) and the slope (○) of Fig. 2 with the content of V_2O_5 in the bismuth–vanadate glass

Anderson and Compton [12], on the basis of the observed absorption of V^{4+} ions in other glass systems in the visible and near infrared [11, 20, 21], have shown that the V^{4+} ions contribute to the broad absorption tail and it is thus suggested that the d–d-type electronic transitions at the V^{4+} ion sites [5] are, in large part, responsible for the broad absorption tail for the bismuth–vanadate glass films. Positional disorder in the system probably broadens the 3d absorption bands causing the broad absorption throughout the tail region. As in other oxide glasses, though the disorder probably does not significantly alter E_{opt} due to nearly identical local order in the crystalline and glassy states [22] and due to the relatively large separation between the 2p and 3d bands [12, 16], the 2p band should tail somewhat into the gap due to disorder, but the 3d states, being relatively localized, are thus less affected by disorder [6, 10, 11, 14]. The glasses probably absorb, in a large part, by a mechanism involving d-like electrons. It is not, however, conclusive that the absorption tail in the vanadate glasses is brought about entirely by V^{4+} ions rather than by more reduced vanadium ions as well. In many crystalline and non-crystalline semiconductors the absorption coefficient depends exponentially on the photon energy [19]. The exponential dependence known as Urbach rule [23] is given by

$$\alpha = K \exp \frac{\hbar\omega}{\Delta E}, \tag{4}$$

where K is a constant and ΔE the width of the tail of the localized states in the band gap.

The origin of the exponential dependence of the absorption coefficient on photon energy is not clearly known. Tauc [24] has suggested that it may arise from an electron transition between localized states where the density of localized states is exponentially dependent on energy. But Davis and Mott [25] have pointed out that this explanation is not valid for all disordered materials, since the slope of the observed exponential behaviour remains unchanged for many crystalline and non-crystalline materials. The variation of ln (α) with the photon energy ($\hbar\omega$) is shown in Fig. 4. The linearity of the curves (Fig. 4) indicate the exponential behaviour of the absorption coefficient (α) on the photon energy in the absorption tail region. The values of ΔE calculated from the slopes of the linear part of the curve are shown in Table 1 and lie between

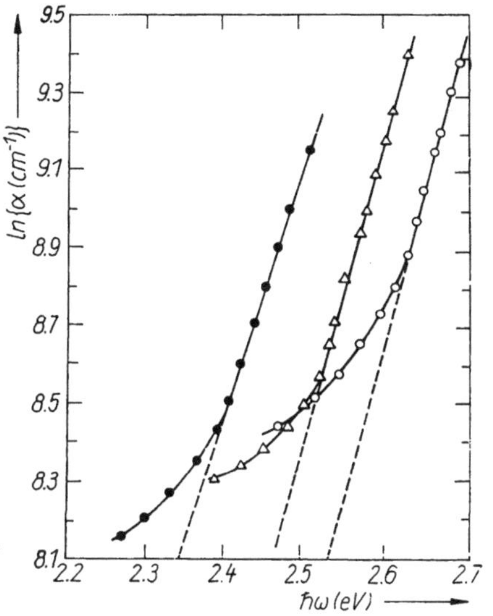

Fig. 4. Variation of the absorption coefficient α, with the photon energy of bismuth–vanadate glass. V_2O_5 content: ● 90; △ 80; ○ 70 mol%

0.51 and 0.92 eV depending on the composition of the glasses. The exponential dependence of α on photon energy suggests that the bismuth–vanadate glass obeys the Urbach rule.

It may be pointed out that, if the glasses behave as quasi-intrinsic semiconductors, then the Fermi-level would be in the middle of the band gap and the electrical energy gap (E_e) would be twice the activation energy (W) for electrical conduction at high temperature and would be expected to be equal to E_{opt}. If $2W$ is much less than E_{opt}, it is clear that the electrical activation process is certainly not across the band gap, but must be between bands and trapping levels. The electrical activation energy values for the bismuth–vanadate glasses have been reported earlier [5]. The values of $E_e = 2W$ are shown in Table 1. It is clear that E_e is much less than E_{opt}, indicating that the electrical activation processes are between the localized states. Fig. 5 shows the variation of E_e with E_{opt} and depicts the following relation between them:

$$E_e = 0.70E_{opt} - 0.93 . \tag{5}$$

Several authors have applied the small polaron theory to the optical absorption spectra of vanadate glasses [26]. The optical absorption by the polaron should occur

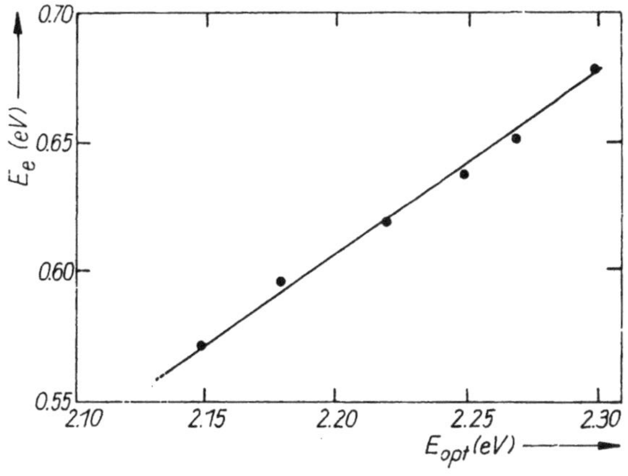

Fig. 5. Variation of electrical energy gap (E_e) with the optical gap (E_{opt}) for bismuth–vanadate glass

at $\hbar\omega \approx 4W$. The values of $4W$ for the present glasses are in the range from 1.14 to 1.36 eV [5]. But in this energy region, any polaron absorption is apparently masked by the intense and broad crystal field absortption band of V^{4+} ions [12, 13].

4. Conclusion

The fundamental absorption edge for the bismuth–vanadate glasses is described by the quadratic power law, $\alpha\hbar\omega \sim (\hbar\omega - E_{opt})^2$. The optical band gap decreases with the increase of V_2O_5 content in the glass. It is suggested that the d–d-type electronic transitions at the V^{4+} ion sites are, in large part, responsible for the broad absorption tail. There is an exponential absorption edge which could be explained to be due to the exponential distribution of localized states in the optical gap.

Acknowledgements

The author wishes to thank Dr. B. K. Choudhury and Prof. U. S. Ghosh, Head of the Department, for their interest in the work.

References

[1] E. P. DENTON, H. RAWSON, and J. E. STANDWORTH, Nature **173**, 1030 (1954).
[2] J. D. MACKENZIE, Modern Aspects of Vitreous State, Vol. 3, Butterworth, Washington 1964 (p. 126).
[3] D. ADLER, Amorphous Semiconductors, Butterworth, London 1971 (p. 88).
[4] C. H. CHUNG, J. D. MACKENZIE, and L. MURAWSKI, Rev. Chim. Min. **16**, 308 (1979).
[5] A. GHOSH and B. K. CHAUDHURI, J. non-crystall. Solids **83**, 151 (1986).
[6] I. G. AUSTIN and N. F. MOTT, Adv. Phys. **18**, 41 (1969).
[7] G. S. LINSLEY, A. E. OWEN, and F. M. HAYATEE, J. non-crystall. Solids **4**, 208 (1970).
[8] G. N. GREAVES, J. non-crystall. Solids **11**, 427 (1973).
[9] M. SAYER and A. MANSINGH, Phys. Rev. B **6**, 4629 (1972).
[10] A. P. SCHMID, J. appl. Phys. **40**, 4128 (1969).
[11] M. SAYER, A. MASINGH, J. M. REYES, and G. F. LYNCH, Proc. Conf. Low Mobility Solids, Eilat (Israel), Taylor and Francis, London 1971 (p. 115).
[12] G. W. ANDERSON and W. D. COMPTON, J. chem. Phys. **52**, 6166 (1970); **54**, 3247 (1971).
[13] G. W. ANDERSON, J. appl. Phys. **44**, 406 (1973).
[14] BH. V. JANAKIRAMA-RAO, J. Amer. Ceram. Soc. **49**, 605 (1966).
[15] V. SADAGOPAN and H. C. GATOS, Mater. Sci. Engng. **2**, 273 (1967/68).
[16] N. KENNY, C. R. KANNEWURF, and D. H. WHITMORE, J. Phys. Chem. Solids **27**, 1237 (1966).
[17] Z. BODO and I. HEVESI, phys. stat. sol. **20**, K45 (1967).
[18] E. J. FRIEBELE, L. K. WILSON, and D. L. KINSER, J. Amer. Ceram. Soc. **55**, 164 (1972).
[19] N. F. MOTT and E. A. DAVIS, Electronic Processes in Non-Crystalline Materials, Clarendon Press, Oxford 1979 (p. 287).
[20] G. J. KAKABADSE and E. VASSILION, Phys. Chem. Glasses **6**, 33 (1965).
[21] W. D. JOHNSTON, J. Amer. Ceram. Soc. **48**, 608 (1965).
[22] R. LANDSBERGER and P. J. BRAY, J. chem. Phys. **53**, 2757 (1970).
[23] F. URBACH, Phys. Rev. **92**, 1324 (1953).
[24] J. TAUC and M. ZANINI, J. non-crystall. Solids **23**, 349 (1977).
[25] E. A. DAVIS and N. F. MOTT, Phil. Mag. **22**, 309 (1970).
[26] I. G. AUSTIN, M. SAYER, and R. S. SUSSMANN, Proc. 5th Internat. Conf. Amorphous and Liquid Semiconductors, Vol. 2, Taylor and Francis, London 1974 (p. 1343).

(Received February 18, 1988; in revised form July 19, 1988)

phys. stat. sol. (a) **110**, 657 (1988)

Subject classification: 73.40 and 78.60; 85; S8.11

Central Institute for Electron Physics, Academy of Sciences of the GDR, Berlin[1])

Measuring on Thin Film Electroluminescent Devices

By

G. O. MÜLLER, R. MACH, B. SELLE, and G. SCHULZ

Nondestructive methods for the evaluation of film thicknesses in the stack of thin film electroluminescence devices are important tools in development, in degradation studies, and even in research concerning the basic processes in these devices. A simple model of the operation of these devices allows for a fitting procedure of the transferred charge/applied voltage characteristic, yielding the capacities of the semiconductor and dielectric films separately. Their thicknesses on the other hand can be assessed by Rutherford backscattering (RBS). With known values of the dielectric constants their comparison results in good agreement assuming a density of the vacuum-deposited films 5% less than the bulk values.

Zerstörungsfreie Meßmethoden für die Dicke der Filme in Dünnfilm-Elektrolumineszenz-Bauelementen sind sowohl für die Kontrolle der Entwurfsziele, in Degradations-Untersuchungen als auch für die Forschung zu Grundprozessen in solchen Strukturen wichtig. Ein einfaches Modell dieses Bauelements ermöglicht die Bestimmung der Kapazitäten der Halbleiter- und Isolatorfilme durch Anpassung der gemessenen Abhängigkeit der transferierten Ladung von der angelegten Spannung. Die Dicken der Einzel-Filme können andererseits aus Rutherford-Rückstreumessungen (RBS) bestimmt werden. Der Vergleich liefert Werte für die Dielektrizitätskonstanten, die von den bekannten Volumenwerten etwas abweichen, aber durch die plausible Annahme einer 5% geringeren Dichte der aufgedampften Filme in Übereinstimmung gebracht werden können.

1. Introduction

The ac driven thin film electroluminescent device (ELD) of the MISIM type (M metal, S semiconductor, I insulator) has matured from the first announcement [1] to industrial production in large volumes of rather complex — 640 × 400 picture elements (pixels) — information displays [2] during the last 15 years. This relates to the monochrome, yellow, ZnS:Mn-based one, which seems to be rather unique in its properties. However multicolor [3] and even full-color [4, 5] have been established recently in the laboratory, and production appears feasible.

The degradation behavior is under study [6, 14], new materials are being introduced, as well for the insulator films [7] as for the active or S-film [4]. The basic physics of the ZnS:Mn based devices is fairly well understood within a "simple model" [8, 9], and it turned out recently that the new materials — the earth alkaline sulphides — are not so far from realizing the same [10].

As with every new device specific measuring techniques are coming up, and have to be developed to a level, high enough to routinely assess the relevant parameters either on the devices themselves or on test samples fabricated in the same run. Not only to understand the basic physics and/or the consequences of technological changes or adjustments, but for routine checks and in degradation studies this is an important issue. The aim of this paper is in this direction, discussing electrical measurements on

[1]) Hausvogteiplatz 5/7, DDR-1086 Berlin, GDR.

ELD test samples. These test samples should be of simple geometric shape, not capacitively coupled to a matrix or other test fields. In a complete matrix not even the capacity of one pixel can be measured in a simple way, as the pixels are meshed and netted capacitively to all other ones on the same matrix [11]. The simplest way to answer the important question towards the capacitance of the MISIM structure is to connect all lines and columns, respectively, in parallel; however, the fill factor (relative active area) must be known accurately or assessed in a tedious manner. So it is highly advisible to have test areas on each glass, providing a nonmeshed MISIM sample of not too high a capacitive load to be driven on a bipolar waveform generator. Electrical measurements undoubtedly possess the advantage to be non-destructive. In aging runs degradation can be followed on the same sample, thereby allowing for much smaller collectives to obtain representative results. In research and development in general it is rather enlightening sometimes to come back to just the sample measured (long) before with other methods or new objectives.

Besides electrical measurements one excellent, non-destructive analytical tool is Rutherford backscattering (RBS). It is highly suited for thin film devices, as mostly the thicknesses of the films do not exceed the range accessible with MeV $^4He^+$ ions mainly used in RBS experiments. Of course specific problems arise from the fact that, stacked on one another, compounds are present in the MISIM the components of which are scattering into the same channels, as the ordering according to geometry overlaps with the ordering by masses [12]. A rather sophisticated evaluation of the measured spectra is necessary if neither the elements or compounds composing the stack nor the film thicknesses are known from the beginning. But the problem much simplifies if elements and compounds are given and the thicknesses have to be determined only. For this case, we have developed a straightforward evaluation method [13] which is described shortly below and gives a very good possibility to check the results of electrical evaluations of film thicknesses and/or to determine dielectric constants.

2. Measuring Transferred Charge

The physical layout of an ELD is shown schematically in Fig. 1, together with the equivalent circuit of this device. The latter consists of capacities of the insulator films, which can be thought of as acting in series, yielding a total capacitance C_i, and of the semiconductor or active film, C_s, respectively. Below the threshold voltage U_{th} of the brightness–voltage characteristic $B(U)$ (Fig. 2) the device is purely capacitive, with a loss factor determined by the non-stoichiometry of the insulator films [14]; the ZnS film being comparatively loss-free. Typical values of tan δ are about 0.005 at 1 kHz. Above the threshold U_{th} the nonlinear resistance R_s becomes increasingly important, and goes to zero or even becomes negative, as will be shown below. In fact, in each half-period of the usually applied bipolar pulse train the S-film shows

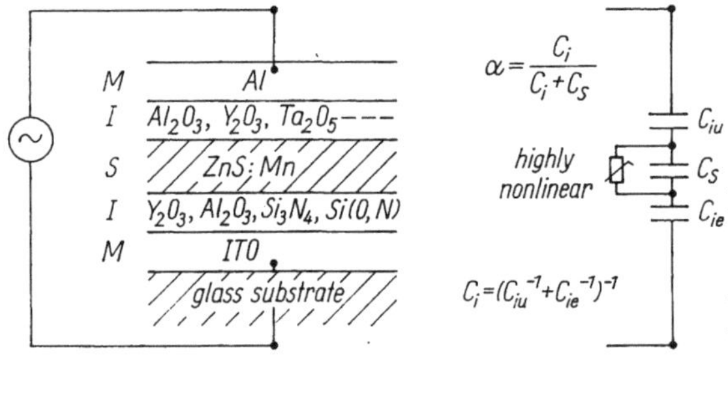

Fig. 1. Schematics of the ELD, which basically is a stack of thin films of metals (M), insulators (I), and semiconductors (S), together with its equivalent circuit, in which R_s is highly nonlinear

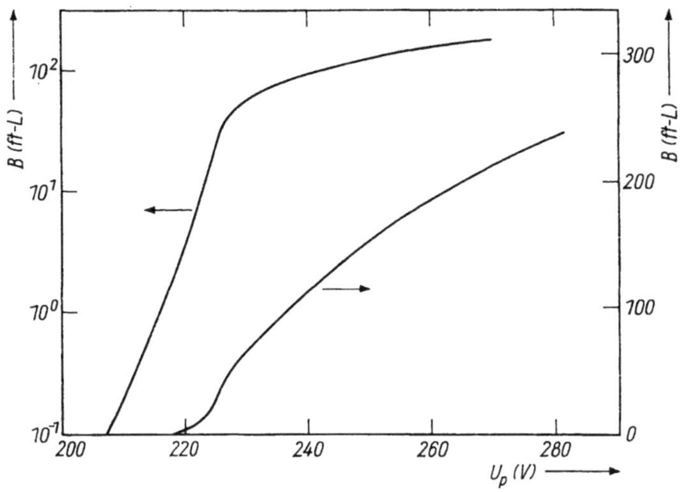

Fig. 2. Typical brightness (B)–voltage (U) characteristic on logarithmic (left) and linear (right) scales measured with bipolar 100 μs pulses of 200 Hz repetition frequency

electrical breakdown [8] necessary for efficient operation. It is stabilised and kept safe by the redistribution of the applied voltage, below threshold capacitively divided according to $\alpha = C_i/(C_s + C_i)$, in such way that any dissipative current through the S-film charges C_i and reduces the voltage fraction over the S-film. So simply the voltage U_s at the S-film is given by

$$U_s(t) = \alpha U(t); \qquad U(t) < U_{th}, \tag{1}$$

$$U_s(t) = \alpha U(t) - \frac{1}{C_i} \int^t i(t')\, dt' + U_0; \quad U(t) > U_{th}, \tag{2}$$

with U_0 being a left-over from the preceding half-cycle.

In a bridge circuit the capacitive current can be compensated, adjusting C below threshold. The dissipative current density measured (preferably done in the same pulse in which $U(t)$ is recorded, too) can be put into (2), and $i(t)$ plotted over $U_s(t)$ in order to judge the transport properties, as e.g. avalanching, more closely. In the following for comparability of different samples the current as well as the voltage across the S-film

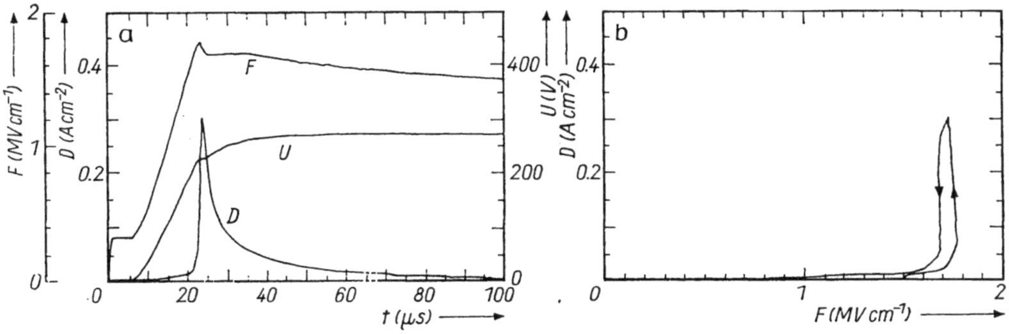

Fig. 3. a) Time behavior of applied voltage U, dissipative current density D, and mean electric field in S-film F; b) current–voltage characteristic $D(F)$ of S-film; sample R31. $Q = 2.5\ \mu C\ cm^{-2}$

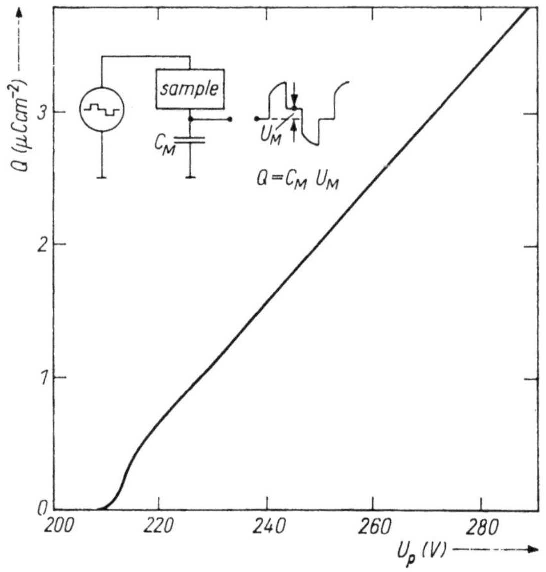

Fig. 4. Transferred charge Q upon applied (peak) voltage U_p of bipolar 100 μs pulses; sample R31, 200 Hz

have been normalized, and $D = i/\alpha$, the current density in the S-film, and $F = U_s/d_s$, the mean field strength in the S-film are used throughout. An example is given in Fig. 3.

In general this mechanism, well-described by (2) accounts for the fact that the breakdown of the S-film, or any other current flow through the S-film, is "charge-controlled", controlled just by the charge $Q = \int i(t)\,dt$, measured in the outer circuit. It is related to the charge transferred through the S-film, Q_s, by

$$Q_s = \frac{Q}{\alpha} . \tag{3}$$

The transferred charge can easily be measured using a so-called Tower-Sawyer capacity, a series capacity, as shown in the inset of Fig. 4, which is large enough to charge up only to a small fraction of the applied voltage. This voltage can be measured using a double box car integrator, sampling before and after the pulse. Using its analog output rather informative traces of $Q(U)$, U being the peak voltage of the applied pulses, can be recorded as shown in Fig. 4.

Assuming for the moment without real justification a simple model of an ELD, which can in a zeroth approximation be regarded as an antiseries connection of two avalanche diodes, sandwiched between two capacitors. If the avalanche diodes realise a vanishing resistance above U_a, the current $i(t)$ is purely charge-controlled, and simple arithmetics gives

$$Q = 2C_i(\alpha U - U_a) , \tag{4}$$

or identifying the avalanche or threshold voltage U_a with $\alpha U_{th} = E_t d_s$ and rewriting in a more general form

$$Q = 2\alpha C_i(U - U_{th}) . \tag{5}$$

(The factor 2 accounts for the charge left on the capacitor from the preceding, opposite polarity pulse.)

A fit of a measured $Q(U)$ curve in its linear portion promptly results in the inclination $\beta = 2\alpha C_i = 2C_i^2/(C_i + C_s)$, and the zero crossing U_{th}. Further arithmetics gives together with the total capacitance C of the sample values of C_i and C_s.

If, as in the case of ZnS the dielectric constant of the S-film is (thought to be) well-known, from C_s the thickness of the S-film, d_s, is obtained and using U_{th} the tunnel threshold $E_t = \alpha U_{th}/d_s$ follows. To use the example of Fig. 4 ($C = 50$ pF/mm²), we get:

$C_i = 285$ pF/mm², $C_s = 60.5$ pF/mm², $d_s = 1256$ nm, $E_t = 1.35$ MV/cm. A comparison of d_s with RBS data gives a rather good agreement, $d_s^{FBS} = (1134 \pm 2)$ nm.

Knowing the dielectric constant of the insulator material, which can be obtained by measuring RBS and the capacity on a test field, which only contains the dielectrics, of course the sum of the thicknesses of both I-films can be calculated from C_i. We are not able to conclude from electrical measurements on the thicknesses of the single I-films.

3. Thickness Analysis by RBS

The RBS spectrum of an electroluminescent stack is presented in Fig. 5. The insert shows the special sequence of films which was subjected to the analysing 1.7 MeV $^4He^+$ ion beam. To make the matter more simple the usual upper metal conducting contact was not included into the analysis because its parameters are not relevant to the considerations of this paper.

The RBS spectrum of this four-layer system, each layer consisting of two components, should be a simple additive superposition of 4×2 peaks. Some light element peaks (oxygen of films 3 and 4) of deep-lying films are lacking in the measured spectrum because the energy losses on the path before and after the scattering are so big that the backscattered ions cannot leave the sample and do not reach the detector. The edge energies of the remaining peaks are indexed in Fig. 5.

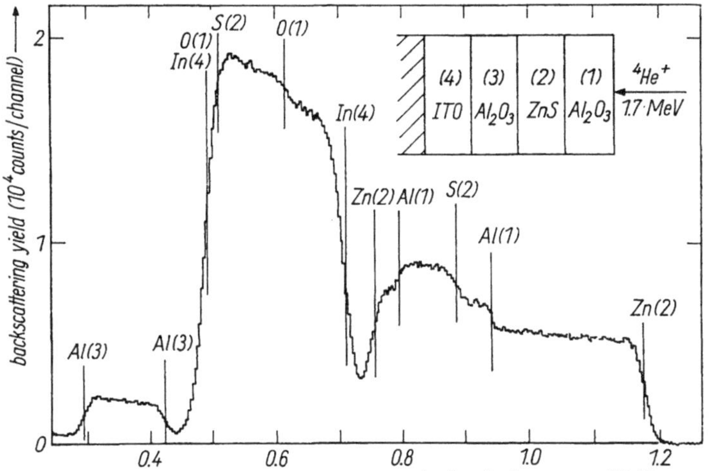

Fig. 5. RBS spectrum of an electroluminescent MISIM structure with the upper metal film lacking (30 μC $^4He^+$ at 1.7 MeV, scattering angle 170°, normal incidence). The edge energies are indexed with the element symbol and the film number

The uppermost ("free") Al_2O_3 layer is represented by the peaks of the elements Al and O which have a high-energy edge E_h at

$$E_h = kE_0 \tag{6}$$

and a low-energy edge E_l at

$$E_l = kE_0 - Nd[\varepsilon]\,. \tag{7}$$

E_0 is the energy of the incident ions, k the kinematic factor, N the atomic density of the layer, d its thickness, and $[\varepsilon]$ the stopping cross-section factor, depending on the special peak element, on the layer composition and — in a note quite simple manner — on the ion energy [15]. The so-called "symmetrical mean energy approximation" [15] can be used for $[\varepsilon]$ to calculate the thickness d with sufficient accuracy from the experimental values of E_h and E_l taken from Fig. 5.

For the deeper layers, the corresponding component peaks are shifted to lower energies due to the electronic energy loss of the incident and backscattered ions when passing the upper layers. An iterative correction procedure [13] can be applied to find now the energy $E_0' < E_0$ of the primary ions impinging on the layer under consideration and the true edge energies $E_h' > E_h$ and $E_l' > E_l$ of the backscattered ions

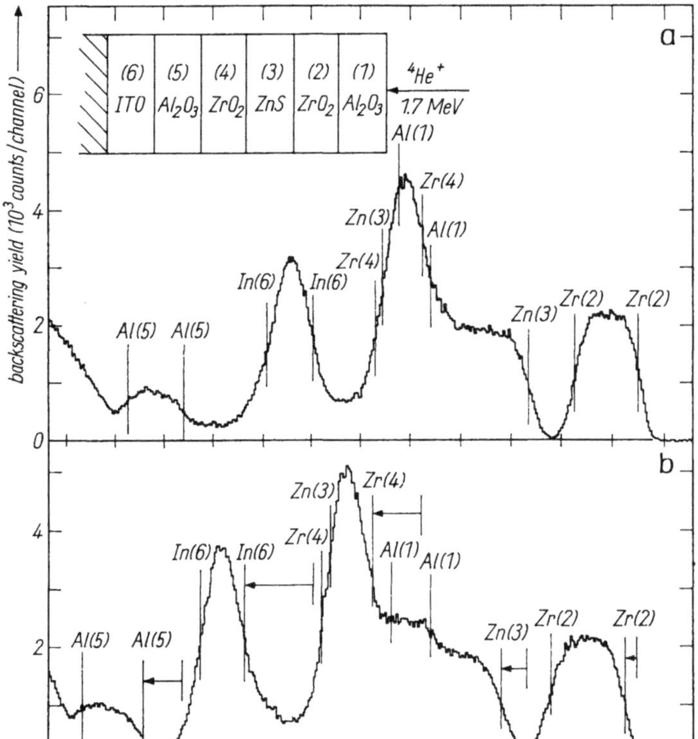

Fig. 6. RBS spectra of an electroluminescent stack with the upper metal layer lacking (10 μC $^4He^+$ at 1.7 MeV, scattering angle 170°) recorded a) at normal incidence and b) with the sample 30° tilted. The edge energies are indexed for the metallic components only

leaving this layer again. This iteration starts from the measured E_h and E_l and converts the covered surface (interface) of a deeper layer mathematically into a "free" one. Equations (6), (7) now can be used to evaluated by inserting the corrected energies E_0', E_h', and E_l''.

For the stack the spectrum of which is shown in Fig. 5 we arrive at the following results:

(i) Al_2O_3 $d = (169 \pm 6)$ nm ,

(ii) ZnS $d = (669 \pm 8)$ nm ,

(iii) Al_2O_3 $d = (168 \pm 13)$ nm ,

(iv) ITO $d = (203 \pm 4)$ nm .

The RBS analysis yields primarily the area densities Nd with good accuracy (see (7)). The film thicknesses given above have been determined by using the bulk atomic densities of these materials which represent only realistic values in the case of film packing densities or fill factors near unity. For packing densities below unity the RBS thickness has to be regarded as a lower limit of the actual thickness.

In comparison with spectrum simulation methods often used in RBS analysis [16] the analytical approach outlined in this section has the advantage that the input of the experimentally measured edge energies leads directly to the desired thicknesses and that time-consuming trial-and-error procedures are not necessary.

Not in all cases, however, the edge energies necessary for computation are as easy to identify as in the example shown in Fig. 5. Recent high-efficiency electroluminescent systems are designed to comprise more complex layer sequences, e.g. by using insulator substacks composed from different materials. The RBS spectrum of such a system, which includes two additional ZrO_2 films in contact with ZnS, is presented in Fig. 6. To facilitate the identification of the energy edges the spectrum was measured under normal incidence (Fig. 6a) and with the sample tilted at 30° to the ion beam direction (Fig. 6b). Tilting the sample has the effect to shift the high-energy edges of all constituent peaks except those on the outer surface (indicated by the arrows in Fig. 6b). Generally, all features of the spectrum are shifted the more, the deeper the constituent producing this feature is located below the outer surface.

This tilting method is also very useful in analysing samples with a certain amount of interface roughness where the energy edges are broadened and do not allow to fix correctly the energies of closely spaced edges. The spectra shown in Fig. 6 was specially selected to illustrate the edge broadening by rough interfaces, too.

4. Discussion

The almost ideal fit of the $Q(U)$ of Fig. 4 to the straight line of (2) allows to ascribe error margins of less than 2% to the C_s-value from which we concluded on the thickness of 1256 nm. This, however, is far outside the error bar of d_s^{RBS}. The most plausible reconcilation of both results is the consideration of the well-known fact that vacuum-deposited films very seldom approach the density of volume or bulk or even monocrystalline material. As was mentioned in Section 3, RBS measures primarily the area density (atoms/cm²), and it is only by the assumption of a specific value of the volume density that it gives a film thickness. Introducing the filling factor f as the ratio of the film density to the bulk density, and — to be consistent — taking into account in the crudest approximation that for a less dense material the permittivity ε should be smaller, and e.g. $(1 + f(\varepsilon - 1))$, we can fit with the parameter f the two

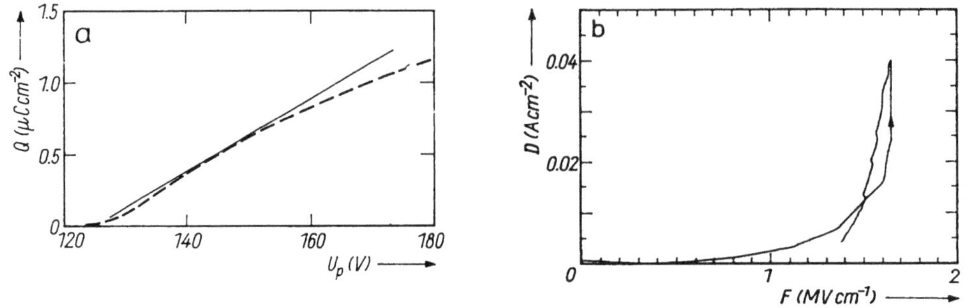

Fig. 7. a) As Fig. 4, sample R61. b) $D(F)$ showing charge-controlled (cc)-behavior in the fitting region of low Q ($Q = 0.52$ μC/ cm^{-2})

differing results. Doing so, and using for ZnS $\varepsilon = 8.6$ we get a perfect agreement for

$$f = 0.95$$

which appears not unreasonable. So summing up the consequences of this correction, we have to increase the thickness, evaluated from the RBS data by 5%, and we have to use 8.2 as the permittivity of the ZnS films.

Not always $Q(U)$ is so simple to evaluate as the one used for demonstration in Fig. 4. For instance Fig. 7 shows a rather different behavior, which at a first glimpse seems to be in extreme disagreement with the simple formula (5) given above. However, having got some experience in doing so, a technician was able to evaluate it, yielding $C_s = 160$ pF/mm², which compares with a later determined C_s^{RBS} ($f = 0.95$) $=$ $= 153$ pF/mm². The fitting interval from which β was derived is the one indicated by the straight line sectioning $Q(U)$ in Fig. 7. The justification of this fitting procedure, and the interrelations to the assumptions made before for the derivation of (5) will become evident below.

A further example in which the resemblance of $Q(U)$ with (5) is not very close, is given in Fig. 8. As Table 1 shows the RBS value of C_s under the assumptions discussed above is 58 pF/mm². The fit in the interval 250 to 260 V results in the value given in Table 1 of 60 pF/mm². Other fits would have been possible and are indicated in Fig. 8. As a guideline in this figure the brightness–voltage characteristic is given. As mentioned in Section 2 the threshold voltage results from the same fit. Comparing its value with the $B(U)$ curve, which because the logarithmic B scale emphasises U_{th}, rules out the two other straight lines drawn, the results of which are given in Table 1, obviously disagreeing more with the experimental value.

The sample used for Fig. 8b is so to speak an "artificial" sample. The behavior of the sample Z44 (Fig. 8a) was modified by a series capacitor, which reduced the total capacitance from 46.5 to 38.2 pF/mm². Of course a shift along the voltage axis results. But this is not interesting in this context. Having the task to fit the curve to a straight line it is very tempting to do so in the really almost straight part 280 to 300 V. In fact, a rather large disagreement of the threshold voltage with $B(U)$ occurs. As again shown in Table 1, the line drawn taking into account a better choice of U_{th} gives better agreement with RBS.

As pointed out recently [6], two different types of behavior of MISIM structures occur, which we termed "charge controlled" (cc) and "interface controlled" (ifc). An example of cc behavior was already given in Fig. 3. Another one is contained in Fig. 9, where it is contrasted to ifc. While in the classical cc case an infinitely steep part of $D(F)$ (current density in S-film D over mean field strength $F = U_s/d_s$) is followed by a negative differential conductance (NDC) region, where the current

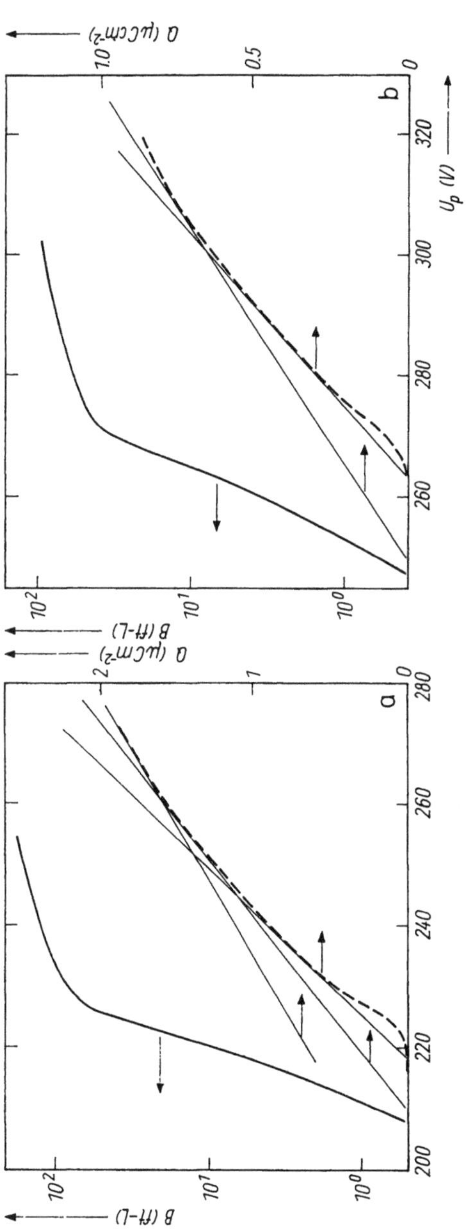

Fig. 8. As Fig. 4. a) sample Z44, b) the same sample as in a), but modified by an extra series capacitor, designated now M44. The straight lines shown represent different possible fits, used in Table 1 and commented in text

increases further in spite of a decreasing mean field. In the ifc case an N-shaped negative conductance is observed instead. Here a further increasing mean field does not increase the current, very probably because of an exhaustion of the interface states, which have to supply the electrons. A rather schematic picture of possible interface densities of states (IF DOS) is shown in Fig. 10, the main feature of which is the density value at the filling margin (quasi-Fermi level), which is supposed to be high to allow for cc, and is presumably much lower in the ifc samples. The reason for S-shaped NDC in cc samples of course is a redistribution of the field strength because

Table 1

Comparison of values for the S-film capacitance C_s, derived from RBS with those from electrical measurements

sample	d_s^{RBS} (nm)	C_s^{RBS} (pF mm^{-2})	C_s^{RBS} (pF mm^{-2}) $f = 0.95$	C_s^{exp} (fit interval (V)) (pF mm^{-2})
R31	1134 ± 2	67.1	60.9	60.5 (220 to 280)
Z44	1195 ± 15	63.7	57.8	56.6 (232 to 238) 60.3 (250 to 260) 64.5 (265 to 270)
M44	1195 ± 15	63.7	57.8	55.8 (285 to 295) 61.4 (305 to 312)
R61	452 ± 16	168.4	152.7	160 (140 to 150) 177 (155 to 160)
AD16	669 ± 8	113.8	103.2	101 (200 to 205) 115 (215 to 220)

of positive space charge accumulation in the bulk of the S-film: its increase near the (momentary) cathode tunnels more primary electrons out of the interface states. So the "opening" (or hysteresis) of $D(F)$ in the cc case has something to do with the amount of space charge piled up mostly near the anode by carrier multiplication, but is counteracted by the ever-present downward shift of the filling margin in the IF

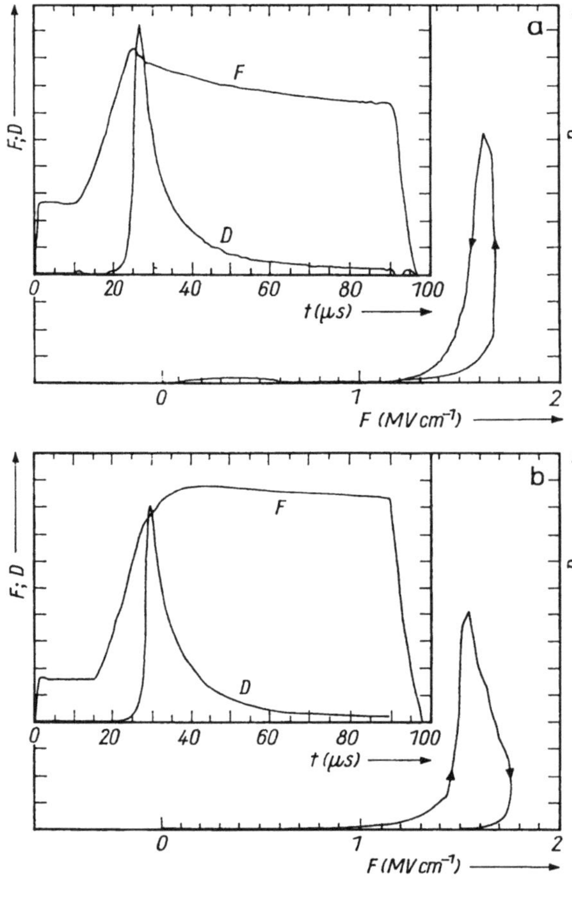

Fig. 9. Two types of electrical behavior of ELDs. a) Charge controlled (c.c.) and b) interface controlled (ifc) current flow, evidenced by the time behavior of field F and current density D and current–voltage characteristics $D(F)$

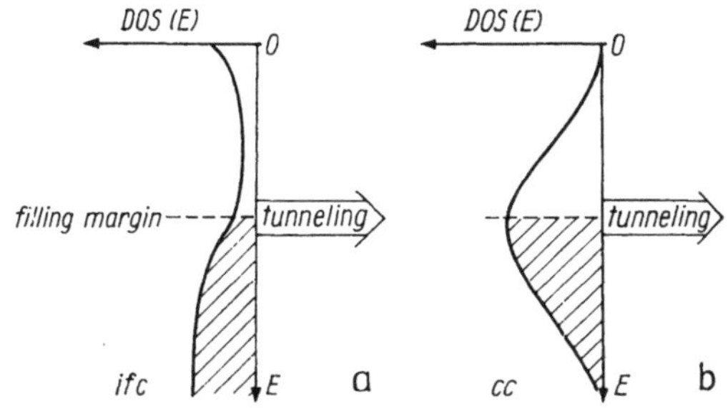

Fig. 10. Hypothetical density of states (DOS) of interface localised levels, responsible for electron supply by tunneling from filled ones. b) In the case cc only a small shift of the filling margin due to the high DOS corresponds to a certain transferred charge Q, but a) in the ifc case rather large shifts would be necessary

DOS. In the same sense in ifc samples the exhaustion of filled IF DOS is counteracted by positive space charge accumulation, raising the cathode field. So the two examples given are two almost "pure" cases, which are surprisingly often realised. However, in-between cases are also numerous.

In much the same way as we argued in correlating (2) and (5), we consider the cc case as the upper margin for the transferable charge for each applied voltage. So if we can assure by measuring $D(F)$ that a cc case is realised we can be quite sure that a fit of a straight line to that portion of $Q(U)$ will give us the correct β, α, C_s, etc. So in some sense a cross-check of the reliability of the fitting procedure can be derived from $D(F)$. In the case of ifc situations caution is necessary: $Q(U)$ might lie below the "true", i.e. cc case, which justifies (5); β might be smaller, and consequently the C_s determined carelessly would be too large.

Using the considerations of the previous paragraph, because of the findings of Fig. 4, we trust completely the evaluation in the case of the sample R31, which we therefore used to determine f.

Fig. 11 providing data of Z44 (Fig. 8a and Table 1) proves rather high reliability in this respect, too.

An intermediate case between ifc and cc might be seen in AD16 (Fig. 12) for which $F(t)$ comes down after some N-shaped NDC, signalising "enough charge transferred".

The strangest case from the simple view developed is R61, which is clearly cc, but with C_s^{exp} definitely above C_s^{RBS} ($f = 0.95$). The most likely explanation being the realisation of an $f > 0.95$, possibly 0.98.

In spite of the positive picture drawn for the design data C_s, C_i, and/or α the situation is much more complex with the evaluation of the absolute values of E_t. In the sense, in which this parameter was introduced here, it is rather well-defined. However, sometimes it is difficult to get a clear-cut confirmation for it from time-resolved measurements. Evaluations are sometimes complicated by a fact, frequently encounter-

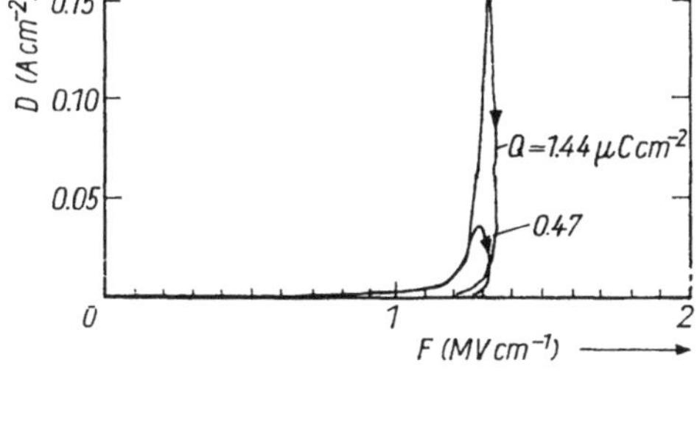

Fig. 11. Interface control is evident in the case of sample Z44, but at higher Q (around 1.4 $\mu C\ cm^{-2}$) it is not a serious one

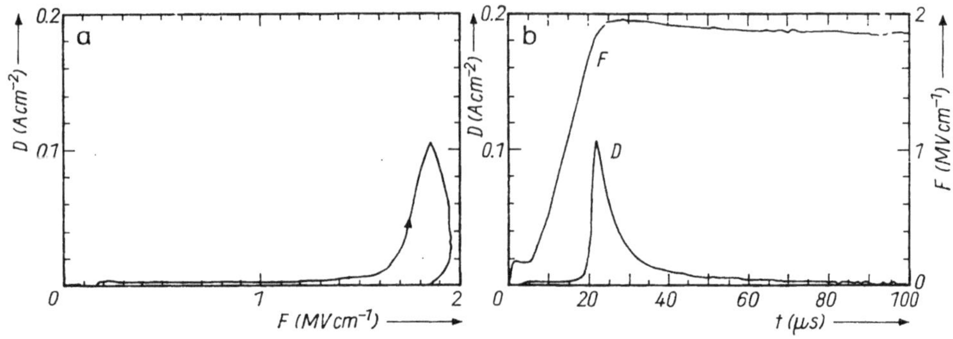

Fig. 12. Sample AD16 realises an intermediate case: After some N-shaped NDC the mean field decreases by voltage transfer to C_i

ed in ELDs — non-equivalent interfaces or asymmetries evident mostly as such in $B(t)$. But even in the mode of the current control — ifc or cc — there is often asymmetry as exemplified by Fig. 13.

Asymmetry in the $D(F)$ characteristics is even a conceptual problem: If an interface is tunnel-injecting electrons prematurely, at much lower voltages than the opposite one, the latter should charge up negatively in order to compensate for the difference. As, however, all the measurements reported have been made in the steady state any self-biasing by such an initial asymmetry in a "starting" condition (which can be thought of as a thermodynamical equilibrium (TDE) only with large question marks) is outside the possibilities of being detected. Consequently the scale or better the zero of the field strengths is not an absolute one. A bias might be superimposed, adding a certain value in one polarity, subtracting the same in the other. A permanent polarisation field stored as an interface charge (in general in the interface, which would emit at higher fields) might be hidden and is of importance, for instance, if considering electrochemical effects, which of course respond to much lower dc fields than to ac field amplitudes. The asymmetry which we measure is the one necessary to establish the steady state condition $Q_- = Q_+$, with Q_k denoting the transferred charges in the different polarities $k = -, +$. All measurements reported here have been done in the following way in order to avoid inaccuracies due to slight asymmetries in the shapes of positive and negative pulses, respectively: the sample was reversely connected to the pulse generator in order to change the polarity, measurement was always done in the negative pulse. Comparing the different polarities at slightly different transferred charges is of no problem, as the behavior does not change abruptly, as confirmed in hundreds of experiments; these slight

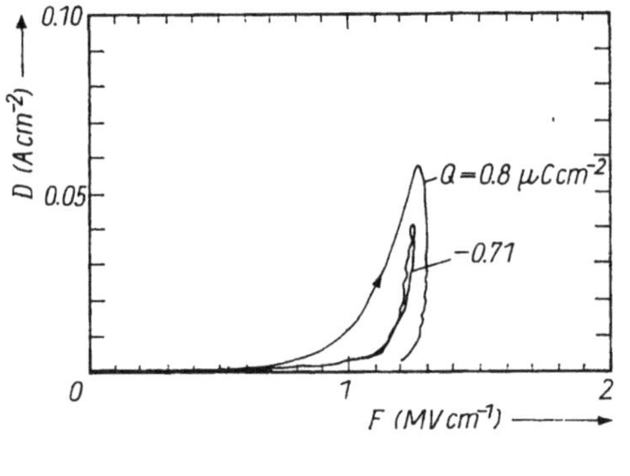

Fig. 13. Exemplifying rather asymmetric conditions by the differences of $D(F)$ for the different polarities; sample E08, not further used in this paper

changes are due to the fact that the voltage was reduced before disconnecting the sample for polarity inversion.

5. Conclusions

It has been shown by a straightforward comparison of model-free evaluations of Rutherford backscattering data with a fitting procedure of easily measurable electrical characteristics, that agreement between the values for thicknesses of s-films buried in a whole stack of up to six films can be obtained. This agreement in it itself gives confidence into the simple electrical measurement, but further support can be obtained from the consistency with other — time-resolving — electrical characterizations. A very likely value for the filling factor f — the ratio of the film density to the bulk density — of $f = 0.95$ was obtained, with some indications that it might be higher in specially prepared samples.

The absolute value of the tunnel fields remains a problem in asymmetric samples.

The methods described here have been successfully applied to a large number of samples followed through degradation studies [14].

References

[1] T. Inoguchi, M. Takeda, Y. Kakihara, Y. Nakata, and M. Yoshida, SID '74 Digest Los Angeles (p. 84).
[2] W. Goede, SID Lecture Notes Anaheim 1988 (p. 12).
[3] N. Yamauchi, H. Kozawaguchi, O. Kogure, and B. Tsujiyama, SID '87 Digest New Orleans (p. 230).
[4] S. Tanaka, H. Yoshiyama, J. Nishiura, S. Ohshio, H. Kawakani, and H. Kobayashi, SID '88 Digest Anaheim (p. 293).
[5] C. King, SID Lecture Notes Anaheim 1988.
[6] G. O. Müller, R. Mach, R. Reetz, and G. U. Reinsperger, SID '88 Digest Anaheim (p. 23).
[7] M. Nishikawa, T. Matsuoka, T. Tohda, Y. Fujita, and J. Kuwata, SID '88 Digest Anaheim (p. 18).
[8] G. O. Müller, SID Lecture Notes 1987 New Orleans (p. 3.1).
[9] G. O. Müller, phys. stat. sol. (a) **81**, 597 (1984).
[10] R. Mach, H. Ohnishi, and G. O. Müller, Proc. 4th Workshop Electroluminescence, Tottori 1988, Springer, Heidelberg, in the press.
[11] P. Pleshko, SID Lecture Notes 1982 Los Angeles (p. 3.2).
[12] E. Rauhala, J. appl. Phys. **62**, 2140 (1987).
[13] R. Groetzschel and B. Selle, in: Physical Problems in Microelectronics, Ed. J. Kassabov, World Scientific, Singapore 1987 (p. 478).
[14] R. Mach and G. O. Müller, see [10].
[15] W. K. Chu, J. W. Mayer, and M.-A. Nicolet, Backscattering Spectrometry, Academic Press, New York 1978.
[16] See e.g. L. R. Doolittle, Nuclear Instrum. and Methods **B9**, 344 (1985).

(Received October 6, 1988)

phys. stat. sol. (a) **110**, 671 (1988)

Subject classification: 73.40 and 85

Department of Electronic Science, University of Delhi South Campus, New Delhi[1])

A Method to Determine Surface Doping and Substrate Doping Profile of n-Channel MOSFETs

By

R. S. Gupta, C. Jagadish[2]), G. S. Chilana[3]), and G. P. Srivastava

A simple method is presented to determine the surface and substrate doping profiles of MOSFET by considering both, the surface inversion and channel depletion case. The dependence of threshold voltage on the substrate bias under surface inversion and channel depletion conditions is used for measuring surface and substrate doping. The surface depletion width and channel depletion width is calculated for different channel doping densities at various back bias voltages.

Es wird eine einfache Methode zur Bestimmung der Oberflächen- und Substrat-Dotierungsprofile von MOSFET angegeben, wobei sowohl die Oberflächeninversion als auch Kanalverarmung berücksichtigt werden. Die Abhängigkeit der Schwellenspannung von der Substratspannung unter Oberflächeninversions- und Kanalverarmungsbedingungen wird zur Messung der Oberflächen- und Substratdotierung benutzt. Die Oberflächenverarmungsweite und Kanalverarmungsweite wird für verschiedene Kanaldotierungsdichten bei verschiedenen Rückseitenvorspannungen berechnet.

1. Introduction

Recently several authors [1 to 6] have described various analytical device models of MOSFETs from the overall point of view of VLSI construction. Some include the carrier velocity saturation effect, while the others describe analytical models for the threshold voltage of small geometry MOSFETs. The various significant characteristics of p-channel and n-channel MOSFETs have been studied and it has been found that the p-channel devices have important advantages over n-channel devices for the realization of sub-μm dimensions in VLSI. The MOSFET threshold voltage variation, being important for decreasing feature size geometries, have extensively been studied. The threshold voltage dependence on channel length of short channel MOSFETs is remodelled [7] taking into consideration the flat band voltage change in the short channel regions. The effect of substrate bias on threshold voltage has been studied recently [8, 9] to extract doping profiles on depletion devices.

In the present paper, an attempt has been made to remodel the dependence of threshold voltage on the substrate bias to determine the substrate doping as well as surface doping profile directly. The surface depletion depth and the channel junction depletion width dependence on various parameters have also been dicussed.

2. Analysis

The basic structure of n-MOS transistor is shown in Fig. 1a in which the channel has just been depleted completely and the potentials are referred to the source. An elemental section of the device at a position along the channel where the channel

---

[1]) Benito Juarez Marg, New Delhi 110021, India.

[2]) Present address: Department of Physics and Electronics, Sri Venkateswara College, University of Delhi, Dhaula Kuan, New Delhi 110021, India.

[3]) Present address: Department of Physics, Ramjas College, University of Delhi, Delhi 110007, India.

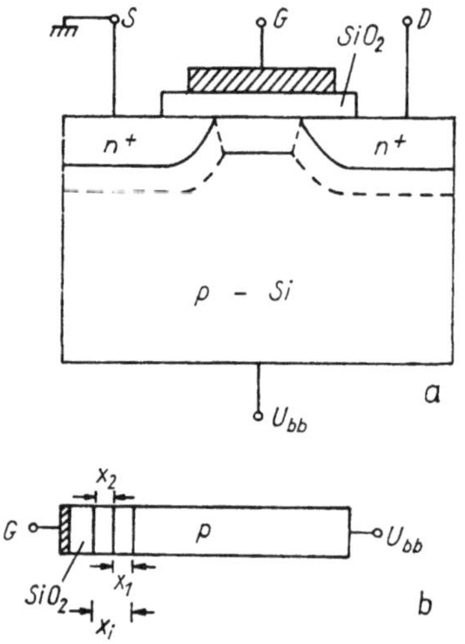

Fig. 1. a) Basic structure of n-channel MOSFET. b) Elemental cross section of the device at a point along the channel where the channel is completely depleted

is completely depleted is illustrated in Fig. 1b, with X_1 and X_2 as the channel junction depletion width due to back bias (U_{bb}) and surface depletion width due to the gate to source voltage, respectively. An analysis of the enhancement device is made in two different cases discussed below:

(i) when the complete channel depletion is prevented by surface inversion and the gate no longer controls the channel current,

(ii) when the channel is just completely depleted and the channel current does no longer flow.

In the surface inversion case an expression for the threshold voltage is given by [10]

$$U_{TIS} = U_{fb} - \frac{(2\varepsilon q N_s)^{1/2}}{C_{ox}} (\Phi_B - U_{bb})^{1/2} - (\Phi_B - U_{bb}) \tag{1}$$

and an expression for the threshold voltage when the channel is just completely depleted is

$$U_T = U_{fb} - \frac{N_E}{N_s} (\Phi_B - U_{bb}) + \frac{(2\varepsilon q N_E)^{1/2}}{C_1} (\Phi_B - U_{bb})^{1/2} - \frac{Q_1}{C_2}, \tag{2}$$

where U_{fb} is the flat band voltage, U_{TIS} the gate to source voltage when the surface inversion occurs, U_T the gate to source voltage when the channel is just completely depleted and the drain current is zero, N_s the surface channel doping, N_B the substrate doping, Φ_B the channel to substrate junction barrier, q the electronic charge, U_{bb} the substrate to source voltage (back bias), C_{ox} the oxide capacitance, $N_E = N_s N_B/(N_s + N_B)$, $C_1 = C_{ox}/(1 + C_{ox}/C_3)$, $C_2 = C_{ox}/(1 + C_{ox}/2C_3)$, $X_i = X_1 + X_2$, $\varepsilon = \varepsilon_0 \varepsilon_r$, $C_3 = \varepsilon/X_i$, $\varepsilon_0 = 8.854 \times 10^{-12}$ F/m, $\varepsilon_r = 12$ for silicon, $Q_1 = qN_sX_i$.

Now considering the threshold voltage variation with substrate to source (back bias) voltage in the surface inversion region, (1) yields

$$\frac{\partial U_{TIS}}{\partial(\Phi_B - U_{bb})^{1/2}} = \frac{-(2\varepsilon q N_s)^{1/2}}{C_{ox}} - 2(\Phi_B - U_{bb})^{1/2}. \tag{3}$$

It can be noted from (3) that $\partial U_{TIS}/\partial(\Phi_B - U_{bb})^{1/2}$ is the slope of the curve between U_{TIS} and $(\Phi_B - U_{bb})^{1/2}$ at a given value of $(\Phi_B - U_{bb})^{1/2}$. Therefore, the slope of the graph at each value of $(\Phi_B - U_{bb})^{1/2}$ can be obtained to plot a graph between $\partial U_{TIS}/\partial(\Phi_B - U_{bb})^{1/2}$ and $(\Phi_B - U_{bb})^{1/2}$. Fig. 2 shows the variation of $\partial U_{TIS}/\partial(\Phi_B - U_{bb})^{1/2}$ with $(\Phi_B - U_{bb})^{1/2}$ and it is a straight line with an intercept of $-(2\varepsilon q N_s)^{1/2}/C_{ox}$. The surface doping can easily be determined from the intercept at the known value of oxide thickness. This also indicates that the oxide thickness of any MOSFET can easily be calculated if the surface doping is known. The figure also shows that the intercept of the line changes with the surface doping, but however, the slope of the line remains constant, thereby allowing an accurate measurement of surface doping.

In the second case when the substrate bias is adjusted such that the channel becomes completely depleted and the channel current does no longer flow, the threshold voltage variation with substrate to source voltage using (2) yields

$$\frac{\partial U_T}{\partial(\Phi_B - U_{bb})^{1/2}} = \frac{-2N_E(\Phi_B - U_{bb})^{1/2}}{N_s} + \frac{(2\varepsilon q N_E)^{1/2}}{C_{ox}}\,. \tag{4}$$

Fig. 3 shows the variation of $\partial U_T/\partial(\Phi_B - U_{bb})^{1/2}$ as a function of $(\Phi_B - U_{bb})^{1/2}$. This plot is a straight line with a slope of $-2N_E/N_s$ and intercept of $(2\varepsilon q N_E)^{1/2}/C_{ox}$. From the intercept the N_E can be calculated for a given oxide thickness. The substitution of this value of N_E in the slope of the line gives the surface doping density N_s. The substrate doping N_B can be calculated using the relation $N_E = N_B N_s/(N_B + N_s)$. The values of N_s calculated from the slope can be used in calculating the substrate doping.

The channel junction depletion width can be calculated using the relation for different back vias voltages using the N_B and N_s values obtained from the measurements

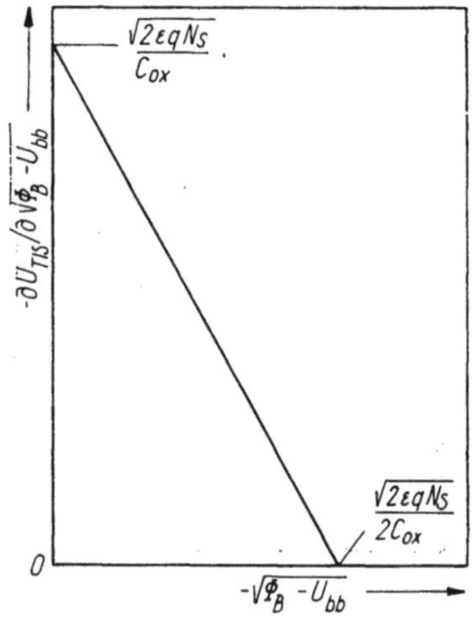

Fig. 2

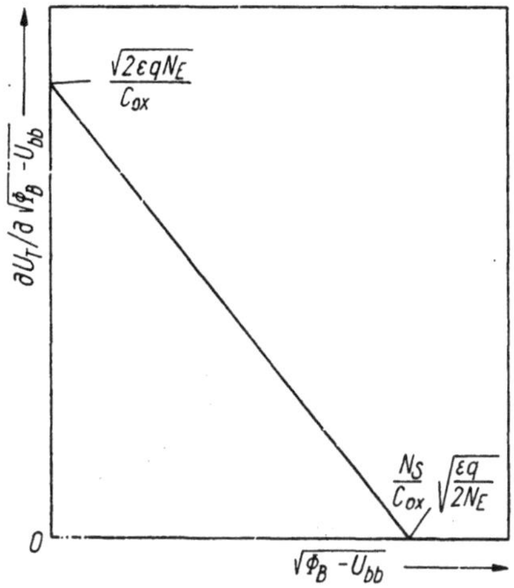

Fig, 3

Fig. 2. Variation of $\partial U_{TIS}/\partial(\Phi_B - U_{bb})^{1/2}$ with $(\Phi_B - U_{bb})^{1/2}$ for surface inversion

Fig. 3. Variation of $\partial U_T/\partial(\Phi_B - U_{bb})^{1/2}$ with $(\Phi_B - U_{bb})^{1/2}$ when the channel is completely depleted of threshold voltage. The channel junction depletion width is given by

$$X_1 = \left(\frac{2\varepsilon N_B}{qN_s(N_s + N_B)}\right)^{1/2} (\Phi_B - U_{bb})^{1/2} \tag{5}$$

and the surface depletion depth X_2 is given by

$$X_2 = \left(\frac{2\varepsilon}{qN_s}\right)^{1/2} \left(U_s - \frac{q\varepsilon N_s}{2C_{ox}^2} - U_g'\right)^{1/2} - \frac{q\varepsilon N_s}{2C_{ox}^2}, \tag{6}$$

where U_g' is the gate voltage less than the flat band voltage.

The total channel depth is the sum of X_1 and X_2 and can be written as

$$X_i = X_1 + X_2 . \tag{7}$$

Equations (5) to (7) show that the surface depletion depth is independent of the back bias while the channel junction depletion width is a function of back bias. When the channel to substrate barrier voltage is equal to the back bias voltage, i.e. $\Phi_B - U_{bb} = 0$, the channel junction depletion width is equal to zero and the total channel depth will then be equal to the surface depletion depth and is given by

$$X_i = \left(\frac{2\varepsilon}{qN_s}\right)^{1/2} \left(U_s - U_g' - \frac{q\varepsilon N_s}{2C_{ox}^2}\right)^{1/2} - \frac{\varepsilon}{C_{ox}} . \tag{8}$$

The variation of channel depletion width as a function of channel doping is shown in Fig. 4 for different values of substrate to source voltage. It can be observed from the figure that the channel depletion width is decreasing with the increase in channel

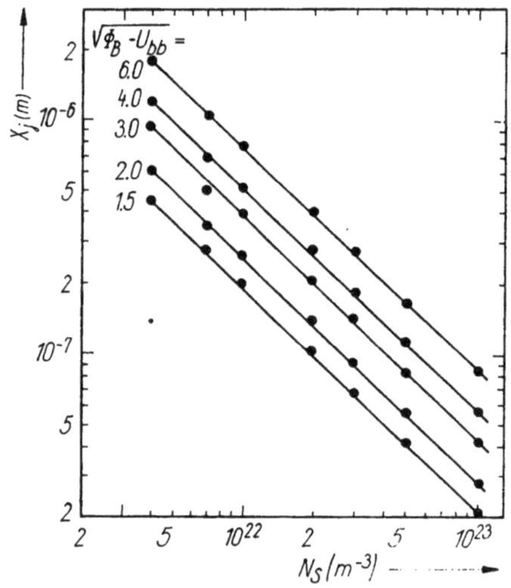

Fig. 4

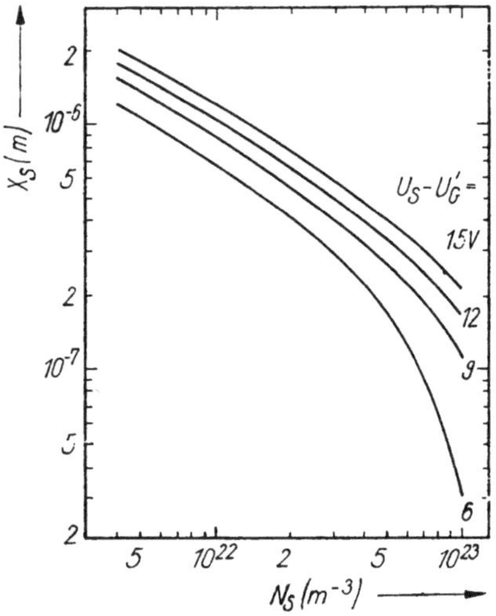

Fig. 5

Fig. 4. Variation of channel depletion width with channel doping ($\lg X_j$ vs. $\lg N_s$) for different values of $(\Phi_B - U_{bb})^{1/2}$ (marked at the curves). $N_B = 1.3 \times 10^{21}\ m^{-3}$

Fig. 5. Variation of surface depletion depth with channel doping ($\lg X_2$ vs. $\lg N_s$) for different values of $U_s - U_g$ (marked at the curves). At the ordinate read X_2 instead of X_s

doping. However, the value of X_1 is increasing with the increase in substrate to source voltage. This suggests that the channel depletion width can be controlled by changing either the channel doping or the substrate to source voltage. It is also observed that the channel depth increases with the increase in substrate doping.

Fig. 5 shows the variation of surface depletion depth with channel doping for different source and gate voltages. The value of surface depletion depth is significantly higher for lower channel dopings whereas this is one to two orders smaller for higher channel dopings. The surface depletion depth is also decreasing with the increase in channel doping. Qualitatively, the channel depth is a function of channel doping and the channel depth decreases with the increase in channel doping.

3. Conclusions

Using this method, the substrate and surface doping can be accurately determined for different values of substrate back bias, while others [8] have taken a linear relationship between threshold voltage and $(\Phi_B - U_{bb})^{1/2}$.

References

[1] I. E. Takeda, Y. Nakagome, H. Kume, N. Suzuki, and S. Assi, IEEE Trans. Electron Devices **30**, 678 (1983).
[2] T. Yamaguchi and S. Morimoto, IEEE Trans. Electron Devices **30**, 559 (1983).
[3] L. M. Dang, H. Iwai, Y. Nishi, and S. Taguchi, Japan. J. appl. Phys. **19**, 107 (1980).
[4] H. Schichijo, IEDM Tech. Digest 219 (1981).
[5] P. K. Chatterjee, W. R. Hunter, I. C. Holloway, and Y. T. Lin, IEEE Trans. Electron Devices, Letters **1**, 220 (1980).
[6] J. S. T. Huang and G. W. Taylor, IEEE Trans. Electron Devices **22**, 995 (1975).
[7] A. Srivastava, phys. stat. sol. (a) **99**, 303 (1987).
[8] R. A. Burghard and Y. A. El. Mansy, IEEE Trans. Electron Devices **34**, 940 (1987).
[9] D. S. Wu, IEEE Trans. Electron Devices **22**, 995 (1980).
[10] T. A. El. Manay, IEEE Trans. Solid State Cirkuits **15**, 331 (1980).

(Received May 11, 1988)

phys. stat. sol. (a) **110**, 677 (1988)

Subject classification: 61.70 and 73.40; 85; S5.11

Department of Electronic Engineering, Kumamoto National College of Technology[1])

Recovery Mechanism of Lattice Defects Formed in the Collector Region for Electron-Irradiated npn Si Transistors

By

H. OHYAMA and K. NEMOTO

The recovery mechanism of lattice defects formed in the collector region of electron-irradiated npn Si transistors is investigated using the DLTS methods for the first time. A center (A–O), E center (V–P), and divacancies with single and double minus charge states (V_2^- and V_2^{--}) are formed in the collector region. The recovery of A center and E center due to pulse aging is enhanced as compared with thermal annealing. This recovery enhancement is increased with increasing the collector peak current and the collector–base voltage. From the experimental results, it is shown that the recovery enhancement due to electron injection under reverse voltage is attributed to the energy transportation from the hot electron to the lattice system.

Erstmalig wird mit DLTS-Methoden der Erholungsmechanismus der Gitterdefekte, die im Kollektorbereich elektronenbestrahlter npn-Siliziumtransistoren gebildet werden, untersucht. Im Kollektorbereich werden A-Zentren (A–O), E-Zentren (V–P) und Doppelleerstellen mit einer und doppelter negativer Ladung (V_2^-, V_2^{--}) gebildet. Die Erholung der A-Zentren und E-Zentren infolge Pulsalterung ist höher als infolge von thermischer Ausheilung. Diese erhöhte Erholung wird mit Zunahme des Kollektorpulsstroms und der Kollektor–Basis-Spannung immer größer. Aus den experimentellen Resultaten wird gefolgert, daß die erhöhte Erholung durch Elektroneninjektion unter Gegenspannung der von den heißen Elektronen in das Gittersystem transportierten Energie zuzuschreiben ist.

1. Introduction

There are many investigations of the lattice defects induced in the semiconductor crystal by irradiation because of the increease of reliability for devices in high-level fluence environment [1 to 5]. The characteristics of the lattice defects (i.e. energy level, capture cross section, and density) have been determined using the deep-level transient spectroscopy (DLTS) technique [6]. Since the DLTS methods were proposed by Lang, the configurations and annealing behavior of the lattice defects could be easily examined.

For the lattice defects induced by electron irradiation in GaAs and GaP p^+n diodes there were reports that the annealing rate by the injection of minority carrier under forward biased voltage is greater than that by thermal annealing, and that the activation energy on the contrary becomes smaller [7, 8]. This is well known as the recombination enhancement effect (R.E. effect) [9]. In this effect, the energy liberated during the nonradiative recombination of minority carriers aided the recovery of the lattice defects. Based on these results, in light emitting diodes made from III–V compound semiconductors, its degradation during performance was thought to be

[1]) Kikuchi-gun, Kumamoto 861-11, Japan.

atrributed to the propagation and motion of the defects resulting from the R.E. effect [9].

On the other hand, Evwaraye and Kimerling indicated from DLTS measurements that several electron capture levels are induced by electron irradiation in n-type Si irradiated to a fluence of about 10^{15} e/cm^2, and that the energy level (E_c–0.18 eV) is associated with the A center (vacancy-oxygen complex), (E_c–0.23 eV) or (E_c–0.39 eV) with a divacancy, and (E_c–0.44 eV) with the E center (vacancy–phosphorous complex), respectively [1, 11]. The recovery property of the E center induced in a p^+n Si diode was studied by experiments such as thermal annealing with reverse biased voltage and hole injection under forward biased voltage [1, 12]. It was shown from these experiments that the recovery of the E center under these conditions is enhanced as compared with purely thermal annealing. The decrease of dislocation density in Si crystals was also reported to be enhanced by laser irradiation [13].

In this way, although annealing behavior and the configurations of the lattice defects induced in the p^+n diodes made from Si and compound semiconductors by irradiation were minutely examined, there have been few detailed considerations of the lattice defects formed in the collector region for npn Si transistors by irradiation. Therefore, we have investigated the characteristics and the recovery properties of the lattice defects formed in the collector region of npn Si transistors by electron irradiation [14].

In this paper, we report the recovery process of the lattice defects formed in the collector region of an npn Si transistor, which was irradiated with 2 MeV electrons to a fluence of 10^{15} e/cm^2. The difference in the recovery properties between purely thermal annealing and pulse aging is also investigated. Because of the use of a transistor structure, the effects of current and voltage on the recovery of the lattice defects could be studied. Based on the experimental results, it was shown that A center (A–O), E center (V–P), and divacancy with single and double minus charge state (V_2^- and V_2^{--}) are formed in the collector region. It was also found that the recovery of A center and E center due to pulse aging is enhanced as compared with purely thermal annealing. This recovery enhancement was also increased with increasing collector peak current and collector–base reverse voltage. The energy transport caused by the interactions between injected electrons and lattice in the collector region was finally discussed as the reason of recovery enhancement for A center and E center due to pulse aging.

2. Experiment

2.1 Sample preparation

The transistor chips used in this study were fabricated by thermal diffusion on n-epitaxial layers grown over n^+ substrates formed from CZ-silicon. The thickness and specific resistance of the n-epitaxial layer were about 12.5 μm and 6 Ω cm, respectively. The concentration of oxygen in the n-type epitaxial layer was estimated to be about 10^{18} cm^{-3}. The junction depths for the boron-diffused base and the phosphorus-diffused emitter were about 2.6 and 1.6 μm, respectively. The dopant (boron and phosphorus) concentrations of the base and collector regions of this sample evaluated from capacitance-voltage measurements were about 4×10^{16} and $1 \times$ $\times 10^{15}$ cm^{-3}, respectively. Ti–Pt–Au multilayer electrodes were formed on the emitter and base contacts. The transistor surface was covered with a silicon nitride film over a silicon dioxide film, and was finally encapsulated in a package (TO-18) consisting of a steel alloy (Kovar) of thickness about 0.3 mm. Furthermore, the minimum collector to base voltage (U_{CBO}) of this transistor was 60 V.

The transistor encapsulated in package was irradiated at room temperature with 2 MeV electrons to a fluence of 10^{15} e/cm^2 in the dynamitron at Takasaki Radiation Chemistry Research Establishment of Japan Atomic Energy Institute.

The irradiation direction was perpendicular to the surface of the transistor.

2.2 *Measurement of the lattice defects formed by electron irradiation*

After irradiation, the electron capture levels formed in the collector region were measured by DLTS methods. The DLTS methods with bipolar rectangular weighting function were used to increase the accuracy of the detection of the electron capture levels [15]. This DLTS measurement was performed using the wafer analyzer system 17D040 (SPC Electronics Corporation) [15]. After DLTS spectra had been measured, the energy level (E_t), the capture cross section (σ_n), and the defect density (N_t) of the electron capture level were automatically evaluated using the personal computer which was connected to the signal disposal apparatus, the capacitance-meter and the temperature measurement apparatus by the GP-IB interface function. The measurement limit of the defect density in this system was given by [15]

$$n = 1 \times 10^{-5} N_D\,, \tag{1}$$

where N_D denotes the dopant concentration in the collector region.

In order to measure the capture levels of majority carriers in the collector region (i.e. electron capture level), the DLTS measurement was performed for base-collector reverse voltage $U_{CB} = 5$ V, pulse width 2 ms, and emission rate 50.83 s^{-1}. Because the impurity density of the base region was about ten times that of the collector region as mentioned above, the depletion layer mainly spread to the collector region. When $U_{CB} = 5$ V, the width of the depletion layer spreading to the collector region was calculated to be about 1.6 μm. The DLTS measurement carried out only up to 300 K previously, was performed to 423 K in the present experiment.

2.3 *Thermal annealing and pulse aging of irradiated transistors*

In order to examine the annealing behavior of the lattice defects formed in the collector region by electron irradiation, thermal annealing and pulse aging of irradiated transistors were performed. Experimental conditions are shown in Table 1. Thermal

Table 1
Test conditions of the electron-irradiated transistor

thermal annealing	pulse aging			
	pulse width (duty ratio)	collector peak current I_{CP} (mA)	collector–base reverse voltage U_{CB} (V)	ambient temperature T_a (°C)
50 to 300 °C	1 ms (2.5%)	a) 100 mA	5	−40 to 265
		b) 200 mA	10	
		c) 400 mA	20	

annealing was carried out in an oven controlled with an accuracy of 1 K. Common base configuration was adopted as the pulse aging circuit to investigate the effects of current and voltage on the recovery process of the lattice defects. In order to minimize the increase of the junction temperature due to pulse aging, the voltage source having pulse width of 1 ms and duty ratio of 2.5% was employed. This pulse voltage source could also provide a maximum current of 1 A. Because of the pulse voltage source, emitter–base forward voltage (U_{EB}) and collector–base reverse voltage (U_{CB}) have the same period. Consequently, collector and base currents become of pulse form: collector peak current (I_{CP}) and base peak current (I_{BP}). I_{CP} was first changed by U_{EB} under constant U_{CB}. U_{CB} was next changed under constant I_{CP}. I_{CP} was evaluated from the voltage drop across the load resistance connected between collector and ground. It was also confirmed that I_{CB} is nearly equal to I_{BP}. This indicated that the common-base current gain (α) is nearly unity.

Although a pulse voltage source was used, an increase of junction temperature (T_p) due to pulse aging was existed. T_p was obtained from the following equation:

$$T_p = R_{th(J\text{-}C)} P_{CI} , \tag{2}$$

where $R_{th(J\text{-}C)}$ shows the transient thermal resistance of the emitter–base junction measured with a pulse width of 1 ms. P_{CI} is the instantaneous collector consumption power, which is defined as

$$P_{CI} = I_{CP} U_{CB} . \tag{3}$$

$R_{th(J\text{-}C)}$ was measured to be about 10 K/W by the measurement apparatus of transistor transient thermal resistance (TH-165, Kuwano Electric Company), and was nearly equal at other P_{CI}. Using (2) and (3), T_p could be calculated to be 10 °C for $I_{CP} = 100$ mA, $U_{CB} = 10$ V (i.e. $P_{CI} = 1$ W).

The wave form of I_{CP} and U_{CB} was observed by the oscilloscope watching the thermal run-away of the transistor. The wave form of I_{CP} and U_{CB} was not changed during pulse aging. This indicated that thermal run-away did not occur and T_p was constant. Consequently, the junction temperature (T) of thermal annealing and pulse aging was given by the following expressions:

$$T = T_a \qquad \text{(for thermal annealing)} , \tag{4}$$

$$T = T_a + T_p \qquad \text{(for pulse aging)} ,$$

where T_a denotes the ambient temperature. In order to compare both recovery properties at the same junction temperature, pulse aging was performed in the oven. After thermal annealing and pulse aging at various junction temperatures had been performed, DLTS measurements were carried out.

3. Results and Discussion

3.1 Defects formed in the collector region by electron irradiation

Fig. 1 shows the DLTS spectra in the collector region for a transistor irradiated with 10^{15} e/cm^2 electron fluence. These DLTS spectra show the electron capture levels in the collector region. As shown in Fig. 1, four levels, E_{15a} (E_c–0.19 eV), E_{15b} (E_c–0.24 eV), E_{15c} (E_c–0.44 eV), and E_{15d} (E_c–0.51 eV) were induced. Table 2 shows the energy level (E_t), capture cross section (σ_n), and defect density (N_t) of these levels.

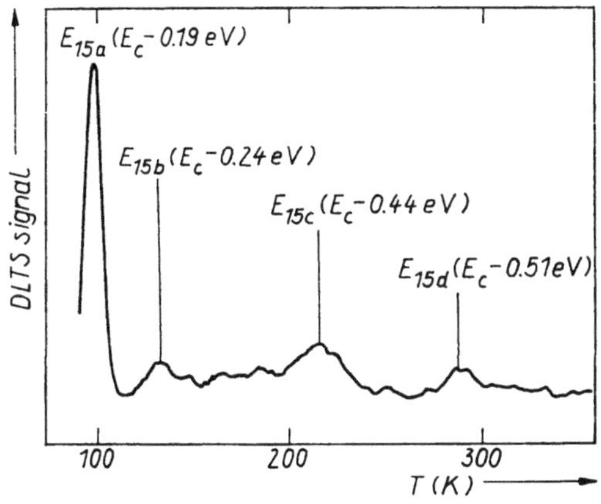

Fig. 1. DLTS spectra in the collector region for an npn Si transistor irradiated with 2 MeV electrons to a fluences of 10^{15} e/cm^2

Table 2

Characteristics of the electron capture levels induced in the collector region by electron irradiacion

level	energy level E_t (eV)	capture cross section σ_n (cm^2)	defect density N_t (cm^{-3})
E_{15a}	$E_c - 0.19$	4.0×10^{-14}	1.5×10^{14}
E_{15b}	$E_c - 0.24$	3.4×10^{-16}	1.8×10^{13}
E_{15c}	$E_c - 0.44$	3.5×10^{-16}	2.0×10^{13}
E_{15d}	$E_c - 0.51$	3.8×10^{-15}	1.2×10^{13}

3.2 *Recovery property of defects due to thermal annealing*

N_t of these electron capture levels was decreased by thermal annealing. This decrease was saturated after 30 min for each annealing temperature [14]. If the decrease of N_t is of first order in the exponential isothermal data until 30 min the unannealed fraction can be described as follows [1]:

$$f = \frac{N_{t_A}}{N_{t_0}} = \exp\left(-\frac{t}{\tau_A}\right), \tag{5}$$

where N_{t_0} and N_{t_A} are the defect densities before and after thermal annealing, respectively, and t is the annealing time. $1/\tau_A$ is the annealing rate for thermal annealing and is related to the activation energy (E_A) for thermal annealing as follows [1]:

$$\frac{1}{\tau_A} = \nu_0 \exp\left(-\frac{E_A}{kT}\right), \tag{6}$$

where ν_0 is the characteristic frequency factor. T denotes the junction temperature and is defined by (4). The temperature dependence of the annealing rate for the electron capture levels are shown in Fig. 2. The following results were obtained from thermal annealing for 30 min.

1. The annealing effect of level E_{15a} was observed from 225 °C and this density became 60% at thermal annealing at 300 °C.

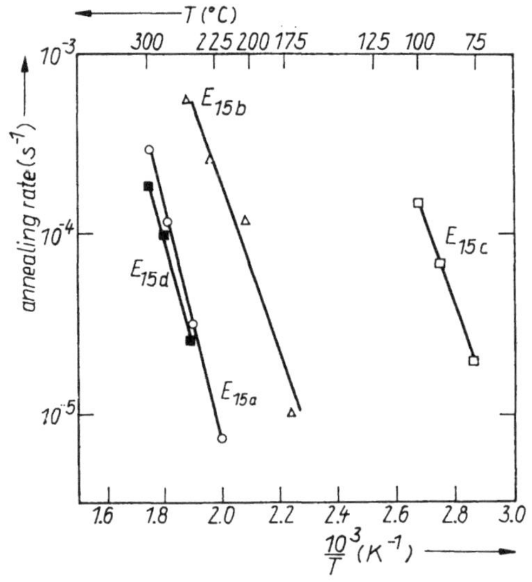

Fig. 2. Temperature dependence of the annealing rate of the electron capture level for an exposure to an electron fluence of 10^{15} e/cm^2

2. The annealing effect of level E_{15b} was observed from 175 °C and could not be observed above 275 °C.

3. The annealing effect of level E_{15c} was observed from 75 °C and could not be observed above 125 °C.

4. The annealing effect of level E_{15d} was observed from 250 °C and this density became as small as the measurement limits at 300 °C.

Using (5) and (6), the activation energies of levels E_{15a}, E_{15b}, E_{15c}, and E_{15d} were estimated to be 1.30, 1.06, 0.90, and 1.05 eV, respectively. The E_A and the characteristics of levels E_{15a} and E_{15c} indicated in Table 2 were approximately in agreement with those of the A center (vacancy–oxygen complex: V–O) and the E center (vacancy–phosphorus complex: V–P), respectively [11]. Therefore, these levels seemed to be associated with A center and E center, respectively [14]. Furthermore, the properties of levels E_{15b} and E_{15d} were also nearly equal to those of divacancy with double minus charge state (V_2^{2-}) and divacancy with single minus charge state (V_2^-), respectively [11]. Therefore, levels E_{15b} and E_{15d} were thought of to relate to V_2^{--} and V_2^-, respectively [14].

3.3 Comparison of recovery properties for defects resulting from thermal annealing and pulse aging

The defect density (N_t) of these lattice defects was decreased by pulse aging. However, V_2^{2-} and V_2^- immediately disappeared for $P_{CI} = 0.5$ W (i.e. $I_{CP} = 100$ mA, $U_{CB} = 5$ V).

The difference of the recovery properties between thermal annealing and pulse aging of A center and E center were minutely investigated. The decrease of the lattice defect density N_t was saturated after 30 min for each junction temperature as well as after thermal annealing [16]. If the decrease of N_t is of first order in the exponential isothermal data until 30 min, the unannealed fraction can be described as follows [1]:

$$g = \frac{N_{t_p}}{N_{t_0}} = \exp\left(-\frac{t}{\tau_p}\right), \tag{7}$$

where N_{t_0} and N_{t_p} are the defect densities before and after pulse aging, respectively, and t is the pulse aging time. $1/\tau_p$ is the annealing rate by pulse aging and is related to the activation energy (E_p) by pulse aging as follows [1]:

$$\frac{1}{\tau_p} = \nu_0 \exp\left(-\frac{E_p}{kT}\right), \tag{8}$$

where ν_0 is the characteristic frequency factor. T denotes the junction temperature.

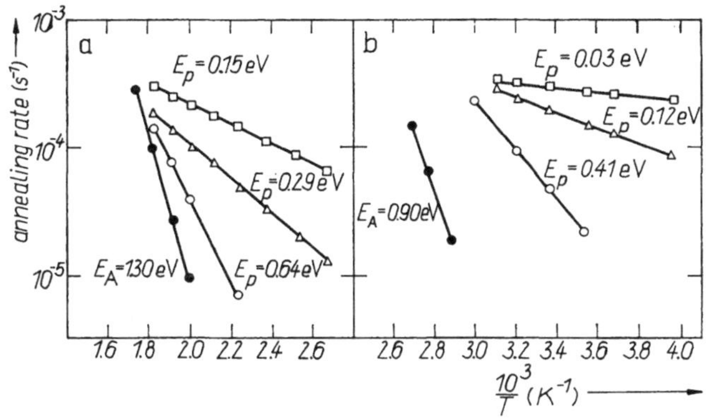

Fig. 3. Dependence of I_{CP} on recovery enhancement for a) A and b) E centers due to pulse aging under $U_{CB} = 10$ V. ● Annealing; ○ 100 mA, △ 200 mA, □ 400 mA

Subsequently, the effects of current and voltage on recovery of A center and E center were investigated. The pulse aging change due to changing I_{CP} under constant U_{CB} was first performed. The junction temperature dependence of the annealing rate for these lattice defects is shown in Fig. 3 a and b, respectively. These figures also compare the rates of thermal annealing and pulse aging for $I_{CP} = 100$, 200, and 400 mA at $U_{CB} = 10$ V. In both A center and E center, $1/\tau_p$ was greater than $1/\tau_A$ at the same T as evidenced in Fig. 3 a and b. Moreover, $1/\tau_p$ increased with increasing I_{CP}. On the other hand, as shown in Fig. 3a, E_p of the A center for $I_{CP} = 100$, 200, and 400 mA was 0.64, 0.29, and 0.15 eV, respectively. Similarly, Fig. 3b indicated that E_p of the E center for $I_{CP} = 100$, 200, and 400 mA was 0.41, 0.12, and 0.03 eV, respectively. In this way, in both A center and E center, E_p was smaller than E_A and became smaller with increasing I_{CP}. A similar current dependence of recovery by pulse aging was obtained under $U_{CB} = 5$ and 20 V.

The effect of voltage on the recovery of A center and E center for constant current was examined next. The relationship between annealing rate and junction temperature for thermal annealing and pulse aging for $U_{CB} = 5$, 10, and 20 V at I_{CP} of 200 mA is indicated in Fig. 4 a and b, respectively. $1/\tau_p$ was greater than $1/\tau_A$ as evidenced in Fig. 4 a and b, and increased with U_{CB}. Furthermore, as indicated in Fig. 4a, E_p of the A center for $U_{CB} = 5$, 10, and 20 V was 0.64, 0.29, and 0.15 eV, respectively. Similarly, Fig. 4b illustrated that E_p of the E center for $U_{CB} = 5$, 10, and 20 V was 0.25, 0.12, and 0.03 eV, respectively. In this way, in both A center and E center, E_p was smaller than E_A and becomes smaller with increasing U_{CB}. The voltage dependence of recovery by pulse aging was similarly observed under $I_{CB} = 100$ and 400 mA.

As mentioned above, it was found for the first time from experiments using a transistor structure that the recovery of A center and E center by pulse aging was enhanced as compared with purely thermal annealing and that this enhancement was dependent on I_{CP} and U_{CB}.

3.4 Recovery mechanism of defect by pulse aging

A and E centers taken in the present study were electron capture levels as mentioned above. The reverse voltage was biased in the base–collector junction so that A center and E center were recovered by the electrons injected into the collector region under reverse biased voltage. Recovery enhancement in this experiment was also dependent on I_{CP} and U_{CB} and increased with increasing I_{CP} and U_{CB}.

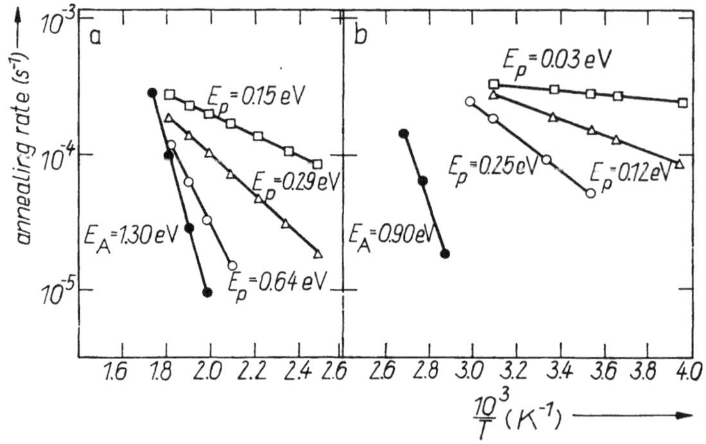

Fig. 4. Dependence of U_{CB} on recovery enhancement for a) A and b) E centers due to pulse aging under $I_{CP} = 200$ mA. ● Annealing, ○ 5 V, △ 10 V, □ 20 V

On the other hand, the electric field in the collector depletion region was estimated to be about 10^4 V/cm. It was reported that the drift velocity of electrons in this electric field is independent of field and approaches a scattering-limited velocity of 1×10^7 cm/s [17]. Moreover, the temperature development of the electron system, when a high electric field was applied, was reported to saturate after about 1 ps [18]. This time was much shorter than the pulse width used in our experiment so that the temperature of electron system seemed to be saturated during electron injection. That is, an electron injected in the collector depletion region turned into a hot electron. Furthermore, it may be possible that the energy transferred from the hot electron to the lattice system in the collector region aids the decomposition of A and E centers and decreases the activation energy of recovery. Based on this speculation, E_p can be described as follows:

$$E_p = E_A - E_l\,, \tag{9}$$

where E_l is the energy transferred from the hot electron to the lattice or a lattice defect in the collector region.

The recovery mechanism considered above of A and E centers due to electron injection under reverse biased voltage will be subsequently confirmed. If all energy supplied to the electrons from the electric field is equally transported to the lattice system, E_l can be described according to the energy transport model as follows [18, 19]:

$$E_l = \frac{J_{CP}\tau_{L\cdot C}F}{Nq}\,, \tag{10}$$

where J_{CP} and F are the collector peak current density and the electric field in the collector depletion region, respectively. N denotes the sum of atomic density of Si and defect density. $\tau_{L\cdot C}$ and q are the energy relaxation time of the lattice system and the magnitude of electric charge, respectively.

In this experiment, the collector peak current cannot be ignored. Consequently, it is necessary to solve the Poisson equation when calculating F under the following hypotheses [20]:

1. Acceptor density of the collector depletion layer can be ignored as compared to donor density.

2. Acceptor density of the base region is much higher than the donor density of the collector region and the base–collector junction is a one-sided step junction.

3. Because of the voltage drop caused by the collector peak current, the reverse voltage across the depletion region is decreased.

From the above assumption, F and the width of the depletion region (W_{dep}) are given by

$$F = \frac{(U_{\text{CB}} - \varrho_{\text{C}} W_{\text{C}} J_{\text{CP}})}{W_{\text{dep}}}, \tag{11}$$

$$W_{\text{dep}} = \left[\frac{2\varepsilon(U_{\text{D}} + U_{\text{CB}} - \varrho_{\text{C}} W_{\text{C}} J_{\text{CP}})}{(qN_{\text{D}} - J_{\text{CP}}/V_{\text{S}})}\right]^{1/2}, \tag{12}$$

where U_{D}, N_{D}, and ϱ_{C} show the diffusion voltage, the donor density, and the specific resistance of the collector region, respectively. W_{C} and V_{S} denote the length of the collector region and the drift velocity, respectively. Substituting (10) and (11) into (9), E_{I} can be reduced to the following expression:

$$E_{\text{I}} = \frac{J_{\text{CP}}\tau_{\text{L}\cdot\text{C}}}{Nq} \frac{(U_{\text{CB}} - \varrho_{\text{C}} W_{\text{C}} J_{\text{CP}})}{\left(\dfrac{2\varepsilon(U_{\text{D}} + U_{\text{CB}} - \varrho_{\text{C}} W_{\text{C}} J_{\text{CP}})}{(qN_{\text{D}} - J_{\text{CP}}/V_{\text{S}})}\right)^{1/2}}. \tag{13}$$

As mentioned earlier, V_{S} under the experimental conditions was presumed to be about 1×10^7 cm/s. $\tau_{\text{L}\cdot\text{C}}$ was also reported to be about 1 ms, [18] and equal to the pulse width used in this study. Consequently, when $\tau_{\text{L}\cdot\text{C}}$, ϱ_{C}, and W_{C} were 1 ms, 6 Ω cm, and 9.9 μm, respectively, E_{I} at I_{CP} of 200 mA and U_{CB} of 20 V was calculated to be 1.14 eV by (13). E_{A} of the A center was 1.30 eV as evidenced in Fig. 2. Using (9), E_{P} of this pulse aging was calcualted to be 0.16 eV. The experimental value of E_{P} for this pulse aging was 0.15 eV as shown in Fig. 4a, and was nearly equal with the calculated value.

Fig. 5 a and b shows the comparison of calculated and experimental values for E_{P} when $E_{\text{A}} - E_{\text{I}} > 0$. If $E_{\text{I}} > E_{\text{A}}$, it is meaningless to compare the recovery properties between thermal annealing and pulse aging. In this case, the lattice defects seem to be mainly recovered by pulse aging. These figures also indicate the dependence on I_{CP} and U_{CB}, respectively. As shown in Fig. 5 a and b, the experimental values were approximately in agreement with the calculated values. Similar results were obtained also at $U_{\text{CB}} = 5$ and 20 V. Furthermore, experimental values were nearly equal to the calculated values for different I_{CP}.

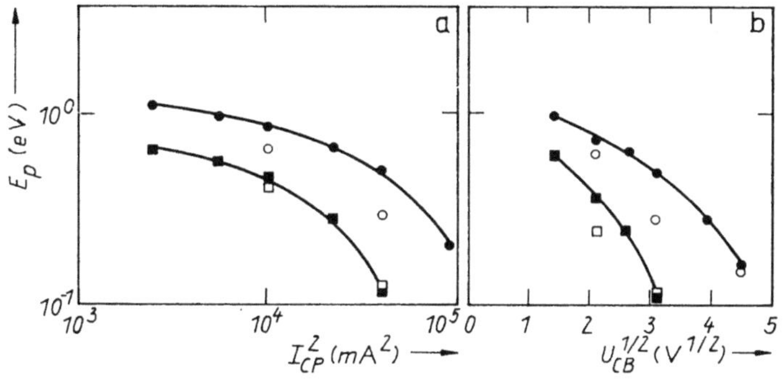

Fig. 5. Comparison of calculated (full symbols) and experimental (open symbols) values of E_{P} a) dependent on I_{CP} ($U_{\text{CB}} = 10$ V); b) dependent on U_{CB} ($I_{\text{CP}} = 200$ mA). A center is marked by circles, E center by squares

These results proved that E_I aids the recovery of A center and E center by pulse aging. Consequently, it was shown that the recovery enhancement of A center and E center due to electron injection under reverse biased voltage is attributed to the energy transport from hot electrons in the collector depletion region to the lattice system.

4. Conclusions

The recovery process of lattice defects formed in the collector region for an npn Si transistor, which was irradiated with 2 MeV electrons to a fluence of 10^{15} e/cm^2, was investigated.

From this study, the following results were obtained:

1. A center (V–O), E center (V–P), divacancy with single and double minus charge state (V_2^- and V_2^{--}) were formed by electron irradiation.
2. The annealing rate of A center and E center by pulse aging is greater than that of purely thermal annealing and increases with increasing collector peak current and collector–base reverse voltage.
3. The activation energy by pulse aging on the contrary is smaller than that by thermal annealing and decreases with increasing collector peak current and collector–base reverse voltage.
4. The recovery enhancement of A center and E center due to electron injection under reverse biased voltage is attributed to the energy transport from hot electrons in the collector depletion region, to the lattice system.

Acknowledgement

The authors would like to express their thanks to Prof. T. Nakamura and Associated Prof. M. Ishida of Toyohashi University of Technology for their helpful discussion.

References

[1] A. O. EVWARAYE, J. appl. Phys. **48**, 734 (1977).
[2] A. O. EVWARAYE and E. SUN, J. appl. Phys. **47**, 3776 (1976).
[3] A. O. EVWARAYE, J. appl. Phys. **48**, 1840 (1977).
[4] Y. TOKUDA and A. USAMI, IEEE Trans. Nuclear Sci. **28**, 3564 (1981).
[5] H. J. STEIN, J. appl. Phys. **37**, 3382 (1966).
[6] D. V. LANG, J. appl. Phys. **45**, 3023 (1974).
[7] D. V. LANG, J. appl. Phys. **47**, 3587 (1976).
[8] L. C. KIMERLING, Solid State Electronics **21**, 1391 (1978).
[9] J. D. WEEKS, Phys. Rev. B **12**, 3286 (1975).
[10] P. M. PETROFF, O. G. LORIMOR, and J. M. RALSTON, J. appl. Phys. **47**, 1583 (1976).
[11] L. C. KIMERLING, IEEE Trans. Nuclear Sci **23**, 1497 (1976).
[12] C. E. BARNES and G. A. SAMARA, Appl. Phys. Letters **48**, 934 (1986).
[13] K. H. KÜSTERS and H. ALEXANDER, Physica (Utrecht) **116B**, 594 (1983).
[14] H. OHYAMA and K. NEMOTO, Extended Abstracts, The 34-th Spring Meeting, 1987, The Japan Society of Applied Physics and Related Societies, (p. 507) (in Japanease).
[15] Y. TOKUDA, N. SHIMIZU, and A. USAMI, Japan. J. appl. Phys. **18**, 309 (1979).
[16] H. OHYAMA and K. NEMOTO, Extended Abstracts, The 46-th Autumn Meeting, 1985, The Japan Society of Applied Physics, (p. 480) (in Japanease).
[17] S. M. SZE, Physics of Semiconductor Devices, John Wiley, 1969.
[18] B. R. NAG, Electron Transport in Compound Semiconductors, Springer-Verlag, 1980.
[19] R. STRATTON, Phys. Rev. **126**, 2002 (1962).
[20] M. KURATA, Y. TAKEISHI, and H. HARA, Behavior Theory of Bipolar Transistor, Kindai Kagaku Sha, Tokyo 1980.

(Received July 25, 1988)

phys. stat. sol. (a) **110**, 687 (1988)

Subject classification: 73.40 and 85

Sektion Physik der Karl-Marx-Universität, Wissenschaftsbereich Halbleiterphysik, Leipzig[1] *(a), Department of Semiconductor Physics, Faculty of Mathematics and Physics, Charles University, Prague*[2] *(b), and Tesla VÚST, Prague*[3] *(c)*

Transients in $p^+\pi n^+$ Photodiodes

By

G. Grummt (a), J. Toušek (b), and B. Tryzna (c)

Expressions for the transient short-circuit current and open-circuit voltage are derived by solving the time-dependent diffusion equation. It is shown, that the electric field in the π region influences the decay time in $p^+\pi n^+$ diodes. The lifetime of carriers can be evaluated by means of open-circuit voltage decay measurements, whereas the short-circuit current decay yields information concerning the internal electric field.

Durch Lösen der zeitabhängigen Diffusionsgleichung werden theoretische Beziehungen für die Relaxation des Kurzschlußstromes und der Leerlaufspannung erhalten. Es wird gezeigt, daß das elektrische Feld im π-Gebiet die Abklingzeit einer $p^+\pi n^+$-Diode beeinflußt. Die Lebensdauer der Ladungsträger kann durch Messung des Leerlaufspannungsabklingens bestimmt werden, das Abklingen des Kurzschlußstromes enthält Informationen über das interne elektrische Feld.

1. Introduction

Measurements of short-circuit current (SCC) and open-circuit voltage (OCV) transients were performed on solar cells already with the aim to extract the lifetime of carriers in the base region and the recombination rate on the back surface of the sample [1, 2]. In the mentioned papers the free carrier transport in the base region could be described by a diffusion mechanism only. On the other hand, it is not possible to neglect the effect of electric field in $p^+\pi n^+$ diodes.

2. Theory

The present method of transient photovoltaic effect investigation consists in the examination of diffusion and drift of minority carriers in the nearly intrinsic region. The effect of both junction regions appears in the boundary conditions.

2.1 Short-circuit current and open-circuit voltage transients

Let us consider a $p^+\pi n^+$ photodiode illuminated by a rectangular pulse of a light emitting or laser diode as on Fig. 1. The change of SCC or of OCV in time is investigated. A sufficiently long light pulse is assumed, so that steady state is established inside the diode. When the irradiation stops, the concentration of excess minority carriers relaxes towards equilibrium according to the continuity equation. In the bulk

[1]) Linnéstr. 5, DDR-7010 Leipzig, GDR.
[2]) Ke Karlovu 5, CS-121 16 Prague 2, Czechoslovakia.
[3]) Novodvorská 994, CS-140 00 Prague 4, Czechoslovakia.

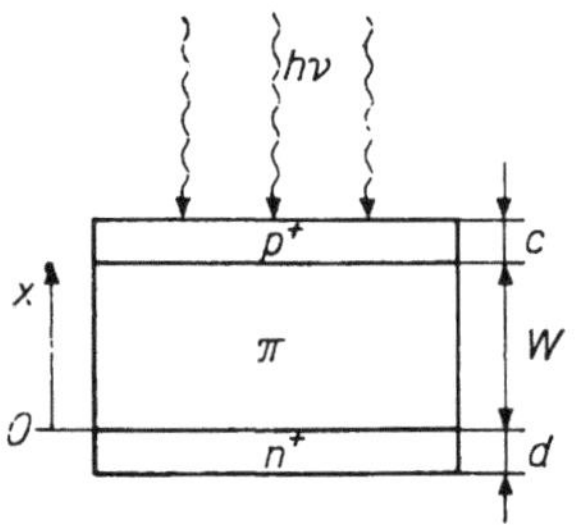

Fig. 1. Schematic representation of the analyzed structure

of the π region the equation

$$\frac{\partial}{\partial t} \Delta n(x, t) = D_n \frac{\partial^2}{\partial x^2} \Delta n(x, t) + \mu_n E \frac{\partial}{\partial x} \Delta n(x, t) - \frac{\Delta n(x, t)}{\tau} \tag{1}$$

is valid in one-dimensional geometry ($\Delta n(x, t)$ is the concentration of excess electrons with diffusion coefficient D_n, mobility μ_n, and lifetime τ; E is the electric field which is supposed to be constant).

The introduction of the electric field into (1) means a generalization of previous approaches (e.g. [1]) and is important in the case of $p^+\pi n^+$ diodes as will be shown further.

Equation (1) is solved with the following boundary conditions:

$$J_n(W, t) = -eS \, \Delta n(W, t) , \tag{2}$$

$$\Delta n(0, t) = 0 \qquad \text{(SCC configuration)}, \tag{3}$$

$$J_n(0, t) = 0 \qquad \text{(OCV configuration)}. \tag{4}$$

$J_n = eD_n(\partial/\partial x) \Delta n + e \, \Delta n \, \mu_n E$ is the density of current, S the recombination velocity at the interface ($x = W$). The point $x = 0$ was chosen at the edge of the depletion region of the $n^+\pi$ junction at the π side. The change of its position and consequently that of the length W with illumination is neglected. There are time boundary conditions, namely

$$\Delta n(x, 0) = \Delta n(x) , \tag{5}$$

$$\Delta n(x, \infty) = 0 . \tag{6}$$

Generally the OCV condition requires that the total current is zero. The boundary condition (4) is valid under the assumptions that the injection ratio is near unity, no generation or recombination occurs in the depletion region and recombination in the n^+ region and on the back surface can be neglected [1]. Let us consider that the concentration can be written in the form: $\Delta n(x, t) = N(x) \, M(t)$. Inserting into (1) we receive

$$\left(\frac{1}{M} \frac{dM}{dt} + \frac{1}{\tau}\right) \frac{1}{D_n} = \frac{1}{N} \frac{d^2N}{dx^2} + 2\varepsilon \frac{1}{N} \frac{dN}{dx} = -\alpha^2 , \tag{7}$$

where $\varepsilon = \mu_n E/2D_n = eE/2k_0T$ and α is a constant. The solution of (7) with regard to M yields

$$M_\alpha(t) = \exp\left[-\left(\alpha^2 D_n + \frac{1}{\tau}\right) t\right] \tag{8}$$

and with regard to N gives

$$N_\alpha(x) = \exp(-\varepsilon x)\,[A_\alpha \sin(x\sqrt{\alpha^2-\varepsilon^2}) + B_\alpha \cos(x\sqrt{\alpha^2-\varepsilon^2})]\,. \tag{9}$$

Let us introduce a parameter $\beta \equiv \sqrt{\alpha^2-\varepsilon^2}$. The values of the parameters α and β are discrete as follows from the boundary conditions. As certain value of $N_\beta(x)$ must ascribed to each parameter β, the general solution of (7) is

$$\Delta n(x, t) = N_\beta(x)\, M_\beta(t). \tag{10}$$

The values of β can be obtained by the solution of the transcendental equations (11), (12), found by inserting (10) into the boundary conditions (3), (4)

$$\beta_j \cot(\beta_j W) = -(\sigma + \varepsilon); \qquad \sigma \equiv \frac{S}{D_n} \tag{11}$$

in the short-circuit regime and

$$\beta_V \cot(\beta_V W) = \frac{\beta_V^2 + \varepsilon^2}{\sigma} + \varepsilon \tag{12}$$

in the open-circuit regime.

There is an infinite number of discrete real solutions of (11) and (12) which will be indicated as β_j^k and β_V^k. The resulting form of the solution is

$$\Delta n_i(x, t) = \sum_{k=0}^{\infty} [A_i^k \sin(\beta_i^k x) + B_i^k \cos(\beta_i^k x)] \exp\left(-\varepsilon x - \frac{t}{\tau_i^k}\right), \tag{13}$$

$$\frac{1}{\tau_i^k} = (\beta_i^k)^2 D_n + D_n \left(\frac{eE}{2k_0 T}\right)^2 + \frac{1}{\tau}, \tag{14}$$

where $i = j$ or V indicates the solutions for SCC and OCV regimes, respectively. Parameters A_i^k and B_i^k are determined by the initial distribution $\Delta n(x, 0)$. Now it is possible to derive easily the relation for the short-circuit photocurrent relaxation. For the minority carrier current in the p-type semiconductor

$$J_{sc}(t) = eA\mu_n E\, \Delta n_j(0, t) + eAD_n \frac{\partial}{\partial x} \Delta n_j|_{x=0}; \tag{15}$$

A is the junction area.

Employing (13) to (15)

$$J_{sc}(t) = eAD_n \sum_{k=0}^{\infty} A_j^k \beta_j^k \exp\left(\frac{-t}{\tau_j^k}\right) \tag{16}$$

is obtained, where

$$\frac{1}{\tau_j^k} = (\beta_j^k)^2 D_n + \varepsilon^2 D_n + \frac{1}{\tau}. \tag{17}$$

The open-circuit voltage can be calculated according to [3]

$$U_{oc}(t) = \frac{\chi k_0 T}{e} \lg\left(1 + \frac{\Delta n_V(0, t)}{n_0}\right), \tag{18}$$

where $\Delta n_V(0, t)$ is the concentration of excess carriers at $x = 0$ in the OCV regime described by (13), χ the quality factor of the diode, and n_0 the dark electron concentration.

The current transient is not simply exponential, it consists of an infinite number of exponential functions as is evident from (16).

The decay of SCC can be considered as simple exponential, with the highest constant τ^k, after the faster relaxation terms in right-hand side of (16) have vanished. According to (17) the decay constant τ^k depends on the parameter β^k, which can be computed with the aid of (11) and (12). If parameters S and E are from the interval $(0, +\infty)$, then the parameter β^k has the value $\pi(k + \delta)/W$, where $\delta \in \langle 0, 1/2 \rangle$ and $k = 0, 1, 2, \dots$ For the sake of simplicity let us take $E = 0$ and then we shall derive τ^k inserting β^k in the above form into (17). We obtain

$$\frac{1}{\tau_k} = \pi^2 \left(\frac{k + \delta}{W}\right)^2 D_n + \frac{1}{\tau} = \frac{1}{\tau}\left[\pi^2 (k + \delta)^2 \left(\frac{L_n}{W}\right)^2 + 1\right]. \tag{19}$$

Let us discuss the following cases:

1. $W \gg L$: If the π region is sufficiently thick, the first term in brackets in (19) can be neglected and the decay has an exponential term with decay constant τ.

2. $W > L$: τ^k decreases slightly with increasing k; many terms in the sum (16) contribute essentially for relatively long time and the decay is strongly non-exponential.

3. $W \leqq L$: There is a strong dependence of τ^k on k. After a time which is approximately equal to $2\tau^1$ after switching off the light, the decay can be described by an exponential function with time constant τ^0, where $\tau^0 = \tau^k$ $(k = 0)$.

4. $W \ll L$: The first term in the sum (16) determines the decay again for $t > 2\tau^1$. In this case we obtain only the first term in brackets in (19), e.g. τ^0 depends no longer on lifetime τ.

With regard to (17) the presence of an electric field can be included by substitution of an effective diffusion length L_{ef} for the diffusion length L according to

$$\frac{1}{L_{ef}} = \sqrt{\frac{1}{L^2} + \left(\frac{eE}{2k_0T}\right)^2}\,. \tag{20}$$

According to the presented considerations there are two cases for which the dependence of J_{sc} on t may be considered as exponential with sufficient accuracy, namely: $W \gg L$ and $W \leqq L$, where L is the minority carrier diffusion length. Then (17) reduces to

$$\frac{1}{\tau_j} = (\beta_j^0)^2 D_n + \varepsilon^2 D_n + \frac{1}{\tau}, \tag{17a}$$

where β_j^0 is the first solution of (11).

The electric field occurring implicitly in (17a), can be formed as a consequence of three effects:

a) inhomogeneous doping of the π-region yielding a constant drift field E_0,

b) appearance of a voltage drop in the π-region in the SCC regime creating an electric field E_p,

c) the applied reverse voltage U produces a field E_V for which holds $E_V = hU$ (h is a constant).

The values of β and W depend on the electric field E much less than the second term in (17a) which follows from the solution of (11). Hence, we can neglect their voltage change. After inserting sequentially $E = E_0 + E_p + hU$ and $E = E_0 + E_p$ $(U = 0)$ into equation (17a) and solving for $\tau_j(U)$ and $\tau_j(0)$, respectively, we obtain

$$\frac{1}{\tau_j(U)} - \frac{1}{\tau_j(0)} = D_n \left(\frac{e}{2k_0T}\right)^2 [2(E_0 + E_p)\, hU + h^2U^2] \,. \tag{21}$$

The values of $(E_0 + E_p)$ and h can be estimated by fitting the function (21) to the measured dependence of $\tau_j(U)$.

The open-circuit voltage decay is calculated from (18), where $\Delta n_V(0, t)$ may be considered as single exponential from (13) with the time constant τ_V, but according to (18) the voltage decay will not be exponential. A strong bias illumination suggested by Moore [3] leading to an excess carrier concentration Δn_I $(\Delta n_I \gg \Delta n_V(t))$ reveals the exponential OCV decay with τ_V.

2.2 Influence of parasitic effects on the relaxation

The measurement of SCC and OCV is usually experimentally realized by the voltage drop on a small and large load resistance, respectively. Let us analyse the simple equivalent circuit shown in Fig. 2. C_0, R_0 represent the capacitance and the resistance of the $n^+\pi$ junction (C_0 includes generally the capacitance of the space charge region and the diffusion capacitance) and R_l is the load resistance. Parasitic capacitances are particularly included into C_0. The series resistance of the π-region and the $p^+\pi$ junction were not taken into account here.

For R_0 holds

$$R_0(U) = \left(\frac{dJ_p}{dU}\right)^{-1} = \frac{\chi k_0 T}{eJ_p(U)} \tag{22}$$

($J_p(U)$ is the dark current flowing through the junction at the applied voltage U).

Now, let us consider the following form of time dependence for the current source (see Fig. 2):

$$\begin{aligned} J(t) &= 1 \qquad &\text{for} \quad t \leqq 0\,, \\ J(t) &= \exp\left(\frac{-t}{\tau}\right) \qquad &\text{for} \quad t > 0\,. \end{aligned} \tag{23}$$

The solution of the equivalent circuit leads to the output voltage

$$U(t) = \frac{R_0R_l}{R_0 + R_l}\,\frac{1}{\tau - \tau_c}\left\{\tau \exp\left(\frac{-t}{\tau}\right) - \tau_c \exp\left(\frac{-t}{\tau_c}\right)\right\}, \tag{24}$$

where

$$\tau_c = C_0 \frac{R_0R_l}{R_0 + R_l}\,. \tag{25}$$

The symbol τ_c will be further called "circuit time constant". The relaxation of the function $U(t)$ in (24) is controlled by that of τ_c or τ which value is higher. In the SCC

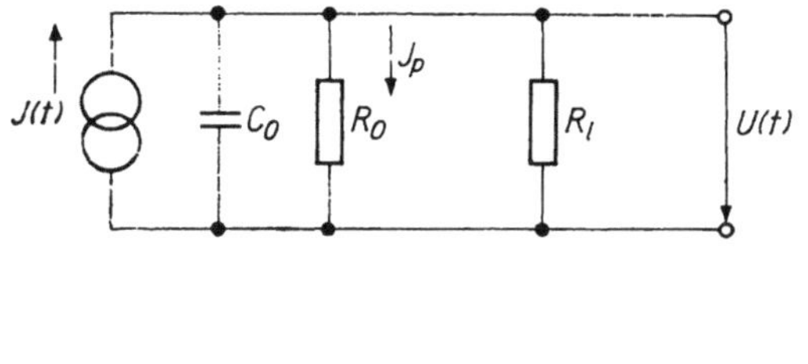

Fig. 2 Equivalent circuit for the analysis of parasitic effects in relaxation

regime the voltage U is zero implying $J_p(U) = 0$ and $R_0 \to \infty$. Thus

$$\tau_c = C_0 R_l \tag{26}$$

is valid. The load resistance R_l should be sufficiently small to suppress the circuit time constant. Including the series resistance R_s of the π-region, the relation

$$\tau_c = C_0(R_s + R_l) \tag{27}$$

is obtained.

In the OCV regime

$$\tau_c = C_0 R_0 \tag{28}$$

holds if the relation $R_0 \ll R_l$ is fulfilled. It is obvious that C_0 and R_0 depend on the open-circuit voltage U_{oc}. At low U_{oc} the value of τ_c is large due to (22) and the relaxation of $U(t)$ yields a constant circuit time. Using dc bias light together with the measuring signal, R_0 falls down due to the rise of current $J_p(U_{oc})$ according to (22). As a result τ_c decreases. If $\tau_c < \tau$ is attained, the output voltage relaxes with the time constant τ. However, the decrease of τ_c is limited by the increase of the junction capacitance with rising dc bias light intensity.

3. Experiment

The kinetics was studied on silicon $p^+\pi n^+$ diodes with a 140 μm thick active region doped to the order of 2×10^{12} cm^{-3}. The sample was placed in a cryostat provided with two windows. The measuring optical pulses from a laser diode ($\lambda = 838$ nm) passed through one of them, whereas the other served for the bias light. Both beams were focused onto the sample. The laser diode was driven by a pulse generator equipped with power amplifier. The light pulses had rectangular form with 25 ns edges. The signal from the sample was amplified by means of a preamplifier with tenfold amplification and bandwidth 50 MHz. An oscilloscope Iwatsu SS 6100 with bandwidth of 100 MHz recorded the transients. For the bias light a halogen 150 W/24 V lamp was employed.

4. Results and Discussion

4.1 Determination of decay constants

The curve of transient decay of OCV (Fig. 3) is approximately exponential in the whole range of the bias light intensities I_0. The bias light intensity was deduced from the magnitude of SCC passing through the sample. As we will show the transient

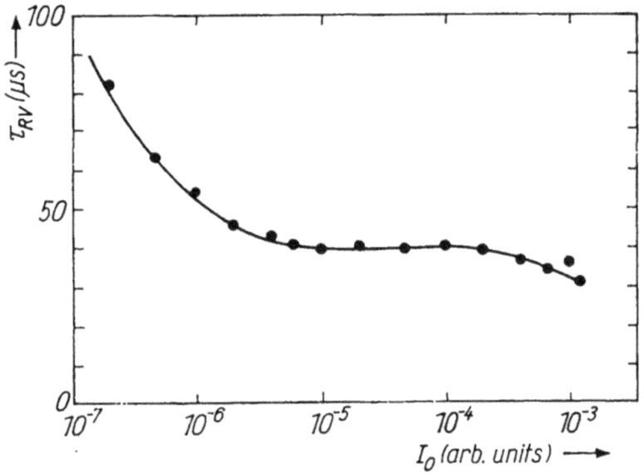

Fig. 3. OCV decay time constant τ_{RV} as a function of the bias light intensity I_0 at room temperature. The intensity is given by the corresponding short-circuit current

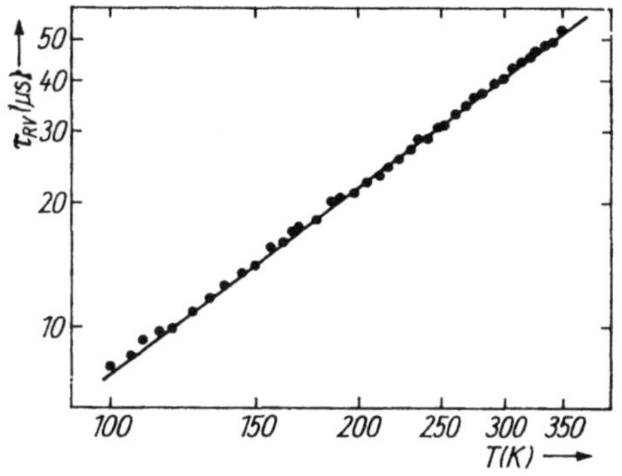

Fig. 4. OCV decay time constant τ_{RV} vs. temperature, for $I_0 = 50\ \mu A$

is not controlled by (18) and the observed decay constant τ_{RV} is not equal to τ_V. The initial decrease of τ_{RV} is connected with a fall of the circuit time constant due to reduction of the junction resistance R_0 (see (22) and (28)). In the region of the plateau which follows, the temperature dependence of τ_{RV} was measured. The second decline τ_{RV} at the highest intensities corresponds to a decrease of the carrier lifetime as a result of high injection. Fig. 4 shows the temperature dependence of the decay constant τ_{RV} which fulfils well the relation $\tau_{RV} \sim T^{1.65}$.

The sample capacitance C_0 and the resistance of the junction R_0 were measured in the dark by means of a multifrequency *LCR* meter HP at suitable applied direct bias and frequency as a function of temperature [5]. It was found that the product C_0R_0 increases proportional to $T^{1.7}$, thus the temperature dependence is approximately the same as that of the circuit time constant from decay experiments. However, provided that $\omega\tau \ll 1$ and $L > W$ (which was probably fulfilled in our experiment), $C_0 = C_d$ and $C_dR_0 = \tau$ [4, 5], τ being the excess carrier lifetime and C_d the diffusion capacitance, which governs the sample capacitance under illumination. Hence the decay constant τ_{RV} has the meaning of the lifetime and the measurement of the decay of OVC under dc illumination represents the method of its determination in the case $L > W$ with low surface recombination velocity.

The measured decay time of SCC is shown on Fig. 5. It can be seen that $1/\tau_j$ increases at first as the intensity of dc illumination I_0 as well as the pulse intensity I_p rise, which is probably caused by a voltage drop in the π-region. According to (21) the increase of $1/\tau_j$ is a consequence of the relevant electric field E_p. At higher light intensities the

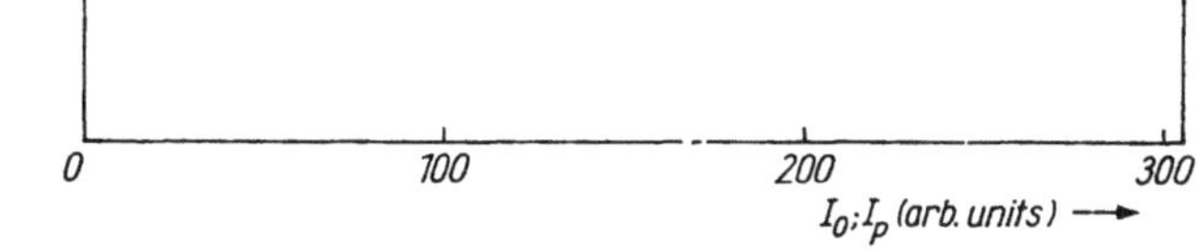

Fig. 5. Inverse SCC decay time conscant $1/\tau_j$ vs. bias light I_0 (full circles) and pulse intensity I_p (open circles)

electric field is screened by free carriers and its influence on the decay constant τ_j decreases because the condition of high generation is fulfilled (ambipolar drift mobility $\mu = 0$). Then the term $2hU(E_0 + E_p)$ in (21) can be omitted; $1/\tau_j$ decreases and finally becomes independent of I_0. The intensity of the measuring pulse has a similar effect. Extrapolating the $1/\tau_j$ versus I_p dependence to $I_p = 0$, we obtain $\tau_j = (1.14 \pm \pm 0.06)$ μs which is the value no longer influenced by the electric field caused by the light pulse I_p.

4.2 Determination of electric field in the π-region

The field in the π-region will be evaluated employing (17) and (11) and the following considerations. From the measurement of the transients of OCV and SCC mentioned above the value of τ and τ_j, respectively, is obtained and it follows that $\tau \gg \tau_j$. As a consequence in (17a) the term $1/\tau$ can be neglected in comparison with $1/\tau_j$. Hence we can write

$$\frac{1}{\tau_j} = \beta_j^2 D_n + \left(\frac{e}{2k_0T}\right)^2 (E_0 + E_p)^2 D_n \,. \tag{17b}$$

Let us now examine (11). The term ε can be evaluated using the value of the electric field which was deduced from the capacitance measurements [5]. The term σ, which characterizes the influence of the $p^+\pi$ junction is usually very small in comparison with ε and could be omitted in (11).

Then we obtain

$$\beta_j \cot(\beta_j W) = -\left(\frac{e}{2k_0T}\right)(E_0 + E_p) \,. \tag{11'}$$

By solution of these equations and with the experimental value $\tau_j = 0.87$ μs ($I_p = = 3$ μA) we get $E_0 + E_p = (435 \pm 30)$ V m^{-1}. Inserting the extrapolated value τ_j (1.14 μs) into (17b) one obtains the field originating from an inhomogeneous doping $E_0 = (290 \pm 30)$ V m^{-1}. A C–U doping profile measurement was performed to prove this. A nearly constant doping gradient was found in the whole π-region, from which one can calculate an electric field $E_0 = 270$ V m^{-1} in surprising agreement with the above value. The field created by the measuring pulse can be simply found then. Moreover we have measured the voltage dependence of the decay constant τ_j. In accordance with (17′a) the applied field accelerates the relaxation, which agrees with the experimental curve on Fig. 6. By fitting the theoretical relation (21) to the experi-

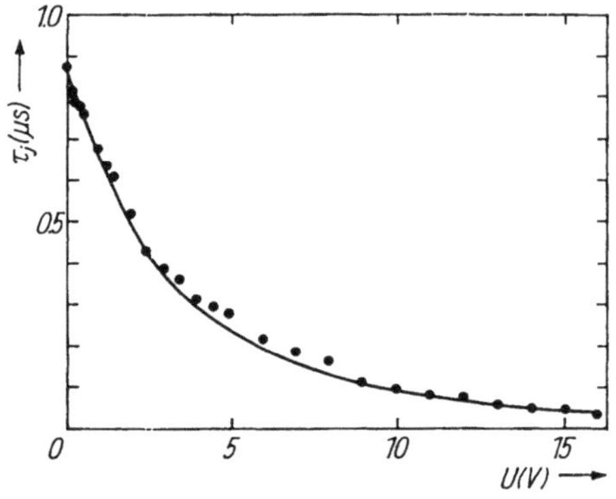

Fig. 6. SCC decay time constant τ_j as a function of applied reverse voltage. The points are experimental data and the solid line is the fit to the data

mental curve, for the same I_p as above, the sum $(E_0 + E_p)$ was determined as 455 V m^{-1}, in good agreement with the above result.

5. Conclusion

The transient of the photovoltaic effect on $p^+\pi n^+$ structures yields some important information. The decay constant depends generally on excess carrier lifetime, recombination velocity in the $p^+\pi$ interface, and electric field in the π-region. The decay is not exponential, nevertheless we can consider it like that in the following cases: $W \gg L$, $W \leqq L$.

a) Short-circuit current decay: From the experimental results it was found that the lifetime and the recombination in the $p^+\pi$ interface had no influence on the SCC decay, thus the transient had been affected mainly by the electric field in the π-region. The field caused by the applied voltage, the inhomogeneity of π-region doping, and the voltage drop after illumination accelerate the relaxation. At high irradiation level, the electric field has no more influence on the relaxation and τ_I is not changed with light intensity.

b) Open-cricuit voltage decay: The theoretical decay constant τ_V does not determine the OCV decay even with bias light. The relaxation is dominated by the circuit time constant which approaches the lifetime τ at intermediate light intensity. Consequently, the method of Rose and Weaver [1, 2] to determine lifetime and surface recombination from τ_I and τ_V is not usable for samples with $L > W$ with low surface recombination velocity. However, for $L > W$ and for low surface recombination the OCV decay time constant under suitable dc illumination was shown to give a good value of the lifetime.

Acknowledgement

The authors thank Prof. E. Klier for useful remarks on the manuscript.

References

[1] B. H. Rose and H. T. Weaver, J. appl. Phys. **54**, 238 (1983).
[2] B. H. Rose, IEEE Trans. Electron. Devices **31**, 559 (1984).
[3] A. R. Moore, RCA Rev. **41**, 549 (1980).
[4] N. G. Franklin and A. Neugroschel, IEEE Trans. Electron. Devices **31**, 413 (1984).
[5] G. Grummt, Thesis, Faculty of Mathematics and Physics, Charles University Prague, 1986.

(Received October 6, 1988)

Short Notes

phys. stat. sol. (a) 110, K61 (1988)
Subject classification: 61.70 and 66.30; S5.11

Leningrad Polytechnical Institute[1)]

Diffusion of Impurities from Implanted Silicon Layers by Rapid Thermal Annealing

By

O.V. ALEKSANDROV, V.V. KOZLOVSKII, V.V. POPOV, and B.E. SAMORUKOV

Experimental investigations of impurity diffusion in implanted layers of semiconductors by rapid thermal annealing (RTA) showed the enhanced diffusion of a number of dopants (B, P, Sb, As) at short time RTA in the order of 5 to 7 s /1 to 3/. To explain the experimental data in /2 to 4/ it was supposed that deep penetration of dopants resulted from the point defects (PD) at RTA. The excess PD, vacancies V in particular, can occur due to dissociation by RTA of complex radiation defects induced by ion dopant implantation.

This note presents an analysis of the diffusion of substitutional dopants from implanted semiconductor layers at short time RTA when the concentration of "radiation" vacancies may substantially exceed the concentration of equilibrium vacancies.

In the analysis dopant diffusion equations considering a nonuniform depth distribution of vacancies were used /5, 6/:

$$\frac{\partial C}{\partial t} = \frac{D_v}{N_s}\left(V\frac{\partial^2 C}{\partial x^2} - C\frac{\partial^2 V}{\partial x^2}\right) , \qquad (1)$$

where V, C, N_s are concentrations of vancies, dopant atoms, and matrix atoms, respectively.

The vacancy depth distribution in the semiconductor was obtained from the continuity equation

$$\frac{\partial V}{\partial t} = D_v \frac{\partial^2 V}{\partial x^2} + \frac{C_k}{\tau_k} - \frac{V}{\tau_v} , \qquad (2)$$

where the first term describes the vacancy diffusion with the diffusion coefficient D_v; the second the vacancy generation caused by radiation defect (RD) dissociation with lifetime τ_k; the third the recombination of vacancies with lifetime τ_v. In solving (1) the $\tau_v(T)$ and $D_v(T)$ dependences from /7, 8/ were used.

---

1) Ul. Politekhnicheskaya 29, SU-195251 Leningrad, USSR.

The energy of elastic losses of striking ions was assumed to follow a simplified Poisson IV distribution which was taken as the initial depth distribution of complex radiation defects /9/:

$$C_{k_o}(x) = \sqrt{\frac{2}{\pi}}\,\frac{I}{\sigma_d}\,\frac{\exp\left[-\frac{1}{2}(x - R_d)^2/\sigma_d^2\right]}{1 + \mathrm{erf}\left[R_d/\sqrt{2}\,\sigma_d\right]}, \qquad (3)$$

where I density of ion flux, σ_d and R_d straggling and mean depth distribution, respectively, of the ion energy due to collisions.

The vacancy generation velocity under these conditions is affected by the annealing rate of RD and at non-steady RTA is a function not only of coordinates but of time as well,

$$G(x, t) = -\frac{\partial C_k(x, t)}{\partial t} = \frac{C_k(x, t)}{\tau_k}, \qquad (4)$$

$$\tau_k(T) = \tau_{k_o} \exp\left(\frac{E_k}{kT}\right), \qquad (5)$$

where E_k is the annealing activation energy.

By solving (4) the depth distribution of the RD concentration as a function of time is obtained,

$$C_k(x, t) = C_{k_o}(x) \exp\left[-\int_o^t \frac{dt}{\tau_k[T(t)]}\right]. \qquad (6)$$

The temperature-time dependence at RTA was obtained from the standard thermal balance equation with the temperature dependence of reflection and absorption coefficients of incident light energy taken into account /10/

$$\frac{dT}{dt} = \frac{1}{\varrho B d}\left[E_o(1 - R)(1 - e^{-Kd}) - 2\varepsilon_{np}\sigma(T^4 - T_o^4)\right], \qquad (7)$$

where ϱ and B density and heat capacity of a semiconductor, d substrate thickness, E_o substrate irradiation intensity, R and K integral reflection and absorption coefficients; ε_{np} reduced irradiation ability of the sample-chamber heating system, σ Boltzmann constant, and T_o chamber wall temperature.

Equation (7) was solved by the Runge-Kutta method of fourth order numerical integration taking into account the temperature dependence of B and K, with the following values: E_o = 35 W/cm^2, R = 0.33, ε_{np} = 0.41, d = 0.3 mm, T_o = = 300 K. The typical T(t) relationship at an annealing duration of 20 s is shown in the inset of Fig. 1.

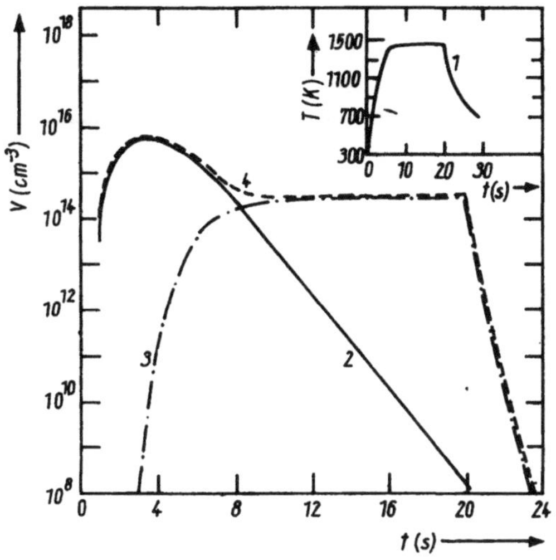

Fig. 1. Time dependence of substrate temperature (1), radiation vacancy concentration (2), thermal equilibrium vacancy (3), and total vacancy concentration (4) at an RTA duration of 20 s and power density 35.5 W/cm^2

The continuity equations for the vacancies and the dopant were solved by the run method using the implicit difference scheme under the following initial and boundary conditions:

$$V(x, 0) = 0; \quad V(0, t) = V_T(T); \quad V(\infty, t) = V_T(T) ; \tag{8}$$

$$C(x, 0) = C_0(x); \quad \frac{\partial C}{\partial x}(0, t) = 0; \quad C(\infty, t) = 0 \quad . \tag{9}$$

The concentration of thermally balanced vacancies in (8) was assumed /11/ to be

$$V_T = 4.5\mathrm{x}10^{28} \exp\left(-\frac{4.28\mathrm{eV}}{kT}\right) \mathrm{cm}^{-3} . \tag{10}$$

The second condition in (9) demonstrates the absence of dopant evaporation from the semiconductor surface during RTA.

Fig. 1 illustrates the calculated results on the kinetic dependences of the vacancy concentration induced by annealing, i.e. radiation vacancies (curve 2) and their total concentrations (curve 4).

In Fig. 1 it can be seen that radiation vacancies dominate at short time intervals in RTA ($\lesssim$ 8 s) whereas at long time intervals thermal equilibrium vacancies dominate. The profiles of the dopant distribution corresponding to short time intervals with (curve 3) and without (curve 2) radiation vacancies taken into account are shown Fig. 2. In the same figure the experimental phosphorus depth distribution profile is represented by dots. This profile was obtained by the differential conductivity method with annealing pulse duration of 6 s, the implantation conditions being: ion energy 75 keV, dose $5\mathrm{x}10^{15}$ cm^{-2}.

As shown in Fig. 2, the consideration of "radiation" vacancies can explain the acceleration of dopant diffusion at short time RTA. The best agreement

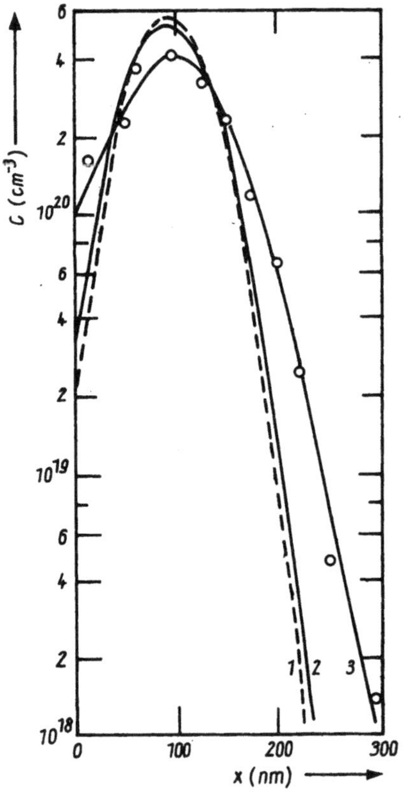

Fig. 2. Depth distribution of ion-implanted phosphorus before (1) and after (curves 2, 3) RTA with 6 s duration. Dots are experiment, solid line is calculated with (3) and without (2) "radiation" vacancies

with the experimental results in the above example could be obtained with the following values of RD: $\tau_{k_0} = 5x10^{-5}$ s, $E_k = 1.3$ eV. It should be mentioned that the simplest defect complexes such as divacancies (V-V), A centres (V-O), and negatively charged E-centres (V-P) caused by ion implantation of P^+ with annealing activation energies approaching the values obtained, can act as radiation defects which are sources of excess vacancies. More complicated multi-particle vacancy clusters, disordered regions, and layers can also be sources of radiation defects /12/. These defects can give rise to simpler defect complexes and PD which enhance dopant diffusion. The most intensive generation of PD occurs at short time intervals of RTA. In the process of annealing the depletion of RD causes a drop of the PD concentration and diffusion speed, these two values tending to approach their equilibrium magnitudes.

Thus, taking into account the consideration of excess PD induced during RTA in the process of dissociation of complex RD having implantation origin makes it possible to explain the acceleration of the substitutional dopant diffusion at short time RTA by using the vacancy mechanism only.

References

/1/ V. E. BORISENKO and V. A. LABUNOV, phys. stat. sol. (a) 72, K173 (1982).
/2/ G.S. OEHRLEIN, S.A. COHEN, and T.O. DEGWICK, Appl. Phys. Letters 45, 417 (1984).
/3/ J. NARAYAN and O.W. HOLLAND, J. appl. Phys. 56, 2913 (1984).

/4/ R.B. FAIR, J.J. WORTMAN, and J. LIU, J. Electrochem. Soc. 130, 2387 (1984).

/5/ Y. MORIKAWA, K. YAMAMOTO, and K. NAGAMI, Appl. Phys. Letters 36, 997 (1980).

/6/ V.V. KOZLOVSKII, V.N. LOMASOV, and L.S. VLASENKO, Radiat. Eff. 744 (1987).

/7/ T. ITO and I. OHDOMARI, J. appl. Phys. 41, 434 (1970).

/8/ B.J. MASTERS and E.F. GOREY, J. appl. Phys. 49, 2717 (1978).

/9/ E.F. MABY, J. appl. Phys. 47, 830 (1976).

/10/ J.R. MEYER, M.R. KRUER, and F.J. BARTOLLI, J. appl. Phys. 51, 5513 (1980).

/11/ B.I. BOLTAKS, Compensated Silicon, Izd. Nauka, Leningrad 1972 (in Russian).

/12/ H.D. GEILER, Radiation Enhanced Diffusion and Annealing Phenomena in Semiconductors, Zentralinstitut Kernforschung der AdW der DDR, Rossendorf, Dresden 1978.

(Received July 15, 1988)

phys. stat. sol. (a) 110, K67 (1988)
Subject classification: 64.70 and 65.50; S1.3

Institute of Solid State Physics, Academia Sinica, Hefei[1)]

A Study of Structural Relaxation in Glassy $Pd_{77.5}Cu_{6.0}Si_{16.5}$ by Microcalorimetric Measurements

By

XINGZHAO DING and YIZHEN HE

The study of structural relaxations which are a common phenomenon in amorphous materials is of great significance in theoretical as well as technical aspects. Chen /1/ has systematically studied the enthalpy relaxations in some (FeNi)-based metallic glasses, and suggested that the enthalpy relaxation processes are different in the low and high temperature regions: (1) sub-sub-T_g region, $T_a < T_g$-100 K (T_a annealing temperature, T_g glass transition temperature) and (2) sub-T_g region, T_g-100K $< T_a < T_g$, respectively. Suzuki et al. /2, 3/ have studied the sub-T_g region enthalpy relaxation in a few metallic glasses, the magnitude of the heat absorbed near T_g was considered as related to a decrease of enthalpy of the amorphous system during annealing.

In this note, the sub-T_g region enthalpy relaxations in a metallic glass $Pd_{77.5}Cu_{6.0}Si_{16.5}$ were studied in detail. The apparent activation energies of the endothermal peak near T_g on the differential scanning calorimetric (DSC) curves of the annealed specimens with different annealing time t_a were obtained by using the Kissinger method. The relaxation phenomena observed were discussed in terms of having solid-like and liquid-like regions in a glassy matrix.

Experimental The amorphous nature of the as-received melt-spun ribbon samples was confirmed by X-ray diffraction and transmission electron microscopy (TEM) experiments. The glass transition temperature (T_g) and crystallization temperature (T_x) were obtained as $T_g \approx 636.8$ K and $T_x \approx \approx 699.0$ K, from DSC measurements (scan rate: 20 K/min). The annealing treatments of the specimens were performed in a vacuum furnace (4×10^{-3} Torr) and the annealing temperature was controlled within 0.5 K. A scanning rate of 80 K/min was used for the calorimetric measurements.

---

1) POB 29, Hefei, People's Republic of China.

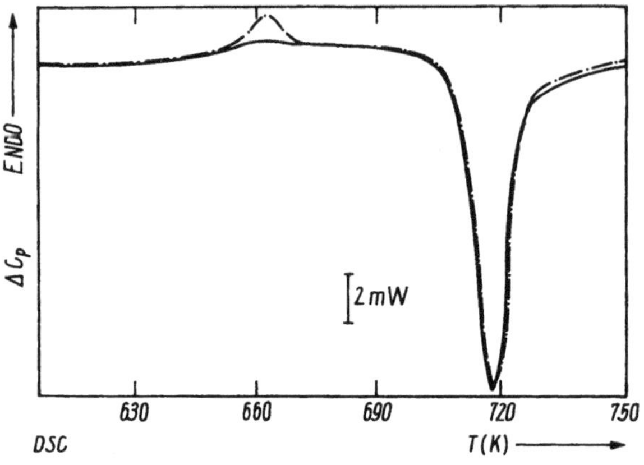

Fig. 1. DSC curves of an annealed specimen at T_a = 563 K of $Pd_{77.5}Cu_{6.0}Si_{16.5}$ and the as-quenched specimen; scan rate 80 K/min, —— as-quenched, — · —annealed, t_a = 16 h

Results An endothermal peak near T_g on the DSC curve of the annealed specimens (dots and dashes) is shown in Fig. 1. This peak is superimposed upon a step that is related to the glass transition and it is defined as an "enthalpy relaxation peak" (ERP). The crystallization peak of the annealed specimen is about the same as that of the as-quenched specimen. The DSC curves of specimens annealed at 563 K for various times were compared in Fig. 2. The amount of heat absorbed has increased with increasing annealing time (t_a) and up to a certain saturated value, meanwhile the temperature of the endothermal peak shifted to a higher value. The apparent activation energies of the enthalpy relaxation peaks of a series of specimens annealed at 563 K were obtained by using five different scanning rates (40, 50, 60, 80, and 100 K/min), see Fig. 3, and the calculated values are listed in Table 1. The dependence of the apparent activation energy on the logarithm of t_a is shown in Fig. 4. It is evident that the apparent activation energy (E) varies with ln t_a differently in two stages, E increased with increasing t_a for the shorter t_a region and decreased with increasing t_a for the longer t_a region.

Discussion In this note, a model of having solid-like and liquid-like regions in a glassy matrix was adopted to explain the relaxation characteristics

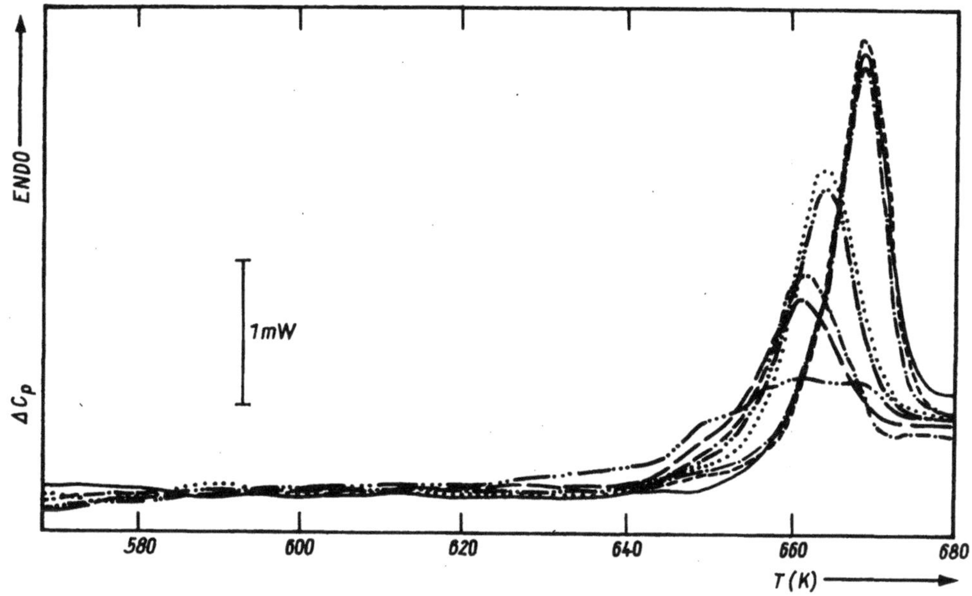

Fig. 2. DSC curves of a series of specimens of $Pd_{77.5}Cu_{6.0}Si_{16.5}$ annealed at 563 K with scan rate of 80 K/min; t_a = 64——, 32---, 28—·—, 16 ···, 8—··—, 4—·-·—, 2 — —, 1 h—···—

observed in the sub-T_g temperature region. The structure of metallic glasses quenched from the melt can be discussed as made up of two components: the liquid-like regions (L-LR) and solid-like regions (S-LR). After the as-quenched samples were annealed at T_a ($< T_g$), both the liquid-like regions and the solid-like regions will relax towards the internal equilibrium state at T_a, and as a result, the enthalpy of the system would be decreased. When the annealed specimens were reheated up to T_g, both the liquid-like regions and the

Table 1

The apparent activation energy of the ERP of the specimens annealed at 563 K for various annealing time t_a

ln t_a(s)	8.882	9.575	10.268	10.961	11.521	11.654	12.348
E (kJ/mol)	470	561	574	533	397	497	422

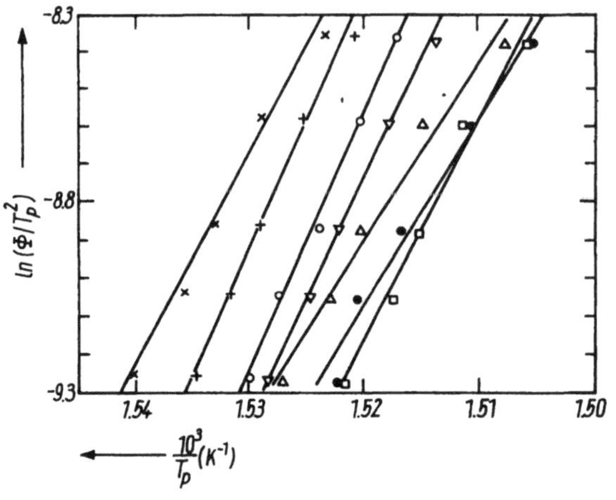

Fig. 3. $\ln(\Phi/T_p^2)$ versus $(1/T_p)$ for the enthalpy relaxation peak at different scanning rates (Φ) of specimens annealed at 563 K. Here T_p is the temperature of the enthalpy relaxation peak. T_a = × 2, + 4, ○ 8, ▽ 16, △ 28, □ 32, ● 64 h

solid-like regions will transform to the equilibrium state of the under-cooled liquid (UCL) and the system will have to absorb some energy, which can be the origin of the endothermal peak observed near T_g.

For the transition from a liquid-like region to the under-cooled liquid state, a smaller amount of energy should be required, so this process will take place earlier and will proceed faster. But for the transition from a solid-like region to the under-cooled liquid state, much more energy will have to be absorbed, so this process will take place at a later time and will proceed more slowly. In the as-quenched samples, most of the glassy matrix is liquid-like,

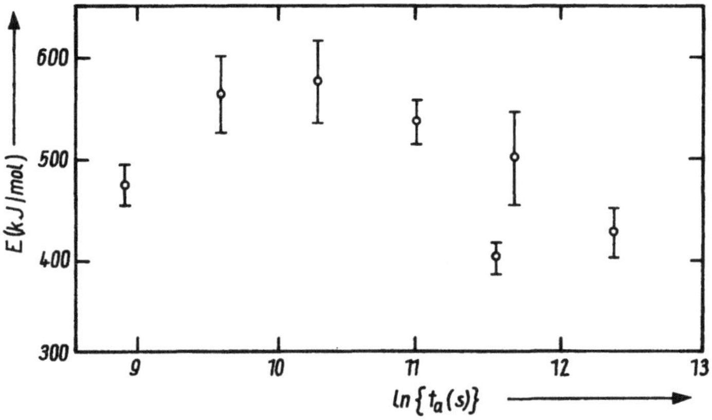

Fig. 4. The variation of apparent activation energy (E) with ln t_a for the enthalpy relaxation peak

but with increasing annealing time t_a the amount of solid-like regions would increase gradually, and both the structures of the solid-like regions and the liquid-like regions will become closer to the internal equilibrium state at T_a. Therefore the amount of heat absorbed during reheating would increase and the ERP would shift to a higher temperature with increasing t_a.

To interpret the time dependence of the apparent activation energy E, one can assume that both the long-range atomic diffusion and a viscous flow process are taking part in the transition from the glassy state to the under-cooled liquid state near T_g.

In the as-quenched samples or in the annealed specimens with shorter t_a, the liquid-like regions dominate in the matrix, and with increasing t_a, the amount of the solid-like regions will increase, the structure becomes denser, and the short-range order in the matrix will be extended, so that the atomic diffusion process becomes more difficult. An increase of the interfaces between the solid-like regions and the liquid-like regions, which should separate the matrix into smaller parts, will make the long-range atomic diffusion more difficult to occur. As a result of what has taken place, the apparent activation energy of the ERP will increase with increasing t_a during the early stage of annealing. With increasing t_a further, the solid-like regions will increase to a fairly high concentration, then the structure of the matrix will become so dense, and viscous flow will be the major process involved in the transition from the glassy state to the under-cooled liquid state. Henceforth, the interfaces between the solid-like regions and the liquid-like regions will no longer increase and might decrease. The interfacial energy would decrease with increasing t_a, the relative motion between the neighbouring regions would become easier, and the viscous flow process would become less difficult to proceed, and this might be directly related to the decrease of the apparent activation energy of the enthalpy relaxation peak.

References

/1/ H.S. CHEN, J. Mater. 20, 2417 (1985).

/2/ R.O. SUZUKI and P.H. SHINGU, J. non-crystall. Solids 61/62, 1003 (1984).

/3/ R.O. SUZUKI, R. KITA, and P.H. SHINGU, Proc. 5th Internat. Conf. Rapidly Quenched Metals, Ed. S. STEEB and H. WARLIMONT, Würzburg 1984 (p. 611).

(Received June 3, 1988; in revised form September 22, 1988)

phys. stat. sol. (a) 110, K73 (1988)
Subject classification: 64.70 and 77.80; S11.1

Institute of Physics, Technical University of Wrocław[1)]

Comments on the Results of Dilatométric Studies of NH_4HSeO_4 Crystals

By
R. POPRAWSKI

In 1983 results of dilatometric studies of NH_4HSeO_4 (AHSe) crystals /1/ were reported. In that work it was found that during a transition from the low-temperature phase to the ferroelectric one at T_{c1} the crystal volume decreases, and the thermal volume expansion coefficient (β) above and below the phase transition temperature is negative. At 298 K a step rise in the crystal volume occurred. After the coefficients related to the pressure-induced phase transition temperature shifts /2/ along with entropy changes due to phase transitions in AHSe crystals /3/ had been determined, it was found that the volume change determined in /1/ had opposite sign to that calculated from the Clausius-Clapeyron (Cl-Cl) equation. That fact made us repeat the dilatometric studies for AHSe crystals in order to explain the disagreement /4/. We found again that the crystal volume change at T_{c2} is negative, whereas β below and above this temperature is positive. Moreover the crystal volume changes at 270 K and no anomaly at 298 K was found. Therefore we claimed that the mentioned disagreement at T_{c1} was due to the crystal cracking during the phase transition. The AHSe crystal cracking has also been observed by other authors, e.g. in /2/.

In order to explain the negative value of β near T_{c1} and the observed volume change at 298 K we shall refer to the results of Sukhovskii et al. /5/ and their earlier works /6 to 9/. It has been said in these works that in AHSe crystals a nucleus formation and new $P2_12_12_1$-symmetry phase growth are possible. The growth of this phase is a very slow process and its rate depends upon various factors, e.g. temperature and strains. This phase may exist in a very wide temperature range including room temperature. The ADSe crystal volume change during the transition from the monoclinic to the orthorhombic khase is of order 10^{-2} /10, 11/. The orthorhombic phase growth inside the monoclinic matrix must cause crystal cracking due to strains resulting from

1) Wybrzeże Stanisława Wyspiańskiego 27, PL-50-370-Wrocław, Poland.

the specific volume difference for the two phases. When measurements were performed on heating (in the presence of water vapour contained in air) the positive thermal deformation was affected with a negative deformation due to a transition from the monoclinic (triclinic) to orthorhombic phase which in turn led to a peculiar relation of the thermal expansion coefficient for ADSe crystals. It results from /6/ that above 271 K the orthorhombic phase is metastable. So above 271 K we deal with a slow transition from the orthorhombic to the monoclinic phase, the crystal volume being this way increased. The temperature of 298 K at which the thermal expansion anomaly is observed is the upper limit of the orthorhombic phase stability. The observed anomaly in thermal expansion of the AHSe crystals has been the first premise indicating a new phase identified as the orthorhombic one. The investigation has shown the possibility of coexistence of the orthorhombic and monoclinic (triclinic, low-temperature) phases. The studies performed under protective atmosphere have confirmed conclusions from /4/ that a moistened crystal surface is a necessary factor for indicating the orthorhombic phase transition. Finally, the interest of the authors of /12/ in our results described in /1/ is here gratefully acknowledged. However, their questioning of our results seems to be unjustified. According to data given in /12/ the crystal volume change at T_{c1} is negative which has not been mentioned in the paper. Moreover, the phase transition at 298 K has been questioned on the basis of their measurements performed in the temperature range of 78 to 245 K!

It is worthwhile to turn our attention to the problem of the incommensurate phase in AHSe crystals. The available data indicate that between the ferroelectric and the paraelectric phase there exists a middle phase. Such conclusion are based on the anomalies reported in /15/ and /16/ of velocity and absorption of elastic waves, as well as the anomalies of the optical properties /17/ at temperatures of about 260 K. Initially, it was suggested that it was an antiferroelectric phase /15, 18/. Aleksandrova and coworkers have published a series of papers, e.g. /19 to 21/, where they attempted to show it was the incommensurate phase. The basic experimental fact supporting the concept of the incommensurate phase is the continuum in the NMR spectrum observed below 262 K. The specific temperature hysteresis of the electric permeability, above 250 K, has also been considered as indicating the existence of the incommensurate phase /24/ as well as the fact that the dependence of the temperature shift of the phase transition T_{c2} on the electric field could be described by $\Delta T = AE\log(B/E)$ where A and B are constants /23/. One has to point out, however, that structural studies of AHSe /11/ crystals have not, as yet, proven the existence of the incommensurate phase. The majority of the experimental results, mentioned above, considered to support the concept of the existence of the incommensurate phase, can well be explained using the classical theory of ferroelectric phase transitions of the first order e.g. the double hysteresis loops observed above T_{c2}, or the $T_{c2}(E)$ behaviour. Recently, even the authors of the concept of the incommensurate phase in AHSe crystals have expressed doubts to its existence. In /24/ they wrote that it was not excluded that this phase was a commensurate with a long period of modulation. The NMR spectra of such a phase would not be distinguishable from those of the incommensurate phase. One can add that an NMR spectrum of similar character can be expected if it is assumed, according to /25/, that nuclei of the orthorhombic phase, arising within the monoclinic phase have a random orientation. Creation of these nuclei may also provide an explanation for the anomalies of absorption of elastic waves at about 262 K. The temperature 262 K would indicate in this case, the stability limit of the orthorhombic phase. Thus the orthorhombic phase would be metastable in the temperature range 262 to 298 K. The problem of existence of the incommensurate phase in AHSe crystals may find its definite explanation in structural (specifically neutronographic) studies.

References

/1/ R. POPRAWSKI, J. DZIEDZIC, and W. BRONOWSKA, Acta Phys. Polon. A63, 601 (1983).

/2/ V.S. KRASIKOV, L.I. ZHEREBTSOVA, and M.P. ZAITSEVA, Fiz. tverd. Tela 23, 289 (1981).

/3/ R. POPRAWSKI, P.S. SMIRNOV, and B.A. STRUKOV, Fiz. tverd. Tela 25, 1339 (1984).

/4/ J. DZIEDZIC and R. POPRAWSKI, Ferroelectrics 77, 1 (1988).

/5/ A.A. SUKHOVSKII, I.P. ALEKSANDROVA, YU.N. MOSKVICH, and O.V. ROSANOV, phys. stat. sol. (a) 105, 145 (1988).

/6/ YU.N. MOSKVICH and O.V. ROSANOV, Preprint 270F, Siberian Branch, Akad. Nauk SSSR, 1984.

/7/ I.P. ALEKSANDROVA, YU.N. MOSKVICH, O.V. ROSANOV, A.F. SADREEV, and A.A. SUKHOVSKII, Zh. eksper. teor. Fiz., Pisma 40, 129 (1984).

/8/ I.P. ALEKSANDROVA, A.A. SUKHOVSKII, O.V. ROSANOV, YU.N. MOSKVICH, and A.F. SADREEV, Ferroelectrics 64, 79 (1985).

/9/ I.P. ALEKSANDROVA, YU.N. MOSKVICH, O.V. ROSANOV, and A.A. SUKHOVSKII, Ferroelectrics Letters 1, 131 (1984).

/10/ K.S. ALEKSANDROV, A.I. KRUGLIK, and M.A. SIMONOV, Kristallografiya 25, 1142 (1980).

/11/ A. ROZYCKI, F. DENYOR, and A. NOVAK, J. Physique 48, 1553 (1987).

/12/ H. PYKACZ, Z. CZAPLA, and J. MRÓZ, phys. stat. sol. (a) 105, K33 (1988).

/13/ H. PYKACZ and Z. CZAPLA, Solid State Commun. 64, 579 (1987).

/14/ H. PYKACZ, Ferroelectrics Letters 7, 61 (1987).

/15/ J. MAJSZCZYK, J. RĄCZKA, and Z. CZAPLA, phys. stat. sol. (a) 67, K123 (1981).

/16/ V.S. KRASIKOV, M.P. ZAITSEVA, L.A. SHABANOVA, T.V. ZRAZHEVSKII, L.I. ZHEREBTSOVA, and K.S. ALEKSANDROV, Ferroelectrics 54, 91 (1984).

/17/ V.G. MARTYNOV and A.T. ANISTRATOV, Fiz. tverd. Tela 24, 213 (1982).

/18/ I.P. ALEKSANDROVA, O.V. ROSANOV, A.A. SUKHOVSKII, and YU.N. MOSKVICH, Fiz. tverd. Tela 24, 1677 (1982).

/19/ I.P. ALEKSANDROVA, O.V. ROSANOV, A.A. SUKHOVSKII, and YU.N. MOSKVICH, Phys. Letters A 95, 339 (1983).

/20/ O.V. ROSANOV, YU.N. MOSKVICH, A.A. SUKHOVSKII, and I.P. ALEKSANDROVA, Izv. Akad. Nauk SSSR 47, 719 (1982).

/21/ I.P. ALEKSANDROVA, YU.N. MOSKVICH, O.V. ROSANOV, A.F. SADREEV, I.V. SERYUKOVA, and A.A. SUKHOVSKII, Ferroelectrics 67, 379 (1986).

/22/ C. PAWLACZYK, Ferroelectrics Letters 5, 149 (1986).

/23/ H. PYKACZ, Ferroelectrics Letters 7, 67 (1987).

/24/ I.P. ALEKSANDROVA, D.H. BUAT, V.I. ZINIENKO, YU.N. MOSKVICH, and A.A. SUKHOVSKII, Izv. Akad. Nauk SSSR, Ser. fiz. 51, 1688 (1987).

/25/ A.A. SUKHOVSKII, I.P. ALEKSANDROVA, YU.N. MOSKVICH, and O.V. ROSANOV, phys. stat. sol. (a) 105, 145 (1987).

(Received May 18, 1988; in revised form August 8, 1988)

phys. stat. sol. (a) 110, K77 (1988)
Subject classification: 68.48; S1.2; S4

Sektion Informationstechnik der Technischen Universität Dresden[1)]

Barrier Behaviour of TiW between Copper and Aluminium

By
H. KRAUTZ, CH. WENZEL, K. BORNKESSEL, and G. BLASEK

Thin TiW films gained significance when used as diffusion barriers in ohmic contacts of integrated VLSI circuits. Investigations have been made on the following systems: TiW-Si /1/, TiW-Al /2/, Si-TiW-Al /3/, TiW-Au /4/, and Al-TiW-Au /5, 6/. The latter system is also applied for chip interconnection by tape automated bonding and its thermal stability has to be mentioned in particular. While no publications concerning the barrier behaviour of thin TiW films in the Al-TiW-Cu system are known, there is a reference of investigation of thin Cu-TiW films /8/. In this note, first results of investigation on the behaviour of a thin TiW film required as a diffusion barrier in the thin film system Al-TiW-Cu are shown. Oxidized silicon (100) wafers have been coated by sputtering successively with Al (1000 nm) and TiW (100 nm) and after a vacuum breaking with Cu (200 nm). The TiW films of a few specimens were reduced in thickness by argon ion bombardment down to 50 nm and then Cu was deposited without a second vacuum breaking. For the composition of the TiW film we have found 65 at% W, 8 at% Ti, 13 at% O, 3 at% N, and 11 at% C (by AES) and 13:87 for Ti to W (by EDX). The specimens were heated up to 500 °C at 10^{-2} Pa, respectively 10^{-6} Pa (in this case in the vacuum chamber of a surface analysing system). The changes within the thin film system were characterized by AES depth profiling and by measuring the short-circuit diffusion using the "first arrival" method.

In Fig. 1, the depth profiles for a) the initial state and b) after heating 450 °C/30 min are shown. Considering the experimental results we can conclude:

1. There are no changes in the thermal stressing of 155 °C/1000 h as well as 300 °C/30 min contrary to the initial stage.

2. For a TiW film with a thickness of 100 nm and a thermal stressing of 450 °C/30 min, a minimum short-circuit diffusion occurs both of Al to the

---

1) Mommsenstr. 13, DDR-8027 Dresden, GDR.

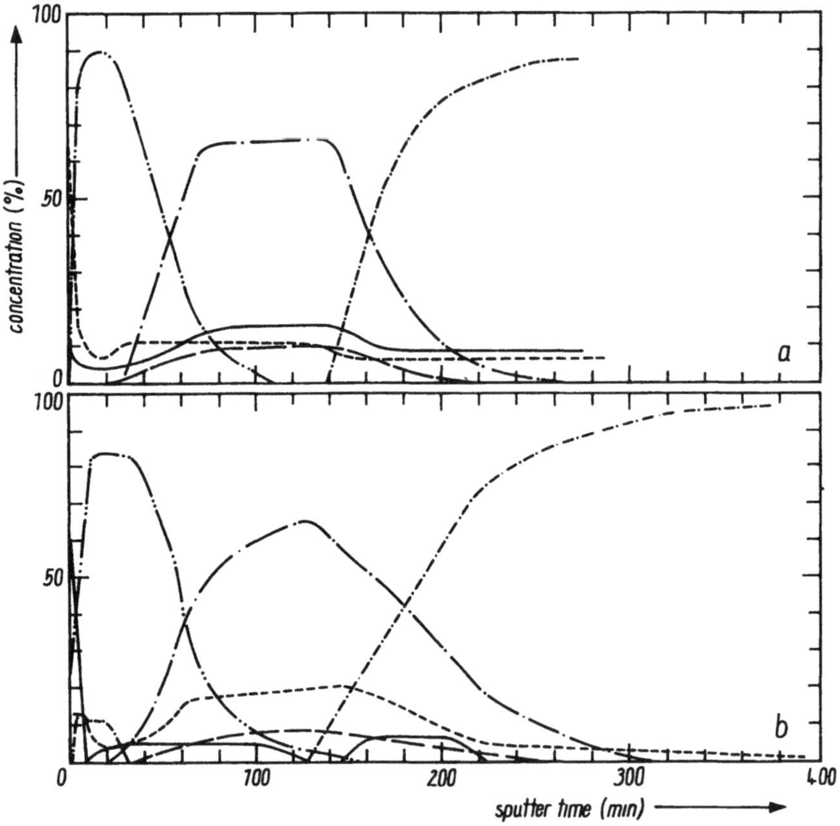

Fig. 1. Depth profiles for a) the initial state and b) after annealing at 450 °C for 30 min of the Al-TiW-Cu film system; —···—Cu,- - - O, —— C, — — —Ti, —·—W,—·—·—Al

copper surface and of Cu in the Al film.

3. The minimum extension of both the interface regions shown in Fig. 1b is influenced by an increasing surface roughness contrary to the initial state above 400 °C and therefore it is not entirely classified as a volume diffusion.

4. No changes in the barrier behaviour are observed for the thinner 50 nm TiW film and for heating to 350 °C/30 min including the possible influence of the cleaning of the TiW surface due to the argon ion bombardment. After step by step thermal treatment (250 °C/60 min - 300 °C/10 min - 350 °C/20 min) neither Al nor W and Ti are detected on the copper surface.

5. A complete layer mixing occurs at 450 $^{o}C/30$ min for 50 nm TiW and at 500 $^{o}C/30$ min for 100 nm TiW.

The thermal stability of the investigated connecting system is expected to be further improved by an optimization of structure, composition, and thickness of the barrier.

References

/1/ J.M. HARRIS, J. Electrochem. Soc. 123, 120 (1976).

/2/ P.B. GHATE, J.C. BLAIR, C.R. FULLER, and G.E. MC GUIRE, Thin Solid Films 53, 117 (1978).

/3/ R.A.M. WOLTERS and A.J.M. NELLISEN, Solid State Technology, February, 131 (1986).

/4/ V. HOFFMANN, Solid State Technology, June, 119 (1983).

/5/ R.S. NOWICKI, J.M. HARRIS, M-A. NICOLET, and I.V. MITCHEL, Thin Solid Films 53, 195 (1978).

/6/ J.M. OPAROWSKI, Thin Solid Films 153, 313 (1987).

/7/ J.-M. CHEYPE, JEEE 0569-5503/86,65.

/8/ C.K. HU, S. CHANG, M.B. SMALL, and J.E. LEWIS, Proc. 3th Internat. JEEE VLSI Multilevel Interconnection Conf., Santa Clara, 9 to 10 June, 1986 (p. 81).

(Received July 8, 1988)

phys. stat. sol. (a) 110, K81 (1988)

Subject classification: 68.60; S8

Zentralinstitut für Kernforschung der Akademie der Wissenschaften der DDR, Rossendorf (a) und
Sektion Physik der Technischen Universität Dresden[1)] (b)

Utilization of Ag:GeSe$_2$ Films as Inorganic Positive Resist

By

R. KLABES (a), A. THOMAS (b), and G. KLUGE (b)

Inorganic Ge-based chalcogenide films coated with a silver layer by a wet dipping process were recently described as a very useful material for inorganic negative resist system /1, 2/. In this negative resist system irradiation with photons, electrons, or ions causes a migration (doping) of silver from the source layer into the underlying chalcogenide. After removing the excess silver from the surface the resist is developed in wet or dry etching media in which the unexposed and therefore undoped regions are preferentially etched.

In an earlier paper we found /2/ that the etching rate is dependent on the silver concentration in a nonlinear manner. Another phenomenon which we observed /2, 3/ is the redistribution of a given silver concentration within the chalcogenide layer under irradiation with ions (ion beam induced diffusion). Based on these findings in the present communication we demonstrate a new variant leading to a positive resist. The proposed resist consists of a thin film of amorphous germanium selenide which is predoped with silver in an extremely thin surface layer. The predoped silver is redistributed by incidence of energetic ions. The positive effect occurs because the time needed for a complete removal of the regions with diluted silver concentration is decreased in comparison to the highly inhomogeneously predoped regions.

Experimental 200 to 250 nm thick GeSe$_2$ films were deposited onto silicon substrates by flash evaporation at a pressure of 10^{-3} Pa and a rate of 5 nm/s. Then elemental silver was thermally evaporated from a tungsten crucible resulting in a very thin Ag-predoped surface layer (predoping due to the kinetic energy of the vapour beam and due to an illumination effect and a related photodoping by the glowing crucible). The total amount of the silver, characterized by Rutherford backscattering spectroscopy (RBS), was adjusted to approximate-

1) Mommsenstr. 13, DDR-8027 Dresden, GDR.

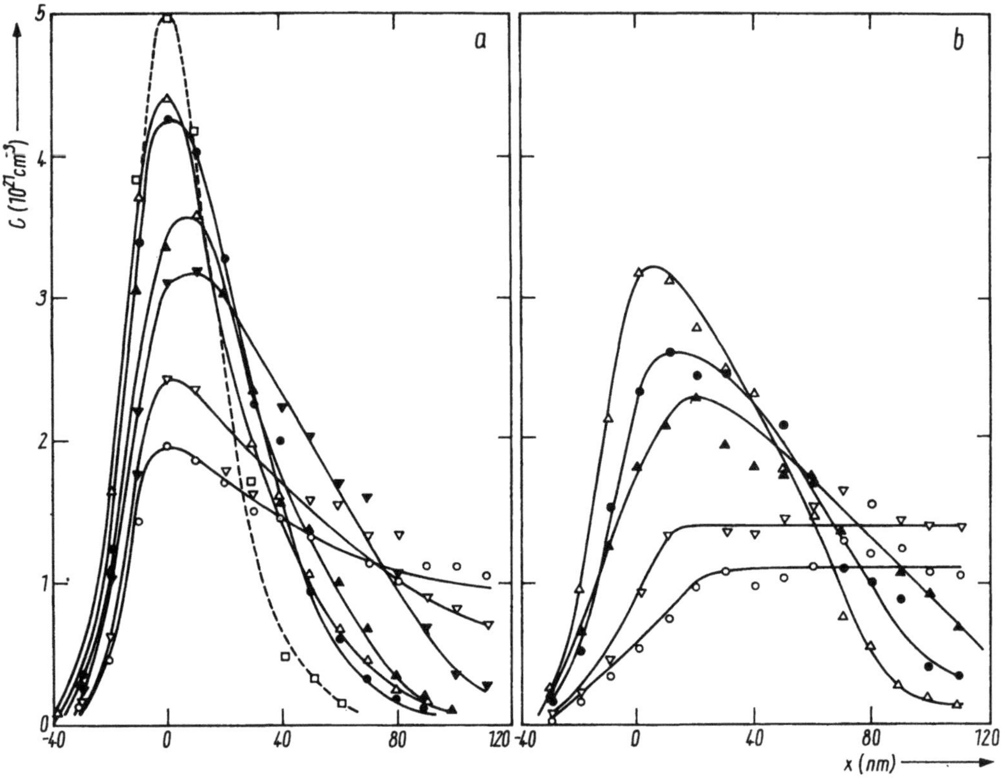

Fig. 1. RBS silver concentration profiles for a) 60 keV He^+ and b) 60 keV Ar^+ ions. △ $1x10^{14}$, ● $5x10^{14}$, ▲ $1x10^{15}$, ▼ $2x10^{15}$, ▽ $5x10^{15}$, ○ $1x10^{16}$ cm^{-2}; --- initial distribution

ly 10^{16} cm^{-2} which is sufficient to ensure a good etch resistance /2/.

The exposure of the samples was carried out by irradiation with 60 keV He^+ and Ar^+ ions. The incident dose was varied between 10^{14} cm^{-2} and 10^{16} cm^{-2}.

In order to observe both the silver redistribution within the $GeSe_2$ layer and to derive the characteristic lithographic curves each of the samples was split into two parts. RBS analysis was performed using He^+ ions with an energy of 1.7 MeV and an incident charge of $1.9x10^{15}$ cm^{-2} to obtain the Ag concentration depth profiles. The second part of the samples was developed by reactive ion etching in CF_4 (pressure 1 Pa; power 0.2 W cm^{-2}). The remaining thickness was measured with an alpha-step device.

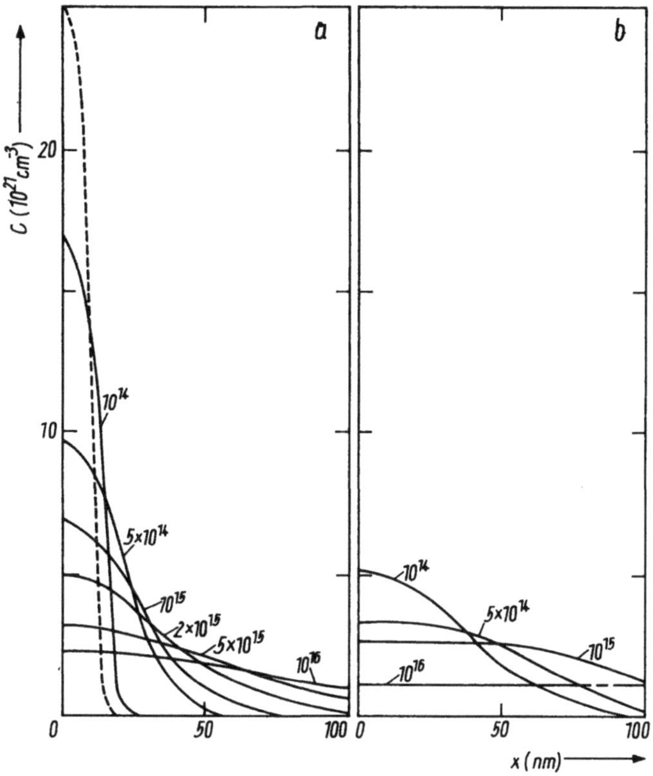

Fig. 2. True silver concentration profiles related to Fig. 1

<u>Results</u> The RBS spectra shown in Fig. 1 clearly demonstrate the redistribution of the predoped silver within the chalcogenide layer under ion bombardment. In order to obtain the true silver depth distribution after 60 keV He^+ and Ar^+ ion bombardment, both the additional influence of the RBS ion beam on the diffusional transport and of the Gaussian spreading caused by the finite detector resolution must be eliminated. That has been done by a mathematical procedure as described in /2/. The results obtained in this manner are given in Fig. 2. It can be seen that the extension of the Ag concentration profiles increases with increasing dose. Incident Ar^+ dose values of more than $5x10^{15}$ cm^{-2} lead to a nearly constant concentration within the depth region of about 100 nm accessible by the RBS measurements.

In Fig. 3 the thicknesses remaining after exposure with He^+ and Ar^+ ions and RIE etching were plotted versus the dose. Nearly no thickness loss was ob-

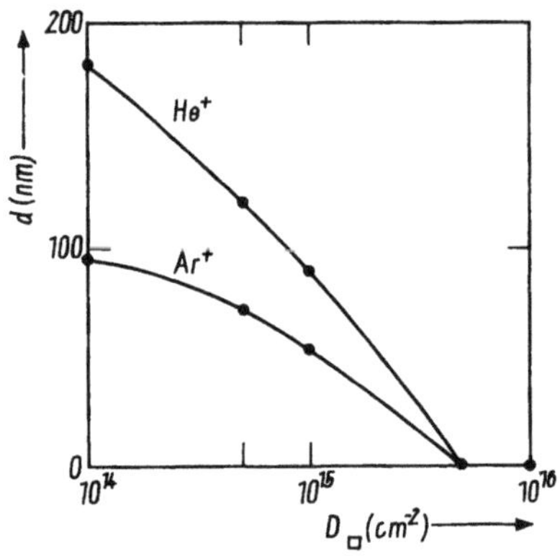

Fig. 3. Characteristic curves of the positive resist after RIE etching in CF_4

served in the unexposed regions. As expected for a positive resist effect, the amount of material removed by reactive ion etching increases with increasing deposited ion beam energy and the related enhanced redistribution of the silver within the chalcogenide. Contrary to the well-known silver-sensitized germanium selenide negative resist the contrast and the sensitivity are still rather poor. However, the purpose of this note was to prove the possibility of obtaining a positive resist effect with the $Ag:GeSe_2$ resist system. Therefore, no attempt has been made yet to optimize the experimental conditions such as deposition, irradiation, and development.

The authors wish to express their gratitude to Prof. P. Süptitz for his interest and encouragement throughout this work.

References

/1/ D.A. DOANE and A. HELLER (Ed.), Proc. Symp. Inorganic Resist Systems, Electrochem. Soc. Proc. 82-9 (1982).

/2/ R. KLABES, A. THOMAS, G. KLUGE, W. BEYER, R. GRÖTZSCHEL, and P. SÜPTITZ, phys. stat. sol. (a) 106, 57 (1988).

/3/ G. KLUGE, R. KLABES, A. THOMAS, R. GRÖTZSCHEL, and P. SÜPTITZ, phys. stat. sol. (a) 101, K9 (1987).

(Received July 18, 1988)

phys. stat. sol. (a) 110, K85 (1988)
Subject classification: 74.70; S10.15

School of Physics, University of Hyderabad[1)]

On the Normal-State Electrical Resistivity of $YBa_2Cu_3O_7$ and Other Oxide Superconductors

By
C.S. SUNANDANA

A linear temperature dependence of the normal-state electrical resistivity ($\rho(T)$) of superconducting $YBa_2Cu_3O_7$ (123) is one among the many unusual properties representative of these high-T_c superconductors. There has been an ongoing debate about the origin of this behaviour insofar as it relates to the mechanism of superconduction in these oxides /1, 2/. The purpose of this short note is to discuss this aspect in the light of (i) the $\rho(T)$ behaviour of related high- and (low-)T_c oxides and (ii) other pertinent experimental facts.

Ignoring the precipitous drop at T_c the normal state ρ of $La_{1.8}Sr_{0.2}CuO_4$, sintered polycrystalline[2)] /3/ 123, Bi-Sr-Ca-Cu-O /4/, and Tl-Ca-Sr-Cu-O /5/ may be represented by the relation

$$\rho(T) = A + B\,T\,, \tag{1}$$

where A is the extrapolated value of ρ at T = 0 and B is the slope of the linear $\rho(T)$. The larger B is, the larger is the residual resistivity ratio (RRR) and the better the sample in terms of monophasicity and density, i.e. (micro)structural homogeneity. Furthermore, the absolute value of B depends on Debye temperature, (average) ionic mass, and electron-phonon interaction. Table 1 presents data on selected materials from published works.

The linear $\rho(T)$ behaviour is not unique to these superconducting oxides alone. Indeed, $LiTi_2O_4$ (in the Li-Ti-O system) /6/, $Ag_{0.76}Sn_{1.24}Se_2$ /7/, and to a restricted extent $BaPb_{0.8}Bi_{0.2}O_3$ /8/ exhibit linear $\rho(T)$. It is to be mentioned that even the "metallic-but-not-superconducting" oxides (in the La-Sr-Cu-O system) tend to exhibit fairly linear resistivity profiles /9/.

---

1) Hyderabad 500134, India.

2) The electrical resistivity behaviour of single crystals is as good as that of pellets, and crystallisation (even more than sintering) entails lot of strains and twinning-induced defects. See /3/.

Table 1

Resistivity parameters of selected high-T_c oxide superconductors

sample	T_c (K)	RRR	ρ_{300K} (μΩcm)	A (μΩcm)	B (μΩ cm/K)	ref.
$YBa_2Cu_3O_{7.00}$	93	6.5	2.3	357	6.57	/10/
$YBa_2Cu_3O_{9.8}$ 1)	92.2	1.9	457	57	4.35	/15/
$YBa_2Cu_3O_{9.8}$ 2)	67.3	1.4	2.7	600	7.0	/15/
$BiSrCaCu_2O_x$	107	28.8	2.2	94	7.0	/4/

1) Slowly cooled, 2) quench cooled.

It would be curious to compare the resistivity profile of the most stoichiometric 123 compound /10/ with that of metallic Cu /11/:

$$\rho_{123}(T) = 357 + 6.57\ T\ , \tag{2a}$$

$$\rho_{Cu}(T) = -0.305 + 6.8855\mathrm{x}10^{-3}\ T\ . \tag{2b}$$

In (2a) and (2b) ρ is in μΩcm. Note that but for a negative sign (2b) is a scaled-down version of (2a). Indeed it is significant that the linear temperature coefficients happen to be nearly the same! However, the RRR values are widely different, 6.5 and 900, respectively. Although the "agreement" in (2a) and (2b) may be fortuitious, this comparison nevertheless underlines the importance of phonons to the normal state resistivity.

In contrast to carefully annealed and slow cooled (in O_2 usually) samples, rapidly quenched 123 (and other) superconductors exhibit near-horizontal resistivity profiles that remind one of "degenerate semiconductors". In this context it would be of interest to look at the resistivity profiles of the spinel system $Li_{1+x}Ti_{2-x}O_4$ ($0 \leq x \leq 1/3$), of which $LiTi_2O_4$ is a superconductor with the highest T_c (13.7 K). Writing this as $Li(Li_xTi_{2-x})O_4$ one notes that as x takes values -0.2, -0.1, 0, 0.1, and 0.2 the resistivity profiles change from linear with distinct positive slope, clear negative slope, increased negative slope followed by two semiconductor-like trends, respectively /6/. This situation has an interesting parallel in $YBa_2Cu_3O_{7-\delta}$ where the profiles change from positive slope through horizontality to nonlinear (semiconductor) behaviour as δ increases /10/. It is noteworthy that just as Ti vacancies control the electrical behaviour of the Li-Ti-O system the oxygen vacancies - which most probably derive from the high-temperature (p-type defect) form of Y_2O_3

/12/ - seem to determine the metallicity and superconductivity of 123 compounds.

This comparison with Li-Ti-O is nontrivial from another view-point (besides cation versus anion disorder), namely the dielectric behaviour. The static dielectric constant of $Li_{4/3}Ti_{5/3}O_4$ is /6/ 11 ± 1 while that of the "metallic" 123 phase is predicted to be 700 at 300 K (f ≈ 500 GHz) from dielectric constant measured on quenched samples /13/. Such large dielectric constants probably arise from large lattice-strain-induced polarizations due to Ba ions in the crystal structure located at such positions as to produce distortions. In any case, it is against this "dielectric background" that the ESR of Cu^{2+} in 123 compound /14/ has to be understood. Although the resistivity experiment reassures us about the metallic nature of $YBa_2Cu_3O_7$, the ESR experiment would have us believe that it is not a metal at all but a polaronic semiconductor! Indeed the ESR spectrum of 123 gives evidence of the presence of a fairly dilute paramagnetic system in which Cu^{2+} sits at the centre of an oxygen square-plane, and, no conduction electron spin resonance has so far been seen. Furthermore, no qualitative changes occur in the spectrum as the sample is cooled through the superconducting transition. Also the ESR hyperfine parameters ($A_{\parallel}$ = 72 G, $A_{\perp}$ = 22 G) give an $A_{isotropic}$ ≈ 39 G which is rather small for an ionic-type compound where A_{iso} ≈ 60 to 70 G is usually observed. A partial replacement of the square-planar oxygen environment by F^- or Cl^- would give us the extent of transferred spin densities and thus ligand covalency. An ESR study of $BaTiO_3:Cu^{2+}$ or $BaCu_2O_3$ would clarify the role of Ba and of the dielectric constant in stabilizing conditions favourable for metallicity and superconductivity in these oxide ceramics. However, the observation of a practically unchanged Cu^{2+} ESR spectrum (but for intensity) through T_c is in accord with the constancy of three important lattice properties: crystal structure, lattice constants, and elastic constants, implying that phonons do play a role - a major one at that - in the formation of the superconducting state.

In conclusion, a qualitative inter-comparison of the electrical resistivity, dielectric, and ESR behaviour of $YBa_2Cu_3O_x$ and allied compounds with those of their metallic and oxide constituents does indicate the involvement of phonons in controlling their normal state behaviour and superconduction.

References

/1/ C. CHEN, Phys. Rev. Letters 60, 750 (1988).
M. GURVITCH and A.T. FIORY, Phys. Rev. Letters 60, 751 (1988).
/2/ P.W. ANDERSON and L. ZHOU, Phys. Rev. Letters 60, 132 (1988).
/3/ G.A. EMELCHENKO et al., Zh. eksper. teor. Fiz., Pisma 46, 204 (1987).
/4/ H.W. ZANDBERGEN et al., Nature 332, 620 (1980).
/5/ R.M. HAZEN et al., Phys. Rev. Letters 60, 1657 (1988).
/6/ D.C. JOHNSTON, J. low-Temp. Phys. 25, 145 (1976).
/7/ D.C. JOHNSTON and H. ADRIAN, J. Phys. Chem. Solids 38, 355 (1977).
/8/ A.W. SLEIGHT, J.L. GILLSON, and P.E. BIERSTEDT, Solid State Commun. 17, 27 (1975).
/9/ J.B. TORRANCE, Y. TOKURA, A. NAZZAL, and S.S.P. PAKIN, Phys. Rev. Letters 60, 542 (1988).
/10/ T.R. DINGER, Proc. EPRI Workshop Utility Applications of High Temp. Superconductivity, Washington (DC), October 1987.
/11/ M.J. LAUBITZ, Canad. J. Phys. 45, 3677 (1967).
/12/ T. NORMY and P. KOFSTAD, 10th Internat. Symp. Reactivity of Solids, Dijon 1984 (p. 466); Abstract No. 3-P-22 (1984).
/13/ L.R. TESTARDI, W.G. MOULTON, H. MATHIAS, H.K. NG, and C.M. REY, Phys. Rev. B 37, 2324 (1988).
/14/ R. BARTUCCI, E. COLAVITA, L. SPORTELLI, G. BALESTRINO, and S. BARBANERA, Phys. Rev. B 37, 2313 (1988).
/15/ S.C. LEE, J.H. LEE, B.J. SUH, S.H. MOON, C.J. LIM, and Z.G. KHIM, Phys. Rev. B 37, 2285 (1988).

(Received July 5, 1988)

phys. stat. sol. (a) 110, K89 (1988)
Subject classification: 64.70 and 74.70; S10.15

Department of Physics (a) and
Center of Materials Analysis (b), Nanjing University[1)]

Tetragonal-to-Orthorhombic Phase Transition of Fe-Substituted $YBa_2Cu_3O_7$

By

HUIMIN SHAO (a), HUAQIN WANG (b), YUANFU HSIA (a),
RONGCHUAN LIU (a), and XIN JIN (a)

The phase structure for the high-T_c superconductor $YBa_2Cu_3O_{7-y}$ has been identified by X-ray diffraction and neutron diffraction studies. The lattice constant measurements on this material above room temperature reveal that it is orthorhombic below about 550 ^oC and transforms to a tetragonal polymorph at higher temperatures /1/.

The influence of Fe substitution on $YBa_2Cu_3O_{7-y}$ has been reported. From the powder X-ray diffraction patterns at room temperature, the orthorhombic structure of the $YBa_2(Cu_{1-x}Fe_x)_3O_{7-y}$ samples with $0 \leqq x \leqq 3\%$ and $y \approx 0$ can easily be recognized, but the structure changes to tetragonal when $x \geqq 4\%$ /2/. The superconductivity is not very sensitive to small amounts of iron. T_c was almost unaffected by substitution up to 4%, and superconductivity was found in the samples with up to 20% Fe /2, 3/.

In this note we report the results of high-temperature X-ray diffraction on the tetragonal-to-orthorhombic transition of $YBa_2(Cu_{1-x}Fe_x)_3O_{7-y}$ with x = 4% and 7% at higher temperatures and the Fe substitution effect on oxygen contents as well.

The $YBa_2(Cu_{1-x}Fe_x)_3O_{7-y}$ samples were synthesized from mixtures of CuO, Y_2O_3, $BaCO_3$, and Fe_2O_3. The detailed description of the procedure can be found elsewhere /2/.

X-ray diffraction studies were performed using a Rigaku D/max-RA rotating anode X-ray diffractometer, and the lattice parameters were calculated by fitting selected peaks.

A thermogravimetric experiment was performed to determine the oxygen contents of the samples. It appears that the oxygen contents are dependent on the Fe substitution.

---

1) Nanjing, People's Republic of China.

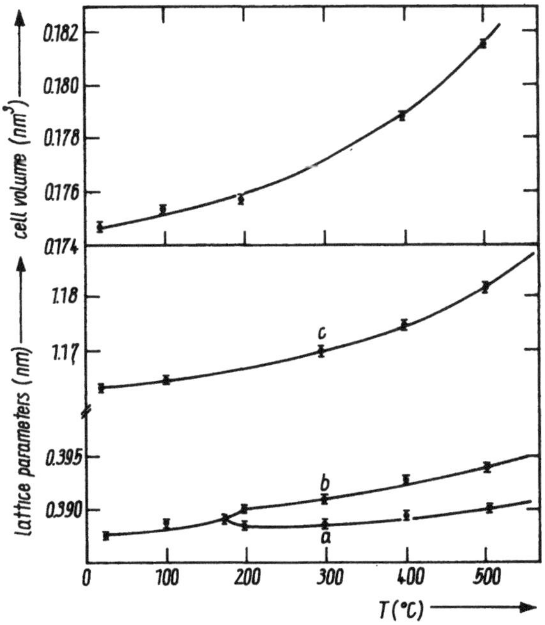

Fig. 1. a) Lattice parameters a, b, and c, and cell volume of $YBa_2(Cu_{0.96}Fe_{0.04})_3O_{7-y}$ as a function of temperature

Fig. 1 shows a plot of the lattice parameters as a function of temperature for $YBa_2(Cu_{0.96}Fe_{0.04})_3O_{7-y}$ in ambient atmosphere.

Below the phase transition at 180 °C, the material is tetragonal, whereas above this temperature it becomes orthorhombic. In this atmosphere, the transition is reversible. For another $YBa_2(Cu_{0.93}Fe_{0.07})_3O_{7-y}$ sample the results were similar.

Fig. 2a shows the results of thermogravimetric measurements. y varies from 0.05 to 0.14 depending on the Fe concentration x but T_c is almost unaffected by substitution up x = 4% (see Fig. 2b). This implies that the oxygen content of $YBa_2(Cu_{1-x}Fe_x)_3O_{7-y}$ is not very sensitive to T_c although the lattice parameters have changed.

The Fe ions are located on the Cu sites, because Fe substitution had a marked influence on the orthorhombicity. It is interesting to observe that the lattice is tetragonal below the transition temperature 180 °C and becomes orthorhombic at higher temperatures. The phase

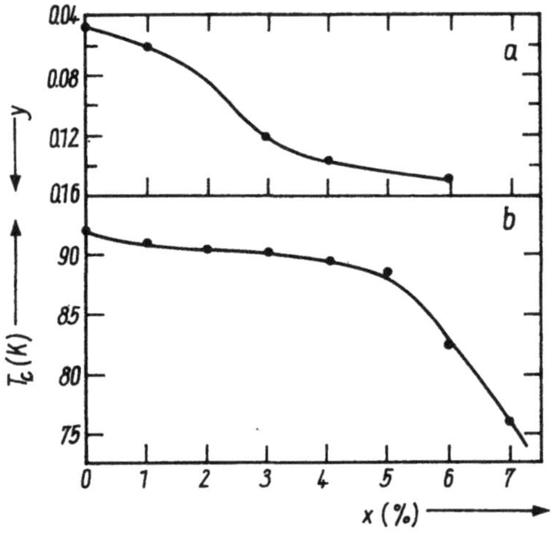

Fig. 2. a) The y value in $YBa_2(Cu_{1-x}Fe_x)_3O_{7-y}$ and b) superconducting transition temperature T_c versus Fe concentration x

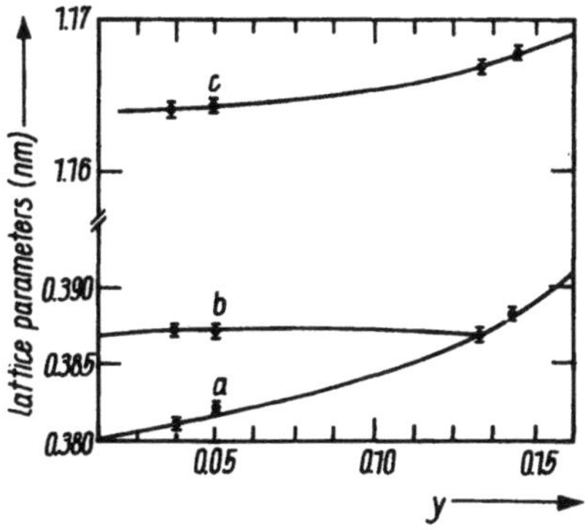

Fig. 3. Lattice parameters a, b, c of the $YBa_2(Cu_{1-x}Fe_x)_3O_{7-y}$ sample as a function of y value

transition differs from the behaviour of undoped $YBa_2Cu_3O_{7-y}$. The relation between lattice parameters and oxygen contents is shown in Fig. 3.

In $YBa_2Cu_3O_{7-y}$, the oxygen vacancies are ordered in b direction, so the lattice parameter b is larger than a. Introduction of Fe^{3+} in square-planar Cu(1) sites should lead to a filling up of the O vacancies in the a-axis. This also would produce a random distribution of O around the Cu(1) sites. Therefore, the result of such effects is an increase of a and decrease of b. Therefore, the orthorhombic phase is not necessary for the superconduction.

We are grateful to Mr. J. Jing, J. Bieg, and Prof. Dr. U. Gonser, Universität des Saarlandes, FRG, for providing us with the samples.

References

/1/ M.O. EATOUGH, D.S. GINLEY, B. MOVOSIN, and E.L. VENTURINI, Appl. Phys. Letters 51, 367 (1987).

/2/ J. JING, J. BIEG, H. ENGELMANN, Y. HSIA, U. GONSER, P. GÜTLICH, and R. JAKOBI, Solid State Commun. 66, 727 (1988).

/3/ J. BIEG, J. JING, H. ENGELMANN, Y. HSIA, U. GONSER, P. GÜTLICH, and R. JAKOBI, Physica 153/155, 952 (1988).

(Received June 23, 1988; in revised form September 19, 1988)

phys. stat. sol. (a) 110, K93 (1988)
Subject classification: 74.70; S10.15

Cryogenic Engineering Centre, Indian Institute of Technology, Kharagpur[1)]

Bismuth Strontium Calcium Copper Oxide High-T_c Superconductors from Nitrate Solutions

By
A.K. PRADHAN, P.C. JANA, and B.K. ROUL

The recent discovery of high-T_c oxide superconductors with the K_2NiF_4 /1/ and $YBa_2Cu_3O_7$ /2/ structure has led to an extensive search for other superconducting oxides. Chu et al. /3/ observed superconductivity upto 120 K in the multi-phase Bi-Al-Ca-Sr-Cu-O and Bi-Ca-Sr-Cu-O systems. The bismuth alkaline earth cuprates constitute the third family of superconducting oxides /4 to 6/. Compared to the compounds of K_2NiF_4 and $YBa_2Cu_3O_7$, they provide a potentially advantageous combination of high critical temperatures and relative environmental inertness /2/. We describe below the preparation of Bi-Sr-Ca-Cu-O compounds with some typical compositions from their nitrate solutions, their resistivity as a function of temperature, and influence of the preparation and excess copper content on T_c. We also discuss the magnetization of the best superconducting sample of the group.

Stoichiometric amounts (in the ratio (2.2/1.7/1.1/2 and excess of copper content ranging from 2 to 4) of AR-grade (99.99% pure) Bi_2O_3, $SrCO_3$, $CaCO_3$, and CuO were dissolved in minimal amount of nitric acid separately and nitrate stock solutions of approximately 0.5 molar in each of these elements were prepared. For complete dissolution, the solution was warmed. The solutions were mixed and diluted to make their aqueous solutions to avoid precipitation of the basic bismuth salt. The solution was heated slowly to form a homogeneous solution. The solution was slowly evaporated out and dried to get black powders. The black powders were ground thoroughly and heated in the furnace at 1123 K for 10 h. The powders were again ground and pressed at $7x10^4$ kg/cm^2 to make pellets of dimension 10 mm diameter and 1 to 2 mm thickness. The pellets were annealed at various temperature ranging from 1123 to 1183 K for different times. We explored a variety of preparatory conditions and obtained our best

1) Kharagpur 721302, India.

result as follows. The calcined pellets of composition $Bi_{2.2}Ca_{1.1}Sr_{1.7}Cu_2O_{8+x}$ were annealed at 1163 K for 12 h and cooled down to room temperature over a period of 12 h yielding black pellets. It was noted that the new materials are also sensitive to their preparatory conditions like in Y-Ba_2-Cu-O systems but not much to the environment.

The compounds were X-rayed and significant differences were marked if annealing time and temperatures were changed. However, the cell parameters of the best superconducting sample of the above composition are found to be closer to the $Bi_2Sr_2Ca_1Cu_2O_{8+y}$ system /5/. The cell parameters for the present sample are a = 0.5405 nm, b = 0.5435 nm, and c = 3.070 nm (orthorhombic system). However, X-ray patterns of samples annealed at 1123 K for 10 h show very less number of reflections and some of them are very poor. Details of the X-ray studies will be discussed elsewhere /7/.

The resistance measurement was carried out using a standard four-probe technique in the temperature region 300 to 4.2 K. The temperature was calibrated with CLTS sensors. Rectangular samples of dimension 8x1x2 mm^3 were cut from the sintered pellets, and pure indium metal was used for electrical contacts. The sample current was of the order of 1 mA.

Fig. 1 shows the variation of resistance with temperature for $Bi_{2.2}Sr_{1.7}Ca_{1.1}Cu_xO_y$ samples prepared under different conditions mentioned in the figure caption. In comparison to the Y-Ba-Cu-O system, the new materials are more sensitive to the preparatory conditions. The transition width

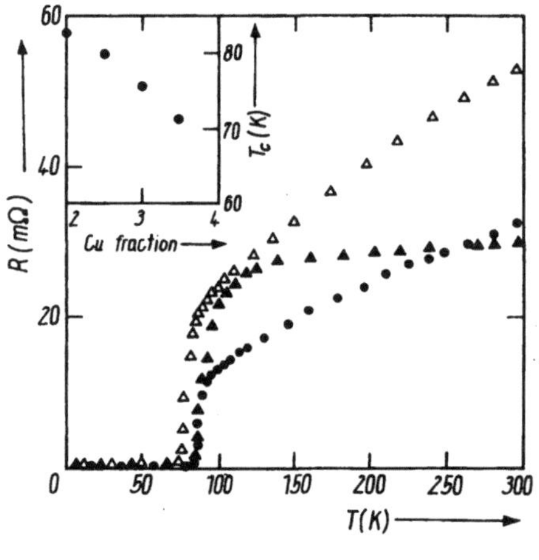

Fig. 1. Variation of resistance as a function of temperature for $Bi_{2.2}Sr_{1.7}Ca_{1.1}Cu_xO_y$ samples; ▲ $Bi_{2.2}Sr_{1.1}Cu_2O_y$ annealed at 1163 K for 12 h and slowly cooled; ● $Bi_{2.2}Sr_{1.1}Cu_2O_y$ annealed at 1153 K for 4 h and slowly cooled; △ $Bi_{2.2}Sr_{1.7}Ca_{1.1}Cu_{3.5}O_y$ annealed at 1163 K for 12 h and slowly cooled. Inset shows variation of T_c with Cu fraction present in these samples

was found to be relatively higher in this system. Several $Bi_{2.2}Sr_{1.7}Ca_{1.1}Cu_xO_y$ samples with copper content x varying from 2 to 4 were prepared by us keeping the same annealing condition. The best superconducting sample as mentioned earlier has x = 2 and T_c was found to be at 83 K. The room temperature resistance was only 28 mΩ and the transition width was comparatively less. But the samples annealed at 1153 K for 4 h and slowly cooled showed a wide transition width with T_c at 83 K. However, 1123 K annnealed (12 h) samples followed by slow cooling showed semiconducting behaviour (not shown in Fig. 1). They did not show any trace of magnetic moment at 77 K, too. The above samples quenched from 1163 K also did not exhibit superconducting behaviour. It is interesting to note that as the Cu fraction increases, the transition temperature decreases (inset of Fig. 1). Annealing at 900 °C for 12 h showed the evidence of partial melting. This becomes more prominent when the Cu fraction decreases. The melt-samples (annealed at 910 °C and slowly cooled) did not show a superconducting transition. However, they showed metallic behaviour at room temperature and became semiconducting as the temperature decreased. This may be due to the shortage of the Bi content which is present in the superconducting phase. The samples were sintered at a temperature very close to the melting point of the Bi component. The above composition with x = 4 annealed at 890 °C for 12 h and slowly cooled showed semiconducting behaviour in the temperature range 4.2 to 300 K.

The critical current density for the best superconducting sample with the described composition was found around 50 A/cm^2 at 4.2 K. But this decreases drastically to 20 A/cm^2 for Cu-rich samples at 4.2 K. The lower critical field' (H_{c1}) was measured in a PAR Vibrating Sample Magnetometer model 150A. A typical magnetization curve is shown in

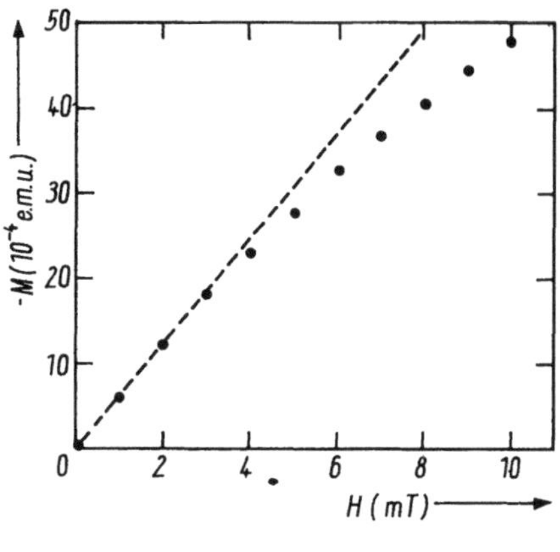

Fig. 2. Magnetization curve for $Bi_{2.2}Cr_{1.7}Ca_{1.1}Cu_2O_{8+y}$ at low magnetic field and at T = 77 K

Fig. 2 for $Bi_{2.2}Sr_{1.7}Ca_{1.1}Cu_2O_{8+y}$ and H_{c1} was defined as the field where M(H) starts to deviate from linearity. At 77 K, H_{c1} is (3 ± 0.2)mT which is comparatively higher than (1 ± 0.2)mT measured for $YBa_2Cu_3O_7$ polycrystalline samples in the same geometry. Detailed magnetization measurements at higher field and still at lower temperature are in progress.

The authors are grateful to Prof. K.L. Chopra, Director of the Institute and Prof. S.K. Dutta Roy, Head of the Centre for their interest and guidance in the present work. The authors are also thankful to Prof. V.R. Kalvey, Prof. S. Sarangi, and Dr. V.V. Rao for their experimental help and encouragement.

References

/1/ J.G. BEDNORZ and K.A. MÜLLER, Z. Phys. B 64, 189 (1986).

/2/ M.K. WU, J.R. ASHBURN, C.J. TORNG, P.H. HOR, R.L. MENG, L. GAO, Z.J. HUANG, Y.Q. WANG, and C.W. CHU, Phys. Rev. Letters 58, 908 (1987).

/3/ C.W. CHU, J. BECHTHOLD, L. GAO, P.H. HOR, Z.J. HUANG, R.L. MENG, Y.Y. SUN, Y.Q. WANG, and Y.Y. XUE, Phys. Rev. Letters 60, 941 (1988).

/4/ H. MAEDA, Y. TANAKA, M. FUKUTOMI, and T. ASANO, Japan. J. appl. Phys. Letters 27, L209 (1988).

/5/ R.M. HAZEN, C.T. PREWITT, R.J. ANGEL, N.L. ROSS, L.W. FINGER, C.G. HADIDIACOS, D.R. VEBLEN et al., Phys. Rev. Letters 60, 1174 (1988).

/6/ M.A. SUBRAMANIAN, C.C. TORADI et al., Science 239, 1015 (1988).

/7/ A.K. PRADHAN, B.K. ROUL, and P.C. JANA, J. Mater. Sci., Letters, to be published.

(Received July 8, 1988)

phys. stat. sol. (a) 110, K97 (1988)

Subject classification: 75.50; S1.2

Instituto de Física, Universidade de São Paulo[1)]

Magnetic Transitions in $Mn_{1-x}Ni_x(HCOO)_2 \cdot 2H_2O$

By

C.H. WESTPHAL and C.C. BECERRA

In this note we show that the inclusion of a small percentage of Ni in the $Mn(HCOO)_2 \cdot 2H_2O$ antiferromagnet leads to a change in the direction of the spin orientation at the Néel temperature T_N and to the suppression of the first-order spin reorientation transition observed in the pure compound at 1.72 K.

The dihydrate formate $M(HCOO)_2 \cdot 2H_2O$ with M = Mn, Ni are isomorphous compounds. They crystallize in the monoclinic system and order antiferromagnetically at T_N = 3.68 and 15.6 K /1/, respectively. The metallic ions are located in two different crystallographic positions (A and B). The A and B ions form alternating sheets parallel to bc plane. Each M^{2+} ion in the A sheet is linked with other four in the same sheet through formate bridges. The ions in the B sheet are not directly linked between them but each of them is connected to two ions in the A sheet through formate groups. It was found that the ions in the A sheet behave as a two-dimensional antiferromagnetic system ordering at T_N and that those in the B sheet act as free spins having thus a paramagnetic behaviour /2, 3/. The Mn compound is the most extensively investigated. The A ions order at 3.68 K along the $\vec{b}$ axis but as the temperature is lowered a first-order transition occurs at T_R = 1.72 K. In this transition the spins reorient to a direction $\vec{d}$ contained in the ac plane and forming an angle of $(63 \pm 5)^\circ$ with the $\vec{a}$ axis /4/. The A ions of the Ni formate order at 15.6 K along a direction in the ac plane very close (within the estimated error) to the $\vec{d}$ direction /1/. The spin reorientation is not observed in the Ni compound. In Fig. 1 we present the data obtained for the ac magnetic susceptibility (χ) taken at 155 Hz and zero applied field, for the Mn formate and for the mixed system $Mn_{1-x}Ni_x(HCCO)_2 \cdot 2H_2O$ with x = 0.04. In both samples the modulation field was applied along the $\vec{b}$ axis of the single crystals. In the Mn compound a small anomaly due to the ordering of the A ions is observed at

1) Caixa Postal 20516, São Paulo 01498, Brazil.

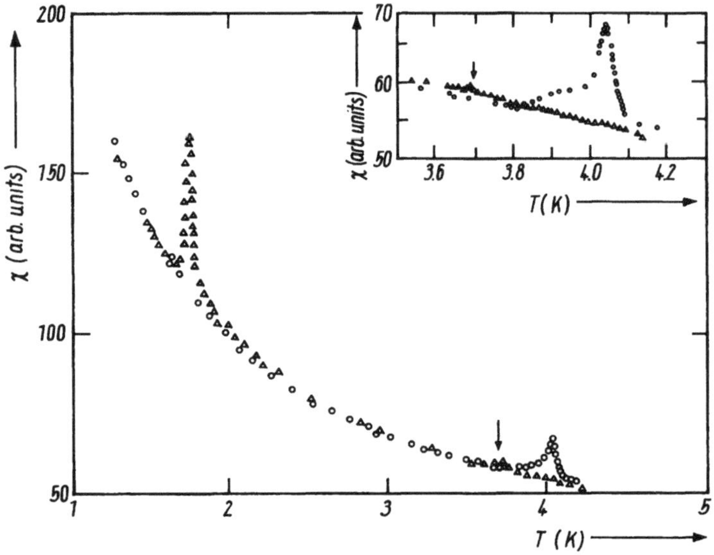

Fig. 1. Magnetic susceptibility as a function of temperature for $Mn(HCOO)_2 \cdot 2H_2O$ (Δ) and $Mn_{1-x}Ni_x(HCOO)_2 \cdot 2H_2O$ (o) with x = 0.04. The insert show a detail of the curve near the ordering temperature of these compounds

3.68 K. The spin reorientation occurring at 1.72 K is evidenced by the strong spike in χ. Both anomalies are superimposed on the Curie-like curve due to the paramagnetic behaviour of the B ions. The obtained data for the mixed compound with 4% of Ni show only a single peak in χ occurring at 4.04 K superimposed in the Curie-like curve of the B ions. This peak is higher in magnitude than that observed in the Mn compound at 3.68 K and we attribute it to the ordering of the magnetic moments of the A ions along the $\vec{d}$ direction. This direction is close to the ordering axis of the Ni formate and is the low temperature reorientation direction of the Mn formate. This explains the absence of the strong spike in the χ versus T curve in the mixed system. A similar suppression of the spin reorientation was formerly observed in the Mn formate with 10% of Fe /1/. We also measured χ as a function of an external magnetic field H applied along the $\vec{b}$ axis and at fixed T (Fig. 2). As it is already known the Mn formate has a rich phase diagram in this configuration /5/ (see

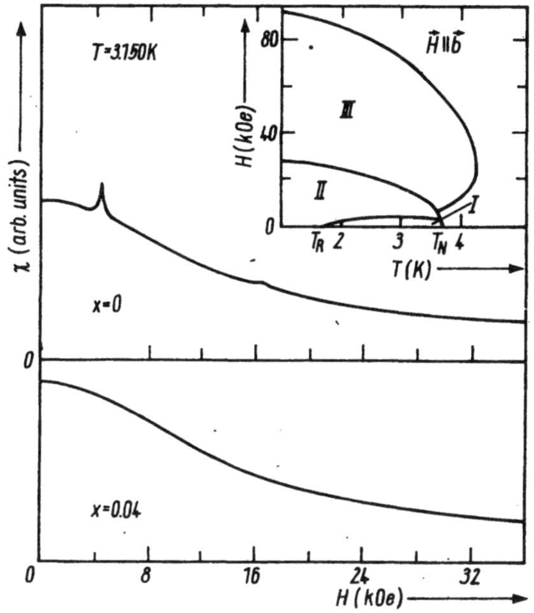

Fig. 2. Magnetic susceptibility for $Mn_{1-x}Ni_x(HCOO)_2 \cdot 2H_2O$ as a function of H applied along the $\vec{b}$ axis and at a fixed temperature (T = 3.150 K). In the insert a schematic view of the phase diagram of $Mn(HCOO)_2 \cdot 2H_2O$ for $\vec{H} \parallel \vec{b}$ based on /5/ is shown

insert of Fig. 2) showing several phase transitions. At low field and between 1.7 and 3.68 an ordered phase (I) with the magnetic moments aligned along the $\vec{b}$ axis is present. Below 1.7 K a second ordered phase (II) with moments perpendicular to the $\vec{b}$ axis is observed. Transitions between phase (I) and (II) are first order and occur only above 1.7 K. The spike observed in the curve of Fig. 2 for the Mn formate corresponds to this transition. A second anomaly is also observed in this curve and marks the transition to a third ordered phase (III) present at higher fields. In the mixed compound the transition (I) → (II) is not observed suggesting that phase (I) (with magnetic moments aligned along the box) has been suppressed. The spin reorientation in the Mn formate has been explained /1/ in terms of the competition of anisotropies due to the dipolar field and the crystalline field, that affect the Mn ion in second order. Both energies are of the order of 1 K. Our experimental data suggest that a small concentration of Ni is sufficient to produce a local anisotropy that favors the $\vec{d}$ direction (common to both compounds at low temperature) and suppresses phase (I) of the phase diagram. A complete study of the phase diagram of the mixed system with lower concentrations of Ni is in progress and will be published elsewhere.

This work is supported by the Brazilian agencies FAPESP, CNPq, and FINEP.

References

/1/ P. BURLET, P. BURLET, J. ROSSAT-MIGNOD, A. DE COMBARIEU, and E. BEDIN, phys. stat. sol. (b) 71, 675 (1975).

/2/ J. SKALVO, Jr., G. SHIRANE, and S.A. FRIEDBERG, Phys. Rev. 188, 1037 (1969).

/3/ M. MATSUURA, H.W.J. BLÖTE, and W.J. HUISKAMP, Physica 50, 444 (1970).

/4/ E.F. BERTAUT, P. BURLET, and P. BURLET, Solid State Commun. 7, 343 (1969).

/5/ J.W. SCHUTTER, J.W. METSELAAR, and D. DE KLERK, Physica 61, 250 (1972).

(Received July 26, 1988)

phys. stat. sol. (a) 110, K101 (1988)
Subject classification: 78.70; S4; S9

Institute of Experimental Physics, University of Wrocław[1)]

Positron Trapping at Ag/Na_3AlF_6 Interface; Life-Time Measurements

By
R. EWERTOWSKI, A. BARANOWSKI, and W. ŚWIĄTKOWSKI[1)]

Introduction The image potential model of interaction of positrons with the metal surface predicts the existence of a potential "trough" just outside the metal surface /1 to 3/. Thus positrons implanted in the metal, which are able to reach the surface can be trapped in the "trough". The analysis of the wave function of positrons trapped in the "trough" as well as of the metal electron wave functions in the region occupied by the trapped positron predicts an anisotropy of the distribution of momenta of photon pairs resulting from the annihilation of the trapped positrons. Moreover one can expect the annihilation rate for the trapped positrons to be smaller than the annihilation rate for positrons diffusing in the bulk. The above predictions are in good agreement with observed changes in the angular correlations of annihilation photons (ACAP) as in positron lifetime spectra when voids are created in metals /4, 5/. A direct verification of the expectation of the anisotropy of photon pair momenta distributions became possible within a short period of time using slow positron beam sources. When positrons with energy of the order of 1 keV enter into the metal most of them can diffuse back to the surface region and be trapped there.

Howell et al. /6/ using a 740 eV positron beam impinging on a clean Cu (121) surface have found that in such a case the observed distribution of annihilation photon pair momenta can be resolved into two components, one associated with energetic positronium emission displaced from zero momentum and a second centered on zero momentum with distinct asymmetry for directions parallel and perpendicular to the surface. However, in other similar experiments /7, 8/ such asymmetry has not been observed.

It seemed likely that the above mentioned image potential effects ascribed to the metal-vacuum surface could take place also in the case of the metal-dielectric interface. The verification of such an expectation could be possible

---

1) Ul. Cybulskiego 36, PL-50-205 Wrocław, Poland.

using directly positrons from a radioactive source and a sample consisting of alternately deposited many thin metal and dielectric layers. Using multilayer metal-dielectric samples (Al/Al_xO_y in first experiment /9/ and Ag/Na_3AlF_6 in the second one /10/) and orienting them in two manners with respect to the pair momentum annihilation spectrometer axis we had observed that the distribution of the annihilation photon pair momentum component being perpendicular to the axis depends on the sample orientation. For both samples ($Al/A_x/O_y$ and Ag/Na_3AlF_6) the distributions appeared to be narrower in the case when the momentum component parallel to the metal-dielectric interface was analyzed. These results appeared thus to be in agreement with the supposition that positrons can be trapped at the metal-dielectric interface. When the trapping of positrons at the metal-dielectric interface takes place, one can expect the appearance of the respective component in the positron lifetime spectrum for a metal-dielectric sample. The intensity of such component should give information on the possibility of positrons implanted in the sample to reach the metal-dielectric interface.

In this note the results of positron lifetime measurements for evaporated Ag sample and a multilayer Ag/Na_3AlF_6 one are presented.

Experimental results The measurements of the positron lifetime spectra were carried out using a spectrometer with time resolution of 350 ps. The first sample was an Ag foil obtained by the evaporation technique. The second one was a multilayer Ag/Na_3AlF_6 foil. This foil consisted as in /10/ of many alternating Ag ($\approx$ 200 nm) and Na_3AlF_6 ($\approx$ 4 nm) layers obtained by evaporation, too. The characteristics of the obtained positron lifetime spectra are collected in Table 1.

Table 1

The results of the lifetime measurements

sample	Ag	Ag/Na_3AlF_6
I_1	76.8 ± 0.7%	56.6 ± 8.5%
τ_1	201 ± 10 ps	210 ± 15 ps
I_2	23.2 ± 0.7%	42.5 ± 8.5%
τ_2	325 ± 20 ps	372 ± 21 ps

As it is seen, there exists a distinct difference between the lifetime spectrum for the Ag foil and that one for the Ag/Na_3AlF_6 foil. Each of the spectra can be resolved practically into two components, the intensity of the second component being much larger in the case of the metal-dielectric foil; the mean lifetime of the second component changed, too. Because of the small dielectric (Na_3AlF_6) content of the metal-dielectric sample such large change in the spectrum when going from the Ag foil to the Ag/Na_3AlF_6 one cannot be ascribed to positrons primary implanted into Na_3AlF_6 layers. The possible explanation is trapping of positrons in the interface region and their annihilation there with an annihilation rate little smaller than that of the second component in the spectrum obtained for Ag. Thus the second component in the spectrum for the Ag/Na_3AlF_6 foil is likely to be due both to the mechanism responsible for the second component in the Ag case and the positron trapping at the interface. From the data in Table 1 for the above interpretation one can obtain a value close to 19% for the probability that the positron implanted in the Ag layer was trapped in the Ag/Na_3AlF_6 interface region. And the value of the mean lifetime for positrons trapped there equals about 430 ps.

Using the procedure as in /11/, one can obtain the positron diffusion length L^+ for evaporated Ag. The respective calculations give L^+ about 20 nm, which is several times smaller than observed in the case of undefected metals. The small value of the positron diffusion length for evaporated Ag can be explained by assuming that the diffusive motion of most positrons ends as a result of positron trapping. The observed two component lifetime spectrum for the evaporated Ag sample indicates that there exist at least two kinds of traps in the sample material, similar as it was suggested for other evaporated metals /13/. For two kinds of traps one should expect the lifetime spectrum to be a three-component one. Two components with longer mean lifetime should be ascribed to annihilation from trapped states and the short-time one connected with annihilation of untrapped positrons. However, for large trapping rates the intensity and mean lifetime of the last mentioned component should be expected to be very small and consequently not observable under our measurement conditions. It should be mentioned that in the above work of Pendrys et al. /13/ such situation takes place, too.

Final remarks The above presented results of positron lifetime measurements seemed to support our earlier suggestion of the possibility of positron trapping at the metal-dielectric interface. It should be mentioned however, that in evaporated metal layers different kinds of defects can exist and the positron diffusion possibility is strongly depressed. What more, one cannot exclude that even small changes in evaporation conditions can lead to essential changes both in concentration and character of defects. Nevertheless the results of our previous work /9, 10/ and those presented here can be simply explained basing on the possibility of positron trapping at the metal-dielectric interface.

At last it should be mentioned that Aravin et al. /12/ had obtained results that also can be explained as an effect of positron trapping at the metal-dielectric interface.

This work was supported by the Ministry of National Education within research project CPBP 01.08.

References

/1/ C.H. HODGES and M.J. STOTT, Solid State Commun. 12, 1153 (1973).

/2/ W. TRIFTSHÄUSER, J.D. MC GERVEY, and R.W. HENDRICKS, Phys. Rev. B 9, 3321 (1974).

/3/ B. ROZENFELD, K. JERIE, and W. ŚWIĄTKOWSKI, Acta Phys. Polon. A54, 93 (1983).

/4/ O. MOGENSEN, K. PETERSEN, R.M.J. COTTERILL, and B. HUDSON, Nature 239, 98 (1972).

/5/ R.M.J. COTTERILL, D.K. MAC KENZIE, L. SMEDSKJAER, G. TRUMPY, and J.H.O.L. TRÄFF, Nature 239, 101 (1972).

/6/ R.H. HOWELL, P. MEYER, I.J. ROSENBERG, and M.J. FLUSS, Phys. Rev. Letters 54, 1698 (1985).

/7/ K.G. LYNN, A.P. MILLS, Jr., R.W. WEST, S. BERKO, K.F. CANTER, and L.O. ROELLING, Phys. Rev. Letters 54, 1702 (1985).

/8/ D.M. CHEN, S. BERKO, K.G. LYNN, A.P.MILLS, Jr., L.O. ROELLING, and R.W. WEST, Bull. Amer. Phys. Soc. 31, 584 (1985).

/9/ R. EWERTOWSKI and W. ŚWIĄTKOWSKI, Physica Scripta 35, 745 (1987).

/10/ R. EWERTOWSKI and W. ŚWIĄTKOWSKI, phys. stat. sol. (a) 102, 555 (1987).

/11/ W. ŚWIĄTKOWSKI, B. ROZENFELD, H.B. KOŁODZIEJ, and S. SZUSZKIEWICZ, Acta Phys. Polon. A47, 79 (1975).

/12/ L.G. ARAVIN, K.A. VALIEV, V.I. GOLDANSKII, YU.A. NOVIKOV,

A.V. RAKOV, M.K. FILIMONOW, and V.P. SHANTAROVICH, Proc. Europ. Meeting Positron Studies of Defects, Wernigerode, March 23 to 27, 1987.

/13/ J.P. PENDRYS, B.T.A. MCKEE, and A.T. STEWART, Appl. Phys. 4, 333 (1974).

(Received July 25, 1988)

phys. stat. sol. (a) 110, K107 (1988)

Subject classification: 78.50; 76.30; S10.15

Institute of Chemistry, Academy of Sciences of the USSR, Gorkii[1] (a), General Physics Institute, Academy of Sciences of the USSR, Moscow[2] (b), and Institute of Nuclear Physics, Moscow State University[3] (c)

UV and γ-Induced Paramagnetic Colour Centers in Germanium-Doped Silica Glass

By

A.N. GURYANOV (a), E.M. DIANOV (b), V.M. KIM (b), V.M. MASHINSKII (b), V.B. NEUSTRUEV (b), V.A. TIKHOMIROV (c), and V.F. KHOPIN (a)

Introduction As is well known Ge(n) (n = 0, 1, 2, 3) paramagnetic centers are formed in the germanium-doped silica preforms and during the fabrication of optical fibers, and by γ-irradiation /1, 2/. Ge-associated absorption bands are also known to be observed in the UV spectral region (185 to 330 nm) /3, 4/, their intensities depending on the technological conditions. The changes in the UV absorption spectrum under γ-irradiation are also reported in /5/. Yet, the individual features of Ge(n) centers have not been observed in the optical spectra. In this note we have studied paramagnetic Ge(n) centers and the induced absorption bands trying to understand mechanisms of their formation.

Experiments Samples of the 10 mol% GeO_2-90 mol% SiO_2 glass were cut out of the optical fiber preform prepared by the MCVD method. ESR and optical absorption spectra were investigated before and after the irradiation by both laser sources (wavelengths 193, 248, 266, and 337 nm or photon energy E_p = 6.4, 5.0, 4.66, and 3.68 eV, respectively) and ^{60}Co γ-source at 300 K. The irradiated samples were subjected to the subsequent temperature treatment at about 1000 K. ESR signals of Ge(n) centers were separated with an accuracy of about 20%.

Results and discussion Fig. 1 shows the spectrum of the initial absorption for the germanium-doped silica glass and the induced absorption after γ- (dose $1.5x10^5$ Gy) and UV irradiation. In the main the induced absorption can be described by the superposition of the increasing bands at 6.2, 5.8, and 4.4 eV

---

1) 38 Vavilov Str., SU-117942 Moscow, USSR.

2) 49 Tropinin Str., SU-603600 Gorkii, USSR.

3) SU-119899 Moscow, USSR.

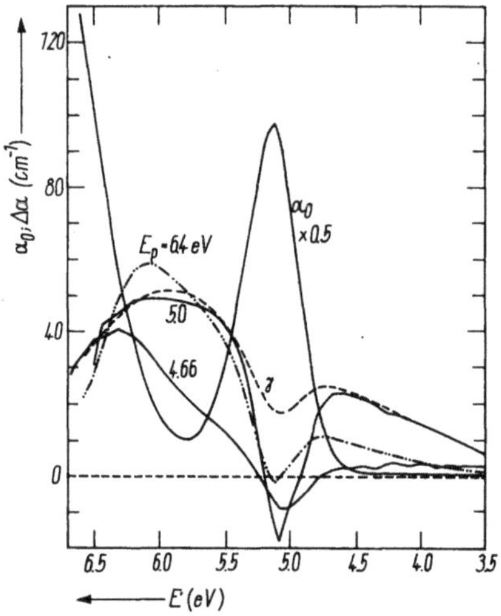

Fig. 1. Spectra of initial and γ- and photo-induced absorption of silica glasses

as well as by the decaying band at 5.12 eV. No changes were observed both in the optical and ESR spectra after 3.68 eV irradiation.

In the initial ESR spectra only the Ge(3) signal was observed with an intensity corresponding to about 3×10^{15} spins/cm^3. All Ge(n) signals are observed after γ-irradiation as well as laser irradiation at E_p = 6.4 or 5.0 eV. According to the isochronal annealing data (Fig. 2) the annealing temperatures of the Ge(1), Ge(2), and Ge(3) centers were determined to be as high as 470, 570, and 870 K, respectively (at a one-half concentration level).

Fig. 3 shows the dependence of the 6.2 and 4.4 eV induced absorption bands and the Ge(n) concentration, on the number of absorbed photons with energy E_p = 5.0 eV. Similar dose dependences were observed under γ- and photo-irradiation with E_p = 6.4 eV.

The correlations between the induced optical absorption bands and Ge(n) concentrations at different kinds of irradiation are shown in more detail in Table 1.

Table 1

Normalized ratios of the induced optical absorption bands and Ge(n) concentrations

E_p(eV)	dose (photons/cm^3 or Gy*))	$\Delta\alpha_{4.4}:\Delta\alpha_{5.8}:\Delta\alpha_{6.2}$	Ge(1):Ge(2):Ge(3)
(1) 4.66	1.1×10^{23}	0.025:0.3:0.675	0:0.19:0.81
(2) 5.0	1.2×10^{22}	0.19:0.38:0.43	0.65:0.08:0.27
(3) 6.4	2.2×10^{23}	0.09:0.38:0.53	0.65:0.15:0.21
(4) 6.4 followed by 4.66	2.2×10^{23} 1.1×10^{23}	0.04:0.34:0.61	0:0.2:0.8
(5) γ, 1.2 MeV	7.4×10^{5}*)	0.17:0.36:0.47	0.56:0.14:0.3

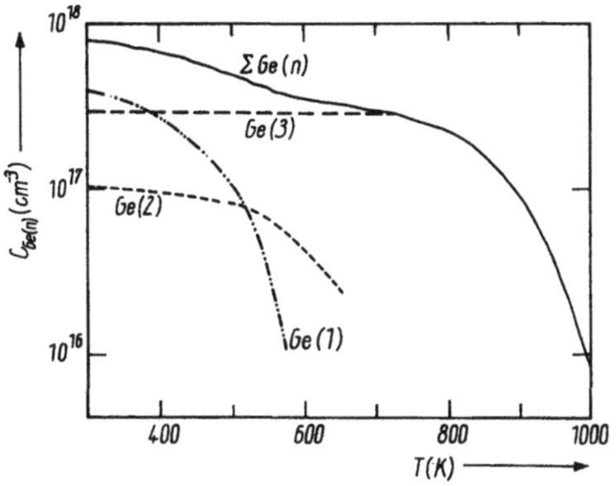

Fig. 2. Annealing behaviour of Ge(n) centers in γ-irradiated MCVD germanium-doped silica glass

When considering all these results we can emphasize the following: a) only irradiation-produced Ge defects produce the absorption in the region 3.5 to 6.5 eV; b) the 6.2, 5.8, and 4.4 eV induced absorption bands are due to different defects; c) only the 6.2 eV band-Ge(3) pair has the higher slope in the dose dependence at doses greater than 10^{20} photons/cm^3; d) the absence of the Ge(1)-4.4 eV optical band as well as photo-bleaching of this pair after the irradiation with 4.66 eV light is evident (see first and fourth lines in Table 1).

So we can suppose that the Ge(1) center is associated with the 4.4 eV induced absorption optical band and the Ge(3) center with the 6.2 eV band. The thermal annealing results showed that these features belonged to different radiation defects. Nevertheless the data on defect formation under γ- and photo-irradiation allow us to assume at least that the radiation defect pair "Ge(1) center-4.4 eV band" is born from one type of precursors and the pair "Ge(3) center-6.2 eV band" - from another one. These precursors seem to be initial germanium defects. As for the Ge(2) center, we did not find any particular feature that associated it with the 5.8 eV induced absorption band except for the similarity of the dose dependences.

In conclusion the analogy between γ- and optical (E_p = 4.6 to 6.4 eV) irradiation in the formation of para-

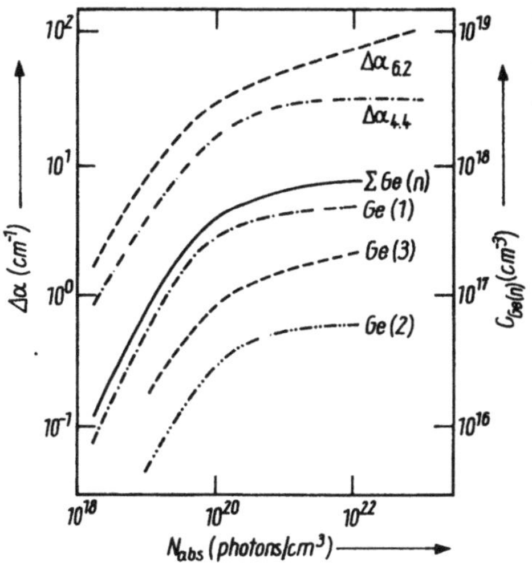

Fig. 3. Dependence of induced absorption and Ge(n) center concentration in MCVD samples under photo-irradiation on the number of absorbed photon (E_p = 5.0 eV)

magnetic colour centers is shown for the germanium-doped silica glass. The association between the Ge(1) center and the 4.4 eV induced absorption band as well as the Ge(3) center and the 6.2 eV band is supposed to be a common feature for each pair precursors.

The authors would like to thank Dr. A.A. Ilyukhin, A.B. Grudinin, D.V. Khaidarov, and V.E. Tukish for their help in the experiments.

References

/1/ E.J. FRIEBELE, D.L. GRISCOM, and G.H. SIGEL, J. appl. Phys. 45, 3424 (1974).

/2/ G. KORDAS, R.A. WEEKS, and P.L. KINSER, J. non-crystall. Solids 69, 293 (1985).

/3/ M.J. YUEN, Appl. Optics 21, 136 (1982).

/4/ A.N. GURYANOV, D.D. GUSOVSKII, E.M. DIANOV, V.M. MASHINSKII, V.B. NEUSTRUEV, and V.F. KHOPIN, Dokl. Akad. Nauk SSSR 264, 90 (1982).

/5/ A.V. AMOSOV and S.F. MALYSHKIN, Fiz. Khim. Stekla 10, 305 (1984).

(Received July 18, 1988)

phys. stat. sol. (a) 110, K111(1988)
Subject classification: 78.30; 71.38; S12

Institute of Chemical Physics, Academy of Sciences of the USSR, Chernogolovka[1)]

Optical Properties of the New Organic Superconductor $(BEDT\text{-}TTF)_2Cu(SCN)_2$

By
M.G. KAPLUNOV, N.D. KUSHCH, and E.B. YAGUBSKII

Bis(ethylenedithio)tetrathiafulvalene (ET or BEDT-TTF) forms conducting ion-radical salts with different anions, some of these salts being organic superconductors /1 to 4/. Recently, it was shown that the compound $(ET)_2Cu(SCN)_2$ is an organic superconductor at ambient pressure with T_c = 10.4 K /4/.

Earlier we have investigated optical properties of some conducting and superconducting ET compounds: α- and β-$(ET)_2I_3$ /5/ and $(ET)_4Hg_3Cl_8$ /6/. The polarization dependence of their reflection spectra shows that the electron structures of these compounds have more or less two-dimensional character. For β-$(ET)_2I_3$ it was shown /7/ that two-dimensionality is higher in the phase with T_c = 8 K than in the phase with T_c = 1.5 K (the optical properties of the latter phase seem to be rather one-dimensional). The closest approach of two-dimensional behaviour was observed for $(ET)_4Hg_3Cl_8$ /6/. The optical properties of ET compounds are also defined by the interaction between conduction electrons and totally symmetric intramolecular vibrations of ET, which manifest itself by specific electron-vibrational bands in the infrared reflection spectra /5, 6/. The comparison of the experimental spectra with calculated model spectra makes it possible to find the values of the electron-vibrational coupling constants /5, 6/ which gives an opportunity to evaluate possible contributions of this coupling to superconductivity /8/.

In the present work we have measured the infrared reflectivity spectra (200 to 8000 cm^{-1}) of single crystals of $(ET)_2Cu(SCN)_2$ at room temperature with polarised light and evaluated the anisotropic parameters of the electronic structure and the values of the electron-vibrational coupling constants.

In Fig. 1a the reflectivity spectra of $(ET)_2Cu(SCN)_2$ from the crystal face parallel to conducting layers are shown for two polarisations: the electric

---

1) SU-142 432 Chernogolovka, Moscow District, USSR.

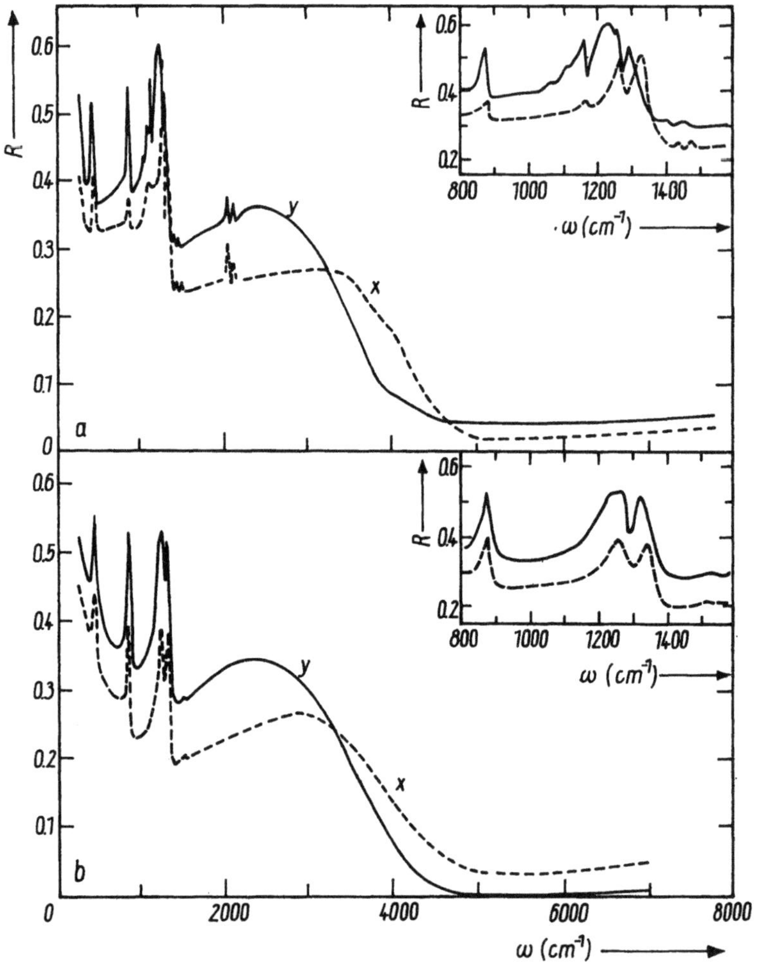

Fig. 1. The reflectivity spectra of $(ET)_2Cu(SCN)_2$ for the two main optical directions x and y; a) experiment, b) theory

vector of light being parallel and perpendicular to the long crystal axis (x and y directions, respectively). The sample consisted of three single crystals arranged with their long axes parallel to each other. The specified x and y polarisations correspond to the two main optical directions for which maximum difference between the two spectra is observed. Our experimental data are similar to that of /9/.

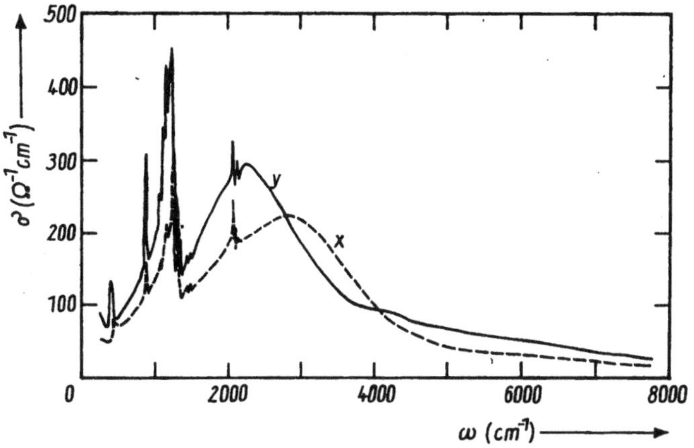

Fig. 2. The optical conductivity spectra of $(ET)_2Cu(SCN)_2$ for the two main optical directions x and y

It must be noted that the anisotropy of the spectra is rather small. In both polarisations there are similar bands present with almost equal intensities: the electronic band with plasma edge and electron-vibrational bands in the region 800 to 1400 cm^{-1}. The small spectral anisotropy makes a difference of $(ET)_2Cu(SCN)_2$ from α- and β-$(ET)_2I_3$ /5/, where the plasma edge is well-defined only for one of the main directions. At the same time the optical properties of $(ET)_2Cu(SCN)_2$ remind of $(ET)_4Hg_3Cl_8$ /6/ where the reflectivity spectra for two main directions are also similar. This similarity of $(ET)_2Cu(SCN)_2$ and $(ET)_4Hg_3Cl_8$ may be due to the similarity of their crystal structures /3, 10/. Thus, according to its optical properties, $(ET)_2Cu(SCN)_2$ is one of the most two-dimensional salts of ET.

In Fig. 2 optical conductivity spectra of $(ET)_2Cu(SCN)_2$ are shown, which are obtained by Kramers-Kronig analysis of the reflectivity spectra. The maximum in $\sigma(\omega)$ in the region 3000 cm^{-1} in both directions may be an evidence of the existence of an energy gap.

To evaluated parameters of the electronic structure and the constants of electron-vibrational interaction we have calculated the reflectivity spectra of $(ET)_2Cu(SCN)_2$ for the two directions in the frame of the phase phonon model taking into consideration the electron damping and the difference in

Table 1

The parameters of the electronic band in the spectra of $(ET)_2Cu(SCN)_2$ for the two main optical directions x and y

	x	y
ε_∞	3.1	2.05
$\sqrt{\varepsilon_\infty}\omega_p(10^3\ cm^{-1})$	6.1	5.0
$2\Delta(10^3\ cm^{-1})$	2.7	1.8
$\Gamma(10^3\ cm^{-1})$	0.6	0.7
V/Δ	0.60	0.55
m/m_o	2.9	4.3
t(eV)	0.095	0.045

vibrational frequencies of the neutral ET molecule and ET^+ cation-radicals. We had used this model for the calculation of the spectra of $(ET)_4Hg_3Cl_8$ in /6/ where the formulae are given. In Fig. 1b the calculated spectra of $(ET)_2Cu(SCN)_2$ for the two directions are shown.

In Table 1, the electronic structure parameters used in the calculation are given for the two main optical directions: dielectric constant at high frequencies ε_∞, plasma frequency ω_p, electron energy gap 2Δ, electronic damping Γ, accepted parameter V/Δ, where 2V is the value which the electronic gap would assume in the absence of electron-vibrational interaction carrier,effective mass m taken from the equation $\varepsilon_\infty\omega_p^2 = 4\pi ne^2/m$ (e is electron charge, n is concentration of carriers accepting $n = 1.19\times10^{21}\ cm^{-3}$ from structure data /10/), and transfer integrals t obtained by numerical integration in analogy with /6, 7/ (periods along main directions are $a_x = 0.422$ nm, $a_y = 0.656$ nm /9, 10/).

In Table 2, the parameters of the electron-vibrational spectrum are given: the frequencies ω_k of totally symmetric vibrational modes of ET molecule and ET^+ cation taken from Raman spectra /5, 11, 12/, their natural linewidths γ_k, accepted values of dimensionless electron-vibrational coupling constants λ_k. The largest constants λ_k are obtained for modes assigned /11/ to the central C=C

Table 2

The parameters of the electron-vibrational spectra of $(ET)_2Cu(SCN)_2$

ω_k (cm^{-1})		λ_k	γ_k (cm^{-1})
ET^0	ET^+		
1552	1455	0.032	40
1494	1431	0.032	40
1285	1294	0.001	40
911	905	0.013	20
486	508	0.031	20

bond of ET and C=C bonds in the rings (1552 and 1494 cm^{-1} for ET^0) and to C=S bonds (911 and 486 cm^{-1}). The total value of the coupling constant $\lambda = \Sigma\lambda_k \approx 0.11$ is less than unity, which corresponds to the case of weak coupling.

Some vibrational bands (for example 1163 cm^{-1}) are not described by our model and may be connected with usual nonsymmetric ET modes. Weak bands at about 2060 and 2110 cm^{-1} correspond to anion CN modes.

Thus the optical data give evidence of two-dimensionality and of considerable electron-vibrational interaction in $(ET)_2CU(SCN)_2$.

References

/1/ E.B. YAGUBSKII, I.F. SHCHEGOLEV, V.N. LAUKHIN et al., Zh. eksper. teor. Fiz., Pisma 39, 12 (1984).

/2/ V.A. MERZHANOV, E.E. KOSTYUCHENKO, V.N. LAUKHIN et al., Zh. eksper. teor. Fiz., Pisma 41, 146 (1985).

/3/ R.N. LYUBOVSKAYA, R.B. LYUBOVSKII, R.P. SHIBAEVA et al., Zh. eksper. teor. Fiz., Pisma 42, 380 (1985).

/4/ H. URAYAMA, H. YAMOCHI, G. SAITO et al., Chem. Letters 55 (1988).

/5/ M.G. KAPLUNOV, E.B. YAGUBSKII, L.P. ROSENBERG, and YU.G. BORODKO, phys. stat. sol. (a) 89, 509 (1985).

/6/ M.G. KAPLUNOV, R.N. LYUBOVSKAYA, M.Z. ALDOSHINA, and YU.G. BORODKO, phys. stat. sol. (a) 104, 833 (1987).

/7/ M.G. KAPLUNOV, E.G. YAGUBSKII, R.P. SHIBAEVA, and YU.G. BORODKO, Khim. Fiz. 6, 760 (1987).

/8/ K. YAMAJI, Solid State Commun. 61, 413 (1987).

/9/ T. SUGANO, H. HAYASHI, H. TAKENOUCHI et al., Chem. Letters, in the press.

/10/ H. URAYAMA, H. YAMOCHI, G. SAITO et al., Chem. Letters 463 (1988).

/11/ M.E. KOZLOV, K.I. POKHODNYA, and A.A. YURCHENKO, Spectrochim. Acta 43A, 323 (1987).

/12/ M.E. KOZLOV, K.I. POKHODNYA, and A.A. YURCHENKO, Spectrochim. Acta, in the press.

(Received July 26, 1988)

Pre-Printed Titles

Papers to be published
in "physica status solidi (a) - applied research"
Vol. 111, No. 1

Original Papers

C.D. LOKHANDE and S.H. PAWAR, Kolhapur
Electrodeposition of Thin Film Semiconductors

Y. MERA, K. MAEDA, and S. HYODO, Tokyo
Microscopic Origin of Electric Conduction Anisotropy in Plastically Deformed CdS Single Crystals

S.S. BAWA, K. SAXENA, and S. CHANDRA, New Delhi
Polarization Reversal Current Behaviour for Parallel and Antiparallel Rubbed Surface Stabilized Ferroelectric Liquid Crystals

T.H. YOUSSEF, F.A. SAADALAH, and I.K. BISHAY, Cairo
Plastic Instability during Combined Torsion-Tension Deformation in Al-0.72% Si Alloy

NGUYEN HOANG LUONG, NGUYEN PHU THUY, LUU TUAN TAI, and THAN DUC HIEN, Hanoi
Rare Earth Magnetocrystalline Anisotropy in $R_2Fe_{14}B$ (R = Sm, Er, Tm, Yb)

A. ROGALSKI and K. JOZWIKOWSKI, Warsaw
The Intrinsic Carrier Concentration in $Pb_{1-x}Sn_xTe$, $Pb_{1-x}Sn_xSe$, and $PbS_{1-x}Se_x$

D. VLADIKOVA, L. ILKOV, and S. KARBANOV, Sofia
Influence of the Microstructure on Some Microwave Properties of Substituted Nickel Ferrites

R.A. YAKSHIBAEV, V.N. ZABOLOTSKII, and R.F. ALMUKHAMETOV, Ufa
Structural Features and Ionic Transport in the Ag_xNbSe_2 Two-Dimensional Superionic Conductor

J.J. BARA and B.F. BOGACZ, Cracow
Investigation of ^{57}Fe-Doped α-$Al_xV_2O_5$ Bronzes

A.I. GUSEV, Sverdlovsk
Structural Stability Boundaries for Nonstoichiometric Compounds

K. YAMAKAWA, B. HOHLER, and H. KRONMÜLLER, Stuttgart
Diffusion of Hydrogen in Nickel Based Alloys

A.F. VLADIMIROV and E.N. MOOS, Ryazan
Development of Eshelby Concept in the Solution of Boundary Problems for Elastic Solids with Defects

S. WAPLAK and T. KRAJEWSKI, Poznań
EPR Studies of $LiNaSO_4$ Cr^{6+} Doped Monocrystals after X-Ray Irradiation

M.M. KAZANIN and V.V. KUTASOV, Leningrad
Fe Impurity in α-, β-Boron and in Some Boron Compounds

V.V. TOKAREV, V.E. BORISENKO, T.M. PYATKOVA, and D.I. ZAROVSKII, Minsk
Argon Ion Implantation Influence on Phase Formation and Structure of Nickel Silicides

M.A. AHMED, Tanta
Electrical Properties of Co-Zn Ferrites

T. KOJIMA, Tokyo
Semiconducting and Thermoelectric Properties of Sintered Iron Disilicide

R. TOVAR, M. QUINTERO, P. GRIMA, and J.C. WOOLLEY, Mérida
Phase Diagram, Lattice Parameter, and Optical Energy Gap Values for the $Zn_{2x}(AgIn)_{1-x}Te_2$ Alloys

A.A. SUKHOVSKII, I.P. ALEKSANDROVA, and YU.N. MOSKVICH, Krasnoyarsk
NMR Confirmation of the Existence of Superstructure in the Ferroelectric Phase of $NH_4Se\,)_4$

J. PASTRNAK, J. OSWALD, I. GREGORA, V. VORLIČEK, and M. BABINSKY, Prague
Laterial Distribution of Si-Related Defects and Carrier Density in GaAs Crystals

E. SCHULZ, G. DRÄGER, and O. BRÜMMER, Halle (Saale)
X-Ray Spectroscopic Investigations of Titanium Oxides and Basic Materials for Ferroelectric Ceramics

U. PÁL, S. SAHA, B.K. SAMANTARAY, H.D. BANERJEE, and A.K. CHOUDHURI, Kharagpur
X-Ray Line Broadening and Electron Microscopic Studies on ZnTe Thin Films

P. SVOBODA, Z. SMETANA, and A.V. ANDREEV, Prague
Magnetic Behaviour of $TmCu_2$ below T_N

H.-J. WÜNSCHE and H. WENZEL, Berlin
Influence of Surface Recombination on the Rate of Electron Leakage in Ridge-Waveguide Injection Lasers

M.V.S. LAKSHMI and K. RAMKUMAR, Bangalore
I-U Characteristics of MOS Structures on Polycrystalline Silicon

J. SUNE, I. PLACENCIA, N. BARNIOL, E. FARRES, and X. AYMERICH, Bellaterra
Degradation and Breakdown of Gate Oxides in VLSI Devices

L.E. SOSHNIKOV and A.U. SHELEG, Minsk
Elastic Properties and Phase Transitions in α-CdP_2

F. HAIDER, Göttingen
Calculation of TEM Contrast of Small Coherent Precipitates

H. PAGNIA and N. SOTNIK, Darmstadt
Metal Influence on Switching MIM Diodes

S.A. AL-KUHAIMI, Riyadh
Optical and Compositional Properties of $CuInSe_2$ Thin Films at High Temperature

Papers to be published
in "physica status solidi (b) - basic research"
Vol. 151, No. 1

Review Article

V.P. ZHUKOV, N.I. MEDVEDEVA, and V.A. GUBANOV, Sverdlovsk
The Electronic Band Structure and Non-Empirical Calculations of Cohesive Properties of Refractory Compounds

Original Papers

S. KUMAR, P.K. AHLUWALIA, and K.C. SHARMA, Shimla
Thermoelectric Power of Dilute Magnetic Alloys in the Presence of CEF and Exchange Scattering

R.G. GABRIELYAN and KH.V. KOTANDIAN, Erevan
Mössbauer Radiation Transfer in the Plane Layer

R. CYWIŃSKI, E. MUGEŃSKI, W. NOWY-WIECHUŁA, and J. WIECHUŁA, Wrocław and Opole
Suzuki Phase Precipitates in Doubly Doped $KCl:Pb^{2+}$, Mn^{2+} Crystals

V.I. GREBENNIKOV and O.B. SOKOLOV, Sverdlovsk
The Local Electron-Spin-Density Fluctuation Mechanism of the Superconductivity in Narrow-Band Metals

G.S. CHADDHA and G.S. KALSI, Ludhiana
Anisotropic Ferromagnet with Biquadratic Exchange

K.R. ALLAKHVERDIEV, S.S. BABAEV, M.M. TAGIEV, and M.M. SHIRINOV, Baku
Low-Temperature IR and Raman Scattering Spectra of $TlInS_2$ Layered Crystal

G.A. MEDVEDKIN, V.D. PROCHUKHAN, YU.V. RUD, and M.A. TAIROV, Leningrad
Anisotropy of the Optical Band Edge Absorption in $CuInTe_2$ Single Crystals

H.N. NAZARENO and J.C. GALLARDO, Santa Barbara, California, and Upton, New York
Bloch Oscillations of an Electron in a Crystal under the Action of an Electric Field

P. NAGY and P. MARKOS, Bratislava
On the Nature of the Self-Trapping of Different Types of Polarons

E. KÖNIGSBERGER and H. SCHRUNNER, Leoben
On Lattice Parameters and Enthalpies of Mixing of Alkali Halide Solid Solutions

I.I. GEGUSIN, V.N. DATSYUK, A.D. YUMATOFF, and R.V. VEDRINSKII, Rostov
Atomic and Solid State Effects in X-Ray Absorption Spectra of Ionic Crystals

J.L. RAO, M.R. REDDY, and S.V.J. LAKSHMAN, Tirupati
Electronic Absorption Spectrum of Ni^{2+} Ions Doped in Zinc Thallium Sulphate Hexahydrate

V.A. BIBIK and N.A. DAVYDOVA, Kiev
Transformation of Excitonic Spectra of Layer Crystals PbI_2 under Order-Disorder-Order Structural Transitions

W. HENRION and H. LANGE, Berlin
Optical Properties of High-Refractory Disilicide Thin Films

K. KOKKO, Turku
Determination of Electron Momentum Density from Electron Number Density

J.-P. ZÖLLNER and G. PAASCH, Dresden
Conditions for the Occurrence of the Electric Quantum Limit in Inversion Layers

U. TRUTSCHEL, M. GOLZ, and M. ABRAHAM, Jena
Generalized Transfer-Matrix Description of the Optical Properties of Spatially Dispersive Periodic Metal Multilayers

G. WEBER and L.E. OLIVEIRA, Campinas
On the Binding Energies of D^- States in Doped Germanium

F. DE MATTEIS, A. SCACCO, F. SOMMA, G. BALDACCHINI, E. GIOVENALE, and U.M. GRASSANO, Roma
Complex Centres in Additively Coloured RbCl: Li^+

C.P. HILTON, J.J. DAVIES, J.E. NICHOLLS, and O. GOEDE, Hull and Berlin
Optically-Detected Magnetic Resonance of Excitons Bound at Te_2 Isoelectronic Centres in $CdS_{1-x}Te_x$

O. GOEDE, D. BACKS, W. HEIMBRODT, and M. KANIS, Berlin
EPR Study of the Antiferromagnetic Phase Transition in (Cd, Mn)S

Substance Classification*)

S1. Transition metals and their alloys and compounds (for oxides see S.10)
- S1.1. Fe
- S1.2. Other 3d metals
- S1.3. Noble metals
- S1.4. Lanthanides
- S1.5. Actinides
- S1.61. Carbides, silicides, germanides
- S1.62. Nitrides, phosphides, arsenides
- S1.63. Sulfides, selenides, tellurides

S2. Alkali metals and their alloys and compounds (for alkali halides see S9.11)

S3. Alkaline earth metals and their alloys and compounds (for alkaline earth halides, see S9.12)

S4. Other metals and their alloys

S5. Nonmetallic elements
- S5.11. Silicon
- S5.12. Germanium
- S5.2. Solidified Gases

S6. Group IV compounds (for transition metal compounds, see S1.61)

S7. Group V compounds (for transition metal compounds, see S1.62)
- S7.1. III–V compounds
- S7.11. Phosphides
- S7.12. Arsenides
- S7.13. Antimonides
- S7.15. Mixed crystals
- S7.16. Higher compounds

S8. Chalcogenides (for transition metal compounds, see S1.63)
- S8.1. II–VI compounds
- S8.11. Sulfides
- S8.12. Selenides
- S8.13. Tellurides
- S8.15. Mixed crystals
- S8.16. Higher compounds

S9. Halides
- S9.11. Alkali halides
- S9.12. Alkaline earth halides
- S9.15. Mixed crystals
- S9.16. Higher compounds

S10. Oxides
- S10.1. Simple oxides
- S10.15. Mixed crystals

S11. Other higher compounds
- S11.1. Salts of oxy-acids
- S11.2. Spinels, garnets, ferrites

S12. Organic compounds
- S12.1. Quasi-one-dimensional systems

Attention

The *Classification Scheme* on cover two is reproduced and used with permission of International Council for Scientific and Technical Information (ICSTI).

Because of the lack of space only the main points of the Classification Scheme are printed here. The *detailed subdivided scheme* is to be found in the January issues of each annual set and in the Index Volumes from 1986 on.

The *Subject Classification* given at the head of each paper begins with the main subject matter of the paper, and this is followed, if necessary, by figures referring to subsidiary subject matter and the type of solids investigated.

The papers published in this issue are listed in the order of their subject classification in the *Systematic List* at the end of the table of contents.

*) Outside the ICSTI Classification for Physics.

Manuscripts and letters for physica status solidi (b) — basic research and physica status solidi (a) — applied research

should be sent to

Leipziger Straße 3—4, P. B. 1233, DDR-1086 Berlin

or

Kunst und Wissen, Erich Bieber OHG, Postfach 330104, D-1000 Berlin 33.

The object of "physica status solidi (b)" and "physica status solidi (a)" is to publish rapidly important and topical results concerning fundamental research in the field of solid state physics and aspects of application and experimental techniques of fundamental solid state research, respectively. The issues of "physica status solidi (b)" and "physica status solidi (a)" will be published on the 1st and the 16th day of each month, respectively. Either series will contain the following categories:

1. Review articles written by arrangement with the editors. They will be remunerated.
2. Original papers selected from those which have not been submitted for publication elsewhere, after being reviewed. (The latest date for acceptance of contributions is the 8th and the 23rd day of each month for "physica status solidi (b)" and "physica status solidi (a)", respectively, 50 days before the delivery of the journal.)
3. Short notes published in the off-set part of the journal. They are, if necessary after being reviewed, selected by editorial staff. (The latest date for acceptance of manuscripts for this part is the 12th and the 27th day of each month for "physica status solidi (b)" and "physica status solidi (a)", respectively, 16 days before the delivery of the journal.)
4. Pre-printed titles of papers to be published in the next issue of both series.

Editorial Information

Manuscripts should be submitted preferentially in English. Besides, papers in German, French, and Russian are admitted. The final decision on the incorporation of the papers in the series (b) or (a) is taken by the editorial staff after consultation of the referee. Papers on liquids are only published in physica status solidi if the problems treated are also relevant for the solid state. Authors are asked to observe the following rules:

1. Manuscripts have to be written as consisely as possible. At most, 20 and 4 pages (28 lines per page; abstracts, tables, references, figures, and captions of figures included, where 3 figures are counted as about one page) are admitted for original papers and short notes, respectively. All manuscripts should be type-written on one side only. Two copies are required. Badly arranged manuscripts are returned.
2. The author's name, initials, and institution should be given.
3. One abstract of about ten lines in English and one in German, French, or Russian are required for original papers.
4. Because of lack of space conclusions (if any) must not have the character of a summary (i. e. substantially a repetition of the contents of the abstract).
5. Formulae should not be written too small and not with pencil. Separate lines for formulae are desirable. Indistinct writing will lead to a return of the manuscript.
6. Drawings should be enclosed on separate sheets of paper, not exceeding the maximum and minimum sizes of 20×30 and $10 \times 15\ cm^2$, respectively. Photographs should be submitted, on white paper, only for review articles and original papers. Duplicates are required.
7. Captions for the figures should be submitted on separate sheets of paper, and special instructions with regard to the size of illustrations should be given on these.
8. References should be numbered in order of appearance and should be given, on a separate sheet, in the form used in physica status solidi.
9. SI-units should be used.
10. Each paper must contain a proposal for the subject classification according to the classification scheme appearing in the January issues of each annual set.

The authors will be sent galley proofs of review articles and original papers. Proof reading of short notes and pre-printed titles will be done by the editorial staff.

For each original paper and short note a total of 75 reprints is available free of charge. The number of reprints free of charge per review article is 25. Requests for additional reprints (against payment) should be given together with the submitted manuscripts. **The deadline for orders of additional reprints is the 15th and the 30th of each month** for "physica status solidi (b)" and "physica status solidi (a)", respectively, two weeks before the delivery of the journal.

Bei Fragen zur Produktsicherheit wenden Sie sich bitte an:
If you have any questions regarding product safety,
please contact:

Walter de Gruyter GmbH
Genthiner Straße 13
10785 Berlin
productsafety@degruyterbrill.com